Climatology and Paleoclimatology

Climatology and Paleoclimatology

Edited by Andrew Hyman

SYRAWOOD
PUBLISHING HOUSE
New York

Published by Syrawood Publishing House,
750 Third Avenue, 9th Floor,
New York, NY 10017, USA
www.syrawoodpublishinghouse.com

Climatology and Paleoclimatology
Edited by Andrew Hyman

International Standard Book Number: 978-1-68286-448-7 (Hardback)

Cataloging-in-publication Data

Climatology and paleoclimatology / edited by Andrew Hyman.
p. cm.
Includes bibliographical references and index.
ISBN 978-1-68286-448-7
1. Climatology. 2. Paleoclimatology. 3. Climatic changes. I. Hyman, Andrew.
QC981 .C55 2017
551.6--dc23

Printed in the United States of America.

TABLE OF CONTENTS

PREFACE

Climatology is the scientific study of weather conditions and patterns with respect to a period of time whereas paleoclimatology focuses on changes in the Earth's climate. Climatology and paleoclimatology plays a crucial role in climate modeling and predicting the climate of the future. This book on climatology and paleoclimatology discusses the various climate models that have evolved over time. This text elucidates the concepts and innovative models around prospective developments with respect to climatology and paleoclimatology. It will be of great help to those researching in the fields of historical climatology, environmental design and meteorology. It aims to equip students and experts with the advanced topics and upcoming concepts in this area.

Over the recent decade, advancements and applications have progressed exponentially. This has led to the increased interest in this field and projects are being conducted to enhance knowledge. The main objective of this book is to present some of the critical challenges and provide insights into possible solutions. This book will answer the varied questions that arise in the field and also provide an increased scope for furthering studies.

I hope that this book, with its visionary approach, will be a valuable addition and will promote interest among readers. Each of the authors has provided their extraordinary competence in their specific fields by providing different perspectives as they come from diverse nations and regions. I thank them for their contributions.

Editor

1

Dietary Specialization during the Evolution of Western Eurasian Hominoids and the Extinction of European Great Apes

Daniel DeMiguel[1]*, David M. Alba[1,2], Salvador Moyà-Solà[3]

1 Institut Català de Paleontologia Miquel Crusafont, Universitat Autònoma de Barcelona, Barcelona, Spain, 2 Dipartimento di Scienze della Terra, Università di Torino, Torino, Italy, 3 ICREA at Institut Català de Paleontologia Miquel Crusafont and Unitat d'Antropologia Biològica (Dept. BABVE), Universitat Autònoma de Barcelona, Barcelona, Spain

Abstract

Given the central adaptive role of diet, paleodietary inference is essential for understanding the relationship between evolutionary and paleoenvironmental change. Here we rely on dental microwear analysis to investigate the role of dietary specialization in the diversification and extinction of Miocene hominoids from Western Eurasian between 14 and 7 Ma. New microwear results for five extinct taxa are analyzed together with previous data for other Western Eurasian genera. Except *Pierolapithecus* (that resembles hard-object feeders) and *Oreopithecus* (a soft-frugivore probably foraging opportunistically on other foods), most of the extinct taxa lack clear extant dietary analogues. They display some degee of sclerocarpy, which is most clearly expressed in *Griphopithecus* and *Ouranopithecus* (adapted to more open and arid environments), whereas *Anoiapithecus*, *Dryopithecus* and, especially, *Hispanopithecus* species apparently relied more strongly on soft-frugivory. Thus, contrasting with the prevailing sclerocarpic condition at the beginning of the Eurasian hominoid radiation, soft- and mixed-frugivory coexisted with hard-object feeding in the Late Miocene. Therefore, despite a climatic trend towards cooling and increased seasonality, a progressive dietary diversification would have occurred (probably due to competitive exclusion and increased environmental heterogeneity), although strict folivory did not evolve. Overall, our analyses support the view that the same dietary specializations that enabled Western Eurasian hominoids to face progressive climatic deterioration were the main factor ultimately leading to their extinction when more drastic paleoenvironmental changes took place.

Editor: Lorenzo Rook, University of Florence, Italy

Funding: This work was funded by the Spanish Ministerio de Economía y Competitividad (CGL2011-28681/BTE, CGL2011-27343, JCI-2011-11697 (to DDM), and RYC-2009-04533 (to DMA)) and the Generalitat de Catalunya (2009 SGR 754 GRC). The funders had no role in study design, data collection and analysis, decision to publish, or preparation of the manuscript.

Competing Interests: The authors have declared that no competing interests exist.

* E-mail: daniel.demiguel@icp.cat

Introduction

Dietary Adaptation and the Hominoid Radiation in Western Eurasia

After an initial radiation in Africa during the Early to Middle Miocene [1], hominoids dispersed into Eurasia, where they diversified into multiple great ape genera from ca. 14 Ma onwards [2–6]. Available data suggest that vicariance and parallel evolution played a significant role in the Eurasian hominoid radiation, with dryopithecines diversifying in Europe, pongines in Asia, and maybe hominines in Africa [3,4]. Dietary adaptations have long been regarded as very significant for understanding the dispersal of hominoids from Africa into Eurasia and their subsequent radiation [4,6–8]. Paleodietary inference is thus paramount for understanding how fossil great apes adapted to changing environmental conditions through time. Unlike in hominins, however, comparatively little dietary research has focused on fossil great apes from Eurasia [7–15]. Previous results suggest that a considerable dietary diversity was present in the Late Miocene, and that such diversification might have taken place during the Middle and Late Miocene [7].

Previous dental microwear analyses were based on a wide array of extinct hominoids from the Miocene of Western Eurasia [10–12]: *Griphopithecus alpani* [16,17] from the early Middle Miocene (MN6, ca. 14.9–13.7 Ma) [3] of Turkey; *Hispanopithecus* (*Rudapithecus*) *hungaricus* [18,19] from the Late Miocene (MN9, ca. 10.0–9.8 Ma) [3] of Hungary; *Ouranopithecus macedoniensis* [20–22] from the Late Miocene (MN10, ca. 9.7–9.0 Ma) [3] of Greece; *Oreopithecus bambolii* [23–25] from the Late Miocene (MN12, ca. 8.3–6.7 Ma) [26] of Italy; and *Hispanopithecus crusafonti* (MN9, ca. 10.4–10.0 Ma) and *Hispanopithecus laietanus* (MN9, ca. 11.1–9.5 Ma) [4] from Spain [3]. Results based on these taxa [10–12] suggested that, from the hard-object feeding plesiomorphic condition displayed by *G. alpani* [12], a progressive dietary diversification and specialization would have taken place through time—with *Hispanopithecus* spp. being inferred as a frugivore [10,11], *Ou. macedoniensis* as a hard-object specialist [10], and *O. bambolii* as an extreme folivore [10].

These analyses had, however, a significant gap, because late Middle Miocene taxa were not included due to the scarcity of fossil specimens from this time span. This situation, however, drastically changed during the last decade thanks to continued and extensive fieldwork in the late Middle Miocene local stratigraphic series of Abocador de Can Mata in els Hostalets de Pierola (Vallès-Penedès Basin) [4,27,28]. The material recovered there has shown an unprecedented diversity of Middle Miocene dryopithecines in Western Europe [3,4], with two new genera and species (*Pierolapithecus catalaunicus* [29] and *Anoiapithecus brevirostris* [30]) being described, and new material of *Dryopithecus fontani* [31] being recovered. Similarly, excavations at several Late Miocene localities of the same basin have led to the recovery of new dental remains of *Hispanopithecus laietanus*, which further confirm the distinction of this species from *Hispanopithecus crusafonti*, recorded at slightly older localities [4]. Additional paleodietary data are therefore required for these taxa in order to better understand the hominoid radiation in Western Eurasia from a dietary viewpoint.

Here we report new dental microwear analyses for five hominoids from the Iberian Peninsula (Figure 1), ranging in age from 12.3–12.2 to 9.7 Ma [3,4]. Together with previous results for other Miocene hominoids from elsewhere in Europe and Turkey [10–12] (see Table S1), these data allow us to re-evaluate dietary diversification during great-ape evolution in Western Eurasia (ca. 14 to 7 Ma) in the light of paleoenvironmental changes.

Paleodietary Inference

Among the various methods of paleodietary inference in primates, dental gross morphology and ultrastructure are useful because teeth are adapted for food processing [32]. Occlusal morphology, which can be quantified by means of shearing crest analysis [9,11,33], offers some clues because folivorous apes possess longer shearing crests on molars than frugivorous ones. However, shearing crest quotients are highly dependent on the particular group being analyzed and the baseline used as a reference for comparison [11,33], so that they might be potentially biased when applied to extinct taxa [34]. Enamel thickness is also generally considered to reflect dietary adaptations to some degree, given the relationship between hard-object feeding and thick enamel [7,8,35–37]. However, enamel thickness is heavily influenced by phylogenetic constraints [38] and there is no threshold value for distinguishing hard-object feeders on this basis alone [39,40], so that overall there is no direct relationship between enamel thickness and diet. All these morphology-based approaches reflect dietary adaptation as well as phylogenetic constraints, and hence probably are more informative about what extinct taxa were able to eat than about what they actually ate [41].

Non-morphological methods of paleodietary reconstruction, based on either dental microwear or stable isotope geochemistry, provide more direct information on the properties of the foods consumed independently from adaptation [41,42]. Geochemical methods, such as stable carbon isotope ratios derived from fossil tooth enamel, are based on the fact that hominoids that consumed grasses and sedges have higher ^{13}C levels than those that fed on fruits and other plants [41–46]. These methods are based on the carbon isotopic distinction between C_3 and C_4 photosynthetic pathways. However, C_4 plants (grasses) did not globally expand until the Late Miocene (ca. 8-6 Ma) [47,48], being present in Eurasia only from 9.4 Ma onwards [49]—i.e., after the extinction of most of the Western Eurasian hominoids studied in this paper. Moreover, isotopic analysis of tooth enamel implies invasive sampling techniques, which are not advisable given the small available dental samples for most of the studied taxa. Dental

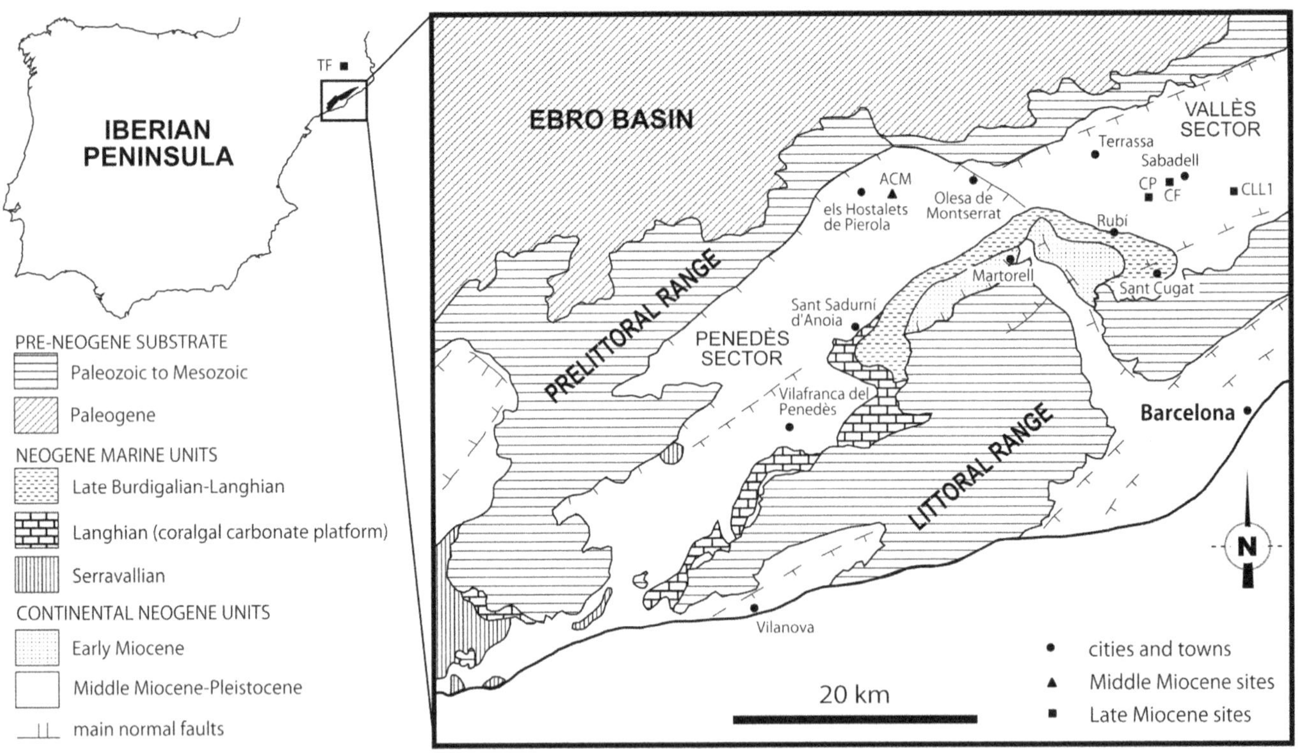

Figure 1. Schematic geological map. Square showing the location of the hominoid-bearing localities that have delivered dental remains studied in this paper, with emphasis on those from the Vallès-Penedès Basin (NE Spain).

microwear analysis similarly provides direct evidence on the type of food items consumed by a particular fossil individual, but unlike isotopic methods it is a non-invasive technique, which relies on the microscopic traces left by foods on the enamel surface [10,15,50].

Dental microwear analysis is thus one of the most powerful methods for inferring dietary behavior in extinct taxa, being based on the strong and consistent association between dental microscopic patterns and the physical properties of the chewed foods [15,50]. While both occlusal morphology and enamel thickness might provide important clues as to what types of food a particular taxon was adapted to consume, microwear features directly reflect what an animal actually ate just prior to death [51]. Hence, since the early 1980s dental microwear analysis has been extensively applied to early hominins and other fossil primates in order to try to determinate their dietary behavior (and seasonal changes thereof), as well as tooth use and masticatory jaw movements [12]. Dental microwear provides direct information on the type of food items consumed shortly prior to an individual's death (days, weeks or months, depending on the nature of the foods being masticated)—a phenomenon referred to as the "Last Supper effect" [52]. The physical properties of both the food items (especially phytoliths from plant taxa) and of exogenous grit (abrasive dirt) being ingested during feeding influence dental microwear patterns. Based on both research in the wild [53] and experimental studies [54,55], some authors have contended that exogenous grit plays an important (or even a primary) role in dental microwear genesis—which might explain why some species have broadly-similar microwear patterns in spite of marked dietary differences [56]. Further experimental research is undoubtedly necessary to better understand the mechanisms responsible of microwear formation, and particularly to determine the role of exogenous grit as a causative agent. However, various studies support the view that food item properties are the main factor determining microwear genesis [57–59]. Thus, work on primates and other mammals has shown a strong relationship between dental microwear features and the types of food consumed, as indicated by different taxa from comparable sites, which exhibit microwear differences that are consistent with their contrasting diets [59–61].

Dental microwear texture analysis [62–65] was recently introduced as an alternative technique to more traditional methods of microwear analysis, being based on 3D surface data and scale-sensitive fractal analysis. Unlike the traditional method, texture analysis does not require the identification of individual features and the analysis is automated—thus being less affected by interobserver error and much less time consuming [51,62]. Contrary to such advantages, microwear texture analysis is a much more costly alternative, because it relies on white-light scanning confocal microscope instead of 2D micrographs taken with a standard Scanning Electron Microscope (SEM). Texture analysis was introduced to increase repeatability and avoid interobserver error [62], but error studies of traditional microwear quantification techniques show that high errors are found only when different methodologies are employed [66]. As long as a consistent technique is employed, such as that offered by the Microware software package, a common microwear database derived by different researchers can be consistently employed [66]. In this sense, using the traditional microwear analysis approach offers the advantage that our new results can be analyzed together with those derived by previous researches for both the extant comparative sample and other extinct hominoids. Whereas traditional microwear data are available for some Western Eurasian hominoids [10,12], no microwear texture data have been thus far published for Miocene apes. As a result, the more traditional approach to microwear analysis followed in this work is still currently used by various researchers [34,67,68].

Methods and Materials

Dental Microwear

Micrographs of the occlusal enamel surfaces of the investigated teeth were taken with an environmental SEM (FEI Quanta 200) at the Serveis Cientificotècnics of the Universitat de Barcelona (Spain), on "Phase II" crushing/grinding facets (9, 10n and x) [20,69]. The standard procedure described in ref. [50] was employed, including 500× magnification and 200 dpi micrograph resolution, in secondary emissions mode and a 20 kV voltage. To avoid interobserver error, an area of 0.02 mm^2 [11,70] was analyzed with Microware 4.02© software by a single author (DDM). Some of the examined specimens showed extensive microscopic damage and were therefore discarded.

Two main microwear features (pits and striations) were distinguished [15,50,60]. Pits are microwear scars that are circular or subcircular in outline. Scratches, in turn, are elongated microfeatures with straight, parallel sides. In this study, pits and scratches were directly categorized by following an arbitrarily-set length to width ratio of 4:1 [50]. The three standard variables customarily quantified in dental microwear analyses were employed [15,50,71]: (1) Percentage of pits (%), i.e., the proportion of pits relative to the total number of microwear features; (2) breadth of striations (in μm); and (3) breadth of pits (in μm). Previous studies have shown that the relative proportion between pits and scratches enables the distinction between frugivores, folivores and hard-object feeders [10,12,15,34,71,72]. Thus, although this is a continuous variable across dietary categories, there is a strong and significant positive correlation between the prevalence of pits and the consumption of hard, brittle foods (such as nuts), as well as between higher scratch frequencies and the consumption of tough items (such as leaves and softer fruits). Microwear feature size is also valuable for further characterizing diets, especially when combined with pitting incidence in multivariate analyses [34]. Moreover, striation breadth has been related to the ratio of exogenous grit versus phytoliths consumed in incisors [73], although this relationship remains to be tested in molar microwear.

Studied Sample

We studied 15 upper and lower molars of the following Miocene hominoids (Table 1 and 2): *Pierolapithecus catalaunicus* from ACM/BCV1 (IPS21350) [29], *Anoiapithecus brevirostris* from ACM/C3-Aj (IPS41712 and IPS43000) [30] and ACM/C1-E* (IPS35027) [74], *Dryopithecus fontani* from ACM/C3-Ae (IPS35026) [31]; *Hispanopithecus* (*Hispanopithecus*) *crusafonti* from CP1 (IPS1820, IPS1818, IPS1812 and IPS1821) and TF (MGSB25314) [75,76]; and *Hispanopithecus* (*Hispanopithecus*) *laietanus* from CF (IPS34753) [77] and CLL1 (IPS1763, IPS1788, IPS1797 and IPS1800) [76,78,79]. The taxonomy employed for hominoids follows ref. [4]. All specimens studied in this paper are housed at the Institut Català de Paleontologia Miquel Crusafont (Sabadell, Spain) and the Museu Geològic del Seminari de Barcelona (Barcelona, Spain). No permits were required for the described study. The ACM specimens come from the late Middle Miocene (MN7+8), whereas those from the remaining localities are Late Miocene (MN9) in age [3,4]. To fully assess the available information and compensate for the small number of individuals in some cases, specimens of a single species from various localities were also analyzed together by using their average values for microwear variables. In particular, we combined specimens of *A. brevirostris* from ACM/

Table 1. Summary results of the microwear analysis.

Taxon	Locality	Age [Ma]	N	Pits [%]			Pit breadth [µm]			Scratch breadth [µm]		
				Mean	SD	Range	Mean	SD	Range	Mean	SD	Range
Pierolapithecus catalaunicus	ACM/BCV1	11.93	1	42.50	—	—	9.59	—	—	3.79	—	—
Anoiapithecus brevirostris	ACM/C1-E*	12.3–12.2	1	19.23	—	—	3.05	—	—	2.87	—	—
Anoiapithecus brevirostris	ACM/C3-Aj	11.94	2	31.18	15.82	20.00–42.37	5.42	1.18	4.59–6.27	2.85	0.52	2.49–3.23
Anoiapithecus brevirostris	average	12.3–11.94	3	27.20	13.14	19.23–42.37	4.64	1.61	3.05–6.27	2.86	0.37	2.49–3.23
Dryopithecus fontani	ACM/C3-Ae	11.85	1	48.81	—	—	3.66	—	—	2.25	—	—
Hispanopithecus crusafonti	TF	10.4–10.0	1	51.39	—	—	4.24	—	—	2.60	—	—
Hispanopithecus crusafonti	CP1	10.4–10.0	4	33.43	5.63	27.84–41.17	4.78	1.37	3.78–6.79	2.37	0.12	2.25–2.54
Hispanopithecus crusafonti	average	10.4–10.0	5	37.02	9.40	27.84–51.39	4.67	1.21	3.78–6.79	2.42	0.14	2.25–2.60
Hispanopithecus laietanus	CF	10.0–9.7	1	34.61	—	—	6.77	—	—	2.44	—	—
Hispanopithecus laietanus	CLL1	9.72	4	29.33	6.30	21.66–37.07	5.08	0.84	4.26–6.25	2.59	0.47	2.07–3.22
Hispanopithecus laietanus	average	10.0–9.7	5	30.39	5.94	21.66–37.07	5.42	1.04	4.26–6.77	2.56	0.41	2.07–3.22

Abbreviations: N, sample size; SD, standard deviation.
Locality abbreviations: ACM, Abocador de Can Mata; BCV1, Barranc de Can Vila 1; C1, Cell 1; C3, Cell 3; CLL1, Can Llobateres 1; CP1, Can Poncic 1; TF, Teuleria del Firal.

C1-E* and ACM/C3-Aj, specimens from *H. crusafonti* from TF and CP1, and specimens of *H. laietanus* from CF and CLL1. This procedure is justified by the close geographic situation and age of these localities (Table 1, see also Figure 1, and Table S1).

Comparative Samples

Our results were compared with those derived from previous authors [15,60,71] for a sample of 11 extant anthropoid primates with well-known diets (extant species samples consisting of 10 specimens, except that of *Papio cynocephalus*, which consists of 16). These studies were selected because they used a sufficiently similar technique to allow comparison with our results. As mentioned above, we analyzed our microwear results together with those previously published for other Western Eurasian Miocene hominoids, including: *Griphopithecus alpani* from Paşalar (MN6) [12,13]; *Hispanopithecus hungaricus* from Rudabánya (MN9) [10,11]; *Ouranopithecus macedoniensis* from Ravin de la Pluie, Xirochori and Nikiti (MN10) [10]; and *Oreopithecus bambolii* from Baccinello, Monte Bamboli and Ribolla (MN12) [10].

Dietary Categories

Three extant dietary categories were employed by attributing each of the extant species to one of these groups defined a priori on the basis of published behavioral data [34]: (1) folivores (FOL); (2) frugivores/mixed feeders (FMF); and (3) hard-object feeders (HOF). Several species were subsumed into a single category of "frugivores/mixed feeders" [34,80], because periods of fruit scarcity may impel many frugivorous primates to exploit alternative, non-preferred food sources (fallback foods), thereby resulting in a somewhat eclectic foraging strategy [34]. It should be also taken into account that the HOF category not only includes specialized hard-object feeders (*Lophocebus albigena* and *Cebus apella*) [60,81], but also orangutans (*Pongo pygmaeus*), which are less specialized hard-object feeders but are not frugivores in a strict sense. All extant hominoids have a preference for ripe fruit, but the emphasis on leaves, soft fruits and hard food items various among the various species [7]—with orangutans consuming on average harder and unripe fruits more often than other great apes [13,32], especially as fallback foods [37].

Statistical Techniques

In order to offer insights into the dietary habits of species, hierarchical, complete-linkage (farthest neighbor method) cluster analyses based on Euclidean distances, and discriminant Canonical Variates Analyses (CVA) were used to analyse the extant and fossil data sets. Cluster analysis was intended to explore the similarities in microwear patterns between extant primates and extinct hominoids by using the above-mentioned three microwear variables. CVA, in turn, was intended to evaluate the reliability of these microwear variables for distinguishing between the various dietary categories defined for extant taxa, as well as to classify fossils to these categories. Extant taxa were thus included a priori in one of the three dietary categories described above, whereas the extinct hominoids were left unclassified and classified a posteriory on the basis of the classification probabilities derived by the analysis from Mahalanobis squared distances to extant group centroids. All statistical analyses were performed using the SPSS v. 11 statistical package.

Technical Considerations

Like every paleobiological approach, dental microwear analysis has its particular drawbacks and limitations. When data derived from different researchers are combined into a single analysis, interobserver error in microwear features is a major concern that can complicate the interpretation of the results [66,82]. This caveat applies to this study and should be borne in mind when interpreting our results. However, it is worth mentioning that several methodological precautions were adopted by us to minimize the error introduced. Thus, all the employed data were obtained through SEM imaging, which is less prone to error bias than other techniques such as light microscopy [83]. Moreover, the data used are based on procedures that, although not identical, are highly comparable because: (1) they were obtained using the same quantitative SEM-based technique; (2) microwear features were analyzed with a semiautomated image analysis procedures (primarily Microware 4.0), which results in lesser error rates [66]—with the exception of data taken from ref. [60], which employed traditional digitizer-based measurements; and (3) all the procedures employed followed standard methodological details, i.e., same wear facets selected for analysis, same instrumental settings (voltage, magnification and specimen detector), same micrograph resolution and analyzed surface, same measured microwear variables, etc. Finally, all the data measured in this analysis or taken from the literature were derived by experienced and highly trained microwear researchers, which diminishes the magnitude of error in microwear measurements [66,84]. Although the use of these methodological precautions cannot fully remove the error bias, they provide a reasonable degree of interobserver consistency, thereby ensuring the comparability of the data employed.

Results

Microwear Features

Among the three analyzed variables (Table 1 and 2, and Figure 2 and S1), pitting incidence best distinguishes among dietary categories [50,71], whereas microwear feature (especially striation) breadth allows to further refine paleodietary inferences [34]. With regard to pitting incidence, only *Pierolapithecus catalaunicus* and *Dryopithecus fontani* (represented by a single individual each) resemble extant HOF such as *Pongo pygmaeus* and *Lophocebus albigena*, which habitually consume hard and brittle items. Most of the remaining taxa are somewhat intermediate between *P. pygmaeus* and extant FMF such as *Pan troglodytes* and *Papio cynocephalus*. Although some differences between taxa/ localities must partly reflect interindividual variation (Table 1 and 2, and Figure S1), the pitting incidences of all the hominoids from Spain suggest some degree of sclerocarpy. This is most clear in *P. catalaunicus*, which further displays wider striations—consistent with a preference for hard foods [34]—than in other extinct taxa, in the range of extant HOF and most closely resembling *L. albigena* (Figure 2 A). The remaining taxa (including *Hispanopithecus crusafonti* and *Hispanopithecus laietanus*) are intermediate between extant HOF and FMF when both pitting incidence and striation breadth are considered simultaneously (Figure 2 A). In contrast, none of the taxa overlaps with extant FOL for any of the studied variables.

Compared with other hominoids from Western Eurasia (Figure 2, Figure S1), only *Griphopithecus alpani* and *Ouranopithecus macedoniensis* surpass the pitting incidences of the hominoids from the Iberian Peninsula (and even that of extant HOF in some instances). The former, however, display much narrower microwear features than extant HOF and *P. catalaunicus*, thus overlapping with the remaining studied taxa (Figure 2 A, B). *Hispanopithecus hungaricus* overlaps to a large extent with other species of *Hispanopithecus* in the various microwear variables, thus being rather intermediate between FMF and HOF, whereas *Oreopithecus bambolii* uniquely falls within the range of extant FMF for most individuals.

Multivariate Analyses

A cluster analysis based on microwear fabrics (Figure 3) yields two main clusters separating FOL and FMF (cluster A) from HOF (cluster B). Among extinct hominoids, the average values of *A. brevirostris* and *H. laietanus* are grouped with the mixed feeder *P. cynocephalus* and the soft-fruit eater *P. troglodytes* in subcluster A1.

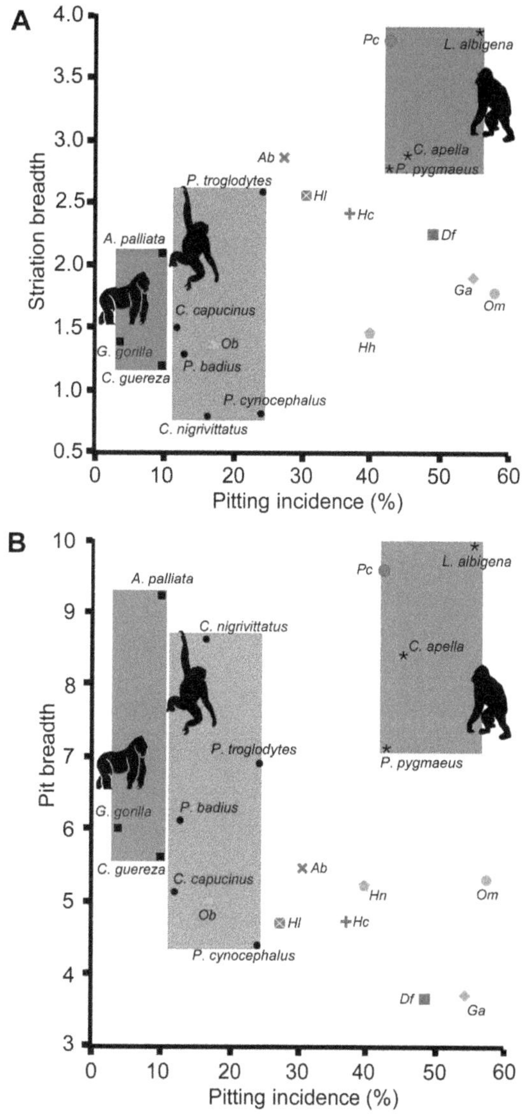

Figure 2. Bivariate plots of microwear feature breadth vs. pitting incidence. (**A**) Striation breadth and (**B**) pit breadth vs. pitting incidence based on species/locality means (Table 1). Fossil taxa abbreviations: *Ab, Anoiapithecus brevirostris; Df, Dryopithecus fontani; Ga, Griphopithecus alpani; Hc, Hispanopithecus crusafonti; Hh, Hispanopithecus hungaricus; Hl, Hispanopithecus laietanus; Ob, Oreopithecus bambolii; Om, Ouranopithecus macedoniensis; Pc, Pierolapithecus catalaunicus.* Different symbols are employed to distinguish each species. For species from multiple localities, only average values are shown (see individual values in Figure S1). The results derived in this study are depicted in red, whereas those taken from previous studies are shown in yellow. The polygons showing the variability of extant dietary categories are depicted in blue (folivores), green (mixed feeders/ frugivores) and magenta (hard-object feeders).

Table 2. Results of the microwear analysis for all the analyzed specimens.

Taxon	Locality	Age (Ma)	Catalog No.	Tooth	% Pits	Pit Breadth	Scratch Breadth
Fierolapithecus catalaunicus	ACM/BCV1	11.93	IPS21350	LM1	42.50	9.59	3.79
Anoiapithecus brevirostris	ACM/C1-E*	12.3–12.2	IPS35027	LM1	19.23	3.05	2.87
Anoiapithecus brevirostris	ACM/C3-Aj	11.94	IPS41712	LM1	20.00	6.27	2.49
Anoiapithecus brevirostris	ACM/C3-Aj	11.94	IPS43000	Lm2	42.37	4.59	3.23
Dryopithecus fontani	ACM/C3-Ae	11.85	IPS35026	LM2	48.81	3.66	2.25
Hispanopithecus crusafonti	TF	10.4–10.0	MGSB25314	Lm2	51.39	4.24	2.60
Hispanopithecus crusafonti	CP1	10.4–10.0	IPS1820	LM2	33.33	4.07	2.37
Hispanopithecus crusafonti	CP1	10.4–10.0	IPS1818	LM1	31.37	6.79	2.54
Hispanopithecus crusafonti	CP1	10.4–10.0	IPS1812	RM2	27.84	3.78	2.25
Hispanopithecus crusafonti	CP1	10.4–10.0	IPS1821	RM2	41.17	4.46	2.32
Hispanopithecus laietanus	CF	10.0–9.7	IPS34753	m1	34.61	6.77	2.44
Hispanopithecus laietanus	CLL1	9.72	IPS1763	Rm1	21.66	4.26	2.46
Hispanopithecus laietanus	CLL1	9.72	IPS1788	RM1	28.78	4.97	2.60
Hispanopithecus laietanus	CLL1	9.72	IPS1797	Rm1	37.07	4.84	2.07
Hispanopithecus laietanus	CLL1	9.72	IPS1800	Lm3	29.82	6.25	3.22

Pit and scratch breadths reported in μm. Estimated ages taken from ref. [3].
Abbreviations: IPS, collections of the Institut Català de Paleontologia Miquel Crusafont; M, upper molar (followed by tooth position); m, lower molar (followed by tooth position); MGSB, Museu de Geologia del Seminari Conciliar de Barcelona; R, right; L, left. See locality abbreviations in Table 1.

The average of *O. bambolii* is in turn included in subcluster A2, together with the remaining extant frugivores and all FOL, which are characterized by lower pitting incidences. The remaining fossil hominoids clump together with HOF in cluster B, displaying higher pit percentages. Average values of *P. catalaunicus, D. fontani, H. crusafonti* and *H. hungaricus* are grouped with *P. pygmaeus* and *C. apella* in subcluster B1, whereas *G. alpani* and *Ou. macedoniensis* cluster with *L. albigena* in subcluster B2 (see Figure S3 for individual variation). Overall, the analyses indicate a hard-object feeding component for many of the fossil hominoids, with the exception of *A. brevirostris, H. laietanus* and *O. bambolii*, which show greater affinities with FMF. None of the taxa clusters with extant FOL.

The CVA (Figure 4 and Table 3; see also Table S2) confirms that the investigated variables provide a satisfactory dietary discrimination (100% of extant taxa correctly classified, 64% in cross-validation). CV1 separates HOF (positive values) from FMF and FOL (negative values) mostly on the basis of pitting incidence, whereas CV2, more influenced by scratch and pit breadths, does not enable a clear distinction among dietary categories. The discriminant analysis (Table S3) based on the CVA classifies most of the taxa as HOF, except the average values of *A. brevirostris, H. laietanus* and *O. bambolii*, which are classified as FMF (Figure 4 A). When individual classifications for extinct taxa are analyzed, *A. brevirostris, H. laietanus* and *H. crusafonti* display some variation in individual classifications between HOF and FMF, whereas several individuals of *O. bambolii* are classified as FOL or HOF instead of FMF (Figure 4 B, Tables 4 and S4). This fact reflects dietary diversity in some of the taxa, which indicates that caution is required when interpreting species (*P. catalaunicus* and *D. fontani*) represented by a single individual. However, based on classification probabilities (Table S4), *P. catalaunicus* falls within the variation of extant HOF, unlike *D. fontani, G. alpani* and *Ou. macedoniensis* ($p<0.05$). *Anoiapithecus brevirostris* and most individuals of *O. bambolii* similarly fit well with extant FMF. In contrast, the classification of *Hispanopithecus* species as either HOF or FMF is not consistent among individuals and not well supported for most of them, suggesting that they were truly intermediate between these categories.

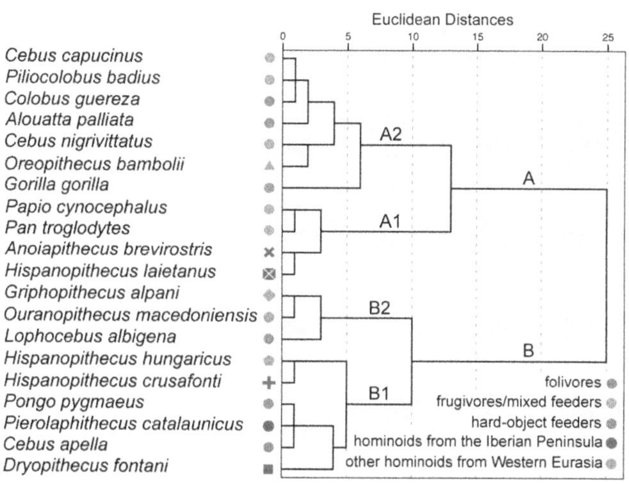

Figure 3. Cluster analysis based on dental microwear features. For species from multiple localities, only average values are shown (see individual values in Figure S2). Symbols and colors as in Figure 2.

Discussion

Miocene Hard-Object Feeders

Only the single individual of *P. catalaunicus* fits well with extant HOF, as shown by its high pitting incidence and broad scratches, thus resembling *L. albigena* and, especially, *P. pygmaeus*. The latter resemble other extant apes in preferring ripe fruits [2], but display thicker enamel as an adaptation to consume harder or unripe fruits, especially as fallback foods [37]. *L. albigena* is also a thick-enameled HOF that consumes fleshy fruits but seasonally forages on hard, brittle objects such as nuts and seeds [85,86]. Although the small sample size precludes a definitive conclusion, our results are consistent with *P. catalaunicus* being a HOF, as previously suggested based on its relatively thick enamel [8] and further confirmed by our multivariate analyses.

Our results also confirm previous inferences, based on pitting incidence, that *Ou. macedoniensis* was a hard-object specialist [10,14] and that *G. alpani* consumed hard fruits at least as often as orangutans [12,13]. These taxa are very similar in microwear features to one another, but differ from *P. catalaunicus* and extant HOF by displaying narrower microwear features. This condition of *P. catalaunicus* is more consistent with being a HOF, which compared to FMF and FOL have wider microwear striations (due to higher occlusal forces) [50] as well as larger pits (due to the higher amount of grit routinely ingested by these taxa) [53,87–89]. Differences in microwear feature size among extinct HOF might reflect their divergent habitats and ecological niches. Thus, the orthograde bodyplan with adaptations for vertical climbing and above-branch palmigrady of *P. catalaunicus* [4,29,90,91] suggests a strong arboreal commitment, as in orangutans and the other extant HOF (*L. albigena* and *C. apella*) [92,93]. In contrast, the postcranials of *G. alpani* suggest a pronograde bodyplan more suitable for semi-terrestrial quadrupedalism [5,94,95]. Similarly, the relatively large body mass [21] and open, pure C_3 environments inferred for *Ou. macedoniensis* based on the associated fauna [96,97] also agree with a semi-terrestrial locomotion. Microwear differences between *P. catalaunicus* and other extinct HOF might be thus attributable to differences in the mechanical properties of the food items found in the canopy as opposed to closer to the ground [15,88], in agreement with previous microwear inferences of a diet primarily based on hard, abrasive items (roots, tubers and/or grasses) for *Ou. macedoniensis* [14]. Alternatively, microwear differences among these taxa might be related to differences in the content of exogenous grit versus phytoliths in the foods consumed, as previously shown for incisor microwear [73]. Abrasive dust particles are more abundant but smaller on average in dry compared to humid environments [88], suggesting that the higher pitting incidences and lower striation breadths of *G. alpani* and *Ou. macedoniensis* might merely reflect their more open and drier habitats compared to both *P. catalaunicus* and extant HOF. Thus, these former taxa might have been predominantly (semi-)terrestrial hard-object feeders, whereas *P. catalaunicus* is best interpreted as an arboreal hard-fruit forager.

Miocene Soft-Fruit Eaters

Among the taxa analyzed, only *O. bambolii* is best interpreted as a soft frugivore, thus contradicting previous interpretations of a specialized folivorous diet [9,10]. Both pitting percentage and striation breadth suggest some dietary diversity (a few individuals show closer microwear resemblances to either FOL and HOF), as further confirmed by the multivariate analyses. *Cebus nigrivittatus*—a mainly frugivorous primate that further consumes a significant proportion of leaves [98]—might be a good analogue of *O. bambolii*, as shown by their almost identical pitting percentages

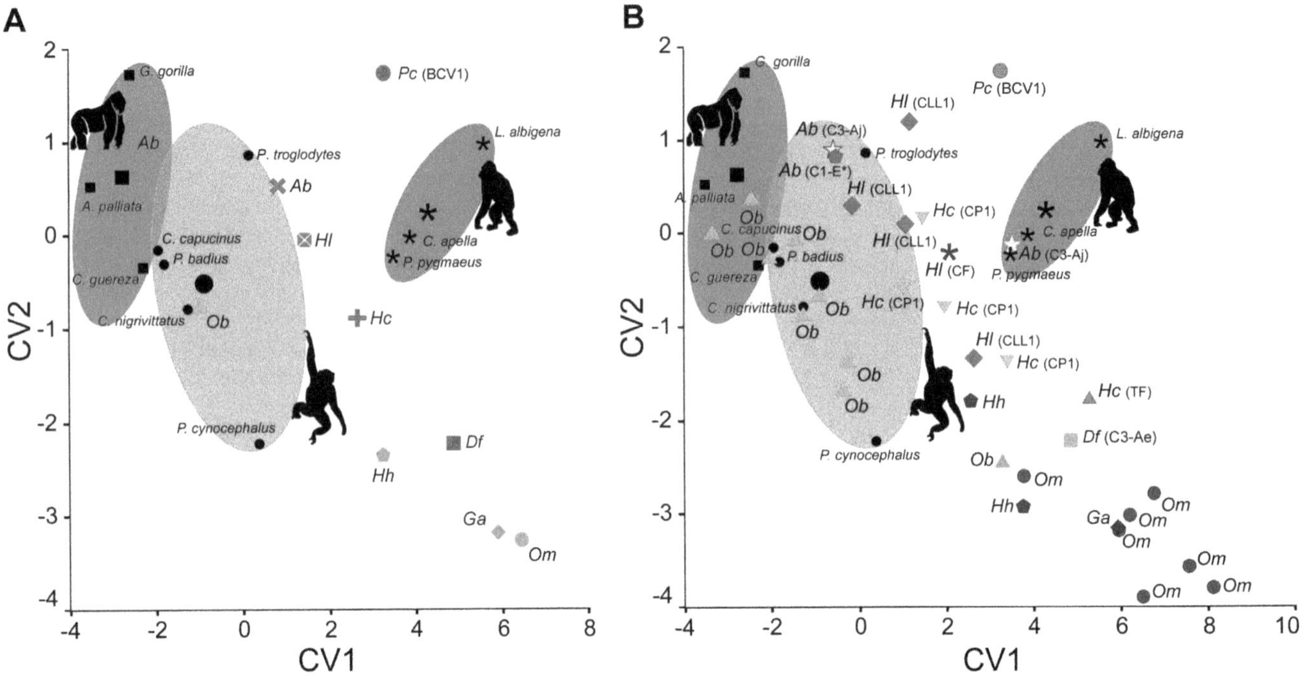

Figure 4. Results of the CVA based on three extant dietary categories and microwear variables. (A) Results based on mean/species locality data and (**B**) on individuals. Colored ellipses denote the morphospace defined by each extant group, black symbols extant taxa (group centroids by larger symbols). Symbols and colors in a as in Figure 2, whereas in B symbols and colors are different to show variability within fossil species.

(16.20% and 16.96%). However, the wide range of pitting incidence displayed by *O. bambolii* does point towards a high dietary flexibility, sporadically including leaves and hard fruits alike, instead of specialized folivory. Although the marked development of molar shearing crests in this taxon was interpreted as a folivorous specialization [9], its pronounced dental relief (with multiple accessory cusps, crests and cingula) more closely resembles that of omnivorous suoids rather than folivores [25,99], in agreement with our microwear results and its moderately thick enamel [100]. *O. bambolii* is thus best interpreted as an eclectic FMF, more clearly relying on soft fruits than other hominoids from Western Eurasia, but further exploiting other resources, possibly due to the trophic restrictions characteristic of insular environments [25,99].

Table 3. Results of the discriminant analysis per taxon/locality and for species average values.

Taxon	Locality	1st group	p	D^2	2nd group	D^2
P. catalaunicus	ACM/BCV1	HOF	0.192	3.296	FRU	22.219
A. brevirostris	ACM /C1-E*	FRU	0.375	1.961	FOL	5.012
A.brevirostris	ACM/C3-Aj	FRU	0.039	6.513	HOF	7.899
A. brevirostris	average	FRU	0.135	4.011	HOF	12.284
D. fontani	ACM/C3-Ae	HOF	0.041	6.403	FRU	36.059
H. crusafonti	TF	HOF	0.078	5.107	FRU	39.672
H. crusafonti	CP1	HOF	0.047	6.129	FRU	8.260
H. crusafonti	average	HOF	0.137	3.970	FRU	12.586
H. laietanus	CF	HOF	0.077	5.120	FRU	8.931
H. laietanus	CLL1	FRU	0.095	4.703	HOF	9.660
H. laietanus	average	FRU	0.067	5.410	HOF	8.612
G. alpani	Paşalar	HOF	0.001	14.400	FRU	53.895
H. hungaricus	Rudabánya	HOF	0.019	7.961	FRU	20.089
O. bambolii	Various	FRU	0.966	0.070	FOL	5.401
Ou. macedoniensis	Various	HOF	0.000	16.847	FRU	61.275

Abbreviations: D^2, Squared Mahalanobis distance; FOL, folivores; FRU, frugivores/mixed-feeders; HOF, hard-object feeders; p, classification probability.

Table 4. Summary results for dietary classification (predicted group) of the fossil individuals studied in this paper according to the discriminant analysis based on microwear features.

Hominoids from the Iberian Peninsula			
Taxon (locality)	**Folivores**	**Frugivores**	**Hard-object feeders**
P. catalaunicus (ACM/BCV1)	0	0	1
A. brevirostris (ACM/C1-E*)	0	1	0
A. brevirostris (ACM/C3-Aj)	0	1	1
D. fontani (ACM/C3-Ae)	0	0	1
H. crusafonti (TF)	0	0	1
H. crusafonti (CP1)	0	2	2
H. laietanus (CF)	0	0	1
H. laietanus (CLL1)	0	3	1
Total	0	7	8
Total (%)	0	46.67	53.33
Other hominoids from Western Eurasia			
Taxon (locality)	**Folivores**	**Frugivores**	**Hard-object feeders**
G. alpani (Paşalar)[1]	0	0	1
H. hungaricus (Rudabánya)[2]	0	0	2
O. bambolii (various locs.)[2]	2	6	1
Ou. macedoniensis (various locs.)[2]	0	0	7
Total	2	6	11
Total (%)	10.52	31.57	57.89

See Table 1 for locality abbreviations, and Tables S2 and S3 for further information of the discriminant analyses.
[1]No individual data available; based on the average value (N = 18) reported in ref. [12].
[2]Individual data taken from ref. [10].

Miocene Mixed Soft/Hard-Fruit Feeders

The remaining hominoids from Western Eurasia do not comfortably fall into any of the extant dietary categories, being somewhat intermediate between FMF and HOF. On average, *A. brevirostris* and *H. laietanus* are classified as FMF, whereas *D. fontani*, *H. crusafonti* and *H. hungaricus* display closer affinities with HOF. Individual values, however, are classified as both FMF and HOF, except for the two individuals of *H. hungaricus* and the single specimen of *D. fontani*. The restricted samples for the latter taxa do not enable to ascertain whether they significantly differed from *H. laietanus* and/or *H. crusafonti*. However, when pitting incidence and striation breadth are considered simultaneously, most of the analyzed individuals are intermediate between extant FMF and HOF, not overlapping with other extinct hominoids here interpreted as HOF. Our analyses therefore suggest that the diet of these taxa might have been intermediate between FMF and HOF (by including both soft and hard fruits to a large proportion), thus lacking an appropriate analog among the comparative sample. Such interpretation differs from previous inferences, based on microwear and shearing crest analyses [9–11], of a mainly frugivorous diet for *Hispanopithecus* species, and clearly discounts a folivorous diet for *H. laietanus* based on buccal microwear [101].

In these taxa, emphasis on hard-object feeding might have varied depending on the species and/or fluctuated depending on environments. Available samples are too small to adequately test whether hard-food items were consumed as fallback foods, depending on seasonal factors influencing resource availability [102], or whether these species displayed an eclectic foraging strategy on a regular, non-seasonal basis. For species of *Hispanopithecus*, paleoenvironmental reconstructions tend to favor the former hypothesis, since these taxa inhabited humid and subtropical to warm-temperate environments [59,75,103,104]. Such environments would have provided soft fruits at least during part of the year. However, the linear enamel hypoplasias frequently displayed by species of *Hispanopithecus* [105,106] indicate repeated episodes of malnutrition due to seasonal fluctuations in resource abundance (due to fruiting cycles) [103,106,107], thus suggesting that they might have consumed hard-food items as fallback foods during the unfavorable season. Whereas postcranial remains are unknown for *A. brevirostris* and very restricted for *D. fontani* [4], the more complete postcranial remains of *H. laietanus* and *H. hungaricus* clearly evidence a high arboreal commitment and more derived suspensory adaptations than in *P. catalaunicus* [4,77,90,95,108–110]. Species of *Hispanopithecus* are thus best interpreted as arboreal feeders, with their enhanced suspensory capabilities enabling a more efficient foraging on terminal branches [77,109].

Evolutionary Implications

Although hominoids are first recorded in Eurasia ca. 17 Ma, coinciding with the beginning of the Miocene Climatic Optimum [3,4,111], additional hominoid dispersal events between Africa and Europe probably took place afterwards [3,4,6]. Kenyapithecines [4,112] extended their range into Eurasia before 14 Ma—*G. alpani* and *Kenyapithecus kizili* [2,17]—and apparently gave rise to the Eurasian hominoid radiation [3,4,30]. Besides their likely semi-terrestrial locomotion [94], the dispersal of kenyapithecines was apparently facilitated by their HOF adaptations [4,7,8], enabling them to occupy subtropical and highly seasonal, single-canopied woodland/forest with abundant ground vegetation and more open areas [2]. This agrees with our results, suggesting that *G. alpani* mainly relied on hard food items. Available evidence suggests that kenyapithecines rapidly spread throughout Eurasia and vicariantly diversified into different clades [4,113], with pongines being recorded in Asia and dryopithecines in the Iberian Peninsula by ca. 12.5 Ma [4].

The European dryopithecines are alternatively interpreted as stem hominids [29,30], hominines [5,6], or the sister-taxon of Asian pongines [4,114]. Between 12.5 and 7 Ma, and despite a climatic trend towards cooling and increased seasonality [4,115], they experienced an adaptive radiation from both taxonomic and ecological viewpoints [4,6]. This radiation, including the acquisition of new locomotor and dietary adaptations, was probably related to the new selection pressures posed by the different biotopes present in Europe during the Middle and Late Miocene, coupled with regional paleoenvironmental differences and changes through time [4,6]. Both the habitat and diet of *G. alpani* from Turkey most closely resemble those of their Middle Miocene relatives from Africa [2,113,116]. In contrast, the habitats encountered by subsequent hominoids in both NE Spain and Central Europe were more humid, less seasonal, and more densely-forested [116–118]. Our results suggest that *P. catalaunicus* retained the ancestral HOF strategy while specializing for arboreal foraging. In contrast, *D. fontani* and *A. brevirostris* apparently displayed a somewhat more frugivorous diet (albeit with some sclerocarpic component), which might be interpreted as an early adaptive response to the new environmental conditions. Competitive exclusion coupled with increased environmental heterogeneity of the plant communities—which favors increased paleobiodiversity by multiplying the number of available ecological niches [119]—would have played a role in the dietary diversification of dryopithecines.

The dietary diversification of hominoids in Western Eurasia was further accentuated during the Late Miocene—as shown by the mixed soft-hard frugivorous condition of *Hispanopithecus*, the more frugivorous but versatile diet of *O. bambolii*, and the specialized hard-object feeding of *Ou. macedoniensis*. The adaptive trend towards increased soft frugivory in *Hispanopithecus* is at odds with concomitant climatic changes toward increased seasonality and lower temperatures. These environmental changes prompted the substitution of evergreen by deciduous trees [120] as well as the fragmentation of habitats suitable for frugivorous hominoids [103]. For some time, *Hispanopithecus* overcame such a paleoenvironmental deterioration thanks to new locomotor adaptations—presumably enabling a more efficient foraging on the canopy [77,109]—instead of exploiting a greater proportion of leafy material and/or foraging on the ground.

The extinction of hominoids in Europe was ultimately related to an increase in environmental uniformity and the resulting loss of suitable habitats [96]. In Western and Central Europe, it has been related to the substitution of (sub)tropical plants by deciduous trees [120]. At least in the Vallès-Penedès, however, this process was gradual, implying the generation of mosaic environments in which (sub)tropical elements became progressively restricted to lowland humid areas [4,103]. *Hispanopithecus* probably had to seasonally recourse to hard food items as fallback foods, until the reduction of its preferred habitat ultimately caused its extinction in Spain and elsewhere in Europe [4,96,103].

In Eastern Europe, *Ou. macedoniensis* survived longer than *Hispanopithecus* (ca. 8.0-7.5 vs. 9.5 Ma), probably thanks to a specialized terrestrial HOF diet adapted to open and arid habitats with a predominantly herbaceous vegetation with abundant C_3 grasses, bushes and herbs [96]. Although its extinction did not coincide with any major climatic shift, it was similarly related to strong seasonal variations [97]. The paleoenvironmental changes leading to the extinction of *Ou. macedoniensis* were apparently opposite to those experienced by *Hispanopithecus*, since in Eastern Europe bushy vegetation expanded over areas previously occupied by grasslands [94], whereas in Western and Central Europe it was rather accompanied by the substitution of (sub)tropical plants by deciduous trees [120].

O. bambolii survived much longer in the Tusco-Sardinian insular ecosystems by displaying a more frugivorous but highly versatile diet, allowing it to opportunistically exploit more fibrous plant materials and harder fruits alike. Its extinction does not offer much insight with regard to that of other apes from mainland Europe, since it is not related to any environmental change [121], but rather to the connection of its insular habitat to the mainland ca. 7 Ma—and the substitution of the endemic associated fauna [26,121].

Conclusions

Contrary to previous interpretations, our microwear analyses show that leaves and stems were not a primary dietary component for any hominoid from Western Eurasia, which are interpreted instead as frugivores (*O. bambolii*) and/or hard-object feeders (e.g., *P. catalaunicus*). Whereas some of the studied taxa fall comfortably within these two dietary categories, many of them (such as *Hispanopithecus* species) seem to be intermediate, suggesting that they have no extant dietary analog in the comparative sample.

From a evolutionary perspective, our results indicate that hominoids from Western Eurasia experienced a progressive dietary diversification between 14 and 7 Ma, from the presumably ancestral condition of (semi-)terrestrial hard-object feeding shown by *G. alpani*. In Western and Central Europe, this diversification might have been triggered by changes in habitat structure (more densely-forested environments), coupled with competitive exclusion and new locomotor adaptations related to arboreal feeding (as shown by *P. catalaunicus*). Other taxa from this area (especially the species of *Hispanopithecus*) apparently combined soft and hard fruits in their diets. The high behavioral plasticity of extant great apes allows them to survive in front of a marked environmental instability (resulting in spatial/temporal uncertainty of preferred fruit resources) [122,123]. Similarly, the suspensory adaptations of *Hispanopithecus* species (enabling a more efficient foraging on terminal branches), coupled with the exploitation of harder food items during the unfavorable season, might have allowed them to temporarily overcome the progressive environmental deterioration. Ultimately, however, the restriction and fragmentation of their preferred habitats would have led to their extinction from Western and Central Europe. In contrast, *Ou. macedoniensis* survived longer in the more open and arid landscapes of Eastern Europe by displaying a more terrestrial trophic niche based on hard food items, whereas *O. bambolii* persisted even longer in the Tusco-Sardinian Paleobioprovince by displaying a versatile frugivorous diet, until its insular ecosystem was connected to the mainland.

The failure by any of these taxa to adapt to folivory in the face of environmental changes towards increased seasonality might be attributable to their specialized (although diverging) trophic niches. The contrasting environmental changes experienced by the respective habitats of *Hispanopithecus* (more deciduous and open forests) and *Ou. macedoniensis* (less open and more bushy habitats), coupled with their strikingly divergent trophic niches, suggest that great-ape vulnerability to environmental change is not attributable to a frugivorous bias per se [122], but rather to the adaptation to whatever hyperspecialized trophic niche.

Supporting Information

Figure S1 Bivariate plots of microwear feature breadth vs. pitting incidence. (A) Striation breadth and (**B**) pit breadth vs. pitting incidence based on individual values reported in Table 2. See Figure 2 for the equivalent plots based on mean species/locality values reported in Table 1. Abbreviations as in Figure 2 (note that symbols and colors are different to show variability within fossil species).

Figure S2 Results of the cluster analysis based on dental microwear features for individual values. Note that symbols and colors are different to those of Figure 2 to show variability within fossil species. See Figure 3 for the results based on mean species/locality data.

Table S1 Sample sizes for the studied extinct species.

Table S2 Results of the CVA based on microwear features.

Table S3 Scores for the two canonical variates in extant and extinct taxa derived by the CVA.

Table S4 Individual results of the discriminant analysis based on the CVA.

Acknowledgments

We thank Isaac Casanovas-Vilar for the map of Figure 1, and Sebastià Calzada for the loan of the TF specimen.

Author Contributions

Conceived and designed the experiments: DDM DMA SMS. Performed the experiments: DDM. Analyzed the data: DDM DMA. Contributed to the writing of the manuscript: DDM DMA. Generated the figures: DDM.

References

1. Harrison T (2010) Dendropithecoidea, Proconsuloidea and Hominoidea (Catarrhini, Primates). In: Werdelin L, Sanders WJ, editors. Cenozoic mammals of Africa. Berkeley, University of California Press. pp. 429–469.
2. Andrews P, Kelley J (2007) Middle Miocene dispersals of apes. Folia Primatol 78: 328–343.
3. Casanovas-Vilar I, Alba DM, Garcés M, Robles JM, Moyà-Solà S (2011) Updated chronology for the Miocene hominoid radiation in Western Eurasia. Proc Natl Acad Sci USA 108: 5554–5559.
4. Alba DM (2012) Fossil apes from the Vallès-Penedès Basin. Evol Anthropol 21: 254–269.
5. Begun DR (2002) European hominoids. In: Hartwig WC, editor. The primate fossil record. Cambridge, Cambridge University Press. pp. 339–368.
6. Begun DR, Nargolwalla MC, Kordos L (2012) European Miocene hominids and the origin of the African ape and human clade. Evol Anthropol 21: 10–23.
7. Andrews P, Martin L (1991) Hominoid dietary evolution. Phil Trans R Soc B 334: 199–209.
8. Alba DM, Fortuny J, Moyà-Solà S (2010) Enamel thickness in Middle Miocene great apes *Anoiapithecus*, *Pierolapithecus* and *Dryopithecus*. Proc R Soc B 277: 2237–2245.
9. Ungar PS, Kay RF (1995) The dietary adaptations of European Miocene catarrhines. Proc Natl Acad Sci USA 92: 5479–5481.
10. Ungar PS (1996) Dental microwear of European Miocene catarrhines: evidence for diets and tooth use. J Hum Evol 31: 335–366.
11. Ungar PS (2005) Dental evidence for the diets of fossil primates from Rudabánya northeastern Hungary with comments on extant primate analogs and "noncompetitive" sympatry. Palaeontograph Ital 90: 97–111.
12. King T, Aiello LC, Andrews P (1999) Dental microwear of *Griphopithecus alpani*. J Hum Evol 36: 3–31.
13. King T (2001) Dental microwear and diet in Eurasian Miocene catarrhines. In: de Bonis L, Koufos GD, Andrews P, editors. Hominoid Evolution and Climatic Change in Europe, Vol. 2. Phylogeny of the Neogene Hominoid Primates of Eurasia. Cambridge, Cambridge University Press pp. 102–117.
14. Merceron G, Blondel C, de Bonis L, Koufos GD, Viriot L (2005) A new method of dental microwear analysis: Application to extant primates and *Ouranopithecus macedoniensis* (Late Miocene of Greece). Palaios 20: 551–561.
15. Teaford MF, Walker A (1984) Quantitative differences in dental microwear between primate species with different diets and a comment on the presumed diet of *Sivapithecus*. Am J Phys Anthropol 64: 191–200.
16. Alpagut B, Andrews P, Martin L (1990) New hominoid specimens from the Middle Miocene site at Pasalar, Turkey. J Hum Evol 19: 397–422.
17. Kelley J, Andrews P, Alpagut B (2008) A new hominoid species from the middle Miocene site of Paşalar, Turkey. J Hum Evol 54: 455–479.
18. Begun DR, Kordos L (1993) Revision of *Dryopithecus brancoi* Schlosser, 1901, based on the fossil hominoid material from Rudábanya. J Hum Evol 25: 271–285.
19. Kordos L, Begun DR (2001) Primates from Rudabánya: allocation of specimens to individuals, sex and age categories. J Hum Evol 40: 17–39.
20. de Bonis L, Koufos GD (1993) The face and the mandible of *Ouranopithecus macedoniensis*: description of new specimens and comparisons. J Hum Evol 24: 469–491.
21. de Bonis L, Koufos GD (1994) Our ancestors' ancestor: *Ouranopithecus* is a Greek link in human ancestry. Evol Anthropol 3: 75–83.
22. de Bonis L, Koufos GD (1997) The phylogenetic and functional implications of *Ouranopithecus macedoniensis*. In: Begun DR, Ward CV, Rose MD, editors. Function, phylogeny and fossils: Miocene hominoid evolution and adaptation. New York: Plenum Press. pp. 317–326.
23. Rook L, Harrison T, Engesser B (1996) The taxonomic status and biochronological implications of new finds of *Oreopithecus* from Baccinello (Tuscany, Italy). J Hum Evol 30: 3–27.
24. Harrison T, Rook L (1997) Enigmatic anthropoid or misunderstood ape? The phylogenetic status of *Oreopithecus bambolii* reconsidered. In: Begun DR, Ward CV, Rose MD, editors. Function, phylogeny and fossils: Miocene hominoid evolution and adaptation. New York: Plenum Press. pp. 327–362.
25. Moyà-Solà S, Köhler M (1997) The phylogenetic relationships of *Oreopithecus bambolii* Gervais, 1872. C R Acad Sci Paris (IIa) 324: 141–148.
26. Rook L, Oms O, Benvenuti M, Papini M (2011) Magnetostratigraphy of the Late Miocene Baccinello–Cinigiano basin (Tuscany, Italy) and the age of *Oreopithecus bambolii* faunal assemblages. Palaeogeogr Palaeoclimatol Palaeoecol 305: 286–294.
27. Alba DM, Moyà-Solà S, Casanovas-Vilar S, Galindo J, Robles JM, et al. (2006) Los vertebrados fósiles del Abocador de Can Mata (els Hostalets de Pierola, l'Anoia, Cataluña), una sucesión de localidades del Aragoniense superior (MN6 y MN7+8) de la cuenca del Vallès-Penedès. Campañas 2002-2003, 2004 y 2005. Est Geol 62: 295–312.
28. Alba DM, Casanovas-Vilar I, Robles JM, Moyà-Solà S (2011) Parada 3. El Aragoniense superior y la transición con el Vallesiense: Can Mata y la exposición paleontological de els Hostalets de Pierola. Paleontol Evol Mem. esp. 6: 95–109.
29. Moyà-Solà S, Köhler M, Alba DM, Casanovas-Vilar I, Galindo J (2004) *Pierolapithecus catalaunicus*, a new Middle Miocene great ape from Spain. Science 306: 1339–1344.
30. Moyà-Solà S, Alba DM, Almécija S, Casanovas-Vilar I, Köhler M, et al. (2009) A unique Middle Miocene European hominoid and the origins of the great ape and human clade. Proc Natl Acad Sci USA 106: 9601–9606.
31. Moyà-Solà S, Köhler M, Alba DM, Casanovas-Vilar I, Galindo J, et al. (2009) First partial face and upper dentition of the Middle Miocene hominoid *Dryopithecus fontani* from Abocador de Can Mata (Vallès-Penedès Basin, Catalonia, NE Spain): Taxonomic and phylogenetic implications. Am J Phys Anthropol 139: 126–145.
32. Ungar PS (1992) Dental evidence for diet in primates. Anthrop Közl 34: 141–155.
33. Kay RF, Ungar PS (1997) Dental evidence for diet in some Miocene catarrhines with comments on the effects of phylogeny on the interpretation of adaptation. In: Begun DR, Ward CV, Rose MD, editors. Function, phylogeny and fossils: Miocene hominoid evolution and adaptation. New York: Plenum Press. pp. 131–151.
34. DeMiguel D, Alba DM, Moyà-Solà S (2013) European pliopithecid diets revised in the light of dental microwear in *Pliopithecus canmatensis* and *Barberapithecus huerzeleri*. Am J Phys Anthropol 151: 573–582.
35. Schwartz GT (2000) Taxonomic and functional aspects of the patterning of enamel thickness distribution in extant large-bodied hominoids. Am J Phys Anthropol 111: 221–244.
36. Lucas P, Constantino P, Wood B, Lawn B (2008) Dental enamel as a dietary indicator. BioEssays 30: 374–385.
37. Vogel ER, van Woerden JT, Lucas PW, Utami Atmoko SS, van Schaik CP, et al. 2008 Functional ecology and evolution of hominoid molar enamel thickness: *Pan troglodytes schweinfurthii* and *Pongo pygmaeus wurmbii*. J Hum Evol 55: 60–74.

38. Olejniczak AJ, Tafforeau P, Feeney RNM, Martin LB (2008) Three-dimensional primate molar enamel thickness. J Hum Evol 54:187–195.
39. Dumont ER (1995) Enamel thickness and dietary adaptation among extant primates and chiropterans. J Mammal 76: 1127–1136.
40. Maas MC, Dumont ER (1999) Built to last: the structure, function, and evolution of primate dental enamel. Evol Anthropol 8: 133–152.
41. Grine FE, Sponheimer M, Ungar PS, Lee-Thorp J, Teaford MF (2012) Dental microwear and stable isotopes inform the paleoecology of extinct hominins. Am J Phys Anthropol 148: 285–317.
42. Ungar PS, Sponheimer M (2013) Hominin diets. In: Begun DR, editor. A companion to paleoanthropology. Chichester, Blackwell Publishing. pp. 165–182.
43. Sponheimer M, Lee-Thorp JA (2007) Hominin paleodiets: the contribution of stable isotopes. In: Henke W, Tattersall I, editors. Handbook of Paleoanthropology. Heidelberg, Springer Verlag. pp. 555–586.
44. Lee-Thorp J, Sponheimer M (2006) Contributions of biogeochemistry to understanding hominin dietary ecology. Yrbk Phys Anthropol 49: 131–148.
45. Lee-Thorp J, Sponheimer M, Passey BH, de Ruiter DJ, Cerling TE (2010) Stable isotopes in fossil hominin tooth enamel suggest a fundamental dietary shift in the Pliocene. Phil Trans Roy Soc B 365: 3389–3396.
46. Sponheimer M, et al. 2013 Isotopic evidence of early hominin diets. Proc Natl Acad Sci USA 110: 10513–10518.
47. Cerling TE, Wang Y, Quade J (1993) Expansion of C4 ecosystems as an indicator of global ecological change in the late Miocene. Nature 361: 344–345.
48. Cerling TE, Harris JM, MacFadden BJ, Leakey MG, Quade J, et al. 1997 Global vegetation change through the Miocene/Pliocene boundary. Nature 389: 153–158.
49. Morgan ME, Kingston JD, Marino BD (1994) Carbon isotope evidence for the emergence of C4 plants in the Neogene from Pakistan and Kenya. Nature 367: 162–165.
50. Teaford MF (1988) A review of dental microwear and diet in modern mammals. Scanning Microsc 2: 1149–1166.
51. Ungar PS, Grine FE, Teaford MF (2008) Dental microwear and diet of the Plio-Pleistocene hominin *Paranthropus boisei*. PLoS ONE 3: e2044.
52. Grine FE (1986) Dental evidence for dietary differences in Australopithecus and *Paranthropus*: a quantitative analysis of permanent molar microwear. J Hum Evol 15: 783–822.
53. Daegling DJ, Grine FE (1999) Terrestrial foraging and dental microwear in *Papio ursinus*. Primates 40: 559–572.
54. Teaford MF, Lytle JD (1996) Diet-induced changes in rates of human tooth microwear: a case study involving stone-ground maize. Am J Phys Anthropol 100: 143–147.
55. Lucas PW, Omar R, Al-Fadhalah K, Almusallam AS, Henry AG, et al. (2013) Mechanisms and causes of wear in tooth enamel: implications for hominin diets. J R Soc Interface 10: 20120923.
56. Strait DS, Weber GW, Constantino P, Lucas PW, Richmond BG, et al. (2012) Microwear, mechanics and the feeding adaptations of *Australopithecus africanus*. J Hum Evol 62: 165–168.
57. Lucas PW, Teaford MF (1995) Significance of silica in leaves eaten by long-tailed macaques (*Macaca fascicularis*). Folia Primatol 64: 30–36.
58. Gügel IL, Gruppe G, Kunzelmann KH (2001) Simulation of dental microwear: characteristic traces by opal phytoliths give clues to ancient human dietary behavior. Am J Phys Anthropol 114: 124–138.
59. Merceron G, Schulz E, Kordos L, Kaiser TM (2007) Paleoenvironment of *Dryopithecus brancoi* at Rudabánya, Hungary: evidence from dental meso- and micro-wear analyses of large vegetarian mammals. J Hum Evol 53: 331–349.
60. Teaford MF (1985) Molar microwear and diet in the genus *Cebus*. Am J Phys Anthropol 66: 363–370.
61. Ungar PS, Grine FE, Teaford MF, El Zaatari S (2006) Dental microwear and diets of African early *Homo*. J Hum Evol 50: 78–95.
62. Scott RS, Ungar PS, Bergstrom TS, Brown CA, Grine FE, et al. (2005) Dental microwear texture analysis shows within-species diet variability in fossil hominins. Nature 436: 693–695.
63. Scott RS, Teaford MF, Ungar PS (2012) Dental microwear texture and anthropoid diets. Am J Phys Anthropol 147: 551–579.
64. Ungar PS, Krueger KL, Blumenschine RJ, Njau JK, Scott RS (2012) Dental microwear texture analysis of hominins recovered by the Olduvai Landscape Paleoanthropology Project, 1995-2007. J Hum Evol 63: 429–437.
65. Merceron G, Scott J, Scott RS, Geraads D, Spassov N, et al. (2009) Folivory or fruit/seed predation for *Mesopithecus*, an earliest colobine from the late Miocene of Eurasia? J Hum Evol 57: 732–738.
66. Grine FE, Ungar PS, Teaford MF (2002) Error rates in dental microwear quantification using scanning electron microscopy. Scanning 24: 144–153.
67. Merceron G, Koufos GD, Valentin X (2009) Feeding habits of the first European colobine, Mesopithecus (Mammalia, Primates): evidence from a comparative dental microwear analysis with modern cercopithecids. Geodiv 31: 865–878.
68. Merceron G, Tafforeau P, Marivaux L (2010) Dietary reconstruction of the Amphipithecidae (Primates, Anthropoidea) from the Paleogene of South Asia and paleoecological implications. J Hum Evol 59: 96–108.
69. Kay RF (1977) The evolution of molar occlusion in the Cercopithecidae and Early Catarrhines. Am J Phys Anthropol 46: 327–352.
70. Grine FE, Ungar PS, Teaford MF, El-Zaatari S (2006) Molar microwear in *Praeanthropus afarensis*: Evidence for dietary stasis through time and under diverse palaeoecological conditions. J Hum Evol 51: 297–319.
71. El-Zaatari S, Grine FE, Teaford MF, Smith AH (2005) Molar microwear and dietary reconstructions of fossil Cercopithecoidea from the Plio-Pleistocene deposits of South Africa. J Hum Evol 49: 180–205.
72. Teaford MF (1985) Molar microwear and diet in the genus *Cebus*. Am J Phys Anthropol 66: 363–370.
73. Ungar PS (1994) Incisor microwear of Sumatran anthropoid primates. Am J Phys Anthropol 94: 339–363.
74. Alba DM, Fortuny J, Perez de los Ríos M, Zanolli C, Almécija S, et al. (2013) New dental remains of *Anoiapithecus* and the first appearance datum of hominoids in the Iberian Peninsula. J Hum Evol 65: 573–584.
75. Begun DR (1992) *Dryopithecus crusafonti* sp. nov., a new Miocene hominoid species from Can Ponsic (Northeastern Spain). Am J Phys Anthropol 87: 291–309.
76. Golpe Posse JM (1993) Los Hispanopitecos (Primates, Pongidae) de los yacimientos del Vallès-Penedès (Cataluña, España). II: Descripción del material existente en el Instituto de Paleontología de Sabadell. Paleontol Evol 26–27: 151–224.
77. Alba DM, Almécija S, Casanovas-Vilar I, Méndez JM, Moyà-Solà S (2012) A partial skeleton of *Hispanopithecus laietanus* from Can Feu and the mosaic evolution of crown-hominoid positional behaviors. PLoS ONE 7: e39617.
78. Begun DR, Moyá-Sola S, Köhler M (1990) New Miocene hominoid specimens from Can Llobateres (Vallès Penedès, Spain) and their geological and paleoecological context. J Hum Evol 19: 255–268.
79. Alba DM, Casanovas-Vilar I, Almécija S, Robles JM, Arias-Martorell J, et al. (2012) New dental remains of *Hispanopithecus laietanus* (Primates: Hominidae) from Can Llobateres 1 and the taxonomy of Late Miocene hominoids from the Vallès-Penedès Basin (NE Iberian Peninsula). J Hum Evol 63: 231–246.
80. Teaford MF, Maas MC, Simons EL (1996) Dental microwear and microstructure in Early Oligocene Primates from the Fayum, Egypt: Implications for diet. Am J Phys Anthropol 101: 527–543.
81. McGraw WS, Pampush JD, Daegling DJ (2012) Brief communication: Enamel thickness and durophagy in mangabeys revisited. Am J Phys Anthropol 147: 326–333.
82. Mihlbachler MC, Beatty BL, Caldera-Siu A, Chan D, Lee R (2012) Error rates in dental microwear analysis using light microscopy. *Palaeontol Electron* 15: 22p.
83. Scott R, Schubert B, Grine FE, Teaford MF (2008) Low magnification microwear: Questions of precision and repeatability. J Hum Evol 28: 139A.
84. Purnell MA, Hart PJB, Baines DC, Bell MA (2006) Quantitative analysis of dental microwear in threespine stickleback: a new approach to analysis of trophic ecology in aquatic vertebrates. J Anim Ecol 75: 967–977.
85. Poulsen JR, Clark CJ, Smith TB (2001) Seasonal variation in the feeding ecology of the grey-cheeked mangabey (*Lophocebus albigena*) in Cameroon. Am J Primatol 54: 91–105.
86. Lambert JE, Chapman CA, Wrangham RW, Conklin-Brittain NL (2004) Hardness of cercopithecine foods: Implications for the critical function of enamel thickness in exploiting fallback foods. Am J Phys Anthropol 125: 363–368.
87. Teaford MF, Oyen OJ (1989) Differences in the rate of molar wear between monkeys raised on different diets. J Dental Res 68: 1513–1518.
88. Ungar PS, Teaford MF, Glander KE, Pastor RF (1995) Dust accumulation in the canopy: A potential cause of dental microwear in primates. Am J Phys Anthropol 97: 93–99.
89. Daegling DJ, McGraw WS, Ungar PS, Pampush JD, Vick AE, et al. (2011) Hard-object feeding in sooty mangabeys (*Cercocebus atys*) and interpretation of early hominin feeding ecology. PLoS ONE 6: e23095.
90. Alba DM, Almécija S, Moyà-Solà S (2010) Locomotor inferences in *Pierolapithecus* and *Hispanopithecus*: Reply to Deane and Begun (2008). J Hum Evol 59: 143–149.
91. Almécija S, Alba DM, Moyà-Solà S (2009) *Pierolapithecus* and the functional morphology of Miocene ape hand phalanges: paleobiological and evolutionary implications. J Hum Evol 57: 284–297.
92. Nakatsukasa M (1996) Locomotor differentiation and different skeletal morphologies in mangabeys (*Lophocebus* and *Cercocebus*). Folia Primatol 66: 15–24.
93. Youlatos D (1998) Positional behavior of two sympatric Guianan capuchin monkeys, the brown capuchin (*Cebus apella*) and the wedge-capped capuchin (*Cebus olivaceus*). Mammalia 62: 351–365.
94. Ersoy A, Kelley J, Andrews P, Alpagut B (2008) Hominoid phalanges from the middle Miocene site of Paşalar, Turkey. J Hum Evol 54: 518–529.
95. Begun DR (1992) Phyletic diversity and locomotion in primitive European hominids. Am J Phys Anthropol 87: 311–340.
96. Merceron G, Kaiser TM, Kostopoulos DS, Schulz E (2010) Ruminant diets and the Miocene extinction of European great apes. Proc R Soc B 277: 3105–3112.
97. Merceron G, Kostopoulos DS, Bonis Ld, Fourel F, Koufos GD, et al. (2013) Stable isotope ecology of Miocene bovids from northern Greece and the ape/monkey turnover in the Balkans. J Hum Evol 65: 185–198.
98. Robinson JG (1986) Seasonal variation in use of time and space by the wedge-capped capuchin monkey, *Cebus oliuaceus*: Implications for foraging theory. Smithsonian Contrib Zool 431: 1–60.

99. Alba DM, Moyà-Solà S, Köhler M, Rook L (2001) Heterochrony and the cranial anatomy of *Oreopithecus*: some cladistic fallacies and the significance of developmental constraints in phylogenetic analysis. In: de Bonis L, Koufos GD, Andrews P, editors. Hominoid Evolution and Climatic Change in Europe, Vol. 2. Phylogeny of the Neogene Hominoid Primates of Eurasia. Cambridge, Cambridge University Press. pp. 284–315.
100. Zanolli C, Rook L, Macchiarelli R (2010) Analyse structurale à haute résolution des dents de *Oreopithecus bambolii*. Ann Univ Ferrara Museol Sci Natural 6: 69–76.
101. Galbany J, Moyà-Solà S, Pérez-Pérez A. (2005) Dental microwear variability on buccal tooth enamel surfaces of extant Catarrhini and the Miocene fossil *Dryopithecus laietanus* (Hominoidea). Folia Primatol 76: 325–341.
102. Teaford MF, Robinson JG (1989) Seasonal or ecological differences in diet and molar microwear in *Cebus nigrivittatus*. Am J Phys Anthropol 80: 391–401.
103. Marmi J, Casanovas-Vilar I, Robles JM, Moyà-Solà S, Alba DM (2012) The paleoenvironment of *Hispanopithecus laietanus* as revealed by paleobotanical evidence from the Late Miocene of Can Llobateres 1 (Catalonia, Spain). J Hum Evol 62: 412–423.
104. Costeur L (2005) Cenogram analysis of the Rudabánya mammalian community: palaeoenvironmental interpretations. Palaeontogr Ital 90: 303–307.
105. Skinner MF, Dupras TL, Moyà-Solà S (1995) Periodicity of linear enamel hipoplasia among Miocene Dryopithecus from Spain. J Paleopathol 7: 195–222.
106. Eastham L, Skinner MM, Begun DR (2009) Resolving seasonal stress in the Late Miocene hominoid *Hispanopithecus laietanus* through the analysis of the dental developmental defect linearenamel hypoplasia. J Vert Paleontol 29: 91A.
107. Skinner MF, Hopwood D (2004) Hypothesis for the causes and periodicity of repetitive linear enamel hypoplasia in large, wild African (*Pan troglodytes* and *Gorilla gorilla*) and Asian (*Pongo pygmaeus*) apes. Am J Phys Anthropol 123: 216–235.
108. Moyà-Solà S, Köhler M (1996) A *Dryopithecus* skeleton and the origins of great-ape locomotion. Nature 379: 156–159.
109. Almécija S, Alba DM, Moyà-Solà S, Köhler M (2007) Orang-like manual adaptations in the fossil hominoid *Hispanopithecus laietanus*: first steps towards great ape suspensory behaviours. Proc R Soc B 274: 2375–2384.
110. Deane AS, Begun DR (2008) Broken fingers: retesting locomotor hypotheses for fossil hominoids using fragmentary proximal phalanges and high-resolution polynomial curve fitting (HR-PCF). J Hum Evol 55: 691–701.
111. Böhme M, Aziz HA, Prieto J, Bachtadse V, Schweigert G (2011) Bio-magnetostratigraphy and environment of the oldest Eurasian hominoid from the Early Miocene of Engelswies (Germany). J Hum Evol 61: 332–339.
112. Ward SC, Duren DL (2002) Middle and Late Miocene African Hominoids. In: Hartwig WC, editor. The primate fossil record. Cambridge, Cambridge University Press. pp. 385–397.
113. Andrews P, Bernor RL (1999) Vicariance biogeography and paleoecology of Eurasian Miocene hominoid primates. In: Agusti J, Rook L, Andrews, P, editors.The evolution of Neogene terrestrial ecosystems in Europe. Cambridge, Cambridge University Press. pp. 454–487.
114. Pérez de los Ríos M, Moyà-Solà S, Alba DM (2012) The nasal and paranasal architecture of the Middle Miocene ape *Pierolapithecus catalaunicus* (Primates: Hominidae): Phylogenetic implications. J Hum Evol 63: 497–506.
115. Utescher T, Bruch AA, Micheels A, Mosbrugger V, Popova S (2011) Cenozoic climate gradients in Eurasia e a palaeo-perspective on future climate change? Palaeogeogr Palaeoclimatol Palaeoecol 304: 351–358.
116. Andrews P (1996) Palaeoecology and hominoid palaeoenvironments. Biol Rev 71: 257–300.
117. Casanovas-Vilar I, Alba DM, Moyà-Solà S, Galindo J, Cabrera L, et al. (2008) Biochronological, taphonomical and paleoenvironmental background of the fossil great ape *Pierolapithecus catalaunicus* (Primates, Hominidae). J Hum Evol 55: 589–603.
118. DeMiguel D, Azanza B, Morales J (2011) Paleoenvironments and paleoclimate of the Middle Miocene of central Spain: A reconstruction from dental wear of ruminants. Palaeogeogr Palaeoclimatol Palaeoecol 302: 452–463.
119. Merceron G, Costeur L, Maridet O, Ramdarshan A, Göhlich UB (2012) Multi-proxy approach detects heterogeneous habitats for primates during the Miocene climatic optimum in Central Europe. J Hum Evol 63: 150–161.
120. Agustí J, Sanz de Siria A, Garcés M (2003) Explaining the end of the hominoid experiment in Europe. J Hum Evol 45: 145–153.
121. Matson SD, Rook R, Oriol O, Fox DL (2012) Carbon isotopic record of terrestrial ecosystems spanning the Late Miocene extinction of *Oreopithecus bambolii*, Baccinello Basin (Tuscany, Italy). J Hum Evol 63: 127–139.
122. Potts R (2004) Paleoenvironments and the evolution of adaptability in great apes. In: Russon AE, Begun DR, editors. The evolution of thought. Evolutionary origins of great ape intelligence. Cambridge, Cambridge University Press. pp. 237–258.
123. Alba DM (2010) Cognitive inferences in fossil apes (Primates: Hominoidea): does encephalization reflect intelligence? J Anthropol Sci 88: 11–48.

2

A New Approach for the Determination of Ammonite and Nautilid Habitats

Isabelle Kruta[1,2]*, Neil H. Landman[1], J. Kirk Cochran[3]

1 Division of Paleontology American Museum of Natural History, New York, New York, United States of America, 2 Department of Geology and Geophysics, Yale University, New Haven, Connecticut, United States of America, 3 School of Marine and Atmospheric Sciences, Stony Brook University, Stony Brook, New York, United States of America

Abstract

Externally shelled cephalopods were important elements in open marine habitats throughout Earth history. Paleotemperatures calculated on the basis of the oxygen isotope composition of their shells can provide insights into ancient marine systems as well as the ecology of this important group of organisms. In some sedimentary deposits, however, the aragonitic shell of the ammonite or nautilid is poorly or not preserved at all, while the calcitic structures belonging to the jaws are present. This study tests for the first time if the calcitic jaw structures in fossil cephalopods can be used as a proxy for paleotemperature. We first analyzed the calcitic structures on the jaws of Recent *Nautilus* and compared the calculated temperatures of precipitation with those from the aragonitic shell in the same individuals. Our results indicate that the jaws of Recent *Nautilus* are secreted in isotopic equilibrium, and the calculated temperatures approximately match those of the shell. We then extended our study to ammonites from the Upper Cretaceous (Campanian) Pierre Shale of the U.S. Western Interior and the age-equivalent Mooreville Chalk of the Gulf Coastal Plain. In the Pierre Shale, jaws occur *in situ* inside the body chambers of well-preserved *Baculites* while in the Mooreville Chalk, the jaw elements appear as isolated occurrences in the sediment and the aragonitic shell material is not preserved. For the Pierre Shale specimens, the calculated temperatures of well-preserved jaw material match those of well-preserved shell material in the same individual. Analyses of the jaw elements in the Mooreville Chalk permit a comparison of the paleotemperatures between the two sites, and show that the Western Interior is warmer than the Gulf Coast at that time. In summary, our data indicate that the calcitic jaw elements of cephalopods can provide a reliable geochemical archive of the habitat of fossil forms.

Editor: Daphne Soares, University of Maryland, United States of America

Funding: IK was funded by AMNH Lerner Gray Postgraduate doctoral fellowship. Funding for isotopic analyses was provided by the N.D. Newell Fund (AMNH). The funders had no role in study design, data collection and analysis, decision to publish, or preparation of the manuscript.

Competing Interests: The authors have declared that no competing interests exist.

* E-mail: ikruta@amnh.org

Introduction

Ammonoids, an extinct class of cephalopods, constitute one of the best documented fossil groups [1]. They are restricted to a marine habitat and exhibit a broad geographic and stratigraphic range. Because the shell is composed of calcium carbonate, ammonites can be used to provide insights into ancient seawater temperatures [2], [3], [4], [5], [6], [7], [8], [9]. Calculation of paleotemperatures also provides information about the ecology (depth distribution and habitat) of this important group of organisms [9], [10], [11]. Studies of the stable isotope composition of the Recent ectocochleate cephalopod *Nautilus* have demonstrated that the $\delta^{18}O$ of the aragonitic shell accurately reflects the temperature of the sea water during the secretion of the shell [12], [13]. [14], [15], [16], [17]. This relationship is assumed to apply as well to externally shelled fossil cephalopods such as ammonites and nautilids [9], [18], [19], [20]. A prerequisite for the use of fossil shell material in such an analysis is that the shell must be well preserved [21]. In some sedimentary deposits, however, the aragonitic shell of the ammonoid or nautilid is poorly or not preserved at all, while the calcitic structures belonging to the jaws are present (e.g., Jurassic Solnhofen Plattenkalk from Germany; Lower Cretaceous Bassin Vocontien from France). Previous studies have used ammonite jaw material as a proxy for paleotemperature [22], [23], but it is unclear in these studies if 1) these structures were secreted in isotopic equilibrium with sea water (defined as the same temperature-dependent fractionation of aragonite/calcite relative to water as that of other aragonite/calcite secreting mollusks) and 2) the state of preservation was sufficiently adequate to retain the original isotopic composition. The aim of this study is to test the hypothesis that unaltered calcitic jaw structures from fossil ammonites can be used to reconstruct paleotemperatures. To begin, we analyzed the calcitic structures on the upper and lower jaws of Recent *Nautilus* and compared the calculated water temperatures with those of the aragonitic shell in the same individuals. The shell and jaws do not exhibit the same mineralogy (aragonite versus calcite), and are not secreted by the same tissue, thus representing independent systems. We then extended our studies to fossil *Baculites* from the Upper Cretaceous of North America. Our studies reveal that both the outer shell and the calcitic jaw elements in fossil ammonites yield reliable sea water temperatures provided that both features are well preserved and retain the original mineralogy and microstructure.

Materials and Methods

Ethics statement

The species of *Nautilus pompilius* (Mollusca: Cephalopoda) is not endangered or protected. The specimens of *Nautilus* were collected with the approval of the Department of Fisheries and Environment Unit of Vanuatu and imported to the American Museum of Natural History with the authorization of the U.S. Fish and Wildlife Service. Copies of the permits are held by the American Museum of Natural History (AMNH) where the specimens are deposited.

Nautilus

We sampled eight specimens of *Nautilus pompilius* from Vanuatu captured in July 2004. All specimens are mature individuals and both sexes are represented. The aragonitic shell and the calcitic coverings of the chitinous jaws were sampled for each individual. The outer shell wall was sampled at the aperture of the body chamber. The rostrum of the upper jaw bears a thick arrowed-shaped calcitic structure called the rhyncholite. The calcitic covering of the lower jaw, which is thinner than the rhyncholite, features calcitic denticules on the oral surface and is called the conchorhynch (Fig. 1). Both calcitic structures were sampled and the isotopic results are listed in Table 1. The rhyncholite and conchorhynch are mostly composed of calcite, but a thin aragonitic layer appears at the contact beween the conchorhynch/rhyncholite and the chitinous part of the jaw [24]. In order to avoid contamination with this aragonitic layer, we sampled the anterior tip of the rhyncholite on the upper jaw and the denticles and the inner part of the conchorhynch on the lower jaw (Fig. 1). The samples were powdered and treated following the method in Allmon *et al.* [25] in order to remove the organic material present in the shell and in the lower jaw. The samples were washed in 15% H_2O_2 for three hours. The H_2O_2 was then pipeted off and flushed three times with methanol (99.9%).

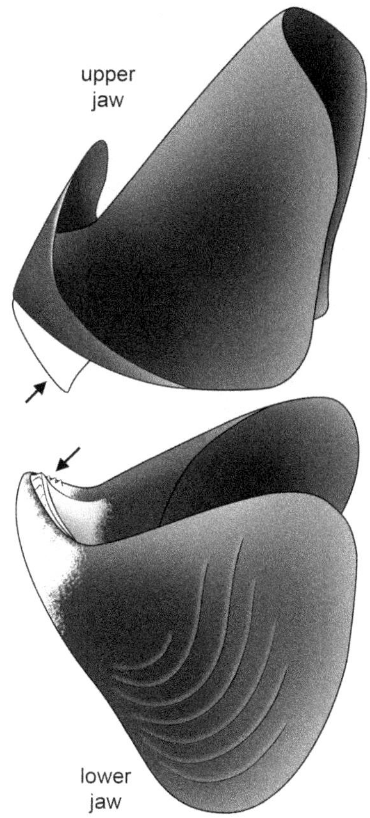

Figure 1. Lateral view of the upper and lower jaws of *Nautilus pompilius*. Calcitic structures are represented in white, while the main chitinous part of the jaws is displayed in black. Arrows indicate where samples were taken for isotopic analyses.

Fossil material

The first set of ammonite samples consists of the outer shell wall as well as jaw elements of *Baculites* sp. (smooth) from the Upper Cretaceous (lower Campanian) Gammon Ferruginous Member of the Pierre Shale, Butte County, South Dakota (Fig. 2) (see [26] for locality information). This locality was selected based on the abundance of extremely well preserved specimens of *Baculites* (Fig. 3A) with pieces of the aragonitic nacreous shell material still attached to the steinkern (composite internal molds) and jaw elements preserved inside the body chamber of the same individuals (*in situ*). While the shell is quite commonly preserved at this locality, jaws inside the body chamber are rare [26] and only seven specimens with both shell and jaw could be sampled for our analysis. The jaw material consists of the calcitic covering of the lower jaw, called the aptychus (Fig. 3B). For comparison, we also analyzed well-preserved shell material from two specimens of *Baculites* sp. (AMNH 78053 and 51754) without jaws inside the body chamber.

The second set of samples consists of aptychi (i.e. the calcitic valves of the lower jaws) preserved loose in the sediment from the time-equivalent lower Campanian Mooreville Chalk, Greene County, Alabama (see [26] for a discussion of the stratigraphic relationships). No outer shell wall material is present at this locality and only the calcitic aptychi are preserved (X-ray diffraction analysis in [26]). The aptychi are attributed to *Baculites* sp. (smooth) by comparison with the jaws preserved inside the body chambers of this species in South Dakota [26]. The calcitic aptychi were sampled for isotopic composition.

Subsamples of the shell were examined under the SEM (Fig. 4) in order to evaluate the microstructure and assign each sample a Preservation Index (PI) ranging from 1 (poor) to 5 (excellent) for the preservation of the nacreous shell wall [21]. We developed a new approach to evaluate the state of preservation of the aptychi. The Preservation Index for the aptychi is based on the quality of preservation of the microstructure. Because Recent material is unavailable for comparison, we selected an example of the best preserved aptychus (AMNH 54277) from the lower Campanian of Alabama. In AMNH 54277, the calcitic increments are identifiable and the main lamellar layer (R1) and the outer layer (R2) are well defined in this specimen (see [27] for a discussion of aptychus microstructure). We assigned a high Preservation Index (5 = excellent preservation) if the two layers (R1 and R2) could be identified and/or if calcitic increments could be observed (for very small pieces of aptychus, the outer layer was not always present). A low Preservation Index indicates specimens with massive calcite without any indication of layers. Aptychi samples for SEM analyses were embedded in epoxy, ground, polished, and etched with EDTA from 2 to 5 minutes. Preservation Index values of the outer shell and aptychus are listed in Table 2. To prepare the specimens for analysis, the surfaces of the aptychi from the Mooreville Chalk were scraped to remove extraneous material and then cleaned with a sonifier, and sampled under the microscope. The fossil material is reposited at the American Museum of Natural History (AMNH). Additional material is reposited at the Black Hills Institute of Geological Research (BHI), Hill City, South Dakota.

Table 1. Isotope data for *Nautilus*.

Recent *Nautilus* (Vanuatu)	Sample	Specimen number	Mineralogy	$\delta^{18}O^{\dagger}$ (‰)	$\delta^{13}C^{\dagger}$ (‰)	T (°C)
Nautilus pompilius	shell	AMNH 310420	aragonite	0.95	2.12	18.3
"	rhyncholite	"	calcite	−0.10	0.06	18.2
"	conchorhynch	"	calcite	−0.02	−0.49	17.8
"	shell	AMNH 310433	aragonite	0.77	1.50	19.1
"	conchorhynch	"	calcite	−0.34	0.36	19.3
"	shell	AMNH 310434	aragonite	0.99	2.22	18.1
"	conchorhynch	"	calcite	−0.23	0.06	18.8
"	shell	AMNH 310435	aragonite	0.89	1.54	18.6
"	rhyncholite	"	calcite	−0.46	0.05	19.8
"	shell	AMNH 310432	aragonite	1.28	1.71	16.7
"	rhyncholite	"	calcite	−0.37	0.3	19.4
"	shell	AMNH 310421	aragonite	1.33	1.47	16.5
"	rhyncholite	"	calcite	−0.07	0.64	18.1
"	shell	AMNH 310422	aragonite	1.47	1.63	15.8
"	rhyncholite	"	calcite	−0.51	0.41	20.0
"	shell	AMNH 310431	aragonite	1.77	1.36	14.4
"	rhyncholite	"	calcite	−0.27	−0.8	19
Mean ± 1σ shell T						*17.2±1.6*
Mean ± 1σ jaw T						*18.9±0.8*

†Values relative to VPDB.

Isotopic analyses

The isotopic analyses were performed at the Keck Paleoenvironmental & Environmental Stable Isotope Laboratory at the University of Kansas (KPESIL). All samples were reacted with phosphoric acid to release CO_2, which was then analyzed for C and O isotopes using a Thermo Finnigan dual inlet MAT253 isotope ratio mass spectrometer (IRMS). Three standards were used- NIST (National Institute of Standards) NBS-18, NBS-19, and an internally calibrated calcite standard-which were included with each run in order to generate a three point calibration curve to the VPBD scale. A fourth standard, NIST 88b (dolomitic limestone), was used for quality control.

The isotopic temperatures were calculated using the equation of Grossman and Ku [28] for aragonite:

$$T(^{\circ}C) = 21.8 - 4.69\,(\delta^{18}O_{arag} - \delta^{18}O_{sw}) \quad (1)$$

and the equation of O'Neil *et al.* [29] for calcite:

$$T(^{\circ}C) = 16.9 - 4.38\,(\delta^{18}O_{calcite} - \delta^{18}O_{sw}) \quad (2)$$

where T is the temperature of the water in which the carbonate precipitated (°C), $\delta^{18}O_{arag}$ and $\delta^{18}O_{calcite}$ are the values of $\delta^{18}O$ for the calcium carbonate (VPDB), and $\delta^{18}O_{sw}$ is the value of $\delta^{18}O$ of the seawater (SMOW), respectively. For the *Nautilus* material we used a value of $\delta^{18}O_{sw}$ of +0.2‰ for seawater at Vanuatu [http://data.giss.nasa.gov/o18data/], and for the fossil material we used a value of $\delta^{18}O_{sw}$ of -1‰ for Cretaceous seawater [30], [31].

Results

Data from Recent Nautilus

The results for *Nautilus pompilius* are reported in Table 1. The values of $\delta^{18}O$ were converted to temperature following equations (1) and (2). The temperatures calculated for a rhyncholite and a conchorhynch from the same specimen are nearly the same, the difference being only 0.4°C. Thus, the two elements are considered as recording the same temperature. The calculated temperatures of the calcitic jaw elements range from 17.8°C to 20°C, averaging 18.9°C±0.8°C (1σ). Compared with the temperatures of the calcitic jaw elements, those of the outer shell are more variable. The temperatures of the outer shell range from 14.4°C to 19.1°C, averaging 17.2°C±1.6°C.

The values of $\delta^{13}C$ are listed in Table 1. The carbon isotope composition of the outer shell ranges from 1.4‰ to 2.2‰. The values of $\delta^{13}C$ of the jaw elements are lighter and range from −0.8‰ to 0.6‰. The conchorhynch and rhyncholite sampled in the same specimen show slightly different values of $\delta^{13}C$ (−0.5‰ and 0.1‰, respectively).

Data from the fossil record

The Preservation Index was assigned to the aragonitic shell of *Baculites* sp. (smooth) following Cochran *et al.* [21] and to aptychi following the criteria outlined in the methods section (Fig. 4). The results are summarized in Table 2. In the specimens of *Baculites* sp. (smooth) with both the outer shell and aptychi from South Dakota, the outer shell shows a wide range of PI from 2 to 4.5 while the aptychi are generally not well preserved (only one specimen with PI = 4). In contrast, the aptychi from the Moorville Chalk are very well preserved (PI = 3.5–4.5).

The calculated temperatures are listed in Table 2. The temperatures of the outer shell in *Baculites* sp. (smooth) from the

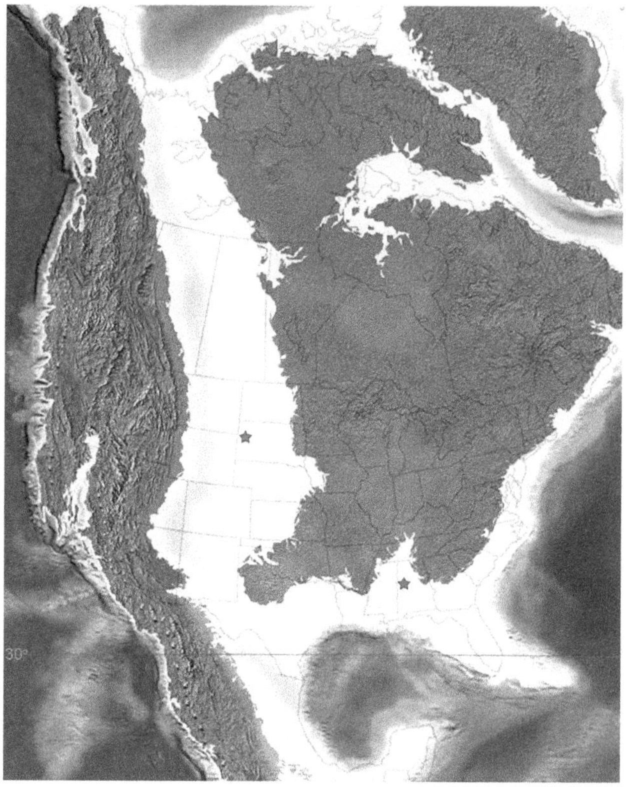

Figure 2. Paleomap of North America during the Late Cretaceous (85Ma, from Blakey, 2011). Fossil localities in South Dakota and Alabama are indicated by red stars. The locality in South Dakota was within the Western Interior Seaway, while the locality in Alabama was on the Gulf Coast. The paleolatitude has been estimated from Smith *et al.* [34].

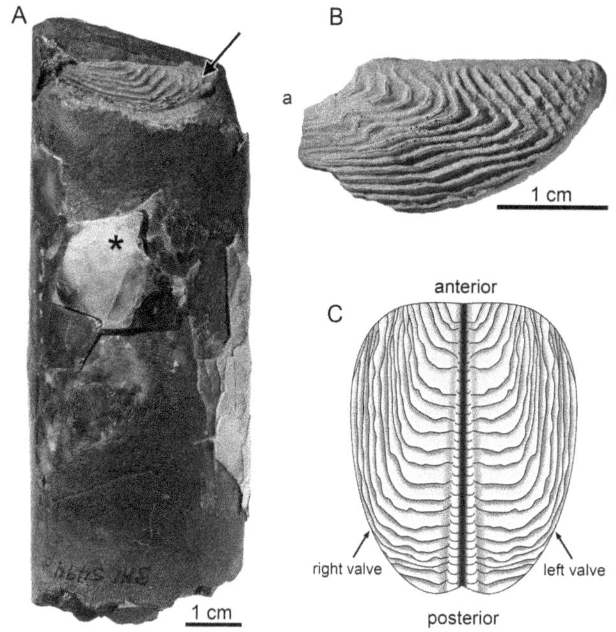

Figure 3. Ammonite shell material and aptychus jaw element. A) *Baculites* sp. (smooth) from the Pierre Shale, Butte County, South Dakota, with aptychus (arrow) preserved in the body chamber of the ammonite. Pieces of the aragonite shell (asterisk) are present on the steinkern. B) Reconstruction of the aptychus type of jaw in *Baculites*. The lower jaw is covered by two calcitic valves that usually separate after death and are found isolated in the sediment. C) Aptychus from the Mooreville Chalk, Alabama. Isolated valves are present in the sediment and are attributed to *Baculites* sp. (smooth). a-anterior, p-posterior.

Pierre Shale range from 23.9°C to 32.5°C. Cochran *et al.* [21] documented changes in $\delta^{18}O$ and $\delta^{13}C$ as preservation declined, with more poorly preserved shells showing lighter values. Samples with Preservation Index >3 (good) recorded the original isotopic composition and thus, paleotemperature [21]. In specimens of *Baculites sp.* (smooth) with PI >3, the mean temperature of the outer shell is 27.5±1.4°C. The calculated temperatures of the aptychi range from 27.9°C to 32.7°C. In the best preserved aptychus (BHI 5146), the calculated temperature is 28°C. The closest match (difference ≤0.6°C) between outer shell and jaw temperatures appears in the two specimens (BHI 5146 and AMNH 51329) with the best preserved outer shell and aptychi (PI for shell = 4.5, PI for aptychi = 3–4). In BHI 5146 the temperatures of the shell and aptychus are 28.6°C and 28°C, respectively. In AMNH 51329 the temperatures of the shell and aptychus are 27.8°C and 27.9°C, respectively. The aptychi of specimens from the Mooreville Chalk are all well preserved (PI>3.5), and the calculated temperatures range from 21.6°C to 24.2°C, with a mean of 22.4±1.1°C.

The values of $\delta^{13}C$ of the outer shell in *Baculites* sp. (smooth) from the Pierre Shale range from −4.0‰ to 0.7‰, with a mean of −1.0‰. The highest value (0.7‰) is from a specimen (BHI 5491) with the poorest preservation (PI = 2). The $\delta^{13}C$ values of the aptychi are generally lighter than those of the outer shell and range from −8.9‰ to −0.8‰. The values of $\delta^{13}C$ of the aptychi from the Mooreville Chalk are heavier than those from the Pierre Shale and range from −2.4‰ to −1.4‰.

Discussion

Isotopic values of Nautilus pompilius

In most of the specimens, the calculated temperatures of the outer shell and calcitic jaw elements match. The average temperature of the outer shell is 17.2±1.6°C and that of the jaw is 18.9±0.8°C. Within the uncertainties, these temperatures are in good agreement. Previous studies have demonstrated that the outer shell is secreted in equilibrium with seawater [15], [16]. Despite the difference in mineralogy, our results suggest that the jaw is also secreted in isotopic equilibrium, and that the temperature calculated for the upper and the lower jaw is the same.

The slight differences in the temperatures calculated between the jaw and outer shell may reflect differences in time averaging. The shell samples are from the aperture and thus represent a finite period of time. Although no data are available on the growth rate of the calcitic jaw elements, the rhyncholite and conchorhynch undoubtedly represent a longer period of time integrating over nearly the entire lifetime of the animal.

Using the temperature-depth profile in Vanuatu (obtained from NOCD database, cruises ID PA-127, PA-165), the depths that correspond to the calculated temperatures of the specimens can be determined. Our results suggest that the outer shell and jaw temperatures correspond to depths of 254–360 m. These values are consistent with habitat depth records of *Nautilus pompilius* elsewhere [32].

The carbon isotopic composition of the outer shell and the jaw elements is more difficult to interpret as several parameters could be involved. The two structures are secreted by independent tissue

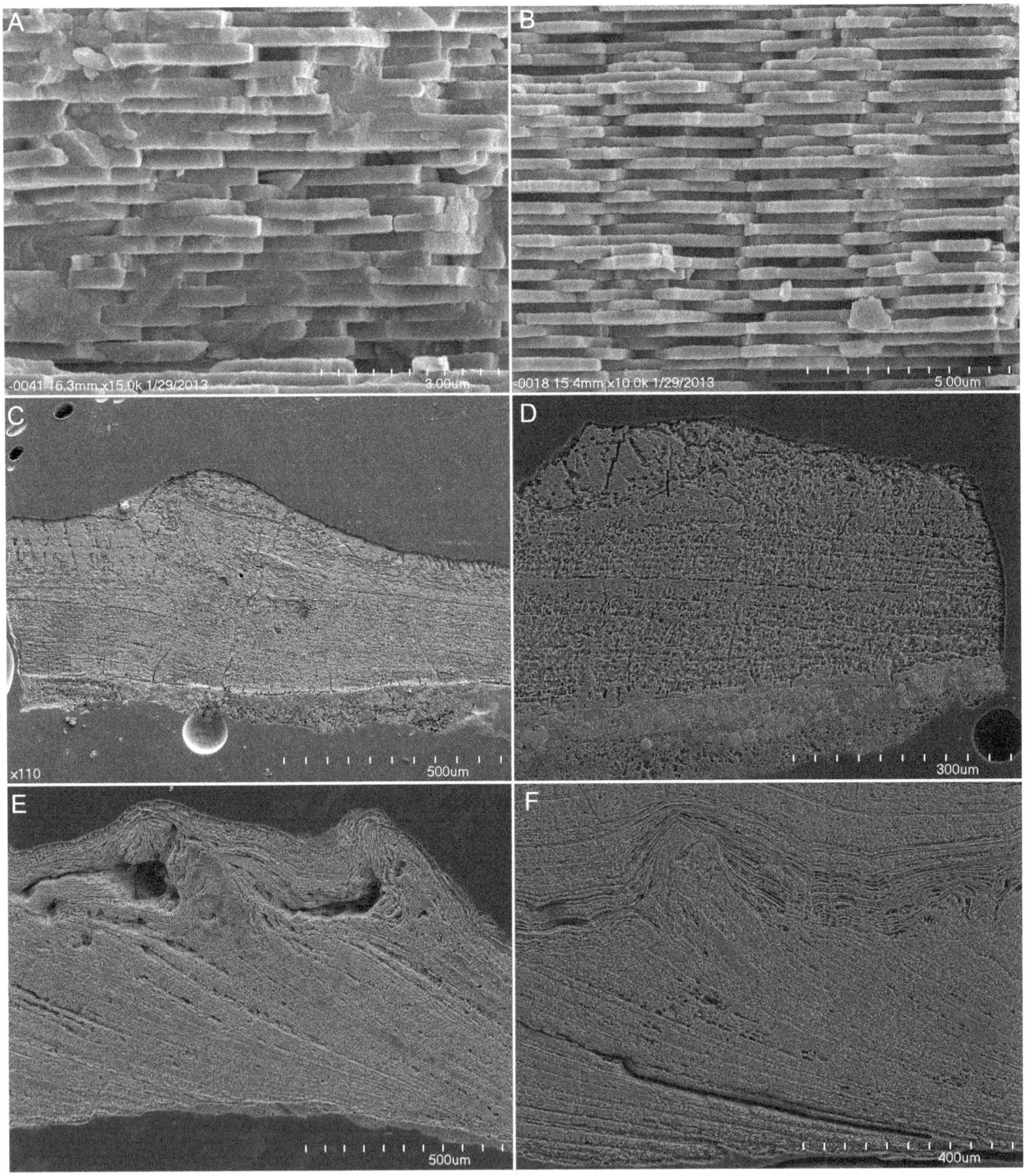

Figure 4. SEM micrographs of the shell and aptychi used to evaluate the Preservation Index. A–B) Nacreous shell in BHI 5146 (A) and AMNH 78053 (B). C–D) Preservation of the aptychi *in situ* in BHI 5146, and close up (D) showing the calcitic increments. E–F) Preservation in the aptychi from the Mooreville Chalk, showing the differentiated layers (R1 and R2) and the calcitic increments in AMNH 66354 (F) and AMNH 66357 (F).

system. The outer shell is secreted by the mantle whereas the jaws are secreted by tissue in the buccal mass. Therefore, different sources of carbon (DIC and possibly diet) could be incorporated during secretion.

Isotopic values of *Baculites* sp. (smooth)

The isotopic values in the fossil material reflect the quality of preservation of the samples. In *Baculites* sp. (smooth) from the Pierre Shale of South Dakota, samples of well-preserved outer shell (PI>3; table 2) yield a mean temperature of 27.5±1.4°C. The sample with the best preserved outer shell and aptychus is BHI 5146. In this specimen, the calculated temperature of the

Table 2. Isotope data for ammonites.

Locality Mooreville Chalk	Sample	Specimen	Mineralogy	$\delta^{18}O^{\dagger}$ (‰)	$\delta^{13}C^{\dagger}$ (‰)	T (°C)	PI
Baculites sp.	aptychus	AMNH 66353	calcite	−2.67	−1.99	24.2	3.5
"	aptychus	AMNH 66354	calcite	−2.07	−2.41	21.6	4
"	aptychus	AMNH 66355	calcite	−2.13	−1.42	21.8	4
"	aptychus	AMNH 66356	calcite	−2.27	−1.7	22.5	4
"	aptychus	AMNH 66357	calcite	−2.16	−1.4	22.0	4.5
Mean ± 1σ jaw T						*22.4±1.1*	
Locality Pierre Shale							
Baculites sp.	shell	AMNH 51329	aragonite	−2.28	−2.15	27.8	4
"	aptychus *in situ*	"	calcite	−3.52	−8.94	27.9	3
"	shell	AMNH 66255	aragonite	−1.71	−0.36	25.1	3.5
"	aptychus *in situ*	"	calcite	−4.62	−6.62	32.7	2
"	shell	AMNH 64489	aragonite	−1.63	−0.04	24.8	3
"	aptychus *in situ*	"	calcite	−3.60	−4.39	28.3	1
"	shell	BHI 5143	aragonite	−1.66	−3.97	24.9	2.5
"	aptychus *in situ*	"	calcite	−3.57	−2.54	28.1	2.5
"	shell	BHI 5146	aragonite	−2.46	−1.73	28.6	4
"	aptychus *in situ*	"	calcite	−3.54	−3.99	28.0	4
"	shell	BHI 5491	aragonite	−3.29	0.69	32.5	2
"	aptychus *in situ*	"	calcite	−3.84	−0.85	29.3	2.5
"	shell	BHI 5494	aragonite	−1.45	−0.60	23.9	3
"	aptychus *in situ*	"	calcite	−4.22	−4.97	31.0	1
"	shell	AMNH 78053	aragonite	−2.45	−1.03	28.6	4.5
"	shell	AMNH 51754	aragonite	−2.23	−0.06	27.6	4.5
Mean ± 1σ shell T (PI>3)						*27.5±1.4*	
Mean jaw T (PI≥3)						*28.0*	

†Values relative to VPDB.

aptychus (28.6°C) is comparable to that of the outer shell (28°C). The calculated temperature of this aptychus also agrees well with the mean temperature for the other shell samples with PI>3, i.e. 27.5±1.4°C. These observations imply that if the aptychus is well preserved, its $\delta^{18}O$-derived temperature matches that of the outer shell. The fossil aptychi from the Mooreville Chalk are generally well preserved (3.5≤PI≤4.5). The two layers (R1 and R2) can be identified as well as the calcitic increments. The calculated temperatures are very consistent, with a mean of 22.4±1.1°C.

Application to Late Cretaceous paleoenvironments

Our study of the temperatures derived from the outer shell and jaw elements of *Baculites* sp. (smooth) in the Western Interior and U.S. Gulf Coastal Plain provides an example of the advantage of using either the carbonate from the shell and the jaw to determine paleotemperatures. In particular during the early Campanian, the Western Interior has abundant well-preserved outer shell material of ammonites, while the Gulf Coastal Plain has well-preserved aptychi but no comparable shell material. Previous studies have suggested that the two localities are time equivalent (see [26] for discussion on the stratigraphic relationships of the two sites) and our results reveal different temperatures for the two sites. In the Western Interior, the temperatures calculated from well-preserved shell and aptychi average 27.5±1.4°C (n = 6) and 28°C (n = 2), respectively. The temperatures calculated from the aptychi on the Gulf Coastal Plain are lower (22.4±1.1°C; n = 5) despite being at a lower paleolatitude (Fig. 2). Similarly elevated temperatures for the Western Interior have previously been reported for the late Campanian of South Dakota [33]. These differences probably reflect variation in paleogeography between the restricted Western Interior Seaway and the open Gulf Coast. Indeed, Dennis *et al.* [31] used clumped isotopes to document cooler temperatures for the open ocean along the Atlantic Coastal margin (Severn Formation, Maryland) relative to the Western Interior Seaway during the late Maastrichtian. However, some of the differences in temperature we observe between the Western Interior and Gulf Coast during the early Campanian may be due to differences in the isotopic composition of the water at the two environments ($\delta^{18}O_w$). Additional work using clumped isotopes might further tease apart the factors responsible for the differences in temperature between the two sites.

Conclusion and Perspectives

Our data demonstrate that the temperatures recorded in the shells of Recent *Nautilus pompilius* match the temperatures of the jaw elements in the same individuals. Calcitic structures of the jaws may thus provide a reliable geochemical archive of the habitat of nautilids and, by extension, ammonites. In ammonites, if the aptychus is well preserved, it records the same temperature as well-preserved outer shells. Data from jaw elements are especially valuable for localities where the aragonitic shell of the ammonites is not preserved. Discrepancies in the calculated temperatures of the outer shell and jaw elements in *N. pompilius* are explainable as reflecting differences in time averaging, with the jaw integrating over a longer portion of the lifetime of the animal while individual samples from the outer shell or aperture are restricted in time. The carbon isotopic composition of the jaws is lighter than that of the shell in *N. pompilius* and may reflect differences related to the source of carbon. If so, such studies on ammonites may yield clues into the diet of these extinct animals.

Acknowledgments

This research was conducted under a permit from the Department of Fisheries and Environment Unit, Vanuatu. We would like to thank Katherine Holmes for collecting the specimens of *Nautilus* in Vanuatu, the Black Hills Institute of Geological Research for lending us the material, and Neal Larson for collecting many of the fossils. We also want to thank Mark Siddall and Joe Wolf for the use of their laboratory equipment and Greg Cane at KPESIL for analyzing the specimens.

Author Contributions

Conceived and designed the experiments: IK NL. Performed the experiments: IK NL. Analyzed the data: IK NL JKC. Contributed reagents/materials/analysis tools: IK NL JKC. Wrote the paper: IK NL JKC.

References

1. House MR (1993) Fluctuations in ammonoid evolution and possible environmental controls. In: House MR editor. The Ammonoidea: environment, ecology and evolutionary change. The Systematics Association Special Volume 47. Oxford: Clarendon Press. pp. 13–34.
2. Tourtelot HA, Rye RO (1969) Distribution of oxygen and carbon isotopes in fossils of Late Cretaceous age, western interior region of North America. Geol Soc Am Bull 80: 1903–1922. doi:10.1130/0016-7606.
3. Rye DM, Sommer MA II (1980) Reconstructing paleotemperature and paleosalinity regimes with oxygen isotopes. In: Rhoads DC, Lutz RA editors. Skeletal Growth of Aquatic Organisms. New York: Plenum. pp. 169–202.
4. Cochran JK, Landman NH, Turekian KK, Michard A, Schrag DP (2003) Paleoceanography of the Late Cretaceous (Maastrichtian) Western Interior Seaway of North America: evidence from Sr and O isotopes. Palaeogeogr Palaeocl 191: 45–64. doi:10.1016/S0031-0182(02)00642-9.
5. He S, Kyser TK, Caldwell WGE (2005) Paleoenvironment of the Western Interior Seaway inferred from ^{18}O and ^{13}C values of molluscs from the Cretaceous Bearpaw marine cyclothem. Palaegeogr Palaeoclim 217: 67–85. doi:10.1016/j.palaeo.2004.11.016.
6. Lécuyer C, Bucher H (2006) Stable isotope compositions of late Jurassic ammonite shell: a record of seasonal surface water temperatures in the southern hemisphere? eEarth 1: 1–7.
7. Wierzbowski H, Joachimski M (2007) Reconstruction of late Bajocian-Bathonian marine palaeoenvironments using carbon and oxygen isotope ratios of calcareous fossils from the Polish Jura Chain (central Poland). Palaeogeogr Palaeoclim 254: 523–540.
8. Wierzbowski H, Rogov M (2011) Reconstructing the palaeoenvironment of the Middle Russian Sea during the Middle-Late Jurassic transition using stable isotope ratios of cephalopod shells and variations in faunal assemblages. Palaeogeogr Palaeoclim 299: 250–264.
9. Landman NH, Cochran JK, Larson NL, Brezina J, Garb MP, et al. (2012) Methane seeps as ammonite habitats in the US Western Interior Seaway revealed by isotopic analyses of well-preserved shell material. Geology 40: 507–510.
10. Landman NH, Rye DM, Shelton KL (1983) Early ontogeny of *Eutrephoceras* compared to Recent *Nautilus* and Mesozoic ammonites - evidence from shell morphology and light stable isotopes. Paleobiology 9: 269–279.
11. Moriya K, Nishi H, Kawahata K, Tanabe K, Takayanagi Y (2003) Demersal habitat of Late Cretaceous ammonoids: Evidence from oxygen isotopes for the Campanian (Late Cretaceous) northwestern Pacific thermal structure. Geology 31: 167–170.
12. Cochran JK, Rye DM, Landman NH (1981) Growth rate and habitat of *Nautilus pompilius* inferred from radioactive and stable isotope studies. Paleobiology 7: 469–480.
13. Taylor BE, Ward PD (1983) Stable isotope studies of *Nautilus macromphalus* Sowerby (New Caledonia) and *Nautilus pompilius* L. (Fiji). Palaeogeogr Palaeoclim 41: 1–16.
14. Crocker KC, DeNiro MJ, Ward PD (1985) Stable isotopic investigations of early development in extant and fossil chambered cephalopods I. Oxygen isotopic composition of eggwater and carbon isotopic composition of siphuncle organic matter in *Nautilus*. Geochim Cosmochim Ac 49: 2527–2532.
15. Landman NH, Cochran JK, Rye DM, Tanabe K, Arnold JM (1994) Early life history of *Nautilus*: evidence from isotopic analyses of aquarium-reared specimens. Paleobiology 20: 40–51.
16. Auclair AC, Lecuyer C, Bucher H, Shepard FP (2004) Carbon and oxygen isotope composition of *Nautilus macromphalus*: a record of thermocline waters of New Caledonia. Chem Geol 207: 91–100.
17. Zakharov YD, Shigeta Y, Smyshlyaeva OP, Popov AM, Ignatiev AV (2006) Relationship between $\delta^{13}C$ and $\delta^{18}O$ values of the Recent *Nautilus* and brachiopod shells in the wild and the problem of reconstruction of fossil cephalopod habitat. Geosci J 10: 331–345.
18. Lukeneder A, Harzhauser M, Mullegger S, Piller WE (2010) Ontogeny and habitat change in Mesozoic cephalopods revealed by stable isotopes (delta O-18, delta C-13). Earth Planet Sc Lett 296: 103–114. doi: 10.1016/j.epsl.2010.04.053.
19. Schlögl J, Chirat R, Balter V, Joachimski M, Hudáč-Ková N, et al. (2011) *Aturia* from the Miocene Paratethys: an exceptional window on nautilid habitat and lifestyle. Palaeogeogr Palaeoclim 308: 330–338. doi: 10.1016/j.palaeo.2011.05.037
20. Seuss B, Titschack J, Seifert S, Neubauer J, Nützel A (2012) Oxygen and stable carbon isotopes from a nautiloid from the middle Pennsylvanian (Late Carboniferous) impregnation Lagerstätte 'Buckhorn Asphalt Quarry'-Primary paleo-environmental signals versus diagenesis. Palaeogeogr Palaeoclim 319–320:1–15.
21. Cochran JK, Kallenberg K, Landman NH, Harries PJ, Weinreb D, et al. (2010) Effect of diagenesis on the Sr, O, and C isotope composition of late Cretaceous mollusks from the Western Interior Seaway of North America. Am J Sci 310: 69–8. doi:10.2475/02.2010.01.
22. Bowen R, Fontes JC (1963) Paléotempératures indiquées par l'analyse isotopique de fossiles du crétacé inférieur des Hautes-Alpes (FRANCE). Experientia 19: 268–275.
23. Price GD, Sellwood BW (1997) "Warm" palaeotemperatures from high Late Jurassic palaeolatitudes (Falkland Plateau). Ecological, environmental or diagenetic controls? Palaeogeogr Palaeoclim 129: 315–327.
24. Lowenstam HA, Traub W, Weiner S (1984) *Nautilus* hard parts: a study of the mineral and organic constituents. Paleobiology 10(2): 268–279.
25. Allmon WD, Jones D, Vaughan N (1992) Observations on the Biology of *Turritella gonostoma* Valenciennes (Prosobranchia: Turritellidae) from the Gulf of California. Veliger 35(1): 52–63.
26. Landman NH, Larson NL, Cobban WA (2007) Jaws and radula of *Baculites* from the Upper Cretaceous (Campanian) of North America. In: Landman NH, Davis RA, Mapes RH editors. Cephalopods present and past: new insights and fresh perspectives. Dordrecht: Springer. pp. 257–298.
27. Kruta I, Rouget I, Landman NH, Tanabe K, Cecca F (2009) Aptychi microstructure in Late Cretaceous Ancyloceratina (Ammonoidea). Lethaia 42: 312–321.
28. Grossman EL, Ku TL (1986) Oxygen and carbon isotope fractionation in biogenic aragonite: temperature effects. Chem Geol 59: 59–74.
29. O'Neil JR, Clayton RN, Mayeda TK (1969) Oxygen isotope fractionation in divalent metal carbonates. J Chem Phys 51: 5547–5558.
30. Shackleton NJ, Kennett JP (1975) Paleotemperature history of the Cenozoic and initiation of Antarctic glaciation: oxygen and carbon isotope analyses in DSDP sites 277, 279 and 281. Initial Rep Deep Sea 29: 743–755.
31. Dennis KJ, Cochran JK, Landman NH, Schrag DP (2013) The climate of the late Cretaceous: New insights from the application of the carbonate clumped isotope thermometer to Western Interior Seaway macrofossil. Earth Planet Sc Lett 362: 51–65.
32. Dunstan AJ, Ward PD, Marshall NJ (2011) Vertical distribution and migration patterns of *Nautilus pompilius*. PLOS One 6(2): e16312.
33. Fatherree JW, Harries PJ, Quinn TM (1998) Oxygen and carbon isotopic 'dissection' of *Baculites compressus* (Mollusca: Cephalopoda) from the Pierre Shale (Upper Campanian) of South Dakota: Implications for paleoenvironmental reconstructions. Palaios 13: 376–385.
34. Smith AG, Smith DG, Funnell BM (1994) Atlas of Mesozoic and Cenozoic Coastlines. Cambridge University Press, Cambridge. 99 pp.

Microrefugia and Shifts of *Hippophae tibetana* (Elaeagnaceae) on the North Side of Mt. Qomolangma (Mt. Everest) during the Last 25000 Years

Lu Xu[1,2¶], Hao Wang[1¶], Qiong La[1,3¶], Fan Lu[1], Kun Sun[2], Yang Fang[1], Mei Yang[1], Yang Zhong[1,3], Qianhong Wu[1], Jiakuan Chen[1], H. John B. Birks[4,5,6], Wenju Zhang[1]*

1 Institute of Biodiversity Science, School of Life Sciences, Fudan University, Shanghai, China, 2 College of Life Sciences, Northwest Normal University, Lanzhou, China, 3 Department of Biology, Tibet University, Lhasa, China, 4 Department of Biology, University of Bergen, Bergen, Norway, 5 Environmental Change Research Centre, University College London, London, United Kingdom, 6 School of Geography and the Environment, University of Oxford, Oxford, United Kingdom

Abstract

Microrefugia at high altitudes or high latitudes are thought to play an important role in the post-glacial colonization of species. However, how populations in such microrefugia have responded to climate changes in alternating cold glacial and warm interglacial stages remain unclear. Here we present evidence to indicate the Rongbuk Valley of the Mt. Qomolangma (Mt. Everest) area, the highest region on earth, had microrefugia for *Hippophae tibetana* and discuss how this low shrub was adapted to the extreme climate fluctuations of the last 25,000 years by shifts. By integrating geological, glaciological, meteorological, and genetic information, we found that the Rongbuk Valley was not only a glacial microrefugium but also an interglacial microrefugium for *H. tibetana*: the former was located on the riverbank below 4800 m above sea level (asl) or lower area and the latter at ~5000 m asl. Our results show that after the Last Glacial Maximum (LGM), *H. tibetana* in the valley has undergone upward and downward migrations around ~5000 m driven by climate fluctuations and the population in the glacial microrefugium has suffered extinction or extreme contraction. Moreover, with the rise of temperature in the last four decades, the upper limit of *H. tibetana* has shifted at least 30 m upward. Combining population history and recent range shift of this species is important in predicting the fate of this endemic species to future climate changes.

Editor: Navnith K. P. Kumaran, Agharkar Research Institute, India

Funding: This research was funded by National Basic Research Program of China (Grant No. 2014CB954103; http://www.973.gov.cn/AreaMana.aspx) and the National Natural Science Foundation of China (Grant No. 41061007 and Grant No. 91131901; http://isisn.nsfc.gov.cn). The funders had no role in study design, data collection and analysis, decision to publish, or preparation of the manuscript.

Competing Interests: The authors have declared that no competing interests exist.

* E-mail: wjzhang@fudan.edu.cn

¶ These authors contributed equally to this work.

Introduction

The alternation between glacial and interglacial stages during the Quaternary has greatly affected species distributions [1], [2], [3]. Many temperate species retreated to refugia at lower latitudes or altitudes during glacials and expanded to higher latitudes or altitudes during interglacials [1], [4], [5], [6]. Apart from these broad-scale refugia, more and more scattered small refugia have been found at high latitudes or in alpine areas in recent decades, with local favourable environmental features outside the species' main distribution area, so-called microrefugia (or cryptic refugia) [3], [4], [5], [7], [8], [9], [10], [11], [12]. Understanding microrefugia is of critical importance in interpreting species genetic diversity and evolutionary processes such as adaptation, speciation, and extinction [1], [2], [3], [9], [13], [14], [15], [16]. For example, the incongruence between estimated post-glacial migration rates and tree dispersal capacity ('Reid's Paradox') can be explained, in part, by the widespread existence of microrefugia [5], [14], [15], [17], [18]. The concept of microrefugia is now well recognized through empirical studies which have mainly focused on identifying microrefugia during the Last Glacial Maximum (LGM) by palaeoecological or phylogeographical methods [4], [5], [10], [17]. However, as Rull (2009) points out, the concept of microrefugia lacks appropriate biogeographical and ecological characterization [9]. Although the importance of landscape physiography and microclimate for plant microrefugia to occur has been highlighted [15], and some potential microrefugia may be deduced according to topoclimate, climate stability, and isolation from the matrix [19], little is known about actual microrefugial situations, their time span, and their exact elevation or location, as well as how species (especially plants) have managed to track a favourable microclimate for their survival [8], [9], [15], [16], [20].

At present, most studies on microrefugia have paid close attention to glacier refugia, especially to the LGM microrefugia, in which some genotypes have been supposed to survive this cold period [4], [5], [7], [8], [9], [10], [14], [17], [21], [22], [23]. But as Hampe et al. (2013) pointed out that, with wide-ranging implications, glacial refugia no longer exist and can hence only be inferred by indirect means [16]. Populations in microrefugia, defined as climate relicts by Hampe et al. (2011) [24], should be the key to understanding the effects of microrefugia. An important fact often has been ignored by many researchers: although populations in the LGM microrefugia were able to persist through this cold period, these populations may have been dramatically changed in interglacial stages because of warm climate (e.g. American pika) [25]. Thus,when we infer glacial refugia by indirect means, especially by genetic information, the dynamics of these climate relics after the LGM are needed to survey at the smaller scale. A meta-population model has been considered to characterize these climate relicts [26]; however, few "climate relict" has been studied carefully in detail.

On the other hand, investigating species' responses to past climate changes is also important in understanding how species might respond to recent and future climate changes [4], [5], [16], [18], [20], [27]. Global temperature has increased 0.6°C in the past three decades and 0.8°C in the past century [28], which has led to both latitudinal and altitudinal shifts in species ranges [29], [30], [31], [32], [33], [34] and even caused some species to be possibly on the brink of extinction [35], [36]. However, present studies on the responses of species to past climate changes and recent warming are very disconnected. The former often focuses on tracking species population history through phylogeographical or paleoecological methods whereas the latter usually involves comparison of well-documented historical records with present distributional data [29], [36], [37]. According to oxygen isotope ($\delta^{18}O$) data from ice cores and sea sediments, past climate oscillations are often more pronounced than recent warming [38], [39], [40], [41], but almost all the species existing today have survived these past oscillations [1], [5]. Thus, we could better predict a species response in the future by combining the present and past responses of the species to climate changes over a range of time scales.

In the present study, we investigate *Hippophae tibetana* Schlecht. (Elaeagnaceae) as a means of characterizing the biogeographical and ecological features of a particular microrefugium and the responses of this species to climate changes since the LGM and to recent warming. *H. tibetana* is a small dioecious shrub propagated by seeds or by horizontal roots [42]. It is endemic to the Qinghai-Tibet Plateau (QTP) and ranges from the west Himalaya to the east-north QTP [43]. In the eastern plateau, *H.tibetana* occurs in the lowlands and in alpine meadows at an altitude below 4000 m, but in the central plateau and the Himalayas, it has a fragmented distribution along several valleys. It is one of the shrubs to occur at the highest altitudes [44], growing up to ~5200 m asl. Our previous study [43] investigated the phylogeography of *H. tibetana* and found that three main lineages (A, B, and C) of the present populations of this species occupy the middle, the western, and the eastern parts of its geographical range, respectively. Based on the distribution of a large number of private haplotypes, we concluded that *H. tibetana* had multiple LGM microrefugia on the Plateau and inferred that the Rongbuk Valley, north of Mt. Qomolangma (Mt. Everest) is a possible microrefugium for *H. tibetana* even though it is the highest region within the geographical range of *H. tibetana*. However, in this valley, where and how *H. tibetana* has survived since the LGM remains unclear. Also, we chose this area for our current study because the area is highly sensitive to global climate change, and the geological and meteorological characteristics of the valley have already been studied well [45], [46], [47], [48]. This area provides an excellent opportunity to consider the favourable local environment required for microrefugia to occur and to study the responses of *H. tibetana* to climate change since the LGM [4], [8] by integrating the available geological, meteorological, and genetic information. In the present study, we hope to clarify two questions: 1) Do microrefugia of the LGM for *H. tibetana* really exist in the Rongbuk valley, one of the highest areas on earth? 2) On the local scale, how has the population of *H. tibetana* in the microrefugia responded to climate changes since the LGM?

Materials and Methods

Ethics Statement

In this study, all field works were carried out in the Rongbuk Valley of Mountain Qomolangma National Nature Preserve (QNNP) and were permitted by QNNP. There is no endangered or protected species involved in this study. The plant species studied in this work, *Hippophae tibetana* (Elaeagnaceae), has a large distribution on the QTP and has not been listed in any protection lists. No animals were used in this study. Coordinate data of sample locations of this study were shown in Table 1.

Study Area

The Rongbuk Valley is located on the northern slopes of Mt. Qomolangma (27.98°N, 86.92°E; elevation 8844 m, the highest mountain on Earth), which lies toward the eastern end of the Himalaya. Three large glaciers occur at the upper end of the valley: West Rongbuk Glacier, Rongbuk Glacier, and East Rongbuk Glacier. These glaciers flow for 13~18 km down to the proglacial plain at 5200 m asl near Everest Base Camp used by many mountaineering expeditions [49], [50]. Glacial meltwater runs through the Rongbuk Valley, which is 0.4−1 km wide, ~91 km long, and ~1500 m fall at altitude (5200 m to 3700 m), and empties into the Pengqu River at 3700 m asl.

The modern Equilibrium Line Altitude (ELA) of these glaciers is above 6000 m asl [51]. In the past, their extents shifted in response to climate change, leaving terminal moraines of different ages in the valley [47], [52], [53], [54], [55] (Fig. 1a). The Far East Rongbuk ice-core only records climate changes for the last 200 years, but these can be matched with the longer Guliya ice-core records [38], [45]. We can thus infer past temperature change over a long time span in the Rongbuk Valley from the Guliya ice-core. An increase (or decrease) of 1‰ in mean annual $\delta^{18}O$ in the Guliya ice-core corresponds to an increase (or decrease) of ~1.5°C in mean annual air temperature [56] as shown in Fig. 1b. There are also detailed studies on the glacial moraines of different ages. From 25 ka B.P. until now, there have been four main glaciations in the valley, and their terminal moraines are the Jilong moraine (~4800 m), Rongbuk moraine (~5000 m), Rongbude moraine (~5100 m), and the Little Ice Age moraine (~5150 m) from the oldest to the youngest (M4, M3, M2, and M1 in Fig. 1a, respectively) [47], [52], [53], [54]. The ages of the moraines have been estimated using optically stimulated luminescence (OSL) and terrestrial cosmogenic nuclide (TCN) dating by Owen *et al.* (2009) [47] and the OSL dates are indicated along the time axis in Fig. 1b.

Main vegetation in study area is alpine steppe vegetation with small shrubs in the bottom of the valley at an altitude between 4400 and 5000 m asl, dominated by *Artemisia wellbyi*, *Stipa purpurea*, *Orinus thoroldii*, *Carex montis-everestii*, *Potentilla parvifolia*. At the higher level, between 5000–5600 m, some alpine sparse and cushion vegetation occurs locally on the upper part of area, such as *Potentilla fruticosa* var. *pumila*, *Androsace tapete*, *Arenaria polytrichoides*, *Kobresia pygmaea*, *Kobresia prainii*, *Saussurea gnaphalodes* and other *Saussurea* sp. In addition, some lichen common species occurs on the leeward side of rocks and gravels, for example *Lecidea auriculata* and *Cloplaca elegans* etc.

Table 1. List of populations (POP) and patches (PAT) analysed in the present study with their sampling localities, number of specimens, coordinates, genetic diversity parameters, and chlorotype composition of each population.

Population No.	*N*	Altitude (m)	Latitude (N)	Longitude (E)	Nuclear microsatellites				Chloroplast Haplotypes		
					F_{IS}	A	H_O	H_E	Haplotypes (Frequencies, %)	*D*	π
POP 1	39	4200	28°24′23″	86°59′14″	−0.061	2.10	0.419	0.390	R1(5), R2(85), R3(10)	0.2780	0.00053
POP 2	23	4400	28°18′49″	86°55′53″	−0.041	2.18	0.430	0.403	R1(44), R2(39), R3(17)	0.6561	0.00148
POP 3	51	4465	28°18′46″	86°53′39″	0.084	2.15	0.328	0.354	R1(39), R2(49), R3(12)	0.6039	0.00127
POP 4	25	4690	28°16′38″	86°48′26″	−0.048	2.28	0.410	0.384	R1(52), R2(36), R3(12)	0.6100	0.00135
POP 5	31	4805	28°14′38″	86°49′05″	0.031	2.25	0.422	0.429	R1(36), R2(61), R4(3)	0.5140	0.00098
POP 6	28	4946	28°12′25″	86°49′19″	−0.850	1.69	0.564	0.304	R1(100)	0.0000	0.00000
POP 7	33	5000	28°10′03″	86°50′23″	0.011	2.15	0.405	0.403*	R1(30), R2(3), R3(18), R4(12), R5(6), R6(31)	0.7879	0.00343
POP 8	35	5035	28°09′42″	86°50′38″	−0.051	2.19	0.429	0.402	R1(57), R2(5), R3(26), R4(11)	0.6084	0.00201
PAT 1	25	>5035	28°09′08″	86°50′52″	−0.778	1.20	0.176	0.099*	R3(100)	0.0000	0.00000
PAT 2	22	5047	28°09′34″	86°50′42″	−0.707	1.40	0.305	0.177*	R3(100)	0.0000	0.00000
PAT 3	15	5046	28°09′35″	86°50′42″	−0.774	1.49	0.373	0.209*	R2(100)	0.0000	0.00000
PAT 4	24	5047	28°09′37″	86°50′42″	−0.378	1.80	0.426	0.305*	R1(100)	0.0000	0.00000
PAT 5	32	5066	28°09′24″	86°50′49″	−1.000	1.20	0.200	0.100*	R2(100)	0.0000	0.00000
PAT 6	16	>5035	28°09′21″	86°50′56″	−0.408	1.78	0.390	0.267*	R2(100)	0.0000	0.00000
PAT 7	15	5058	28°09′32″	86°50′52″	−0.647	1.60	0.373	0.224*	R1(100)	0.0000	0.00000

The exact elevations of PAT 1 and PAT 6 are not given because the GPS data of the two spots are not very accurate. *N*, number of individuals analysed; Nuclear microsatellites data, including F_{IS}, fixation index; A, allelic richness; H_O, observed heterozygosity; H_E, expected heterozygosity;
*significant Hardy–Weinberg disequilibrium. cpDNA haplotypes and their frequencies, as well as estimates of gene diversity (*D*) and nucleotide diversity averaged across loci (π) of the populations and patches studied.

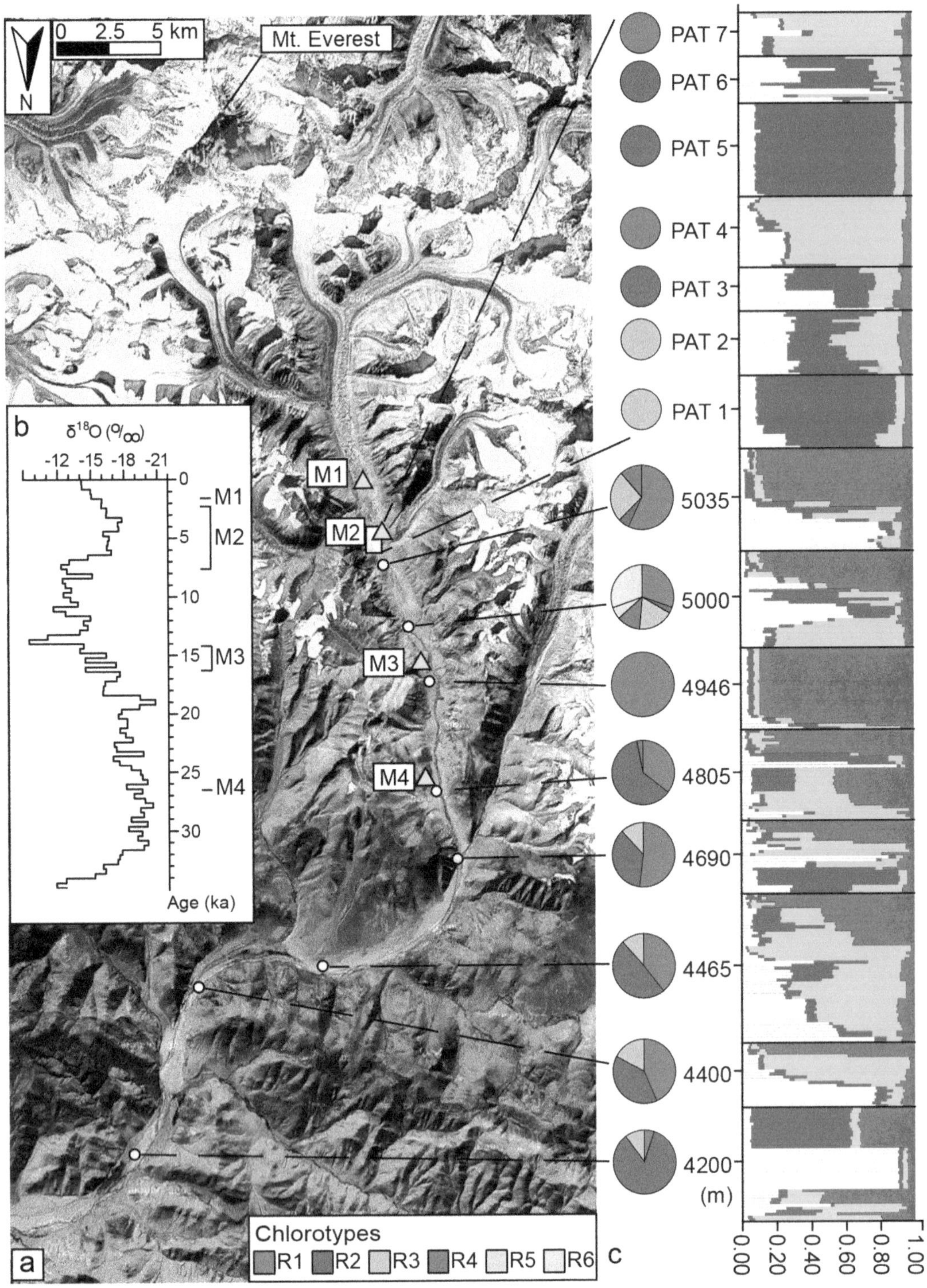

Figure 1. Sampling locations of *Hippophae tibetana* and the genetic composition of each population. (*a*) Map showing the sampling locations along the Rongbuk Valley, north of Mt. Qomolangma. Populations are represented by yellow circles and the patches are in the yellow rectangular area. Blue triangles represent the end-moraine of each glaciation [47], [52], [54] (M1, Little Ice Age moraine; M2, Rongbude moraine; M3 Rongbuk moraine; M4, Jilong moraine). (*b*) Temperature change revealed in the Guliya ice-core $\delta^{18}O$ record over the past 35 ka [38]. OSL ages of the moraines [47] are indicated along the time axis. (*c*) Elevation of each population, chlorotype composition, and proportion shown as pie charts (left column) and the population assignment test results with the software STRUCTURE (right column).

Climate Conditions

In terms of climate conditions, the Himalaya and the QTP are located in the inter-tropical convergence zone. Mt. Qomolangma is influenced by the mid-latitude westerlies and the south Asian monsoon and the climate in the Rongbuk valley is semi-arid and cold [57], [58]. According to data obtained from 2008 to 2011, annual mean air temperature at Qomolangma Station (4300 m alt.) is 4.3°C and average annual precipitation is 213.4 mm [58]. The rainy season is from June to August in summer, when moisture comes with the monsoon from the Indian Ocean and the Bay of Bengal. Although it is cold in the valley, air temperature has increased in the last four decades at a rate of 0.302°C per decade (from 1971 to 2004), reflecting a high sensitivity to recent global warming [46]. This warming trend has contributed to the retreat of glaciers in recent decades [55]. In this valley, mean air temperature drops 0.7–0.8°C with a 100 m increase in altitude as a result of changes in atmospheric pressure and surface conditions [48].

Sample Collection and Field Investigation

In the Rongbuk valley, *H. tibetana* extends from ~4200 m to ~5200 m asl along the banks of the river and is one of dominant species in the valley. There are some large colonies (area>1000 m^2) between 4200~5035 m asl, but only a few small colonies of 20–250 m^2 areas are scattered above 5035 m (>5035 m but < 5100 m asl). We are not sure that all the individuals of *H. tibetana* in this valley consist of one or some populations, so when we collected samples, we treated large colonies at different altitudes as different populations (POP) and a small colony above 5035 m as a patch (PAT). Leaves of *H. tibetana* were collected in the summers of 2007–2010. Population samples (POP) were collected along the elevational gradient from 4200 m to 5035 m every 200 m at most, and individuals are at least 10 m apart (Fig. 1a). We also surveyed the area from 5035 m to 5200 m in a ~150-metre-wide belt transect, and collected leaf samples from all the patches of the shrub that we found (Table 2; Fig. 1a). In the patches, the spacing of individuals is not always over 10 m because of the small area of some patches. Leaves were dried in silica gel. In total, we analyzed 414 individuals from 8 populations and 7 patches, and their GPS data are shown in Table 1. Field measurements were conducted in January 2011 to determine the colonization times of the patches. Areas, crown diameters, and colonization times of the patches were estimated. The five biggest individuals in each patch and the highest large colony (POP 8 at 5035 m) were measured. The basal stems of the biggest individual in each patch and POP8 at 5035 m were cut (~5 cm length) and conserved in plastic bags. Transverse section toward root was smoothed in the laboratory and annual rings were identified by using a dissecting microscope.

Genetic Analysis

Total genomic DNA was extracted using the modified CTAB method [43]. Both the chloroplast trnT-trnF sequences and nuclear microsatellites are used in the present study.

Based on the sequences from our previous phylogeographic study [43], we selected partial *trn*T-*trn*F sequences of chloroplast DNA to identify chlorotypes. Amplification and sequencing were carried out using primers INP1 5′ TAGATCGTTCAAGTATT-CAAAATA 3′ and INP2 5′ CAGGTCGTCATTAATCATTTT-CAGA 3′, following the procedures of Wang *et al.* (2010). The DNA sequences of *H. tibetana* individuals were aligned using the program CLUSTAL X with subsequent manual adjustments, and assigned to different haplotypes using DnaSP 5.10 [59]. Our previous study has shown that all the *trn*T-*trn*F haplotypes of populations in the Mt. Qomolangma area and adjoining region belong to the B-lineage [43] To evaluate the evolutionary relationships among chlorotypes, the chlorotypes obtained in the present study and all the B-lineage chlorotypes from Wang *et al.* (2010) were used to construct a network by the program NETWORK version 4.5.1.0 (http://www.fluxus-engineering.com) which uses the median joining approach to combine a minimum spanning tree within a single network and then, by the criterion of parsimony, median vectors are added to the network (Fig. 2). The indices of unbiased genetic diversity (*D*) and nucleotide diversity (π) [60] were calculated for each population using the program ARLEQUIN version 3.1 [61]. Using part of the *trn*T-*trn*F sequences that we had obtained for all individuals, pairwise mismatch distributions were carried out with ARLEQUIN version 3.1 for populations containing different chlorotypes to infer their demographic history. The sum of the squared differences (SSD) was used as a statistical test to accept or reject the hypothesis of sudden population expansion. The raggedness index and its significance were calculated to quantify the smoothness of the observed mismatch distribution [62].

Microsatellite loci were developed using 5′-anchored PCR [63], [64] and five microsatellite loci exhibiting polymorphism among individuals from the Rongbuk Valley (Table S1 in File S1) were amplified following the procedures of Song *et al.* (2003) to analyze the genetic variation [65]. All the primers had reliable scoring and the interpretation of electropherograms was performed by the same person in the same laboratory for all samples. For each population, the intra-population genetic diversity was evaluated (Table 2). Fixation index (F_{IS}) and allelic richness (A, mean number of alleles per locus based on the minimal sample size) were calculated with FSTAT version 2.9.3.2 [66] using five microsatellite loci, and H_E and H_O were calculated with GENETIX version 4.03 (www.genetix.univ-montp2.fr). For tests of deviation from the Hardy-Weinberg equilibrium and genotypic disequilibrium, the *p*-values obtained (with a 0.05 significance threshold) were adjusted in FSTAT version 2.9.3.2 by applying sequential Bonferroni corrections to avoid false-positives [66]. Pairwise *F*-statistics (F_{ST}) among populations and patches were calculated using five microsatellite loci with GENETIX version 4.03, the significance of which was tested by comparison of the 95 and 99% confidence intervals derived from 1000 bootstrap permutations

Table 2. Area, crown diameter, and annual ring of 5 patches and the highest population.

Patch No.	Altitude (m)	Area (m^2)	Crown diameter (cm)[a]	Annual rings[b]
POP 8	5035	>5000	56.6±2.70	40
PAT 4	5047	~250	26.6±3.21	37
PAT 3	5046	~50	17.0±2.34	30
PAT 2	5047	~30	14.4±6.95	15
PAT 7	5058	~30	8.8±1.30	21
PAT 5	5066	~20	8.4±1.14	15

[a]the mean value ± S.D. of crown diameters of five individuals with the biggest crow; [b]the annual rings of the individual with the largest basal stem.

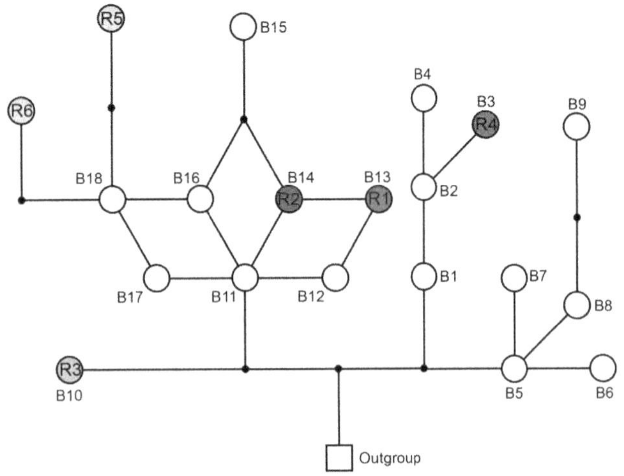

Figure 2. Chlorotype network based on the *trn*T-*trn*F sequences of *H. tibetana*. Circles with letters inside indicate chlorotypes found in the Rongbuk Valley. The other circles represent the remaining B-lineage chlorotypes found in the western part of the species' geographical range and the outgroup consists of the A-lineage chlorotypes that occupy the central part of the range [43]. Genbank accession numbers of these haptotypes were listed in Table S2 in File S1.

individuals across sites. To illustrate the shortest dispersal route between populations, we connected populations on a non-metric multidimensional scaling (NMDS) of F_{ST} using a minimum-spanning tree (MST) [67]. The bottleneck test was performed using BOTTLENECK version 1.2.02 [68] with five microsatellite loci. We employed the Wilcoxon sign-rank test because of the small number of loci [69], [70]. The estimates are based on 1000 simulations performed under both the strict one-step stepwise mutation model (SMM) and the two-phase model (TPM: 70% single-step mutation and 30% multi-step mutation) assumptions. A Bayesian clustering approach was used to infer population structure with STRUCTURE version 2.2 [71]. We used the admixture model and correlated allele models without any prior information. Twenty independent runs with a burn-in of 10,000 and 100,000 Markov chain Monte Carlo simulations for each value of K (from 1 to 15) were used to evaluate the genomic composition of the populations. We then used the ΔK statistics to evaluate the change in likelihood [72].

Results

Size of Patch and Annual Rings of the Biggest Individual

The sizes of the patches above 5035 m asl and the annual rings of the biggest individual in each patch are shown in Table 2. Five out of seven patches were found in the field investigation of 2011 according to the previous GPS data, but PAT 1 and PAT 6 could not be re-found. The topography of the sample region is rugged and some patches are small, thus we guess that the inaccurate GPS data of some patches and the harsh weather in January 2011 led us not to re-find these two patches.

The basal stems of all the cut individuals have clear annual rings when viewed under a dissecting microscope. The age (annual ring) of the biggest individual of the large colony (POP8) at 5035 m is 40 years and that of the largest patch at 5047 m is 37 years. The patch with the smallest size and crown diameter, PAT 5, grows at the highest altitude, 5066 m, and has the youngest age (15 years). The crown diameter of patch is related negatively to its altitude ($R^2 = 0.69$, $F = 8.83$, $p = 0.041$), and the same relationship exists between the annual ring of the biggest individual and its altitude ($R^2 = 0.56$, $F = 5.14$, $p = 0.086$).

Chlorotypes and Distribution

Chloroplast DNA *trn*T-*trn*F sequences of 414 individuals from 8 populations (large colonies) and 7 patches (small colonies over 5035 m asl) at different elevations in the Rongbuk Valley were sequenced in this study and six chlorotypes were identified (Fig. 1c and Table 1). Genebank accession numbers of the chlorotypes are R1 - GU561447, R2 - GU561448, R3 - GU561444, R4 - GU561441, R5 - JF268789, and R6 - JF268790. Among these six chlorotypes, R5 and R6 were first found in this study and the others (R1–R4) were reported in our previous study [43]. Comparing with all the chlorotypes found in *H.tibetana* [43], four of these six haplotypes, R1, R4, R5 and R6, are endemic to this valley, i.e. private haplotypes. The chlorotype compositions of the populations at different elevations are different (Fig. 1c). All the populations below 4700 m contain the same chlorotype component: three chlorotypes (R1, R2, R3), while the population at 5000 m has all six chlorotypes (Fig. 1c); the patches over 5035 m also have three chlorotypes (R1, R2, R3), but each patch consists of only one (Fig. 1c). Unbiased genetic diversity (D) within the populations in the Rongbuk Valley ranges from 0 to 0.7879, and nucleotide diversity (π) from 0 to 0.00343 (Table 1). A network of the six chlorotypes and other chlorotypes occurred in the west part of the species' geographical range (the B-lineage chlorotype, sequences from Wang et al. 2010) shows that chlorotypes found in the Rongbuk Valley are scattered within the network (Fig. 2).

Genetic Structure

Five microsatellite loci exhibiting polymorphism among individuals from the Rongbuk Valley were obtained. Their primer sequences are listed in Table S1 in File S1. Bayesian clustering of the information from the five microsatellite loci demonstrates that the model with $K = 4$ explains the data satisfactorily based on the ΔK statistics. Each patch contains genetically similar individuals revealed by the STRUCTURE analysis, whereas populations usually have more complex components except for the population at 4946 m (Fig. 1c). All the patches and the 4946 m-population also exhibit large negative F_{IS} values and a significant deviation from the Hardy-Weinberg equilibrium (Table 1). Pairwise F_{ST} shows little differentiation between populations except for the 4946 m-population (Table 3). The NMDS projection accounts for 88.5% (stress = 0.115) of the variance in the F_{ST} matrix (Table 3, Fig. 3). The 4690 m-population and the 5000 m-population are at the centre of the MST; populations higher than 4690 m are all linked to the 5000 m-population while others are linked to the 4690 m-population. All the patches are connected to populations below 4690 m in the MST (Fig. 3).

Population Expansion and Population Bottleneck

Results of the Wilcoxon sign-rank test and the mismatch distribution analysis are given in Table 4. The Wilcoxon test values under both the strict one-step stepwise mutation model (SMM) and the two-phase model (TMP) are significant in POP 1, 2, 5, 7, and 8, indicating that these five populations have been through recent bottlenecks. Furthermore, when performing mismatch analysis, the SSD p-value and raggedness index for POP 1, 2, 4, 5, and 7 are not significant at the 95% significance level, suggesting that these five populations have experienced demographic expansion (Table 4).

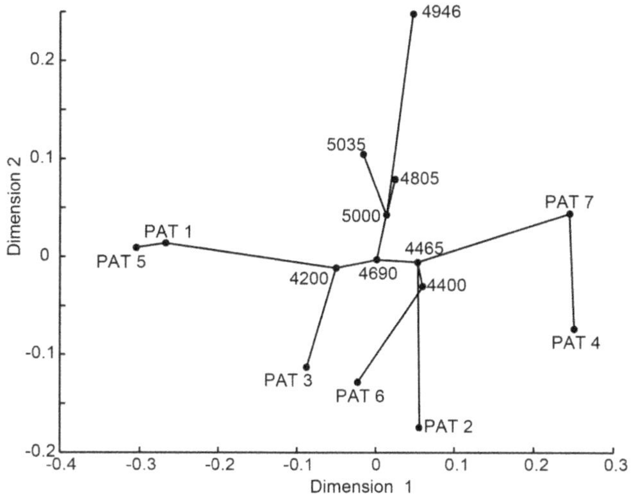

Figure 3. Non-metric multidimensional scaling of pairwise F_{ST} and a minimum-spanning tree linking populations and patches. Populations are indicated by their altitudes (m).

Discussion

The Microrefugia of *H. tibetana* in the Rongbuk Valley

In previous study, we inferred that the Rongbuk Valley may have been a LGM microrefugium for *H.tibetana* based on private chlorotypes [43]. In the present study, we analysed 414 individuals from this valley using chloroplast DNA *trn*T-*trn*F sequences and five nuclear microsatellite loci. More private chlorotypes were found and their distribution in the valley was revealed. By comparing chlorotypes here to those of *H. tibetana* elsewhere, including the two known nearest populations in the Mt. Qomolangma area (Dingri population and Nielamu population, see reference 43), we found that four (R1, R4, R5, R6) of six haplotypes found in the Rongbuk Valley are endemic to this valley, i.e. private haplotypes. In addition, these six chlorotypes are scattered in the chlorotype network (Fig. 2), and their divergence time could date back to 1.52±0.814 million years ago [43]. Considering the Rongbuk Valley is so high and had dramatically influenced by glaciations during the LGM [47], [49], the fact that so many private chlorotypes occur here is surprising. These private haplotypes only have three potential origins: origination from common haplotypes (R2 and R3) after the LGM, coming from other areas after the LGM, and survival through the LGM in the valley. As Opgenoorth et al. discussed in the previous study on *Juniperus tibetica* complex [22], these private chlorotypes are very unlikely to have originated from common haplotypes after the LGM according to the substitution rate of cpDNA *trn*T-*trn*F regions (from the fastest, 8.24×10^{-9} substitutions per site per year, to the lowest, 4.87×10^{-10}) [73]. Also, they are very unlikely to have come recently from other areas because in these other areas, including regions very near this site, the above private chlorotypes have not found according to our previous study [43]. Thus, the third origin, survival in the valley through the LGM is the only acceptable hypothesis. These results provide strong evidence that the Rongbuk valley was a refuge for *H.tibetana* during the LGM and possibly earlier glaciations. According to the definition of microrefugium [9], it is a microrefugium for *H.tibetana* because the Rongbuk valley is far away from its main range and a very small part of its whole range.

Shifts of *H. tibetana* in the Last 25,000 Years

As the 5000 m-population has all six chlorotypes and the highest genetic diversity (*D* and π in Table 1), including two endemic types (R5 and R6) in this population, it seems that in the valley, the microrefugium for *H. tibetana* in the past was located in the area at 5000 m. However, the integration of geological [47], [52], [54], glaciological [38], [45], [55], meteorological [46], [48], and genetic information reveals that the survival process may have been complex. From 25 ka B.P. until today, there have been four major periods of glacier advance that formed four terminal moraines in the Rongbuk valley (M1-M4, Fig. 1a) [52]. A terminal- or end-moraine, forms at the end of the glacier tongue and marks the maximum advance of the glacier. During the LGM, the valley area over 4800 m was covered by the glacier tongue, which consists of ice and gravel and would prevent *H. tibetana* from colonizing. The valley, except riversides and riverbed, is very arid today because of its special topography (46) and *H. tibetana* is not able to survive on the xeric side slopes in the valley. During the glacial stages, the QTP became more arid [74], [75], and the valley would be more arid. Thus, the xeric side slopes in the valley were not suitable habitats for *H. tibetana*. In fact, all populations of *H. tibetana* found in our field investigation and reported by other collectors did not occur over glacier tongues. These results support that the upper limit of *H. tibetana* in the valley must have been below the glacier tongue.

According to data from the Guliya ice-core (Fig. 1b), temperature during the LGM was ~9°C lower than present [38], [56], and the glacier tongue dropped to ~4800 m [47], [54] (M4 in Fig. 1a, b), so the upper limit of *H. tibetana* must then have been below 4800 m. According to the altitudinal difference between the recent glacier tongue and the recent upper limit of *H. tibetana* in this valley (~200 m), we infer that the upper limit of *H. tibetana* during the LGM might have been about 4600 m. It means that the microrefugium for *H. tibetana* during the LGM was located below 4800 m, and all six chlorotypes of *H. tibetana* should occur in the area below 4800 m. But now, though we checked all the populations below 4800 m, three private chlorotypes (R4, R5, and R6) could not be found in this area. This fact indicates that after the LGM, populations of *H. tibetana* in the LGM microrefugium, i.e. climate relict according to Hampe & Jump (2011) [24], have already undergone dramatic contraction or extinction. From the LGM to the Younger Dryas (YD, a cold episode around 12 ka ago), temperature fluctuated several times [38]. Annual average temperature at ~14.2 ka BP may have been up to 7°C higher than present, but it decreased dramatically by ~10°C around 12 ka BP, and suddenly increased by 4.5°C about 11 ka BP according to data from the Guliya ice-core (Fig. 1b). The glacier formed the conspicuous Rongbuk end-moraine complex (5000 m, M3 in Fig. 1a). After the YD, there was a ~4 ka warm period (~11 to 7 ka ago, Fig. 1b) and temperature in this period was ~3°C higher than present (Fig. 1b). Although these rapid and dramatic climate fluctuations were very likely to drive *H. tibetana* to shift up and down in the valley, the history of this species from the LGM to the YD in this valley is difficult to be proved. But at least during the warm period after the YD (~11 to 7 ka ago, Fig. 1b), *H. tibetana* had migrated upward to 5000 m. The reasons are as follows: 1) only the 5000 m-population now has the highest genetic divergence with all six chlorotypes, 2) two of the six chlorotypes, R5 and R6, are endemic to this 5000 m-population, strongly indicating that not less than these two endemic chlorotypes had colonized at this altitude during this period, and 3) these two endemic chlorotypes are unlikely to originate here from other chlorotypes in such a short period (Fig. 2). It means that from the LGM to the warm time after the YD, the upper limit of *H. tibetana*

Table 3. Genetic differences (*Fst*) among populations and patches using data from five microsatellite loci (Table S1).

FST	POP2	POP3	POP4	POP5	POP6	POP7	POP8	PAT1	PAT2	PAT3	PAT4	PAT5	PAT6	PAT7
POP1	0.0603^{ns}	0.0928**	0.0303^{ns}	0.0884^{ns}	0.1950**	0.0488^{ns}	0.0631^{ns}	0.2147**	0.1671**	0.0403^{ns}	0.2473**	0.2336**	0.0552^{ns}	0.2564**
POP2		0.0130^{ns}	0.0158^{ns}	0.0632^{ns}	0.2196**	0.0377^{ns}	0.0860^{ns}	0.2948**	0.0863**	0.1308**	0.1425**	0.3274**	0.0436^{ns}	0.1157**
POP3			0.0119^{ns}	0.0525**	0.1609**	0.0200^{ns}	0.0834**	0.2351**	0.0715**	0.1225**	0.1542**	0.2664**	0.0897**	0.0751^{ns}
FOP4				0.0274^{ns}	0.1638**	−0.0006^{ns}	0.0548^{ns}	0.1820**	0.0876**	0.0718^{ns}	0.1757**	0.2096**	0.0512^{ns}	0.1592**
POP5					0.1503**	0.0068^{ns}	0.0345^{ns}	0.2583**	0.1864**	0.1882**	0.1994**	0.2866**	0.1632**	0.1923**
POP6						0.1244**	0.1254**	0.4405**	0.3579**	0.2944**	0.3667**	0.4813**	0.3480**	0.3050**
POP7							0.0253^{ns}	0.2024**	0.1357**	0.1138**	0.1767**	0.2330**	0.1144**	0.1450**
POP8								0.2835**	0.2057**	0.1794**	0.2997**	0.3171**	0.1759**	0.2464**
PAT1									0.3075**	0.2777**	0.4712**	0.0054^{ns}	0.2863**	0.5185**
PAT2										0.2148**	0.3252**	0.3430**	0.0785^{ns}	0.2545**
PAT3											0.3123**	0.2995**	0.0710^{ns}	0.3420**
PAT4												0.5007**	0.2666**	0.1225**
PAT5													0.3066**	0.5612**
PAT6														0.2920**

¯he *F*-statistics (F_{ST}) values were calculated with GENETIX version 4.03, the significance of which was tested by comparison of the 95 and 99% confidence intervals derived from 1,000 bootstrap permutations individuals across sites.
$^{ns}p>0.05$,
$^{*}p\leq0.05$,
$^{**}p<0.001$.

Table 4. Bottleneck tests and mismatch distribution analysis.

Population No.	Altitude (m)	Wilcoxon Test		Mismatch distribution analysis	
		SMM	TPM	SSD p-value	Raggedness index
POP 1	4200	0.0469*	0.0312*	ns	0.273^{ns}
POP 2	4400	0.0312*	0.0156*	ns	0.164^{ns}
POP 3	4465	0.3125^{ns}	0.3125^{ns}	0.028	0.195*
POP 4	4690	0.5000^{ns}	0.3125^{ns}	ns	0.148^{ns}
POP 5	4805	0.0469*	0.0312*	ns	0.218^{ns}
POP 6	4946	0.0625^{ns}	0.0625^{ns}	–	–
POP 7	5000	0.0312*	0.0312*	ns	0.067^{ns}
POP 8	5035	0.0312*	0.0312*	ns	0.476*
PAT 1	>5035	0.2500^{ns}	0.2500^{ns}	–	–
PAT 2	5047	0.1250^{ns}	0.1250^{ns}	–	–
PAT 3	5046	0.1250^{ns}	0.1875^{ns}	–	–
PAT 4	5047	0.0625^{ns}	0.0625^{ns}	–	–
PAT 5	5066	0.2500^{ns}	0.2500^{ns}	–	–
PAT 6	>5035	0.1250^{ns}	0.1250^{ns}	–	–
PAT 7	5058	0.1250^{ns}	0.1250^{ns}	–	–

($^{ns}p>0.05$, $*p\leq0.05$, $**p<0.001$).

shifted at least 200 m upward, and most probably over 400 m (from ~4600 m to 5000 m asl or higher). Meanwhile, the lower limit of this species here might upward shift to 4800 m or higher in this warm period, and at least populations in the LGM microrefugium have undergone extreme contraction, which resulted in the loss of three chlorotypes (R4, R5, R6) in the LGM microrefugium. These results demonstrate that the area about 5000 m plays an interglacial refuge and that some area below 4800 m was merely a glacial refuge.

When populations shift up and down repeatedly, populations will undergo genetic bottlenecks and expansions [76]. These genetic events have been observed in most populations based on microsatellite and chlorotype data, respectively (Table 4). Analysis of microsatellites provides us with more information. Non-metric multidimensional scaling of pairwise F-statistics (Table 3) and a minimum-spanning tree show that the 4690 m-population and the 5000 m-population are at the centre of the net (Fig. 3), and that populations higher than 4690 m are all linked to the 5000 m-population while others are linked to the 4690 m-population. These results imply that three present populations below 4690 m are the result of the expansion of populations at higher altitude, and that the lower limit of *H. tibetana* moved upward to at least 4690 m when climate became warmer after the LGM, especially after the YD.

Information from both chlorotypes and microsatellites strongly demonstrate that the genetic structure of populations of *H. tibetana* in the LGM microrefugium has been changed greatly after the LGM. If we infer glacial microrefugia only based on genetic information, we are likely to be misled, because present populations in glacial microrefugia may not be true climate relicts of glacial stages. Similar patterns have been found at a broader scale. The study on chloroplast DNA variation in 22 widespread European trees and shrubs sampled in the same forests has shown that the genetically most diverse populations were not located in glacial refugia in southern Europe but at intermediate latitudes [77]. The authors proposed that it was a likely consequence of the admixture of divergent lineages colonizing the continent from separate refugia. We suspect that present populations in glacial refugia in southern Europe have not been true glacial relics.

Recent Shifts of *H. tibetana*

In the last 100 years, altitudinal ranges of many plants have shifted upward, presumably in response to global temperature increases [29], [32], [33], [37], [78]. The Mt. Qomolangma area is one of the most vulnerable areas to global change [79], [80], [81], and in the last four decades (from 1971 to 2004), annual mean temperature has increased here at a rate of 0.302°C per decade and from 1976 to 2006, the ends of Central Rongbu glacier has retreated about 500 m in length [55], while precipitation has stayed at a constant level [46]. In our field investigation, we only found seven scattered patches in the region over 5035 m (Table 1). These patches are mostly located on the Rongbude terminal moraine (M2, Fig. 1a) and have not been covered by landslides. No dead patches could be found in this region and few or no dead individuals could be found in these patches. These results show that these patches are not the relict of an old large colony but new small colonies, and that the annual ring of the largest individual of a patch can indicate the successful colonization age of this patch. These patches have a colonization age of 15~37 years and show a significant negative relationship between colonization time and altitude, i.e. the higher the altitude of the patch, the younger the patch (Table 2), demonstrating that these patches must have colonized here in the last 40 years and that these patches shifted step by step from lower to higher sites. The highest large colony (POP 8) is located at 5035 m, and the highest patch (PAT 5) at 5066 m. In fact, the annual ring of PAT 5 at 5066 m is 15 years, indicating that *H. tibetana* had arrived at 5066 m 15 years ago and that the shift of *H. tibetana* from 5035 m to 5066 m only spent about 25 years. This rate is slower than the mean rate of shift of alpine plants above tree-line in the European Alps and of forest plants in western Europe (27.8 m per decade; 29.4 m per decade) [33], [82]. The shift of *H. tibetana* in the valley during the past 40 years is also smaller than the inferred value (~100 m) based on an average decrease of 0.7~0.8°C per 100 m increase in altitude and an average increase of 0.302°C per decade in the Qomolangma area [46], [48]. Lenoir *et al.* (2008) showed that species with different ecological properties displayed different rates of shift and that small grasses moved faster than large woody plants [33]. We suspect that the high altitude (over 5000 m) of our study area may influence the upward rate of shift of plants. Grabherr *et al.* (1994) found that mountain plants that grew above 3000 m in the central part of the European Alps have slower rates of shift (<4 m per decade) than those that grow below 3000 m [29]. Besides climate warming and glacier shrinkage, precipitation and other ecological factors have not obviously changed during the past 40 years in the Rongbuk Valley, and human activity was very infrequent in this valley 20 years ago [52], [46]. Thus, global warming is likely to be the main reason for the recent upward shift of *H. tibetana.*

This study also highlights the colonization pattern of primary habitats. Each patch contains only one chlorotype and consists of highly genetically similar individuals inferred from the microsatellite data (Fig. 1c), indicating that all the patches expanded mainly by clonal reproduction, which then resulted in a deviation from the Hardy-Weinberg equilibrium and high F_{ST} values (Table 1). Although the 4946 m-population area is much larger than the patches, it is on a landslide and almost entirely consists of small and genetically identical individuals (Fig. 1c, Table 1), implying it is also a new clonal population after the landslide formed.

Factors Determining the Survival of *H. tibetana* in the Rongbuk Valley

Plant survival depends on a favourable local environment [9], [15], and extreme conditions, stable climates, and distinct differences from the surrounding matrix are the basis for microrefgia [19]. According to detailed topoclimate, Ashcroft et al. (2012) deduced that a mountain area may have not only warm refugia but cool refugia for striking differences in climate among different locations [19]. In the Rongbuk Valley, mountains play an important role in the ability of *H. tibetana* to withstand climate changes. Here, the elevational range of the valley is nearly 1500 m (from 3700 m to 5200 m asl) and mean temperature drops 0.7–0.8°C with each 100 m increase in altitude in this area [48], implying that there is nearly a ~11°C temperature gradient, which not only provides *H. tibetana* with a wide range within which to shift its occurrence in response to climate change, but also provides a series of fine-scale ecological niches with different microclimates that can buffer against temperature fluctuations [15], [83], [84]. In fact, with the retreat of glaciers in warmer periods, the area over 5200 m in the valley is likely to become a suitable habitat for *H. tibetana.*

In addition, glaciers are critical for *H. tibetana.* Historically when it became cold on the QTP, widespread aridity also occurred [74], and the special topography of the Mt. Qomolangma area caused the Rongbuk Valley to receive lower rainfall and experience more droughts. Glacial meltwater have thus provided almost all the moisture for *H. tibetana.* Our field investigation also confirms the importance of glaciers. In the semi-arid and arid areas in the QTP, all the eleven populations of *H. tibetana* were located in the valley below large glaciers except for the Ritu population [43], which was once near glacier [85] but the population is now shrinking as the local glacier disappears. Besides *H. tibetana,* there are many other plants which mainly depend on glacial meltwater for their moisture sources in the Rongbuk Valley [52]. We thus think that this valley may have been a favourable microhabitat for some of these plants in both cold and/or warm stages. Furthermore, in arid areas on the QTP, water is a limiting factor for the growth of many plants [86]. Similar to the Rongbuk Valley, water supply depends largely on glaciers in many places on the Plateau [87]. In these areas, glaciers become a crucial resource for the growth of many plants not only in warm periods but also in cold glacial stages.

Taken together, we present evidence to indicate the Rongbuk Valley, Mt. Qomolangma is a microrefugium of the LGM for *H. tibetana,* and how this plant here has responded to climate changes in the last 25 ka. By combining historical and recent range shifts, we have derived a better understanding of the rates of range shift, the dynamics of populations in the microrefugium after the LGM, and the mode of colonization of *H. tibetana,* all of which are important in inferring the potential locations of microrefugia for *H. tibetana* and predicting the response of this endemic species in relation to future climate warming.

Supporting Information

File S1 Tables S1–S2. Table S1. Characterization of polymorphic microsatellite loci in this study. PCR products were electrophoresed on 7.5 m urea 6% polyacrylamide gel, sized with the DNA ladder pUC19/MspI (Fermentas Life Sciences) and visualized by silver stain as in ref 36. All the primers had reliable scoring. Ta, annealing temperature; NA, number of alleles; HO, observed heterozygosity; HE, expected heterozygosity. HO and HE were calculated with Genetix version 4.03 (www.genetix.univ-montp2.fr). When POP 7 and the seven patches were not counted because they are most probably new clone populations, no significant Hardy–Weinberg disequilibrium was detected for each locus (for the seven remaining populations each or as a whole). Table S2. Genbank accession numbers of the haptotypes in Fig. 2. Haptotype B1–B18 and the haptotypes of outgroup were found in our previous work (reference 43). The outgroup consists of Haptotype A1–A6. R5 and R6 were found in the present study.

Acknowledgments

We thank Ou La and Zhiping Song for their help in sample collection. We are indebted to Cathy Jenks, Wen Fan and Xinji Li for editorial help.

Author Contributions

Conceived and designed the experiments: WZ. Performed the experiments: WZ QL FL HW YZ YF LX. Analyzed the data: MY LX HW WZ. Contributed reagents/materials/analysis tools: LX HW MY YZ KS JC QW. Wrote the paper: WZ LX HW HJBB. Obtained materials: WZ KS HW FL.

References

1. Hewitt GM (2000) The genetic legacy of the Quaternary ice ages. Nature 405:907–913.
2. Shafer A, Cullingham CI, Cote SD, Coltman DW (2010) Of glaciers and refugia: a decade of study sheds new light on the phylogeography of northwestern North America. Mol Ecol 19: 4589–4621.
3. Stewart JR, Lister AM, Barnes I, Dalen L (2010) Refugia revisited: individualistic responses of species in space and time. Philos T R Soc B 277: 661–671.
4. Bennett KD, Provan J (2008) What do we mean by 'refugia'? Quaternary Sci Rev 27: 2449–2455.
5. Birks HJB, Willis KJ (2008) Alpines, trees, and refugia in Europe. Plant Ecol Divers 1: 147–160.
6. Keppel G, Van Niel KP, Wardell-Johnson GW, Yates CJ, Byrne M, et al. (2012) Refugia: identifying and understanding safe havens for biodiversity under climate change. Global Ecol Biogeogr 21: 393–404.

7. Abbott RJ, Brochmann C (2003) History and evolution of the arctic flora: in the footsteps of Eric Hultén. Mol Ecol 12: 299–313.
8. Holderegger R, Thiel-Egenter C (2009) A discussion of different types of glacial refugia used in mountain biogeography and phylogeography. J Biogeogr 36: 476–480.
9. Rull V (2009) Microrefugia. J Biogeogr 36: 481–484.
10. Binney HA, Willis KJ, Edwards ME, Bhagwat SA, Anderson PM, et al. (2009) The distribution of late-Quaternary woody taxa in northern Eurasia: evidence from a new macrofossil database. Quaternary Sci Rev 28: 2445–2464.
11. Rull V (2010) On microrefugia and cryptic refugia. J Biogeogr 37: 1623–1625.
12. Provan J, Bennett KD (2008) Phylogeographic insights into cryptic glacial refugia. Trends Ecol Evol 23: 564–571.
13. Stewart IT, Cayan DR, Dettinger MD (2005) Changes toward earlier streamflow timing across western North America. J Climate 18: 1136–1155.
14. Anderson LL, Hu FS, Nelson DM, Petit RJ, Paige KN (2006) Ice-age endurance: DNA evidence of a white spruce refugium in Alaska. P Natl Acad Sci U S A 103: 12447–12450.
15. Dobrowski SZ (2011) A climatic basis for microrefugia: the influence of terrain on climate. Global Change Biol 17: 1022–1035.
16. Hampe A, Rodriguez-Sanchez F, Dobrowski S, Hu FS, Gavin DG (2013) Climate refugia: from the Last Glacial Maximum to the twenty-first century. New Phytol 197: 16–18.
17. McLachlan JS, Clark JS, Manos PS (2005) Molecular indicators of tree migration capacity under rapid climate change. Ecology 86: 2088–2098.
18. Pearson RG (2006) Climate change and the migration capacity of species. Trends Ecol Evol 21: 111–113.
19. Ashcroft MB, Gollan JR, Warton DI, Ramp D (2012) A novel approach to quantify and locate potential microrefugia using topoclimate, climate stability, and isolation from the matrix. Global Change Biol 18: 1866–1879.
20. Graham CH, VanDerWal J, Phillips SJ, Moritz C, Williams SE (2010) Dynamic refugia and species persistence: tracking spatial shifts in habitat through time. Ecography 33: 1062–1069.
21. Previsic A, Walton C, Kucinic M, Mitrikeski PT, Kerovec M (2009) Pleistocene divergence of Dinaric *Drusus* endemics (Trichoptera, Limnephilidae) in multiple microrefugia within the Balkan Peninsula. Mol Ecol 18: 634–647.
22. Opgenoorth L, Vendramin GG, Mao KS, Miehe G, Miehe S, et al. (2010) Tree endurance on the Tibetan Plateau marks the world's highest known tree line of the Last Glacial Maximum. New Phytol 185: 332–342.
23. Moeller DA, Geber MA, Tiffin P (2011) Population genetics and the evolution of geographic range limits in an annual plant. Am Nat 178: S44–S61.
24. Hampe A, Jump AS, Futuyma DJ, Shaffer HB, Simberloff D (2011) Climate relicts: past, present, future. Annu Rev Ecol Evol S 42: 313–333.
25. Galbreath KE, Hafner DJ, Zamudio KR (2009) When cold is better: climate-driven elevation shifts yield complex patterns of diversification and demography in an alpine specialist (American pika, *Ochotona princeps*). Evolution 63: 2848–2863.
26. Mosblech N, Bush MB, van Woesik R (2011) On metapopulations and microrefugia: palaeoecological insights. J Biogeogr 38: 419–429.
27. McLaughlin BC, Zavaleta ES (2012) Predicting species responses to climate change: demography and climate microrefugia in California valley oak (*Quercus lobata*). Global Change Biol 18: 2301–2312.
28. Hansen J, Sato M, Ruedy R, Lo K, Lea DW, et al. (2006) Global temperature change. P Natl Acad Sci U S A 103: 14288–14293.
29. Grabherr G, Gottfried M, Pauli H (1994) Climate effects on mountain plants. Nature 369: 448.
30. Walther GR, Post E, Convey P, Menzel A, Parmesan C, et al. (2002) Ecological responses to recent climate change. Nature 416: 389–395.
31. Parmesan C (2006) Ecological and evolutionary responses to recent climate change. Annu Rev Ecol Evol S 37: 637–669.
32. Colwell RK, Brehm G, Cardelus CL, Gilman AC, Longino JT (2008) Global warming, elevational range shifts, and lowland biotic attrition in the wet tropics. Science 322: 258–261.
33. Lenoir J, Gegout JC, Marquet PA, de Ruffray P, Brisse H (2008) A significant upward shift in plant species optimum elevation during the 20th century. Science 320: 1768–1771.
34. Lenoir J, Svenning JC (2013) Latitudinal and elevational range shifts under contemporary climate change. Encyclopedia of Biodiversity Vol. 4. San Diego, CA: Academic Press. 599–611p.
35. Thomas CD, Cameron A, Green RE, Bakkenes M, Beaumont LJ, et al. (2004) Extinction risk from climate change. Nature 427: 145–148.
36. Parmesan C, Yohe G (2003) A globally coherent fingerprint of climate change impacts across natural systems. Nature 421: 37–42.
37. Pauli H, Gottfried M, Reiter K, Klettner C, Grabherr G (2007) Signals of range expansions and contractions of vascular plants in the high Alps: observations (1994–2004) at the GLORIA master site Schrankogel, Tyrol, Austria. Global Change Biol 13: 147–156.
38. Thompson LG, Davis ME, Mosley-Thompson E, Sowers TA, Henderson KA, et al. (1998) A 25,000-year tropical climate history from Bolivian ice cores. Science 282: 1858–1864.
39. Thompson LG, Yao T, Davis ME, Henderson KA, Mosley-Thompson E, et al. (1997) Tropical climate instability: The last glacial cycle from a Qinghai-Tibetan ice core. Science 276: 1821–1825.
40. Stott L, Poulsen C, Lund S, Thunell R (2002) Super ENSO and global climate oscillations at millennial time scales. Science 297: 222–226.
41. Bush MB, Silman MR, Urrego DH (2004) 48,000 years of climate and forest change in a biodiversity hot spot. Science 303: 827–829.
42. Zhang ZR (1983) Elaeagnaceae, genus *Hippophae*. Delectis Florae Reipublicae Popularis Sinicae. Beijing: Science Press. 61–64p.
43. Wang H, Qiong L, Sun K, Lu F, Wang YG, et al. (2010) Phylogeographic structure of *Hippophae tibetana* (Elaeagnaceae) highlights the highest microrefugia and the rapid uplift of the Qinghai-Tibetan Plateau. Mol Ecol 19: 2964–2979.
44. Chen XL, Ma RJ, Sun K, Lin YS (2003) Germplasm resource and habitat types of Seabuckthorn in China. Acta Botanica Boreali-Occidentalia Sinica 23: 5.
45. Kang SC, Qin DH, Mayewski PA, Wake CP, Ren JW (2001) Climatic and environmental records from the Far East Rongbuk ice core, Mt. Qomolangma (Mt. Everest). Episodes 24: 176–181.
46. Yang XC, Zhang YL, Zhang W, Yan YP, Wang ZF, et al. (2006) Climate change in Mt. Qomolangma region since 1971. J Geogr Sci 16: 11.
47. Owen LA, Robinson R, Benn DI, Finkel RC, Davis NK, et al. (2009) Quaternary glaciation of Mount Everest. Quaternary Sci Rev 28: 1412–1433.
48. Yang XG, Zhang TJ, Qin DH, Kang SC, Qin XA (2011) Characteristics and changes in air temperature and glacier's response on the north slope of Mt. Qomolangma (Mt. Everest). Arct Antarct Alp Res 43: 147–160.
49. Xie ZC, Su Z (1975) Developing conditions, amounts and distributions of glaciers in the Mt. Qomolangma area. Dissertation of the Scientific Expedition to Mt. Qomolangma (1966–1968), Modern Glacier and Topography. Beijing: Science Press. 1–14p.
50. Pecci M, Pignotti S, Smiraglia C, Mortara G (2010) Geomorphology of the central and frontal Rongbuck glacier area (Mt. Everest, Tibet). Geogr Fis Din Quat 33: 79–92.
51. Burbank DW, Kang JC (1991) Relative dating of Quaternary Moraines, Rongbuk Valley, Mount Everest, Tibet - Implications for an ice-sheet on the Tibetan Plateau. Quaternary Res 36: 1–18.
52. Team of Scientific Expedition to Mt. Qomolangma C (1962) Report of the Scientific Expedition to Mt. Qomolangma. Science Press: Beijing.
53. Zheng BS, Shi YF (1976) A study of Quaternary glaciations of the Mount Qomolangma Region. Report of the Scientific Expedition to Mt. Qomolangma (1966–1968), Quaternary Geology. Beijing: Science Press. 29–62p.
54. Hu E, Yi CI, Li YJ (2010) Observation and evolution investigation of the moraine geomorphology in the Rongbuk Valley of Mount Qomolangma. J Glaciol Geocryol, 32, 316–324.
55. Nie Y, Zhang YL, Liu LS, Zhang JP (2010) Glacial change in the vicinity of Mt. Qomolangma (Everest), central high Himalayas since 1976. J Geogr Sci 20: 667–686.
56. Yao TD, Thompson LG, Qin DH, Tian LD, Jiao KQ, et al. (1996) Variations in temperature and precipitation in the past 2000a on the Xizang (Tibet) Plateau - Guliya ice core record. Sci China Ser D 39: 425–433.
57. Yang XG, Qin DH, Zhang TJ, Kang SC, Qin X (2012) Characteristics of the air temperature and humidity on the north slope of Mt. Qomolangma. Acta Meteorol Sin 70: 12.
58. Wang ZY, Ma YM, Liu JS, Han CB (2013) Characteristic analyses on hydrological and related meteorological factors on the north slope of Mt. Qomolangma. Plateau Meteorol 32: 7.
59. Rozas J, Sanchez-DelBarrio JC, Messeguer X, Rozas R (2003) DnaSP, DNA polymorphism analyses by the coalescent and other methods. Bioinformatics 19: 2496–2497.
60. Nei M (1987) Molecular Evolutionary Genetics. New York: Columbia University Press.
61. Excoffier L, Laval G, Schneider S (2005) Arlequin (version 3.0): An integrated software package for population genetics data analysis. Evol Bioinform 1: 47–50.
62. Harpending HC (1994) Signature of ancient population-growth in a low-resolution mitochondrial-DNA mismatch distribution. Hum Biol 66: 591–600.
63. Fisher PJ, Gardner RC, Richardson TE (1996) Single locus microsatellites isolated using 5′ anchored PCR. Nucleic Acids Res 24: 4369–4371.
64. Zhou YB, Wang H, Yang M, Chen JK, Zhang WJ (2009) Development of microsatellites for *Scirpus mariqueter* Wang et Tang (Cyperaceae) and cross-species amplification in *Scirpus planiculmis* F. Schmidt. Mol Ecol Resour 9: 370–372.
65. Song ZP, Xu X, Wang B, Chen JK, Lu BR (2003) Genetic diversity in the northernmost *Oryza rufipogon* populations estimated by SSR markers. Theor Appl Genet 107: 1492–1499.
66. Goudet J (1995) FSTAT (Version 1.2): A computer program to calculate F-statistics. J Hered 86: 485–486.
67. Geffen E, Waidyaratne S, Dalen L, Angerbjorn A, Vila C, et al. (2007) Sea ice occurrence predicts genetic isolation in the Arctic fox. Mol Ecol 16: 4241–4255.
68. Cornuet JM, Luikart G (1996) Description and power analysis of two tests for detecting recent population bottlenecks from allele frequency data. Genetics 144: 2001–2014.
69. Pope LC, Estoup A, Moritz C (2000) Phylogeography and population structure of an ecotonal marsupial, *Bettongia tropica*, determined using mtDNA and microsatellites. Mol Ecol 9: 2041–2053.
70. Hampton JO, Spencer P, Alpers DL, Twigg LE, Woolnough AP, et al. (2004) Molecular techniques, wildlife management and the importance of genetic population structure and dispersal: a case study with feral pigs. J Appl Ecol 41: 735–743.
71. Pritchard JK, Stephens M, Donnelly P (2000) Inference of population structure using multilocus genotype data. Genetics 155: 945–959.

72. Evanno G, Regnaut S, Goudet J (2005) Detecting the number of clusters of individuals using the software STRUCTURE: a simulation study. Mol Ecol 14: 2611–2620.
73. Richardson JE, Pennington RT, Pennington TD, Hollingsworth PM (2001) Rapid diversification of a species-rich genus of neotropical rain forest trees. Science 293: 2242–2245.
74. Gasse F, Arnold M, Fontes JC, Fort M, Gibert E, et al. (1991) A 13,000-year climate record from western Tibet. Nature 353: 742–745.
75. Shi YF, Liu SY (2000) Estimation on the response of glaciers in China to the global warming in the 21st century. Chinese Sci Bull 45: 668–672.
76. Ruedi M, Castella V (2003) Genetic consequences of the ice ages on nurseries of the bat *Myotis myotis*: a mitochondrial and nuclear survey. Mol Ecol 12: 1527–1540.
77. Petit RJ, Aguinagalde I, de Beaulieu JL, Bittkau C, Brewer S, et al. (2003) Glacial refugia: Hotspots but not melting pots of genetic diversity. Science 300: 1563–1565.
78. Harsch MA, Hulme PE, McGlone AS, Duncan RP (2009) Are treelines advancing? A global meta-analysis of treeline response to climate warming. Ecol Lett 12: 1040–1049.
79. Hou SG, Qin DH, Wake CP, Mayewski PA, Ren JW, et al. (2000) Climatological significance of an ice core net-accumulation record at Mt. Qomolangma (Everest). Chinese Sci Bull 45: 259–264.
80. Ren JW, Qin DH, Kang SC, Hou SG, Pu JC, et al. (2004) Glacier variations and climate warming and drying in the central Himalayas. Chinese Sci Bull 49: 65–69.
81. Liu XD, Cheng ZG, Yan LB, Yin ZY (2009) Elevation dependency of recent and future minimum surface air temperature trends in the Tibetan Plateau and its surroundings. Global Planet Change 68: 164–174.
82. Walther GR, Beissner S, Burga CA (2005) Trends in the upward shift of alpine plants. J Veg Sci 16: 541–548.
83. Körner C (2000) Why are there global gradients in species richness? Mountains might hold the answer. Trends Ecol Evol 15: 513–514.
84. Scherrer D, Körner C (2011) Topographically controlled thermal-habitat differentiation buffers alpine plant diversity against climate warming. J Biogeogr 38: 406–416.
85. Li B, Li J, Cui Z (1991) Quaternary Glacial Distribution Map of Qinghai-Xizang (Tibet) Plateau. Beijing: Science Press (Map).
86. Tilman D, Lehman C (2001) Human-caused environmental change: Impacts on plant diversity and evolution. P Natl Acad Sci U S A 98: 5433–5440.
87. Yao TD, Pu J, Lu A, Wang Y, Yu W (2007) Recent glacial retreat and its impact on hydrological processes on the Tibetan plateau, China, and surrounding regions. Arct Antarct Alp Res 39: 642–650.

4

Migration Patterns of Subgenus *Alnus* in Europe since the Last Glacial Maximum

Jan Douda[1]*, Jana Doudová[2], Alena Drašnarová[1,2], Petr Kuneš[3], Věroslava Hadincová[2], Karol Krak[2], Petr Zákravský[2], Bohumil Mandák[1,2]

1 Department of Ecology, Faculty of Environmental Sciences, Czech University of Life Sciences Prague, Prague, Czech Republic, **2** Institute of Botany, Academy of Sciences of the Czech Republic, Průhonice, Czech Republic, **3** Department of Botany, Faculty of Science, Charles University in Prague, Prague, Czech Republic

Abstract

Background/Aims: Recently, new palaeoecological records supported by molecular analyses and palaeodistributional modelling have provided more comprehensive insights into plant behaviour during the last Quaternary cycle. We reviewed the migration history of species of subgenus *Alnus* during the last 50,000 years in Europe with a focus on (1) a general revision of *Alnus* history since the Last Glacial Maximum (LGM), (2) evidence of northern refugia of *Alnus* populations during the LGM and (3) the specific history of *Alnus* in particular European regions.

Methodology: We determined changes in *Alnus* distribution on the basis of 811 and 68 radiocarbon-dated pollen and macrofossil sites, respectively. We compiled data from the European Pollen Database, the Czech Quaternary Palynological Database, the Eurasian Macrofossil Database and additional literature. Pollen percentage thresholds indicating expansions or retreats were used to describe patterns of past *Alnus* occurrence.

Principal Findings: An expansion of *Alnus* during the Late Glacial and early Holocene periods supports the presence of alders during the LGM in southern peninsulas and northerly areas in western Europe, the foothills of the Alps, the Carpathians and northeastern Europe. After glaciers withdrew, the ice-free area of Europe was likely colonized from several regional refugia; the deglaciated area of Scandinavia was likely colonized from a single refugium in northeastern Europe. In the more northerly parts of Europe, we found a scale-dependent pattern of *Alnus* expansion characterised by a synchronous increase of *Alnus* within individual regions, though with regional differences in the times of the expansion. In southern peninsulas, the Alps and the Carpathians, by contrast, it seems that *Alnus* expanded differently at individual sites rather than synchronously in whole regions.

Conclusions: Our synthesis supports the idea that northern LGM populations were important sources of postglacial *Alnus* expansion. The delayed *Alnus* expansion apparent in some regions was likely a result of environmental limitations.

Editor: Jerome Chave, Centre National de la Recherche Scientifique, France

Funding: This study was supported by grant no. P504/11/0402 from the Grant Agency of the Czech Republic (http://www.gacr.cz/en/) and grant CIGA no. 20124201 from the Czech University of Life Sciences Prague (http://ga.czu.cz/) and as part of the long-term research development project no. RVO 67985939 (http://www.msmt.cz/). The funders had no role in study design, data collection and analysis, decision to publish, or preparation of the manuscript.

Competing Interests: The authors have declared that no competing interests exist.

* E-mail: douda@fzp.czu.cz

Introduction

The recent distribution of species in the Northern Hemisphere has been significantly influenced by processes occurring in the last Quaternary cycle, during the last glacial period and subsequent Holocene warming [1], [2]. The 'classic' paradigm states that during the Last Glacial Maximum (LGM, i.e., from 26.5 to 19 to 20 thousand years before present (kyr BP) [3]), temperate plant species, particularly climate-sensitive trees, were harboured in low-latitude refugia. In Europe, southern peninsulas (i.e., Iberian, Italian and Balkan) served as refugial areas for many species [4], [5].

Recently, new palaeoecological records supported by molecular analyses and palaeodistributional modelling have provided more comprehensive insights into plant behaviour during the last Quaternary cycle [6], [7]. In eastern Europe, more northerly distributions of many temperate and boreal plants during the last glacial period have been confirmed, although fossil records directly from the LGM are scarce. The eastern Alps, northern Dinaric Alps, the Carpathians and the Pannonian region probably served as northern refugia for many temperate tree species, namely *Abies alba*, *Carpinus betulus*, *Fagus sylvatica*, *Taxus baccata* and *Ulmus* [6], [8], [9], [10]. Open taiga and hemiboreal forests dominated by *Larix*, *Pinus*, *Picea* and *Betula* likely occurred in the northern Carpathians, Belarus and the northwestern Russian plains [11], [12], [13].

The early postglacial expansion of trees in northern areas thus need not reflect migration from southern regions but may be the result of the population growth and expansion of small tree populations persisting in scattered refugia relatively close to the margin of the ice sheet [14], [15], [16]. Climatic warming has been determined as the most important driver initiating the expansion of trees [1]. However, regional differences in climatic and environmental conditions recorded for the Late Glacial and early Holocene periods could have resulted in nontrivial species-specific and regionally dependent patterns of expansion [17], [18], [19].

Taxonomic Status

The genus *Alnus* Mill. belongs to the family Betulaceae [20], [21]. The oldest macrofossil records assigned to *Alnus* have been reported from the middle Eocene, but fossil pollen grains of *Alnus* from the Late Cretaceous have also been found [22]. The genus *Alnus* comprises about 29 to 35 species of monoecious trees and shrubs distributed throughout the Northern Hemisphere and along the Andes in South America [23]. In Europe, three species of subgenus *Alnus* (i.e. *Alnus cordata*, *A. glutinosa* and *A. incana*) and one species of subgenus *Alnobetula* (*A. alnobetula* (Ehrh.) K. Koch) occur [24]. It has been estimated using molecular methods that the subgenera *Alnus* and *Alnobetula* diverged in the Eocene, 48.6 million years (Myr) BP [25]. *A. cordata* separated from the *A. glutinosa-incana* complex in the Oligocene (22.9 Myr BP), and *A. glutinosa* and *A. incana* diverged in the Pliocene (7.9 Myr BP) [25].

Upper Pleistocene and Holocene *Alnus* History

Pleistocene pollen and macrofossil data indicate repeated population increases and decreases of *Alnus* in Europe, reflecting climate oscillations between glacial periods and interglacials, particularly noticeable in the Middle and Upper Pleistocene [5], [26], [27]. The majority of Upper Pleistocene pollen profiles support a common presence of *Alnus* in the Eemian interglacial (Marine Isotope Stage, MIS 5e–5d) throughout Europe (i.e., 125–115 kyr BP) and its disappearance after the start of the last glacial period [i.e., Hollerup (DK) – [28]; Tenaghi Philipon (GR) – [26]; Valle di Castiglione (IT) – [29]; Les Echets (FR), La Grande Pile (FR) – [30], [31]; Praclaux Crater (FR), Ribains (FR) – [32]; Ioannina (GR) – [5]; Jammertal (DE) – [33]].

In their classic study, Huntley and Birks [1] assumed that the main source refugia for the *Alnus* expansion after the LGM lay in the eastern Alps, the Carpathians and the Ukrainian lowlands. Other LGM refugia were located in Corsica, western France, northern Spain and northwestern Russia. The authors supposed that the Holocene migration of *Alnus* likely began somewhere in eastern Europe and continued by the northward expansion of *A. glutinosa* (L.) Gaertn. and *A. incana* (L.) Moench to the Baltic and Scandinavia and by the westward expansion of *A. glutinosa* along the southern shore of the North Sea as far as the British Isles [1].

A large-scale genetic survey that included Europe and Turkey and focused on the postglacial history of *Alnus glutinosa* accepted Huntley and Birks' migration patterns [34]. King and Ferris [34] revealed 13 cpDNA haplotypes of *A. glutinosa* mainly associated with southern European peninsulas. They suggested that two of these haplotypes colonized northern and temperate Europe from LGM refugia located in the Carpathians. While the first haplotype expanded primarily into western Europe, the second mainly colonized northern Europe [34]. However, the presence of only two haplotypes in the northern part of Europe limits a more detailed determination of *A. glutinosa* migration patterns. Surprisingly, no follow-up study focusing on postglacial migration pattern of *A. glutinosa* has since been published. No phylogeographic study has been performed for *A. incana* thus far, either.

Since the seminal studies of Huntley and Birks [1] and King and Ferris [34] were published, many palaeoecological studies have presented new knowledge about the history of *Alnus* in Europe during the last glacial period and Holocene. We reviewed the migration history of species of subgenus *Alnus*, including *A. glutinosa* and *A. incana*, during the last 50,000 years in Europe based on large numbers of pollen records and macrofossil remains. Pollen of different species of subgenus *Alnus* (further collectively referred to as "*Alnus*") is indistinguishable in palaeoecological studies, but alder species can be identified based on macrofossil remains. In particular, we focused on (1) a general revision of *Alnus* history since the LGM, (2) evidence of northern refugia of *Alnus* during the LGM and (3) the specific history of *Alnus* in particular European regions.

Alnus species are keystones of alluvial and wetland habitats [35], [36] distributed through the European forest zones from the northern treeline to the Mediterranean. Understanding their last glacial occurrence and postglacial migration pattern may shed light upon the resistance and resilience of wetland forest habitats in the course of global climate change. The results of our study allow us to propose guidelines for the sampling design and interpretation of a future detailed phylogeographic and population-genetic survey of *Alnus* species in Europe.

Materials and Methods

Study Species

Two common tree species of *Alnus* grow natively in Europe [24]. Black alder (*A. glutinosa*) is considered a temperate tree. It commonly occurs in the lowlands and mountains across Europe except Scandinavia, where it is associated with a coastal oceanic climate in southern areas [37] (Figure 1). The cold-climate limitation also likely affects its distribution in high-elevation mountainous areas, where black alder populations are often absent. Scarce distributions are found in the Mediterranean region and in the arid Great Hungarian plains, the Ukraine and the Russian steppe zone. Outside Europe, the distribution extends as far as western Siberia and the mountains of Turkey, Iran and North Africa [24], [38]. In Corsica and southern Italy, *A. glutinosa* grows sympatrically with *A. cordata* (Loisel.) Duby.

Grey alder (*Alnus incana*) is considered a boreal and mountain tree. Similar to Norway spruce (*Picea abies*), the range of *A. incana* is divided into a northern and a southern area, which meet in the Polish lowlands. In northern Europe, *A. incana* continuously covers the east Baltic region and all of Scandinavia with a northern margin at latitudes greater than 70°N [24], [37]. In northern Scandinavia, the nominal subspecies grows sympatrically with *A. incana* subsp. *kolaensis* (Orlova) Á. Löve & D. Löve [24]. The distribution of grey alder continues eastwards across European Russia to western Siberia, which contrasts with its patchy mountain occurrence in the southern part of the range linked to the Alps, the northern Apennines, the Hercynian Mountains, the Carpathians, the Bulgarian Mountains, the Dinaric Alps, the Caucasus and Turkey [24].

Alnus glutinosa and *A. incana* dominate in floodplain and swamp forests. These species are indifferent to soil nutrient conditions, except for extremely poor peat bogs. Seeds are dispersed effectively by water, while wind dispersal is commonly limited to the vicinity of the parent tree [38]. Under unfavourable environmental conditions, such as in cold climates, *A. incana* is able to survive and reproduce by clonal growth [39]. Compared with the relatively short-lived *A. incana* (c. 20–50 years), *A. glutinosa* is a long-lived tree (c. 100–120 years), although the age of reproduction is similar for the two species (i.e., 10–20 years) [37], [38].

Pollen Data

This systematic review follows the PRISMA (Preferred Reporting Items for Systematic Reviews and Meta-Analyses) statement as a guide [40] (see Checklist S1). We compiled freely available data

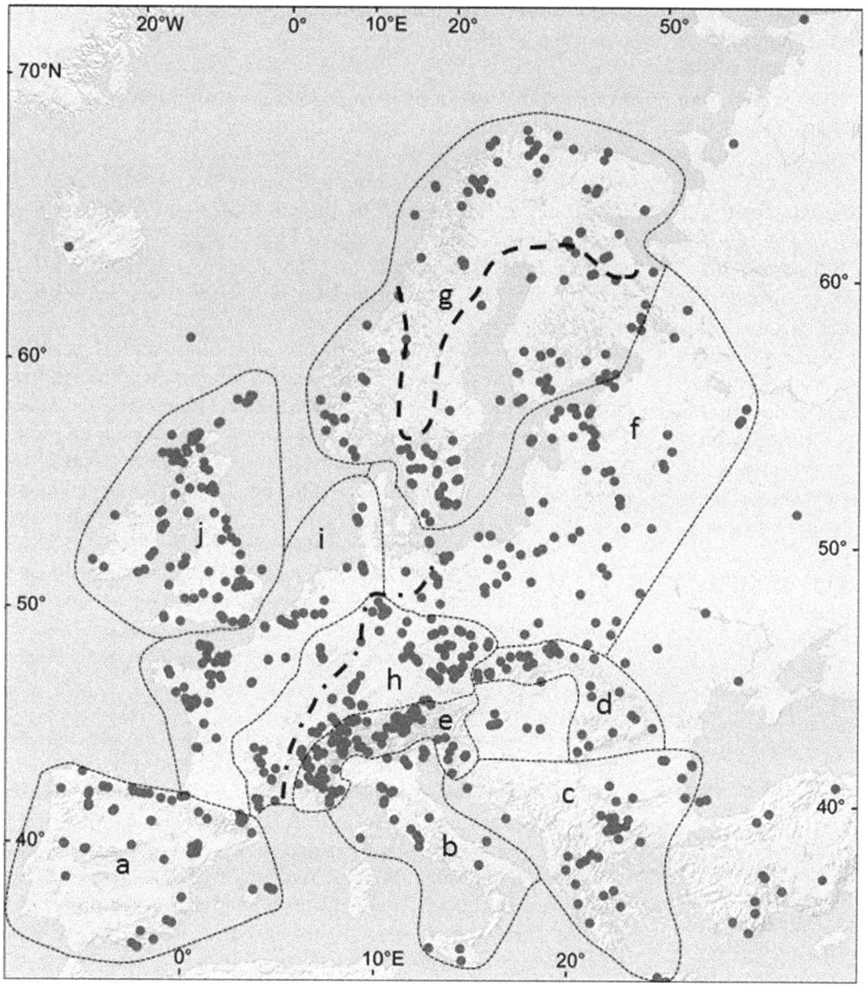

Figure 1. Pollen sites (dots) and European regions (dotted lines) included in this study. a, Iberian region; b, Italian region; c, Balkan region; d, the Carpathians; e, the Alps; f, Baltic and northeastern European plains; g, Scandinavia; h, Hercynian Mountains and Massif Central; i, western European plain; j, British Isles. Bold dashed and dashed-dotted lines show the northern boundary of *Alnus glutinosa* in Scandinavia and the western range boundary of *A. incana* in Western Europe, respectively [24].

from the European Pollen Database (EPD, http://europeanpollendatabase.net, [41]), the Czech Quaternary Palynological Database (PALYCZ, http://botany.natur.cuni.cz/palycz, [42]) and additional literature (Figure 1). The search for additional literature was performed in September 2011 in Web of Science and augmented by Google Scholar. The search included combinations and derivations of the following terms: radiocarbon dates, pollen, wood remains, macrofossils, glacial, vegetation, LGM, Holocene and Europe. To guarantee the chronological accuracy of changes in *Alnus* distribution, we used only pollen data with radiocarbon dating. In total, we used 553 and 258 pollen profiles from databases and the literature, respectively. The list of original publications and sites available in September 2011 is provided in Tables S1 and S3.

Age-depth models were constructed and radiocarbon dates were calibrated (cal.) for all profiles in the EPD [43], PALYCZ and publications using the CLAM code [44] in R [45]. The age-depth models were constructed using smoothing-spline fitting with a default smoothing factor of 0.3 or linear interpolation with preferences for a smoothing spline. Possible errors in the pollen diagram chronology were minimised in several ways. We excluded parts of the chronology outside the marginal ^{14}C dates. Additionally, to determine the oldest unquestionable time of expansion, we checked parts of the pollen diagrams indicating the start of the *Alnus* expansion (i.e., ≥2.5% pollen threshold) to determine whether i) the nearest radiocarbon date is closer than 2,000 years to time of the expansion, ii) there is no presence of reworked pollen or iii) the expansion does not start at the end of the previous 1,000-year interval. Reworking was assumed when an isolated pollen spectrum with *Alnus* ≥2.5% was recorded or when the basal spectra of *Alnus* ≥2.5% were followed by a steep decrease in pollen.

To describe the temporal patterns of *Alnus* occurrence at particular sites, we recorded the pollen percentage of *Alnus* at 1,000-year intervals in the time period from the present to 26 cal. kyr BP (i.e., the start of the LGM) and at 5,000-year intervals in the time period preceding the LGM. The average percentage of pollen of *Alnus* at the site in each time interval (1,000 and 5,000 year) was calculated by dividing the *Alnus* pollen count by the total pollen sum in each sample after excluding aquatic species, cryptogam spores and indeterminable pollen. We excluded pollen of subgenus *Alnobetula* from the total *Alnus* pollen count. Due to their specific pollen morphology, pollen grains of species of subgenus *Alnobetula* are identified and counted separately in palaeoecological studies [46], [47]. We chose 5,000-year intervals before 26 kyr BP because the pollen records were fragmentary and the rare radiocarbon dates do not sufficiently cover the pollen profiles. The total pollen sum was calculated in most literature sources in the same way, allowing us to determine past *Alnus* pollen value for each time period by simple visual inspection of pollen diagrams.

Because alders are high pollen producers, they are generally overrepresented in pollen diagrams [1]. Moreover, *Alnus glutinosa* and *A. incana* often dominate in swamps, at lake and stream shores and at the margins of peat bogs in close vicinity to sample sites [48]. To record the regional presence of *Alnus* from pollen diagrams, several thresholds ranging from 0.5 to 8% have been used in the literature [49], with greater agreement for 2–3% [1], [50], [51], [52]. We used the 2.5% threshold suggested for *Alnus* in a recent study comparing modern pollen data with European tree species distribution [49]. This 2.5% threshold corresponds to the presence of *Alnus* within approximately 50 km of a pollen site [49]. We also incorporated the threshold of 0.5% as an indicator of possible scarce regional occurrence despite the risk of contamination by long-distance pollen transport. Lisitsyna et al. [49] still found strong agreement between pollen presence defined by the 0.5% threshold and the regional occurrence of *Alnus*. Pollen values greater than the 10% threshold are assumed to correspond to the occurrence of an *Alnus*-dominated forest at the site [1], [37], [51]. In summary, four percentage categories were used to describe the patterns of past *Alnus* occurrence, where less than 0.5% indicates the regional absence of a species, 0.5–2.5% may be the result of long-distance pollen transport but could also capture the presence of relatively small populations in the region, 2.5–10% indicates a species' presence within the region, and values greater than 10% indicate the local presence of a species at the site. The description of *Alnus* distribution in the results and discussion section is based on ≥2.5% pollen records to eliminate possible misinterpretation based on the 0.5% threshold.

Macrofossil Data

To obtain macrofossil evidence (e.g., cones, fruits, male catkins, twigs, wood pieces), we used free data available from the Eurasian Macrofossil Database (NEMD, http://oxlel.zoo.ox.ac.uk/reference-collection, [13]) and additional published records. In total, we used macrofossil data from 14 sites in the database and 54 sites in the literature (Tables S2 and S3). Macrofossils of *Alnus glutinosa* and *A. incana* were determined at 38 and 15 sites, respectively. Macrofossil records were assigned according to 1,000- or 5,000-year pollen intervals based on constructed age-depth models (see Pollen data chapter). We interpreted only the *Alnus* presence, as it is problematic to evaluate data regarding the absence or abundance of macrofossils [9].

Pollen and Macrofossil Maps

The pollen and macrofossil maps indicate *Alnus* occurrence at particular time periods during the last 50,000 years. We merged 5,000- and 1,000-yr intervals with a limited number of records to logical periods of the last glacial period; 50–26 cal. kyr BP includes the period preceding the LGM, 26–20 cal. kyr BP the period of the LGM and 20–15 cal. kyr BP the period after the LGM, also known as the Oldest Dryas. The macrofossil remains and maximum pollen thresholds recorded during the merged periods were plotted in maps. We also marked changes in the pollen percentages between the time periods, indicating the expansion, stability or decrease of *Alnus*. The term "*Alnus*" indicates macrofossils that were not assigned to individual species in original studies whereas the names "*Alnus glutinosa*" and "*A. incana*" refer to those that were.

Regional Differences in Late Glacial and Holocene History

To determine the specific postglacial history of *Alnus* in individual European regions, we delimited 10 regions based on different environmental conditions in the last glacial period and the Holocene (Figure 1). The Iberian, Italian and Balkan regions include areas considered southern LGM refugia of trees (Figure 1, regions a–c). The Baltic and northeastern European plains, Scandinavia and the British Isles are regions that were largely covered by the Scandinavian ice sheet during the LGM (Figure 1, regions f, g, j). The Carpathians and Alps covered areas of potential LGM refugia for some temperate and many boreal trees (Figure 1, regions d, e). The Hercynian Mountains, the Massif Central and highlands located to the north of the Alps were mostly ice-free regions (Figure 1, region h). Ice-free lowland areas of the Western European plain were influenced by the oceanic climate (Figure 1, region i). We determined the proportion of pollen sites in each region and time period that reached the 0.5%, 2.5% and 10% thresholds. Only time intervals with more than 10 sites available in particular regions were considered in the analysis. The region of the Great Hungarian plains was excluded from all analyses because fewer than 10 pollen sites had been found there.

Results

Pre-LGM *Alnus* Distribution (50–26 cal. kyr BP)

In southern Europe, *Alnus* exceeds the 2.5% pollen threshold in the Pyrenees Mountains [53] and at several Italian sites [54], [55], [56], [57] (Figure 2A). Other pollen records exceeding 2.5% have been obtained from northwestern France [58] and the western Russian plains [59]. In western Russia and Belarus, the occurrence of *A. glutinosa* and *A. incana* is supported by macrofossil remains [60], [61], [62]. *Alnus* macrofossil records are present along the northern border of the Pannonian lowlands in the Czech Republic and the northeastern foothills of the Carpathians in Romania [10] (Figure 2A).

LGM *Alnus* Distribution (26–20 cal. kyr BP; Figure 2B)

At the LGM, the *Alnus* pollen values decrease in Italy and France (Figure 2B). The only 2.5%-threshold pollen evidence for *Alnus* occurrence in southern-European peninsulas was detected in the Pyrenees Mountains [53]. Further north in Europe, *Alnus* pollen values exceed the 2.5% threshold at two sites in the Bodmin moor in Cornwall [63] and in the Timan Ridge in Arctic Russia [64] (Figure 2B).

Late Glacial *Alnus* Distribution (20–12 cal. kyr BP; Figure 2C, 2D, 3A and 3B)

Between 20 and 15 cal. kyr BP, the 2.5%-threshold pollen evidence of *Alnus* continues in southern England and Arctic Russia (Figure 2C). In southern Europe, only one new 2.5%-threshold pollen record has emerged in the Rila Mountains in Bulgaria [65]. Macrofossil remains of *Alnus* occur in the southwestern foothills of the Alps [66], the Thracian plain in Bulgaria [67] and southern Lithuania [61] (Figure 2C).

Between 15 and 12 cal. kyr BP, several pollen sites exceed the 2.5% *Alnus* threshold in the southwestern and western parts of the Alps (Figs 2D, 3A and 3B). Moreover, macrofossil remains of *A. glutinosa* occur there [66] (Figure 3B). South of the Alps, *Alnus* pollen increases and reaches more than 2.5% in Corsica [68] (Figure 3A), the northern Apennines [69] (Figure 3B) and central Italy [70] (Figure 3B). In the Carpathians, *Alnus* pollen records

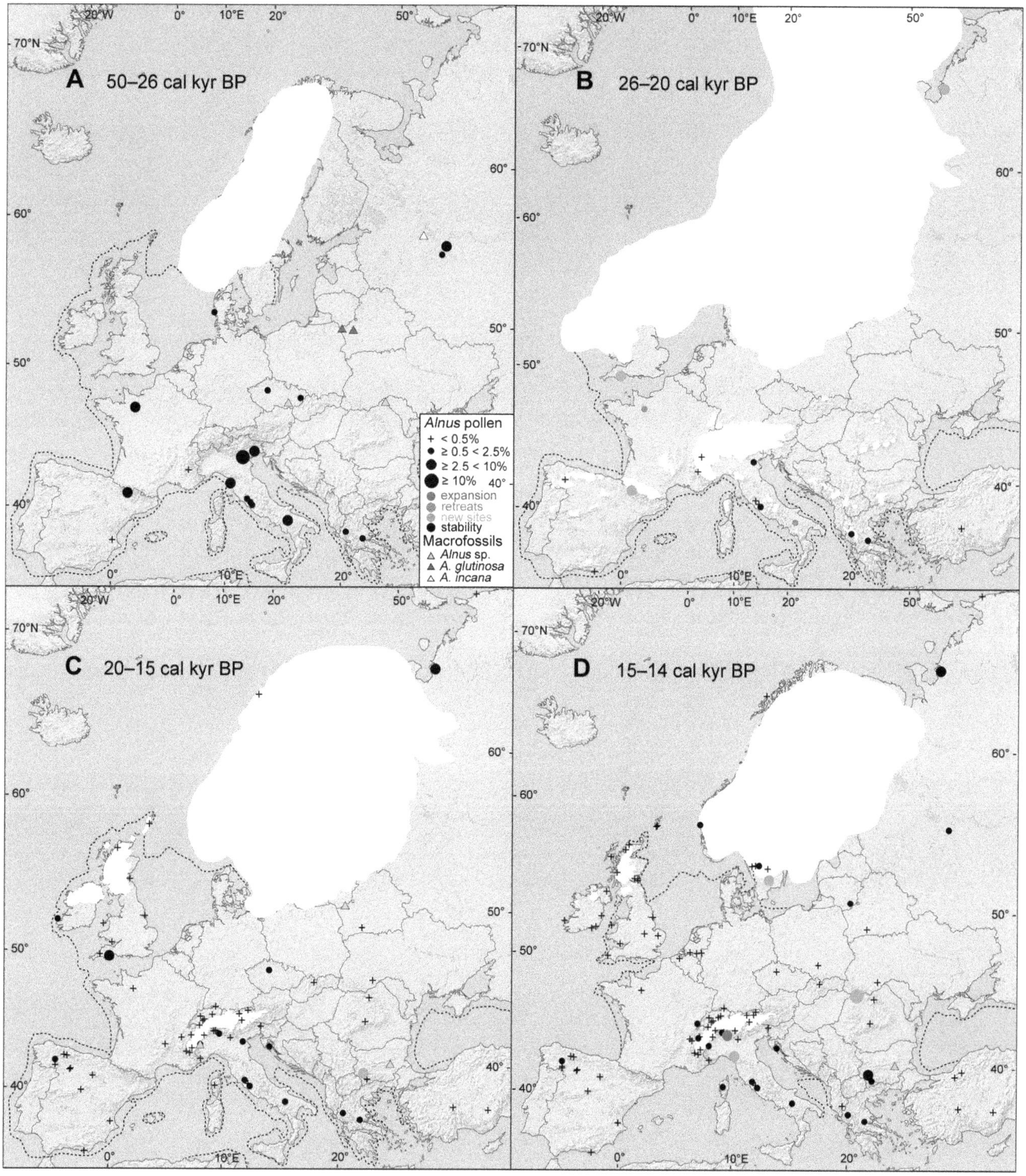

Figure 2. Last glacial period distribution (50–14 cal. kyr BP) of *Alnus* pollen sites. According to four classes of the percentage of *Alnus* pollen and macrofossil remains. The dot colour indicates changes compared with the previous period: red, expansion, *Alnus* pollen <2.5% in preceding period; blue, retreat, *Alnus* pollen ≥2.5% in preceding period; orange, new pollen sites of *Alnus* pollen ≥2.5%; black, stability; the course of deglaciation (white) and changes in coastline (dotted lines).

exceeding 2.5% are present in the Gutaiului Mountains in northwestern Romania [71] (Figure 2D). Sites with evidence of more than 2.5% of *Alnus* pollen emerge in southern Scandinavia [72] (Figure 2D), Estonia [73], [74] (Figure 3A) and northwestern and western Russia [75], [76] (Figure 3A). Macrofossil remains occur in Poland [77] (Figure 3A), Lithuania [61], [78] (Figure 3A and 3B) and Belarus [62] (Figure 3B).

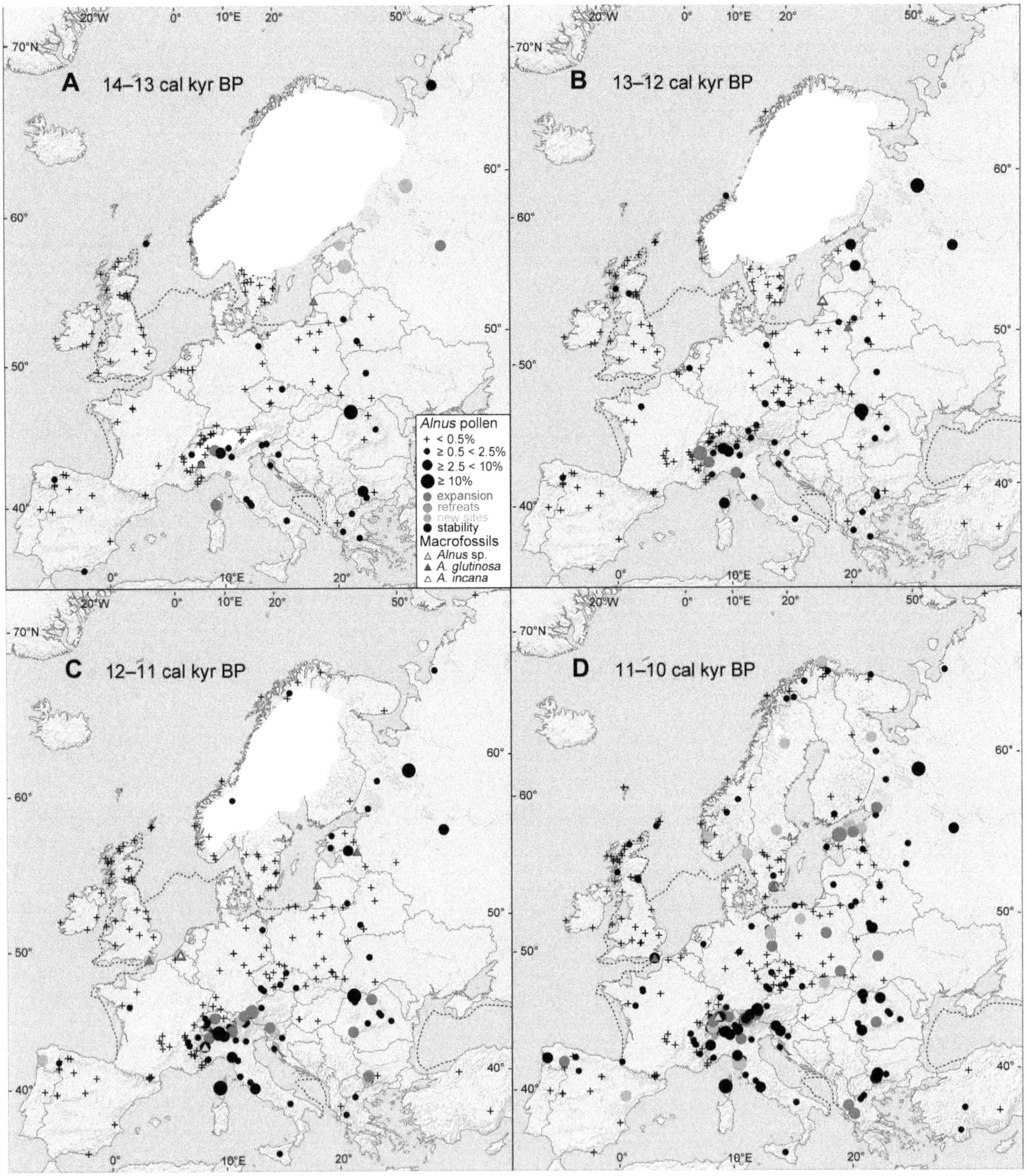

Figure 3. Late Glacial and early Holocene distribution (14–10 cal. kyr BP) of *Alnus* pollen sites. According to four classes of the percentage of *Alnus* pollen and macrofossil remains; for details, see Figure 2.

Holocene *Alnus* Distribution (12–0 cal. kyr BP; Figure 3C, 3D and 4, Figure S1 and S2)

At the beginning of the Holocene (i.e., 12–11 cal. kyr BP, Figure 3C), a continual increase in the number of sites with at least 2.5% *Alnus* pollen is apparent across the Alps, with the exception of the western areas. In western Europe, macrofossils of *A. glutinosa* and *A. incana* are present at the Kreekrak site in southwestern Netherlands [79] and *A. glutinosa* in Pannel Bridge, East Sussex [80]. Several pollen sites exceed the 2.5% pollen threshold in the Romanian Carpathians [81] and the Rila and Pirin Mountains in Bulgaria [82], [83], [84]. The first piece of evidence since the

LGM of more than 2.5% of *Alnus* pollen has been recorded in the Iberian peninsula [85] (Figure 3C).

Between 11 and 10 cal. kyr BP (Figure 3D), many sites reach at least 2.5% of *Alnus* pollen in a large area of the Polish lowland, the northern Carpathians and Scandinavia, including its northern part [86], [87]. An increase of sites exceeding the 2.5% pollen threshold is also evident in the Iberian and the Balkan peninsula (Figure 3D).

Between 10 and 9 cal. kyr BP, the majority of localities in the Carpathians and the Baltic region, including southern Scandinavia, exceed the 2.5% *Alnus* pollen threshold (Figure 4A). Macrofossil remains of *A. glutinosa* occur in the northern border of its recent distribution in central Sweden [14]. By contrast, few sites with more than 2.5% *Alnus* pollen are present in a large zone running from the Bohemian Massif and the northern foothills of the Alps through the Massif Central and the French Alps to western Europe and the British Isles (Figure 4A).

Between 9 and 8 cal. kyr BP, the increase of sites with more than 2.5% of *Alnus* pollen is apparent over the British Isles, the northern foothills of the Alps, the Bohemian Massif, northern Scandinavia and likely in the western European plain (Figure 4B). During the next two millennia (i.e., 8–6 cal. kyr BP), many sites with 2.5% *Alnus* evidence emerge in the French Alps, northern Scotland, Ireland and all southern peninsulas (Figure 4C and 4D). Finally, the number of sites exceeding 2.5% of *Alnus* pollen increases in the Massif Central and the remaining unoccupied areas of France between 7 and 6 cal. kyr BP (Figure 4D).

During the period between 6 and 0 cal. kyr BP, a decrease in the number of sites with more than 2.5% *Alnus* pollen is present in large areas of Europe, likely except in the southern peninsulas and the Carpathians (Figure 5; Figure S1 and S2). After 6 cal. kyr BP, *Alnus* enters a period of retreat in northern Scandinavia and continues southward up to the present (Figure 5G; Figures S1 and S2). In other regions, a decrease is apparent during approximately the last three millennia. A relatively strong decrease appears in the Alps (Figure 5E), Hercynian Mountains (Figure 5H), the western European plain (Figure 5I) and the British Isles (Figure 5J) whereas a weak decrease is apparent in the Baltic region (Figure 5F).

Regional Differences at the Beginning of the *Alnus* Expansion

In the southern peninsulas, the Alps and the Carpathians, there is an increase in the number of sites exceeding the 2.5% pollen threshold beginning in the Late Glacial period and increasing gradually during most of the Holocene (Figure 5A–5E). In more northerly regions, the number of sites with more than 2.5% pollen evidence rises abruptly after the beginning of the Holocene. Specifically, an increase in the number of sites in the Baltic region (Figure 5F) and Scandinavia (Figure 5G) starts between 11 and 10 cal. kyr BP and over three thousand years reaches more than 80% of occupied sites. In Hercynian Mountains (Figure 5H), the western European plain (Figure 5I) and the British Isles (Figure 5J), the expansion starts between 10 and 9 cal. kyr BP, and 80% of sites are occupied after four thousand years.

Discussion

Northern LGM Refugia

For the Last Glacial Maximum, there are two records with more than 2.5% *Alnus* pollen (Figure 6) from the periglacial landscape of the Scandinavian ice sheet in southern England [63] and Arctic Russia [64], but they are likely influenced by wind pollen transport from more distant sites. This is indicated by a low concentration of *Alnus* pollen and the presence of steppe taxa in pollen profiles [64], [88].

Because of the absence of reliable records from the LGM, we used pollen sites and macrofossils from the Late Glacial and early Holocene periods as indicators of possible *Alnus* LGM refugial areas (Figure 6). These sources indicate the presence of *Alnus* during the LGM in western Europe, the northern foothills of the Alps, the Romanian Carpathians and a large area of northeastern Europe (Figure 6). Evidence of more than 2.5% pollen from sites located in northeastern Europe from the Late Glacial period are generally interpreted as a reworking of earlier climatically favourable periods or long-distance dispersal [72], [73], [74], [75], [76] but macrofossil remains found in Poland, Belarus, Lithuania and Latvia support the occurrence of *Alnus* in this area (Figure 6).

The ability of alder trees to tolerate the climatic conditions of the LGM in northern areas has been supported in several ways. Kullman [39] showed a high tolerance of *Alnus incana* to cold climates by assessing its regeneration patterns in a subalpine forest of central Sweden. He suggested that *A. incana* could have survived the last glacial period in northern areas because it has high vegetative survivability far above its generative limit. Palaeodistributional modelling based on the climatic tolerance of trees has suggested the possible existence of *A. incana* in the proximity of the ice sheet, including southern England, northern France, Central Europe, the northern Carpathians and Belarus, but this modelling has also suggested that *A. incana* was absent from the northwestern Russian plains [7]. The northern occurrence of *A. glutinosa* reached France and the northern foothills of the Alps, but the species was absent from the northern Carpathians, Belarus and the northwestern Russian plains [7]. The survival of *Alnus* species in the North throughout the LGM might be supported by their occurrence in floodplains, which were moister and more sheltered sites than the typical dry habitats of the surrounding uplands with the occurrence of permafrost [13].

Southern LGM Refugia

Surprisingly, the 2.5% threshold does not support the Mediterranean peninsulas as LGM refugial areas for *Alnus* with the sole exception of the Pyrenees [53]. This finding contradicts the conclusions of a phylogeographic study on *A. glutinosa* that detected specific cpDNA haplotypes for particular southern peninsulas [34]. A recent population-genetic study of Lepais et al. [89], supported also by pollen data [90], [91] highlights the behaviour of rear-edge stable populations of *A. glutinosa* in North Africa. They found that tetraploid *A. glutinosa* populations in Morocco have diverged for a long-time without contribution of gene flow of Algerian or Tunisian diploid populations. This supports to the idea that *Alnus* survived in the Mediterranean area at mesoclimatically favourable sites (e.g., in foothill valleys) in sparse and isolated populations, which are generally hard to detect by pollen analyses [4], which possibly explains the low percentage of *Alnus* pollen.

Holocene *Alnus* Expansion in Northern Regions

The expansion of *Alnus* began in the Baltic region and Scandinavia between 11 and 10 cal. kyr BP (Figure 3D). The absence of *Alnus* evidence in most of central and northwestern Europe indicates that populations in northeastern Europe were predominant sources for the colonisation of Scandinavia. The delayed expansion of *Alnus* in the British Isles between 10 and 8 cal. kyr (Figure 4A and 4B) appears to have originated in a western European refugium [92], [93], [94], [95] rather than in eastern Europe, as suggested by Huntley and Birks [1]. However, eastern populations, which colonised the Baltic states and

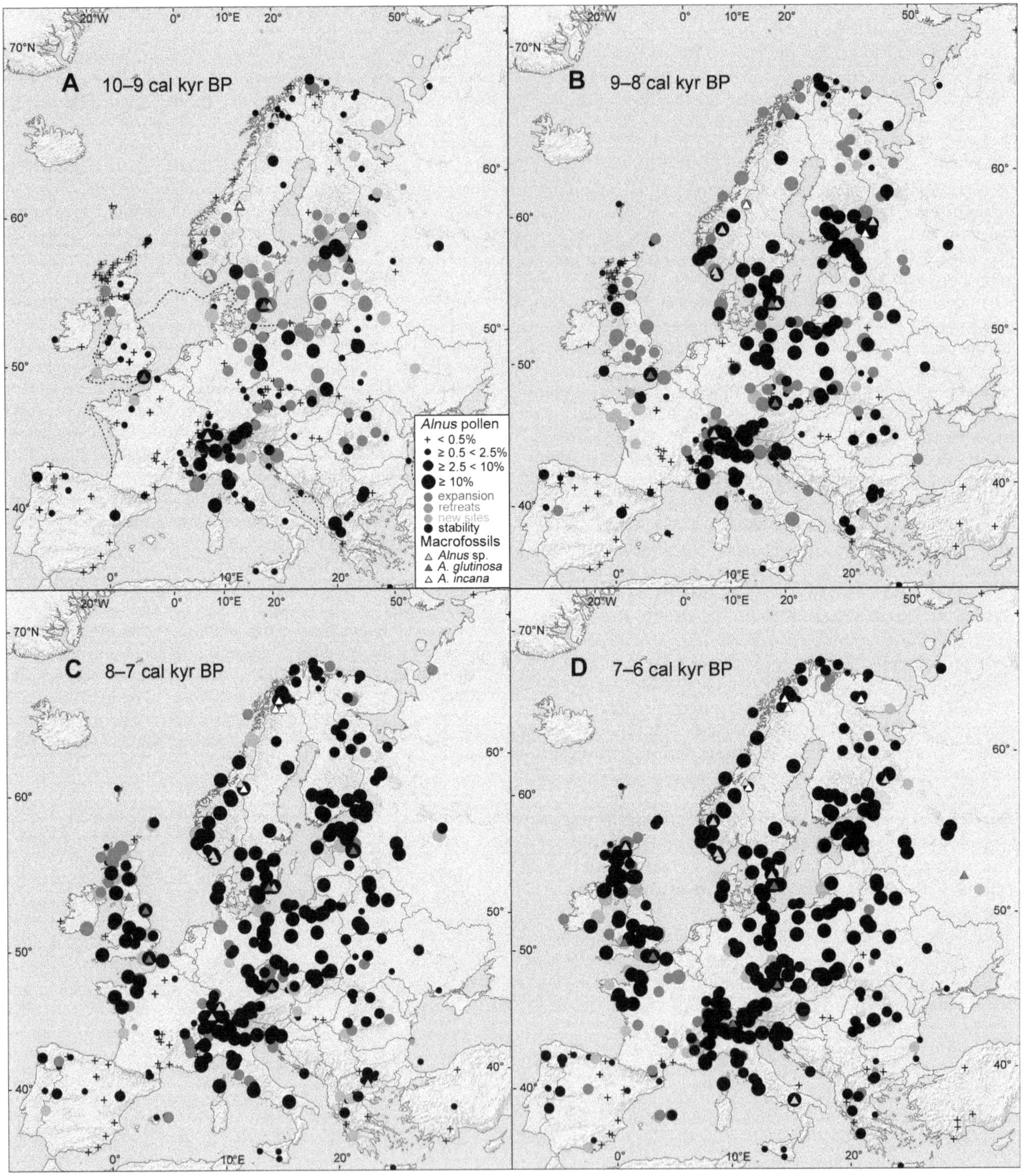

Figure 4. Holocene distribution (10–6 cal. kyr BP) of *Alnus* pollen sites. According to four classes of the percentage of *Alnus* pollen and macrofossil remains; for details, see Figure 2.

Scandinavia, could have spread southwest and mixed with western populations [93]. Synchronously with the rise of *Alnus* in the British Isles, alders expanded in Hercynian Mountains, but it is impossible to tell whether Baltic, Carpathian, Alpine or local alder populations contributed to this expansion (Figure 4A and 4B). Source populations are also unknown for the *Alnus* expansion in the Massif Central and the remaining unoccupied area of France between 7 and 6 cal. kyr BP (Figure 4D).

Scale-dependent pattern of alnus expansion. In northern areas, the *Alnus* expansion shows a scale-dependent pattern characterised by a synchronous increase of *Alnus* within individual regions, but with regional differences in the times of the expansion.

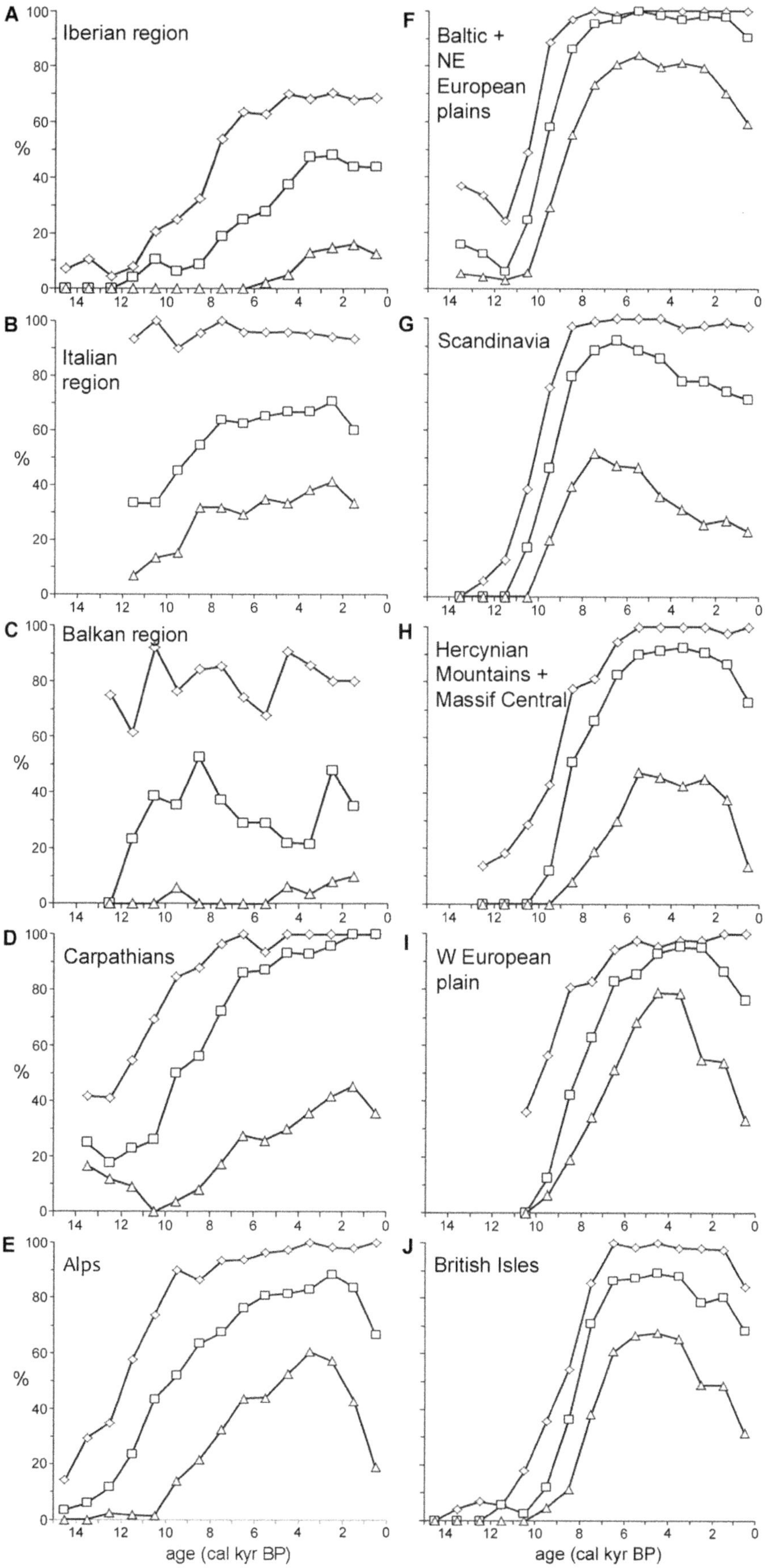
A
Iberian region
B
Italian region
C
Balkan region
D
Carpathians
E
Alps
F
Baltic + NE European plains
G
Scandinavia
H
Hercynian Mountains + Massif Central
I
W European plain
J
British Isles
%
age (cal kyr BP)

Figure 5. Regional proportions of *Alnus* pollen sites during the Late Glacial and Holocene periods. Pollen thresholds: 0.5% (diamonds), 2.5% (squares) and 10% (triangles). Only time intervals with more than 10 sites available in particular regions were considered.

At the scale of hundreds to a thousand kilometres within individual regions, we recorded little or no directional pattern in the *Alnus* expansion, i.e., sites with *Alnus* evidence initially occurred across the whole region, and then the number of sites increased equally. We recorded this pattern in all northern regions, including the deglaciated area of Scandinavia, corroborating the descriptions of Bennett and Birks [95] for the British Isles and Giesecke et al. [18] for the Baltic area. Such a general absence of spatial coherence of the *Alnus* expansion within large areas seems to be very specific in comparison with the generally observed "stepping stone" character of expansions commonly recorded for other European trees [95]. This pattern suggests that the delayed *Alnus* expansion apparent in some regions was likely a result of environmental limitations rather than the effect of slow colonization.

Giesecke et al. [18] suggested that the climate is an important factor affecting regional differences in the expansion of *Alnus*. Global warming is generally assumed to be a trigger of the rapid *Alnus* expansion that began at the turn of the Late Glacial and Holocene periods [1]. However, an arid climate in some regions could have limited the onset of the *Alnus* expansion. The ecological requirements of *Alnus* and their recent distribution indicate that *Alnus* occurrence significantly declines in areas with an arid climate [37]. *Alnus glutinosa* is currently absent from large, arid areas of the Hungarian, Romanian and Ukrainian lowlands and the Iberian peninsula (http://euforgen.org). Increased oceanicity and rising sea levels after the separation of the British Isles from the continent possibly drove the *Alnus* expansion at approximately 9 cal. kyr BP in the British Isles, as suggested by Godwin [96] and Chambers and Elliott [94]. Similarly, the early *Alnus* expansion in the Baltic area could be accelerated by the large area of the Ancylus Lake (i.e., the Baltic sea).

Alnus Expansion in Southern Peninsulas, the Alps and the Carpathians

In southern regions, *Alnus* began its expansion in the Late Glacial and early Holocene periods. It seems that *Alnus* expanded

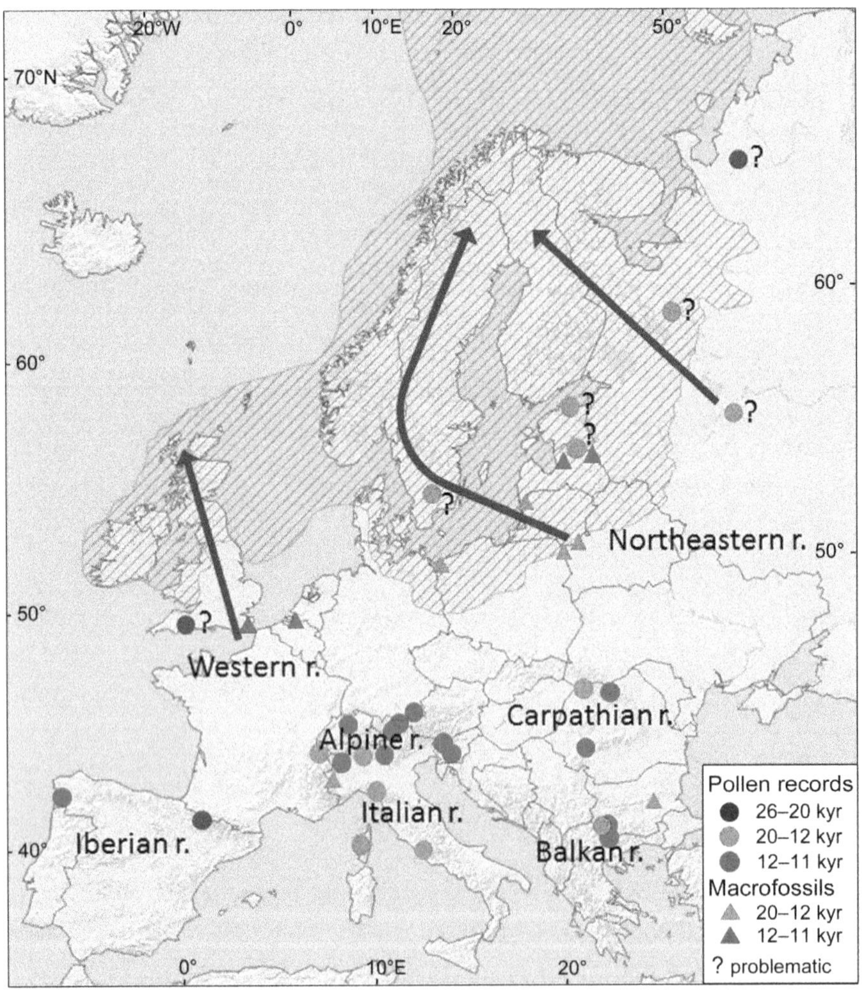

Figure 6. Putative Last Glacial Maximum refugia and directions of postglacial *Alnus* migration. The triangles and dots indicate macrofossil and pollen (≥2.5%) records from the LGM (blue), Late Glacial (green) and early Holocene (red). Arrows indicate directions of *Alnus* migration after northern deglaciation; question marks show problematic pollen records – possible reworking or long-distance pollen dispersal; hatching indicates the maximal extent of the ice sheet during the LGM.

at individual sites rather than synchronously in whole regions. We assume that the arid climate of the Mediterranean, which was temporarily and spatially variable during the Holocene [97], possibly limited the establishment of new populations and locally caused population decreases. Similarly, a harsh, unstable mountain climate [98], [99] possibly drove a relatively slow expansion in the Alps and Carpathians.

During the second part of the Holocene, between 6 and 0 cal. kyr BP, *Alnus* retreats took place in most regions of Europe (Figure 5). In Scandinavia, the northward-southward direction of its population decrease is positively correlated with climate cooling and ombrogenous peat formation, which are likely the main factors initiating this process [37], [39]. Human activity in floodplains resulting in deforestation could be an additional factor contributing to thinning [100].

Species-specific History of *Alnus glutinosa* and *A. incana* Based on Macrofossils

Differences in the LGM refugia of *Alnus glutinosa* and *A. incana* could have significantly affected the time of the *Alnus* expansion in particular regions. For example, the earlier *Alnus* expansion in the Baltic area could be an expansion of the more cold-tolerant *A. incana*. Available macrofossils, however, do not support such differences between *A. glutinosa* and *A. incana*, although results may be influenced by their relative scarcity. It seems that both *Alnus* species colonised Scandinavia from the area of the northeastern refugium. Macrofossil evidence of *A. incana* from Netherlands also supports its Late Glacial occurrence in western Europe, i.e., outside its recent range [79] (Figure 3C).

Drawbacks of the Approach

Different factors may influence the proportion of *Alnus* pollen at individual sites and potentially underestimate or overestimate *Alnus* occurrence in the past. The recorded pollen proportion of species depends on pollen production and dispersal of other species in the vegetation [49]. It has been shown that the occurrence of trees (e.g. *Betula*) in areas with low pollen production, such as borders of tundra and taiga, may be overestimated in comparison to forest zones [49]. Recent studies have shown that the size of sedimentary basins, including bogs and lakes, importantly influences the source area of pollen coming from surrounding vegetation [101], [102]. Small sedimentary basins reflect the composition of surrounding vegetation at the expense of regional vegetation patterns and, thus, may underestimate regional species occurrence [101], [102].

One important factor influencing the representation of *Alnus* pollen is its dispersal ability. *Alnus* has small and light pollen grains (fall speed 0.021 ms^{-1}, according to Eisenhut [103]) effectively dispersed by wind over large distances. Studies have shown that *Alnus* pollen may occur in quite high relative quantities (4%) in remote areas thousands of kilometres from its closest occurrence in the vegetation [104]. An unstable climate and strong winds in the last glacial period likely facilitated long-distance dispersal of *Alnus* pollen, biasing pollen records in generally treeless landscapes with low pollen production.

The level of taxonomic resolution of the pollen spectra may bias interpretations when considering occurrences of *Alnus glutinosa* and *A. incana*. May and Lacourse [105] pointed out problems with the identification of three species, *A. rubra* (analogous to *A. glutinosa*), *A. incana* and *A. alnobetula*, in pollen spectra based on a dataset from North America. They concluded that if all three species were present in the vegetation, it would be statistically impossible to determine their pollen at the species level. This makes it difficult to distinguish *A. alnobetula* pollen from the other two and complicates the interpretation of pollen records. In southern Italy and Corsica, *Alnus* pollen records may also include pollen grains of *A. cordata*, which grows there sympatrically with *A. glutinosa* in alluvial habitats. Similarly, we cannot fully exclude the presence of pollen transported over long distances belonging to other species of subgenus *Alnus* such as *A. djavanshirii* Zare, *A. dolichocarpa* Zare, *A. orientalis* Decne. and *A. subcordata* C. A. Mey, all recently growing in the Eastern Mediterranean area and Iran [106], [107]. Despite the above-mentioned facts, the accordance of macrofossils with the pollen records confirms the robustness of the relative pollen data used in this study.

Comparison with Huntley and Birks, and King and Ferris

Using a much larger pollen dataset, we broadly confirmed LGM refugial areas and the general pattern of the postglacial expansion of *Alnus* as presented in the Huntley and Birks [1] "Pollen Maps", thus supporting the robustness and actuality of their work. The main differences between our study and the conclusions of Huntley and Birks [1] lie in the interpretation of the importance of northern LGM refugial areas for the *Alnus* expansion. Based on our dataset, the refugium in northeastern Europe appears to be more important for the *Alnus* expansion than was proposed by Huntley and Birks [1]. Huntley and Birks [1] mentioned this area only as a possible LGM refugium of *A. incana* subsp. *kolaensis*. We also support that the western refugium rather than eastern European one was the source for the expansion in the British Isles.

King and Ferris [34] have suggested the Carpathians as possible source areas for the expansion of *Alnus glutinosa* in the northern part of Europe. Our study also supports northeastern and western Europe. However, some conclusions of King and Ferris [34] seems to be based on the work of Huntley and Birks [1] rather than on molecular data. Only two largely distributed haplotypes, the first occurring across all northern parts of Europe and the second in the Alps, the Carpathians, western Europe and Scandinavia, were recorded by King and Ferris [34]. The presence of two weakly spatially structured haplotypes in the northern part of Europe may reflect the postglacial expansion of genotypes from the Carpathians [34] but may also correspond to the fragmentation of the continual *A. glutinosa* range during cold phases of the last glacial period. Similarly, some tree species most likely surviving the last glacial period in the northern part of Europe, such as *Betula pendula*, *B. pubescens*, *Populus tremula* and *Salix caprea*, exhibit a low level of phylogeographic structure [108], [109], [110]. To shed light on the last glacial period and Holocene history of *A. glutinosa* in northern Europe, future molecular studies should combine several approaches. For example, more variable chloroplast DNA markers [111] and microsatellites capable of determining the demographic history of *A. glutinosa* [112] in a particular region using approximate Bayesian computation [113] could be employed. A similar study is needed for *A. incana*, for which molecular studies are still missing.

Huntley and Birks [1] postulated two questions concerning the expansion pattern of *Alnus* in Europe. First, they asked why *Alnus* delayed its expansion north of the Alps. They hypothesised that this delay could have been caused by the occurrence of only cold-demanding *A. incana* and *A. alnobetula* in the Alpine LGM refugium. These species were unable to colonise the upland and lowland areas north of the Alps. This answer remains plausible, but the macrofossil finding of *A. glutinosa* in the southern foothills of the Alps in the Late Glacial period makes their interpretations less probable. Second, they posed a question about the importance of a western refugium for the *Alnus* expansion, which appears to be the source for the *Alnus* expansion in the British Isles in our study. However, only future phylogeographic studies can bring progress towards answering the following additional questions: (i) Are there any distinctions among northern LGM refugial areas of *A. glutinosa* and *A. incana* that could influence regional differences at the beginning of the *Alnus* expansion? (ii) Was Scandinavia colonised only from the northeastern refugium, or were there other sources of colonisation located, for example, in western Europe? (iii) What is the origin of *A. incana* subsp. *kolaensis*, whose range has recently been limited to the north of Scandinavia? Within this context, the large area of northwestern Russia and the Baltic states appears to be crucial for future molecular sampling.

Supporting Information

Figure S1 Holocene distribution (6–2 cal. kyr BP) of *Alnus* pollen sites. According to four classes of percentage of *Alnus* pollen and macrofossil remains. The colour of dots indicates changes compared to the previous period; red, expansion, *Alnus* pollen < 2.5% in preceding period; blue, retreat, *Alnus* pollen ≥2.5% in preceding period; orange, new pollen sites of *Alnus* pollen ≥2.5%, respectively; black, stability; the course of deglaciation (white) and changes in coastline (dot lines).

Figure S2 Holocene distribution (2–0 cal. kyr BP) of *Alnus* pollen sites. According to four classes of percentage of *Alnus* pollen and macrofossil remains; for details see Figure S1.

Table S1 Location of the pollen sites from EPD, PALYCZ and the literature (Lit.).

Table S2 Location of the macrofossil sites from NEMD and the literature (Lit.).

Table S3 References of the pollen and macrofossil sites from EPD, PALYCZ, NEMD and the literature (Lit.).

Checklist S1 PRISMA checklist.

Acknowledgments

We wish to thank all contributors to EPD, NEMD and PALYCZ for making their data available and, thus, making this study possible. We would like to thank F. Krahulec and H. Svobodová-Svitavská for insightful comments on previous drafts of this manuscript and American Journal Experts and Frederic Rooks for improving the English of this paper.

Author Contributions

Conceived and designed the experiments: J Douda J Doudová AD BM. Analyzed the data: J Douda J Doudová AD PK VH PZ KK. Wrote the paper: J Douda J Doudová AD PK VH PZ KK BM.

References

1. Huntley B, Birks HJB (1983) An atlas of past and present pollen maps for Europe: 0–13000 years ago. Cambridge: Cambridge University Press. 688 p.
2. Soltis DE, Morris AB, McLachlan JS, Manos PS, Soltis PS (2006) Comparative phylogeography of unglaciated eastern North America. Mol Ecol 15: 4261–4293.
3. Clark PU, Dyke AS, Shakun JD, Carlson AE, Clark J, et al. (2009) The Last Glacial Maximum. Science 325: 710–714.
4. Bennett K, Tzedakis P, Willis K (1991) Quaternary refugia of north European trees. J Biogeogr 18: 103–115.
5. Tzedakis PC (1994) Hierarchical biostratigraphical classification of long pollen sequences. J Quaternary Sci 9: 257–259.
6. Magri D, Vendramin GG, Comps B, Dupanloup I, Geburek T, et al. (2006) A new scenario for the quaternary history of European beech populations: palaeobotanical evidence and genetic consequences. New Phytol 171: 199–221.
7. Svenning J-Ch, Normand S, Kageyama M (2008) Glacial refugia of temperate trees in Europe: insights from species distribution modelling. J Ecol 96: 1117–1127.
8. Litynska-Zajac M (1995) Anthracological analysis. In: Hromada J, Kozlowski J, editors. Complex of upper Palaeolithic sites near Moravany, western Slovakia. Krakow: Jagellonian University Press. 74–79.
9. Willis KJ, Rudner E, Sumegi P (2000) The full-glacial forests of central and south-eastern Europe. Quaternary Res 53: 203–213.
10. Willis KJ, van Andel TH (2004) Trees or no trees? The environments of central and eastern Europe during the Last Glaciation. Quat Sci Rev 23: 2369–2387.
11. Jankovská V, Pokorný P (2008) Forest vegetation of the last full-glacial period in the Western Carpathians (Slovakia and Czech Republic). Preslia 80: 307–324.
12. Kuneš P, Pelánková B, Chytrý M, Jankovská V, Pokorný P, et al. (2008) Interpretation of the last-glacial vegetation of eastern-central Europe using modern analogues from southern Siberia. J Biogeogr 35: 2223–2236.
13. Binney HA, Willis KJ, Edwards ME, Bhagwat SA, Anderson PM, et al. (2009) The distribution of late-Quaternary woody taxa in northern Eurasia: evidence from a new macrofossil database. Quat Sci Rev 28: 2445–2464.
14. Kullman L (1998) Non-analogous tree flora in the Scandes Mountains, Sweden, during the early Holocene - macrofossil evidence of rapid geographic spread and response to palaeoclimate. Boreas 27: 153–161.
15. Väliranta M, Kaakinen A, Kuhry P, Kultti S, Salonen JS, et al. (2011) Scattered late-glacial and early Holocene tree populations as dispersal nuclei for forest development in north-eastern European Russia. J Biogeogr 38: 922–932.
16. Parducci L, Jørgensen T, Tollefsrud MM, Elverland E, Alm T, et al. (2012) Glacial survival of boreal trees in northern Scandinavia. Science 335: 1083–1086.
17. Giesecke T, Hickler T, Kunkel T, Sykes MT, Bradshaw RHW (2007) Towards an understanding of the Holocene distribution of *Fagus sylvatica* L. J Biogeogr 34: 118–131.
18. Giesecke T, Bennett KD, Birks HJB, Bjune AE, Bozilova E, et al. (2011) The pace of Holocene vegetation change – testing for synchronous developments. Quat Sci Rev 30: 2805–2814.
19. Henne PD, Elkin CM, Reineking B, Bugmann H, Tinner W (2011) Did soil development limit spruce (*Picea abies*) expansion in the Central Alps during the Holocene? Testing a palaeobotanical hypothesis with a dynamic landscape model. J Biogeogr 38: 933–949.
20. Furlow JJ (1979) The systematics of the American species of *Alnus* (Betulaceae). Rhodora 81: 151–248.
21. Stevens PF (2001 onwards) Angiosperm Phylogeny Website. Version 12, July 2012 [and more or less continuously updated since]. Available: http://www.mobot.org/MOBOT/research/APweb/.
22. Chen ZD, Manchester SR, Sun HY (1999) Phylogeny and evolution of the Betulaceae as inferred from DNA sequences, morphology, and paleobotany. Am J Bot 86: 1168–1181.
23. Chen Z, Li J (2004) Phylogenetics and biogeography of *Alnus* (Betulaceae) inferred from sequences of nuclear ribosomal DNA ITS region. Int J Plant Sci 165: 325–335.
24. Jalas J, Suominen J (1976) Atlas Florae Europaeae. Distribution of vascular plants in Europe. 3. Salicaceae to Balanophoraceae. Helsinki: The Committee for Mapping the Flora of Europe and Societas Biologica Fennica Vanamo. 128 p.
25. Rochet J, Moreau P-A, Manzi S, Gardes M (2011) Comparative phylogenies and host specialization in the alder ectomycorrhizal fungi *Alnicola*, *Alpova* and *Lactarius* (Basidiomycota) in Europe. BMC Evol Biol 11: 40.
26. Wijmstra TA (1969) Palynology of the first 30 metres of a 120 m deep section in Northern Greece. Acta Botanica Neerlandica 18: 511–527.
27. West RG (1980) Pleistocene forest history in East Anglia. New Phytol 85: 571–622.
28. Andersen ST (1965) Interglacialer og interstadialer i Danmarks Kvartaer. Meddelelser fra Dansk Geologisk Forening 15: 496–506.
29. Follieri M, Giardini M, Magri D, Sadori L (1998) Palynostratigraphy of the last glacial period in the volcanic region of Central Italy. Quatern Int 47–48: 3–20.
30. de Beaulieu J-L, Reille M (1984) A long Upper Pleistocene pollen record from Les Echets, near Lyon, France. Boreas 13: 111–132.
31. de Beaulieu J-L, Reille M (1992) The last climatic cycle at La Grande Pile (Vosges, France) a new pollen profile. Quat Sci Rev 11: 431–438.
32. Reille M, de Beaulieu J-L, Svobodova H, Andrieu-Ponel V, Goeury C (2000) Pollen analytical biostratigraphy of the last five climatic cycles from a long continental sequence from the Velay region (Massif Central, France). J Quaternary Sci 15: 665–685.
33. Müller UC (2000) A Late-Pleistocene pollen sequence from the Jammertal, south-western Germany with particular reference to location and altitude as factors determining Eemian forest composition. Veg Hist Archaeobot 9: 125–131.

34. King RA, Ferris C (1998) Chloroplast DNA phylogeography of *Alnus glutinosa* (L.) Gaertn. Mol Ecol 7: 1151–1161.
35. Douda J, Čejková A, Douda K, Kochánková J (2009) Development of alder carr after the abandonment of wet grasslands during the last 70 years. Ann For Sci 66: 712.
36. Douda J (2010) The role of landscape configuration in plant composition of floodplain forests across different physiographic areas. Journal of Vegetation Science 21: 1110–1124.
37. Tallantire PA (1974) The palaeohistory of the grey alder (*Alnus incana* (L.) Moench.) and black alder (*A. glutinosa* (L.) Gaertn.) in Fennoscandia. New Phytol 73: 529–546.
38. McVean DN (1953) Biological flora of the British Isles: *Alnus glutinosa* (L.) Gaertn. J Ecol 41: 447–466.
39. Kullman L (1992) The ecological status of grey alder (*Alnus incana* (L.) Moench) in the upper subalpine birch forest of the central Scandes. New Phytol 120: 445–451.
40. Moher D, Liberati A, Tetzlaff J, Altman DG (2010) Preferred reporting items for systematic reviews and meta-analyses: the PRISMA statement. Int J Surg 8: 336–341.
41. Fyfe RM, de Beaulieu J-L, Binney H, Bradshaw RHW, Brewer S, et al. (2009) The European Pollen Database: past efforts and current activities. Veg Hist Archaeobot 18: 417–424.
42. Kuneš P, Abraham V, Kovařík O, Kopecký M, Contributors P (2009) Czech Quaternary Palynological Database – PALYCZ: review and basic statistics of the data. Preslia 81: 209–238.
43. Giesecke T, Davis B, Brewer S, Finsinger W, Wolters S, et al. (2013) Towards mapping the late Quaternary vegetation change of Europe. Veg Hist Archaeobot. doi: 10.1007/s00334-012-0390-y.
44. Blaauw M (2010) Methods and code for "classical" age-modelling of radiocarbon sequences. Quat Geochronol 5: 512–518.
45. R Development Core Team (2012) R: A language and environment for statistical computing. Vienna: R Foundation for Statistical Computing.
46. Punt W, Blackmore S, Hoen PP, Stafford PJ (2003) The Northwest European Pollen Flora, Volume 8. Elsevier. 194 p.
47. Leopold EB, Birkebak J, Reinink-Smith L, Jayachandar AP, Narváez P, et al. (2012) Pollen morphology of the three subgenera of *Alnus*. Palynology 36: 131–151.
48. Douda J (2008) Formalized classification of the vegetation of alder carr and floodplain forests in the Czech Republic. Preslia 80: 199–224.
49. Lisitsyna OV, Giesecke T, Hicks S (2011) Exploring pollen percentage threshold values as an indication for the regional presence of major European trees. Rev Palaeobot Palynol 166: 311–324.
50. Tallantire PA (1992) The alder [*Alnus glutinosa* (L.) Gaertn.] problem in the British Isles: a third approach to its palaeohistory. New Phytol 122: 717–731.
51. Montanari C (1996) Recent pollen deposition in alder woods and in other riverine plant comunities. Allionia 34: 309–323.
52. Ralska-Jasiewiczowa M, Latałowa M, Wasylikowa K, Tobolski K, Madeyska E, et al. (2004) Late Glacial and Holocene history of vegetation in Poland based on isopollen maps. Cracow, Poland: W. Szafer Institute of Botany, Polish Academy of Sciences. 444 p.
53. González-Sampériz P, Valero-Garcés BL, Carrión JS, Peña-Monné JL, García-Ruiz JM, et al. (2005) Glacial and Lateglacial vegetation in northeastern Spain: New data and a review. Quatern Int 140–141: 4–20.
54. Watts W, Allen JRM, Huntley B (1996) Vegetation history and palaeoclimate of the last glacial period of Lago Grande di Monticchio, southern Italy. Quat Sci Rev 15: 133–153.
55. Lucchi M (2008) Vegetation dynamics during the Last Interglacial–Glacial cycle in the Arno coastal plain (Tuscany, western Italy): location of a new tree refuge. Quat Sci Rev 27: 2456–2466.
56. Pini R, Ravazzi C, Donegana M (2009) Pollen stratigraphy, vegetation and climate history of the last 215ka in the Azzano Decimo core (plain of Friuli, north-eastern Italy). Quat Sci Rev 28: 1268–1290.
57. Pini R, Ravazzi C, Reimer PJ (2010) The vegetation and climate history of the last glacial cycle in a new pollen record from Lake Fimon (southern Alpine foreland, N-Italy). Quat Sci Rev 29: 3115–3137.
58. Barbier D (1999) Histoire de la vegetation du nord-mayennais de la fin du Weichselien à l'aube du XXIème siècle: mise en évidence d'un Tardiglaciaire armoricain; interactions homme-milieu. Univ. de Nantes (Thesis).
59. Aleshinskaya ZV, Gunova VS (1976) History of Lake Nero as reflection on the surrounding landscape dynamics. In: Kalinin GP, Klige RK, editors. Problemy Paleohydrologii. Moscow: Nauka. 214–222.
60. Zarrina YP, Spiridonova YA, Arslanov KA, Kolesnikova TD, Simonova GF (1973) New profile of Middle Valday deposits near the village Shenskoe (Mologo-Sheksna basin). In: Subakov VA, editor. Pleistocene Chronology and Climatic Stratigraphy. Leningrad: Nauka Press, Lenigrad Division. 160–167.
61. Voznyachuk LN, Valchik MA (1978) Morphology, structure and history of the valley of the Neman in Pleistocene and Holocene. Minsk: Nauka i Tekhnika.
62. Gaigalas AI, Dvaretskas VV, Banis YY, Davaynis GA, Kibilda ZA, et al. (1981) Radiocarbon age of river terraces in the southern Baltic. Isotopic and geochemical methods in biology, geology and archeology. Tartu. 28–32.
63. Kelly A, Charman DJ, Newnham RM (2010) A Last Glacial Maximum pollen record from Bodmin Moor showing a possible cryptic northern refugium in southwest England. J Quaternary Sci 25: 296–308.
64. Paus A, Svendsen JI, Matiouchkov A (2003) Late Weichselian (Valdaian) and Holocene vegetation and environmental history of the northern Timan Ridge, European Arctic Russia. Quat Sci Rev 22: 2285–2302.
65. Tonkov S, Possnert G, Bozilova E (2006) The lateglacial vegetation and radiocarbon dating of Lake Trilistnika, Rila Mountains (Bulgaria). Veg Hist Archaeobot 16: 15–22.
66. Finsinger W, Tinner W, van der Knaap WO, Ammann B (2006) The expansion of hazel (*Corylus avellana* L.) in the southern Alps: a key for understanding its early Holocene history in Europe? Quat Sci Rev 25: 612–631.
67. Magyari EK, Chapman JC, Gaydarska B, Marinova E, Deli T, et al. (2008) The "oriental" component of the Balkan flora: evidence of presence on the Thracian Plain during the Weichselian late-glacial. J Biogeogr 35: 865–883.
68. Reille M, Gamisans J, de Beaulieu J-L, Andrieu V (1997) The late-glacial at Lac de Creno (Corsica, France): a key site in the western Mediterranean basin. New Phytol 135: 547–559.
69. Lowe JJ, Watson C (1993) Lateglacial and early Holocene pollen stratigraphy of the northern Apennines, Italy. Quat Sci Rev 12: 727–738.
70. Kelly MG, Huntley B (1991) An 11 000-year record of vegetation and environment from Lago di Martignano, Latium, Italy. J Quaternary Sci 6: 209–224.
71. Björkman L, Feurdean A, Wohlfarth B (2003) Late-Glacial and Holocene forest dynamics at Steregoiu in the Gutaiului Mountains, Northwest Romania. Rev Palaeobot Palynol 124: 79–111.
72. Björck S, Möller P (1987) Late Weichselian environmental history in southeastern Sweden during the deglaciation of the Scandinavian ice sheet. Quaternary Res 28: 1–37.
73. Saarse L, Niinemets E, Amon L, Heinsalu A, Veski S, et al. (2009) Development of the late glacial Baltic basin and the succession of vegetation cover as revealed at Palaeolake Haljala, northern Estonia. Estonian Journal of Earth Sciences 58: 317–333.
74. Amon L, Veski S, Heinsalu A, Saarse L (2012) Timing of Lateglacial vegetation dynamics and respective palaeoenvironmental conditions in southern Estonia: evidence from the sediment record of Lake Nakri. J Quaternary Sci 27: 169–180.
75. Wohlfarth B, Filimonova L, Bennike O, Björkman L, Brunnberg L, et al. (2002) Late-Glacial and Early Holocene Environmental and Climatic Change at Lake Tambichozero, Southeastern Russian Karelia. Quaternary Res 58: 261–272.
76. Wohlfarth B, Tarasov P, Bennike O, Lacourse T, Subetto D, et al. (2006) Late Glacial and Holocene Palaeoenvironmental Changes in the Rostov-Yaroslavl' Area, West Central Russia. J Paleolimnol 35: 543–569.
77. Latałowa M, Borówka RK (2006) The Allerød/Younger Dryas transition in Wolin Island, northwest Poland, as reflected by pollen, macrofossils, and chemical content of an organic layer separating two aeolian series. Veg Hist Archaeobot 15: 321–331.
78. Stančikaitė M, Šinkūnas P, Šeirienė V, Kisielienė D (2008) Patterns and chronology of the Lateglacial environmental development at Pamerkiai and Kašučiai, Lithuania. Quat Sci Rev 27: 127–147.
79. Bos JAA, Huisman DJ, Kiden P, Hoek WZ, van Geel B (2005) Early Holocene environmental change in the Kreekrak area (Zeeland, SW-Netherlands): A multi-proxy analysis. Palaeogeogr Palaeoclimatol Palaeoecol 227: 259–289.
80. Waller MP, Marlow AD (1994) Flandrian vegetational history of south-eastern England. Stratigraphy of the Brede valley and pollen data from Brede Bridge. New Phytol 126: 369–392.
81. Farcas S, de Beaulieu J-L, Reille M, Coldea G, Diaconeasa B, et al. (1999) First 14C datings of Late Glacial and Holocene pollen sequences from Romanian Carpathes. Comptes Rendus de l'Académie des Sciences - Series III - Sciences de la Vie 322: 799–807.
82. Bozilova ED, Tonkov S, Pavlova D (1986) Pollen and plant macrofossil analyses of the Lake Sucho Ezero in the south Rila mountains. Annual of Sofia University, Faculty of Biology 80: 48–57.
83. Bozilova ED, Tonkov SB (2000) Pollen from Lake Sedmo Rilsko reveals southeast European postglacial vegetation in the highest mountain area of the Balkans. New Phytol 148: 315–325.
84. Tonkov S, Panovska H, Possnert G, Bozilova E (2002) The Holocene vegetation history of Northern Pirin Mountain, southwestern Bulgaria: pollen analysis and radiocarbon dating of a core from Lake Ribno Banderishko. Holocene 12: 201–210.
85. López-Merino L, Silva Sánchez N, Kaal J, López-Sáez JA, Martínez Cortizas A (2012) Post-disturbance vegetation dynamics during the Late Pleistocene and the Holocene: An example from NW Iberia. Glob Planet Change 92–93: 58–70.
86. Seppä H (1996) Post-glacial dynamics of vegetation and tree-lines in the far north of Fennoscandia. Fennia 174: 1–96.
87. Carcaillet C, Hörnberg G, Zackrisson O (2012) Woody vegetation, fuel and fire track the melting of the Scandinavian ice-sheet before 9500 cal yr BP. Quaternary Res 78: 540–548.
88. Scourse J (2010) Comment: A Last Glacial Maximum pollen record from Bodmin Moor showing a possible cryptic northern refugium in southwest England. (Kelly et al., 2010). J Quaternary Sci 25: 826–827.
89. Lepais O, Muller SD, Ben Saad-Limam S, Benslama M, Rhazi L, et al. (2013) High genetic diversity and distinctiveness of rear-edge climate relicts maintained by ancient tetraploidisation for *Alnus glutinosa*. PLoS ONE 8: e75029. doi: 10.1371/journal.pone.0075029.

90. Ben Tiba B, Reille M (1982) Recherches pollenanalytiques dans les montagnes de Kroumirie (Tunisie septentrionale): premiers résultats. Ecol Mediterr 8: 75–86.
91. Stambouli-Essassi S, Roche E, Bouzid S (2007) Evolution of vegetation and cimatic changes in North-Western Tunisia during the last 40 millennia. Internat J Trop Geol Geogr Ecol 31: 171–214.
92. Chambers FM, Price S-M (1985) Palaeoecology of *Alnus* (alder): early post-glacial rise in a valley Mire, north-west Wales. New Phytol 101: 333–344.
93. Bush MB, Hall AR (1987) Flandrian *Alnus*: expansion or immigration? J Biogeogr 14: 479–481.
94. Chambers FM, Elliott L (1989) Spread and expansion of *Alnus* Mill in the British Isles: timing, agencies and possible vectors. J Biogeogr 16: 541–550.
95. Bennett KD, Birks HJB (1990) Postglacial history of alder (*Alnus glutinosa* (L.) Gaertn.) in the British Isles. J Quaternary Sci 5: 123–133.
96. Godwin H (1975) History of the British flora. 2nd ed. Cambridge: Cambridge University Press. 580 p.
97. Magny M, Vannière B, Calo C, Millet L, Leroux A, et al. (2011) Holocene hydrological changes in south-western Mediterranean as recorded by lake-level fluctuations at Lago Preola, a coastal lake in southern Sicily, Italy. Quat Sci Rev 30: 2459–2475.
98. Wick L, Tinner W (1997) Vegetation changes and timberline fluctuations in the Central Alps as indicators of Holocene climatic oscillations. Arct Antarct Alp Res 29: 445–458.
99. Magyari EK, Jakab G, Bálint M, Kern Z, Buczkó K, et al. (2012) Rapid vegetation response to Lateglacial and early Holocene climatic fluctuation in the South Carpathian Mountains (Romania). Quat Sci Rev 35: 116–130.
100. Brown A (1988) The palaeoecology of *Alnus* (alder) and the postglacial history of floodplain vegetation. Pollen percentage and influx data from the West Midlands, United Kingdom. New Phytol 110: 425–436.
101. Sugita S (1994) Pollen representation of vegetation in Quaternary sediments: theory and method in patchy vegetation. Journal of Ecology 82: 881–897.
102. Gaillard M-J, Sugita S, Bunting MJ, Middleton R, Broström A, et al. (2008) The use of modelling and simulation approach in reconstructing past landscapes from fossil pollen data: a review and results from the POLLANDCAL network. Veg Hist Archaeobot 17: 419–443.
103. Eisenhut G (1961) Untersuchungen über die Morphologie und Ökologie der Pollenkörner heimischer und fremdländischer Waldbäume (translated into English by Jackson TS and Jaumann P, 1989). Hamburg: Paul Parey. 68 p.
104. van der Knaap W, van Leeuwen JF, Froyd CA, Willis KJ (2012) Detecting the provenance of Galapagos non-native pollen: The role of humans and air currents as transport mechanisms. Holocene 22: 1373–1383.
105. May L, Lacourse T (2012) Morphological differentiation of *Alnus* (alder) pollen from western North America. Rev Palaeobot Palynol 180: 15–24.
106. Yaltrik Y (1982) Betulaceae. In: Davis PH, editor. Flora of Turkey, Vol. 7. Edinburgh: Edinburgh University Press. 688–694.
107. Zare H, Amini T (2012) A review of the genus *Alnus* Gaertn. in Iran, new records and new species. Iran J Bot 18: 10–21.
108. Palmé AE, Semerikov V, Lascoux M (2003) Absence of geographical structure of chloroplast DNA variation in sallow, *Salix caprea* L. Heredity 91: 465–474.
109. Petit RJ, Aguinagalde I, de Beaulieu J-L, Bittkau C, Brewer S, et al. (2003) Glacial refugia: hotspots but not melting pots of genetic diversity. Science 300: 1563–1565.
110. Maliouchenko O, Palmé AE, Buonamici A, Vendramin GG, Lascoux M (2007) Comparative phylogeography and population structure of European *Betula* species, with particular focus on *B. pendula* and *B. pubescens*. J Biogeogr 34: 1601–1610.
111. Shaw J, Lickey EB, Beck JT, Farmer SB, Liu W, et al. (2005) The tortoise and the hare II: relative utility of 21 noncoding chloroplast DNA sequences for phylogenetic analysis. Am J Bot 92: 142–166.
112. Lepais O, Bacles CFE (2011) De novo discovery and multiplexed amplification of microsatellite markers for black alder (*Alnus glutinosa*) and related species using SSR-enriched shotgun pyrosequencing. J Hered 102: 627–632.
113. Csilléry K, Blum MGB, Gaggiotti OE, François O (2010) Approximate Bayesian Computation (ABC) in practice. Trends Ecol Evol 25: 410–418.

5

Quantifying Phytogeographical Regions of Australia Using Geospatial Turnover in Species Composition

Carlos E. González-Orozco[1*¤], Malte C. Ebach[2], Shawn Laffan[2], Andrew H. Thornhill[1,5], Nunzio J. Knerr[1], Alexander N. Schmidt-Lebuhn[1], Christine C. Cargill[1], Mark Clements[1], Nathalie S. Nagalingum[3], Brent D. Mishler[4], Joseph T. Miller[1]

1 Centre for Australian National Biodiversity Research, Commonwealth Scientific and Industrial Research Organisation, Plant Industry, Canberra, Australian Capital Territory, Australia, **2** School of Biological, Earth and Environmental Sciences, University of New South Wales, Sydney, New South Wales, Australia, **3** National Herbarium of New South Wales, Botanic Gardens Trust, Sydney, New South Wales, Australia, **4** University and Jepson Herbaria, Department of Integrative Biology, University of California, Berkeley, Berkeley, California, United States of America, **5** Australian Tropical Herbarium, James Cook University, Cairns, Queensland, Australia

Abstract

The largest digitized dataset of land plant distributions in Australia assembled to date (750,741 georeferenced herbarium records; 6,043 species) was used to partition the Australian continent into phytogeographical regions. We used a set of six widely distributed vascular plant groups and three non-vascular plant groups which together occur in a variety of landscapes/habitats across Australia. Phytogeographical regions were identified using quantitative analyses of species turnover, the rate of change in species composition between sites, calculated as Simpson's beta. We propose six major phytogeographical regions for Australia: Northern, Northern Desert, Eremaean, Eastern Queensland, Euronotian and South-Western. Our new phytogeographical regions show a spatial agreement of 65% with respect to previously defined phytogeographical regions of Australia. We also confirm that these new regions are in general agreement with the biomes of Australia and other contemporary biogeographical classifications. To assess the meaningfulness of the proposed phytogeographical regions, we evaluated how they relate to broad scale environmental gradients. Physiographic factors such as geology do not have a strong correspondence with our proposed regions. Instead, we identified climate as the main environmental driver. The use of an unprecedentedly large dataset of multiple plant groups, coupled with an explicit quantitative analysis, makes this study novel and allows an improved historical bioregionalization scheme for Australian plants. Our analyses show that: (1) there is considerable overlap between our results and older biogeographic classifications; (2) phytogeographical regions based on species turnover can be a powerful tool to further partition the landscape into meaningful units; (3) further studies using phylogenetic turnover metrics are needed to test the taxonomic areas.

Editor: Ulrich Joger, State Natural History Museum, Germany

Funding: MCE's contribution was supported under Australian Research Council's 'Future Fellow' funding scheme (project number FT0992002). MCE and SWL also wish to thank the University of New South Wales, Australia for two Goldstar Grants (RG114989 & RG124461). The authors would like to thank Bernd Gruber for supporting the spatial overlap analyses. BDM thanks CSIRO (Australia) for a Distinguished Visiting Scientist Award. The authors acknowledge the support received from the Cancer Research UK and EPSRC Cancer Imaging Centre in association with the MRC and Department of Health (England) grant C1060/A10334, also NHS funding to the NIHR Biomedical Research Centre, MRC-funded studentship, also Chang Gung Medical Foundation (Taiwan) grants CMRPG370443 and CMRPG3B1922. MOL is an NIHR Senior Investigator. The authors thank Alice Warley at the Kings College London Centre for Ultrastructural Imaging (CUI) for assistance with electron microscopy. The funders had no role in study design, data collection and analysis, decision to publish, or preparation of the manuscript

Competing Interests: The authors have declared that no competing interests exist.

* E-mail: carlos.gonzalez-orozco@canberra.edu.au

¤ Current address: Institute for Applied Ecology and Collaborative Research Network for Murray-Darling Basin Futures, University of Canberra, Canberra, Australian Capital Territory, Australia

Introduction

The definition of biogeographical regions (also referred to as bioregions) is fundamental for understanding the distribution of biodiversity [1]. Bioregionalizations are important because they allow us to classify organisms into fundamental geographic units, at different scales, to be used to establish global conservation agreements and to make diversity assessments [2,3,4,5]. The characteristics and terms used to define areas in biogeography are not always used consistently (Table 1). For example, biomes are defined by both the climate and the types of organisms that have adapted to it and floristic zones are defined only by the types of vegetation they contain. However, they are sometimes used interchangeably. Bioregions (phytogeographical and zoogeographical regions) are defined on the distributions of specific taxonomic groups, and therefore are simpler to understand and to use comparatively.

The history of Australian bioregionalization spans 190 years [6] and may be divided into the colonial, post-federation, ecogeographical and systematic periods. The first attempt at a bioregionalization classification of Australia was by Ferdinand von Mueller in 1858, using vegetation types rather than taxic distributions [7]. In contrast, the naturalist Ralph Tate in 1889 produced the first bioregionalization using a combination of taxic distributions and climate, coining the terms Eremaean and Euronotian, both of which are still in use [8] and are also used here. In 1933, during the Post-federation period, the

Table 1. Glossary of terms.

Term name	Description
Area	The region of distribution of any taxonomic unit (species, genus, family) of the plant (or animal) world (Wulff 1950: 25) [51]. An endemic area is the geographical area to which a taxon or biota is understood to be native. (Parenti and Ebach 2009: 253) [50].
Biome	Bioclimatic Zone. The geographical area defined by climate and the types of organisms that have adapted to it (e.g., mesic, arid).
Biota	A group of taxa (organisms), the combined distribution of which occupies a common set of geographical limits. (Parenti and Ebach 2009: 252) [50].
Biotic Area	The geographical area inhabited by a biota. Limits of taxon distribution specify limits of the area (Parenti and Ebach 2009: 251) [50].
Biogeographical Region	Bioregion or phytogeographical and zoological regions. The geographical area based on the distributions of specific taxonomic groups (e.g., plant or animal taxa).
Vegetative (Floristic) Zone	The geographical area defined by a particular type of vegetation (e.g., savannah, tundra, Mulga Scrub).

The terms, regions, areas, and vegetation are often used inter-changeably, however, they do have specific meanings that we use herein with the following definitions.

zoogeographer G.E. Nicholls presented the first combined regionalization of Australian terrestrial flora and fauna [9]. Rather than adopt a combined bioregionalization, however, many Australian plant and animal geographers chose instead to keep animal and plant distributions separate. Most important was Nancy Burbidge, who by the Ecogeographical Period developed her own regionalization consisting of areas that were largely based on Tate's work [8] as well as her expert knowledge of flora distributions. Burbidge's classification of Australia flora used the concept of interzones, that is, areas of overlap (including the MacPherson-Macleay overlap zone), along with three floristic zones (Temperate, Tropical and Eremaean) [10]. Possibly the last continental bioregionalization of Australia's flora was made by Dutch botanist Henk Doing in 1970, who created the first hierarchical classification of Australian flora according to vegetation type, communities and climate [11].

Other studies attempted to regionalize Australia through using numerical, evidence-based data, such as Barlow's 33 botanical regions derived from herbarium specimens [12]. By the 1980s regionalization was done at the regional rather than continental level and many classifications ignored biotic areas in favour of vegetative or climatic zones (i.e., biomes) [13]. However, by the time of the Systematic Period, particularly in the early 1990s, there was a resurgence of biotic area classification [14,15,16,17]. Of these new classifications, very few were in agreement, leaving phytogeography with more areas and area names than ever before [18]. In summary, there has to date been an accumulation of, often conflicting, area classifications, none of which were quantitatively produced or assessed. Modern advances in the development of large databases of georeferenced specimen observations, allied with concurrent improvements in spatial analysis tools, means that it is now possible to quantitatively define and assess biotic regions [19,20].

A key concept to define biogeographical regions (herein phytogeographical regions) is species turnover, which is the rate of change in species composition between sites [21]. Species turnover has been used to generate classifications of bioregions [19,20,22,23,24,17] and there have been some studies, for example, in sub-Saharan regions of Africa, where multiple taxa were used to successfully partition the continent into phytogeographical regions [25]. In other cases, species turnover and multivariate statistical methods effectively diagnosed bioregions on different biological groups [26,27,28,29,30]. However, no large dataset of geo-referenced taxa has been used to identify phytogeographical regions of plants across an entire continent such as Australia. Moreover, there has been no quantitative attempt to test existing phytogeographical regions that have been in use since the late 19th century [8].

The method we apply to quantifying phytogeographical regions is similar to recent zoogeographical studies [1,24]. Limited access to large nationally digitized spatial datasets is a likely reason why large studies do not exist on this topic. Australia is an exception because of the existence of Australia's Virtual Herbarium AVH [31], which has digitized most specimens housed in herbaria around Australia (http://www.avh.ala.org.au/). This amalgamated database of herbarium records for an entire continent makes it possible to investigate large scale patterns of plant distributions. In this paper we used quantitative methods to prepare a phytogeographical classification for the entire Australian continent using species turnover of nine major plant groups (Table 1) and to test the validity of an existing classification of Australia's three regions and 18 sub-regions. Our dataset contains representatives from diverse land plant groups (bryophytes, ferns, and angiosperms).

By first developing a species turnover-based phytogeographical classification, using taxonomic groups instead of climate, we are then able to test how the environmental patterns, such as of climate and soil types [19,20], fit the observed phytogeographical regions.

The aims of this study were to:

1. Quantify and map the plant regions of Australia through spatial analyses of modern databases of georeferenced specimen data and compare these with the current phytogeographical regionalization.
2. Identify what the major environmental drivers of species turnover are for each of these phytogeographical regions.
3. Test the validity of previously proposed regions and sub-regions from the last 190 years, and propose an improved classification of the phytogeographical regions of Australia.

Methods

Spatial dataset

Table 2 summarizes the taxa and number of occurrence records examined in this study. A total of 802,273 records were downloaded from *Australia's Virtual Herbarium* (AVH) [31]. This dataset does not contain absence records, of course. Collections were mostly curated to the accepted taxonomy of the *Australian Plant Census* [32]. We did not consider any infra-specific taxa in the analysis. For each species, spatial errors were removed using ArcGIS 9.2 [33]. The spelling and consistency of scientific names across taxa were corrected using Google Refine. Spatial outliers and all records without a geographic location were deleted. Records that fell in the ocean or outside continental Australia were excluded. After the correction process, 750,741 records remained for use in our final analyses. The geographic coordinates of each record were projected into an Albers equal area conic conformal coordinate system to avoid the latitudinal biases of geographic coordinate systems. The records were imported and aggregated to 100 km×100 km grid cells (870 in total) using BIODIVERSE 0.18 [34] (http://purl.org/biodiverse). We calculated the ratio of species records to number of samples per grid cell to measure redundancy, as an indicator of sample coverage [35]. We found that 70% of the analyzed grid cells had a good level (60%) of species record redundancy within each grid cell (see Appendix S1).

Taxonomic dataset

The dataset included Australian representatives from a diverse set of major terrestrial plant groups (bryophytes, ferns and several large angiosperm genera and families) that represent a wide geographic distribution across Australia [53,54]. For example, *Acacia* and eucalypts are the most abundant canopy and sub-canopy woody plants in Australia [36,37,53]. One of the problems when gathering a large dataset is the reliability of the taxonomic identification. The more taxa included, the more challenging it becomes to achieve high taxonomic and spatial reliability of herbarium data. The strategy we applied was to utilize as many taxa as possible that combine a strong taxonomic tradition and wide geographical ranges. Despite having experts on each group involved in the cleaning process checking for taxonomic and spatial errors in the sampled groups, we expect there to remain some low degree of taxonomic uncertainty.

Spatial analysis: bioregionalization

All spatial analyses were conducted using BIODIVERSE [34]. A matrix of species turnover was generated for all pair-wise combinations of grid cells (757,770 pairs). Simpson's beta (β_{sim}) was used as the turnover measure because it corrects for species richness differences between sites (Equation 1).

$$\beta_{sim} = 1 - \frac{a}{a + \min(a,c)} \quad \text{(Eq1.)}$$

Where a refers to the number of species common to cells i and j, b is the number found in cell i but not cell j, and c is the number found in cell j but not cell i. A low β_{sim} value indicates that many taxa are shared between two grid cells (low dissimilarity) and a high β_{sim} means a small number of shared taxa (high dissimilarity).

An agglomerative cluster analysis of the βsim turnover matrix was used to generate a WPGMA hierarchical cluster diagram in BIODIVERSE. Current literature suggests that dissimilarity clustering algorithms are exposed to topology biases [55]. In order to reduce biases in our analyses, the cluster analysis included a tie-breaker condition [18,19] such that, when more than one pair of sub-clusters had the same turnover score, the pair which had the highest Corrected Weighted Endemism (CWE) score was selected for merging. This approach guarantees the same cluster result would be generated each time the model is run, as well as increasing the endemicity of the resultant clusters. We identified the phytogeographical regions from the clusters based on two criteria: (1) a phytogeographical region is preferably represented by a group of contiguous, or near-contiguous, grid cells; (2) each cluster that represents a phytogeographical region needs to be clearly separated from its children (immediate descendent nodes) or parent (immediate ancestral node) in the dendrogram.

Spatial overlap analysis

Our phytogeographical regions were visually compared with several previous bioregionalizations of Australia. We also conducted a formalized comparison of the degree of overlap between our phytogeographical regions and the terrestrial phytogeographical sub-regions proposed in the Australian Bioregionalization Atlas (ABA). This comparison used only the ABA because it is the existing classification that best reflects a historical viewpoint of geographical regionalizations of flora for Australia. Both classifications were converted to raster format with a resolution of 100×100 km. Because we aimed to identify phytogeographical regions at the continental scale, a coarse grid cell size was applied. The effect of changes in grid cell size on the bioregions was explored in prior studies with eucalypts across Australia [19]. In that study we found no significant implications on the identifica-

Table 2. The plant groups used in this study, number of occurrence points, and the number of species per group, with the totals.

Taxon name	Number of records	Number of species
Acacia	165,518	1,020
Asteraceae	105,692	823
Eucalypts (*Angophora, Corymbia and Eucalyptus*)	202,736	791
Ferns	58,774	356
Hornworts	370	13
Liverworts	16,502	735
Melaleuca	41,092	282
Mosses	79,210	835
Orchids	80,847	1,188
TOTAL	**750,741**	**6,043**

tion of the major regions but some changes in the delimitation of the regions boundaries occurred. The ABA sub-regions were then overlaid on our phytogeographical regions using ArcMap 9.2 [33]. The overall spatial agreement between both classification schemes (which differ in the number of classes and their extent) was calculated by the count of ABA sub-region cells that overlapped with only one of our regions and then divided by the total number of cells in all ABA sub-regions. This value was summed up and expressed as a percentage of overlap.

Environmental correlates

Eleven environmental variables were used in this study (Table 3). A correlation matrix available on the spatial portal of the Atlas of Living Australia was used to select variables which represented different environmental traits and demonstrated minimal correlation. The spatial resolution of the layers was 1 km (approximately 0.01 degrees). The environmental layers were re-projected into the same Alber's conic conformal coordinate system as the species data using the R software [38] and aggregated to 100 km×100 km grid cells using BIODIVERSE. The environmental variables were developed using ANUCLIM [39,40]. We also included four layers related to soils and topography, sourced from the National Land & Water Resources Audit [41]. The mean value for each environmental variable within each 100 km×100 km grid cell was calculated using BIODIVERSE.

Relative environmental turnover (RET) aims to identify phytogeographical regions based on species turnover and investigate their environmental correlates. It has been shown to be a useful method to partition the continent into meaningful phytogeographic regions in *Acacia* and eucalypts [18,19]. RET was derived from a framework to delineate biogeographic regions initially proposed by Kreft & Jetz [23]. Previous studies used the term environmental turnover to explore rates of change of dissimilarity in vertebrates and their relationship to environment depending on the geographic distance [24]. RET is different from previous approaches because it is not geographic distance based, but instead combines grid cell analyses with ordinations. RET consist of two parts, one is an ordination and the second is a gridded analysis. Here, we only used the gridded component of RET. The key question addressed is what are the main environmental differences among the phytogeographic regions that were inferred from species turnover?

The gridded approach consists of three steps. First, summary statistics were calculated for 100 km×100 km grid cells of the eleven environmental variables in BIODIVERSE. Then, each of the grid cells of the environmental variables were spatially linked to each of the grid cells corresponding to the phytogeographic regions. Finally, the association between the phytogeographical regions and the environmental variables were calculated using Getis-Ord Gi* hotspot statistic spatial statistics in BIODIVERSE. We used the Getis-Ord Gi* hotspot statistic to assess if the environmental values within the clusters (phytogeographic regions) were significantly different from those for the country as a whole [42,43]. The Gi* statistic is expressed as a z-score indicating the degree to which the values of a subset of grid cells, in this case the cells comprising a cluster, are greater or less than the mean of the dataset. Those clusters with Gi* values greater than 2 or less than −2 represent sets of cells that have environmental values significantly different from expected ($\alpha<0.05$).

Results

Phytogeographical regions

We found and propose six phytogeographical regions: Northern Region (Euronotian, Monsoonal Tropics and Monsoon *sensu* [57,18,19]), Northern Desert Region (Eremaean North *sensu* [18,19], Eremaean (Eremaean South *sensu* [18,19]), Eastern Queensland (South-eastern Temperate and southeast *sensu* [14,18,19] and South-Western (Southwest *sensu* [8]) (see Fig 1). The names of the proposed regions are aligned to correspond to the Australian Bioregionalisation Atlas [17].

Overall, the spatial arrangement of the phytogeographical regions follows a distinctive north to south pattern, with an east to west pattern at the sub-regional level. The phytogeographical regions are nested in geographically related pairs. The first split is of the Northern and Northern Desert regions (branch length (bl) of 0.09; dendrogram in Fig 1) from the other four regions. The Northern is on a long branch (bl = 0.12) and readily subdivided east (bl = 0.09) to west (bl = 0.03; Fig 2). The region with the highest species similarity to the Northern is the Northern Desert

Table 3. Environmental variables used in our analyses.

Environmental variable	Description
Annual precipitation	Monthly precipitation estimates (mm)
Annual mean temperature	The mean of the week's maximum and minimum temperature (°C)
Annual mean radiation	The mean of all the weekly radiation estimates (Mj/m^2/day)
Precipitation of coldest quarter	Total precipitation over the coldest period of the year
Radiation seasonality	Standard deviation of the weekly radiation estimates expressed as a percentage of the annual mean (Mj/m^2/day)
Precipitation seasonality	Standard deviation of the weekly precipitation estimates expressed as a percentage of the annual mean (mm)
Temperature seasonality	Standard deviation of the weekly mean temperatures estimates expressed as a percentage of the annual mean (°C)
Ridge top flatness	Metric of the topographic flatness derived from a surface of 9 second grid cells (dimensionless)
Rock grain size	Lithological property of the bedrocks related to the mean grain size (0–10 units)
Sand	Content of sand on the top 30 cm of soil layer estimated from soil maps at a resolution of 1 km (%)
Clay	Content of clay on the top 30 cm of soil layer estimated from soil maps at a resolution of 1 km (%)

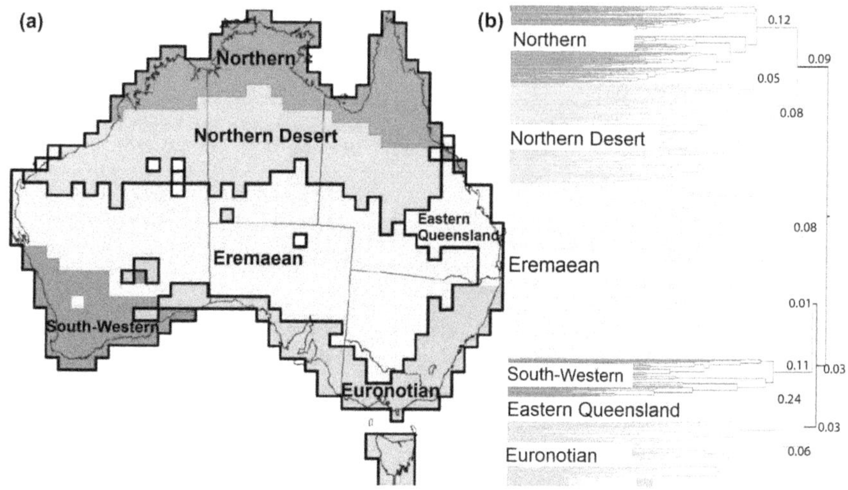

Figure 1. Phytogeographical regions of Australian terrestrial flora (a) as defined by the corresponding dendrogram (b). The colors of the regions in the map correspond to those used to plot the dendrogram. The dendrogram is a representation of the spatial relationship of dissimilarities in species composition among regions.

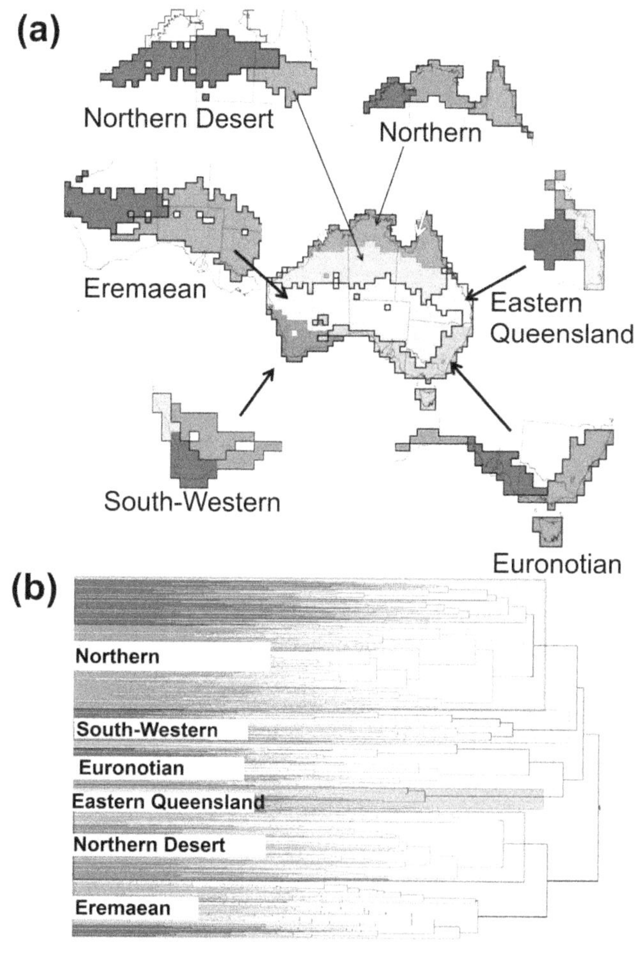

Figure 2. Phytogeographical regions and new subregions proposed for Australia (a), and their corresponding dendrogram (b). Note that the colors of the dendrogram clusters correspond to the colors of the subregions. Shaded colors indicate relationships: light blue and dark blue cluster together before clustering with brown colours.

Region (bl = 0.08; Fig 1). This phytogeographical region also has a clear east (bl = 0.05) to west subdivision (bl = 0.08; Fig 2).

The other major cluster has a short branch (bl = 0.03). The largest clustering of grid cells is the Eremaean phytogeographical region which shares a very short branch (bl = 0.01) with the South-Western phytogeographical region (Fig 1). The Eremaean is on a long branch (0.08) and subdivides east (bl = 0.07) to west (bl = 0.04; Fig 2). The South-Western phytogeographical region is also on a long branch (bl = 0.11) and is subdivided into three coastal (bl = 0.02) subregions and an inland (bl = 0.04) subregion (Fig 2).

The Euronotian and Eastern-Queensland phytogeographical regions cluster together (bl = 0.03). The Eastern Queensland phytogeographical region (bl = 0.24) runs along the eastern coast of Queensland from the Wet Tropics to the New South Wales border and inland into south central Queensland (Fig. 1). This region subdivides into northern and southern coastal subregions (bl = 0.03) and an inland (bl = 0.07; Fig 2) subregion. The Euronotian phytogeographical region (bl = 0.06) has a strong east subregion (bl = 0.16) to central-west subregion (bl = 0.12) structure (Fig. 2).

Environmental correlates of the phytogeographical regions

The environmental correlates of the six phytogeographical regions are shown in Table 4. The most extreme Gi* score was a precipitation trait for four of the six phytogeographical regions, while a temperature trait was the most extreme for the other two (see underlined values in Table 4). Species distribution and bioregions of *Acacia* and eucalypts in Australia are strongly influenced by annual precipitation and seasonal temperatures as well [18,19].

Precipitation and temperature seasonality are the environmental variables that better correlate with turnover of the Northern phytogeographical region (which could be termed the "monsoonal region" environmentally). The Eremaean phytogeographical regions are differentiated by seasonality traits in the Northern Desert Region and annual precipitation in the Eremaean, reflecting a possible Tropic of Capricorn division [10]. For the Euronotian phytogeographical region, the amount of solar radiation and precipitation during the coldest quarter of the year

Table 4. Gi* spatial statistics for the six phytogeographical regions of Australian flora. Bolded means statistically significant ($\alpha = 0.05$).

Environmental variable	Northern (N = 141)	Northern Desert (N = 185)	Eremaean (N = 317)	Eastern Queensland (N = 43)	Euronotian (N = 114)	South-Western (N = 70)
Annual mean radiation	**3.37**	**9.04**	**4.91**	**−2.03**	**−14.90**	**−6.75**
Annual mean temperature	**12.69**	**11.66**	**−2.55**	−0.82	**−17.02**	**−8.45**
Annual mean precipitation	**17.69**	**−4.89**	**−15.45**	**5.01**	**6.86**	−1.78
Clay	**−2.28**	1.16	**2.05**	**3.85**	0.37	**−5.81**
Precipitation coldest quarter	**−7.31**	**−7.76**	**−5.51**	1.83	**16.26**	**9.71**
Precipitation seasonality	**17.38**	**13.23**	**−12.83**	−1.61	**−12.59**	**−3.84**
Radiation seasonality	**−14.83**	**−11.44**	**7.22**	**−2.44**	**15.81**	**6.86**
Ridge Top flatness	−1.90	**3.29**	**3.60**	**−3.16**	**−5.74**	0.89
Rock grain size	1.52	**−4.71**	−0.20	−0.89	−1.31	**7.70**
Sand	**3.31**	0.78	−2.26	**−2.79**	−1.84	**2.85**
Temperature seasonality	**−18.07**	0.87	**18.42**	-0.86	**−4.80**	**−2.77**

N = number of grid cells per region. Underlined values are the most extreme scores for each region.

(winter) are the main environmental drivers. In the South-Western phytogeographical region precipitation in the coldest quarter, temperature, and landscape properties are the main drivers.

Spatial comparison of the new phytogeographical regions to the ABA terrestrial sub-regions

Our phytogeographical regions resemble the nomenclature proposed by the terrestrial phytogeographical sub-regions of the ABA (Fig. 3c) [17] (Table 5). In numerical terms, the spatial agreement, between our phytogeographical regions (Fig 3a) and the ABA classification scheme is 65% (Fig 3c). This result represents a high level of agreement but there are still major gaps among many of the ABA subregions that we were able to fill using the species turnover approach.

Discussion

Our data suggest that the Northern region overlaps with the ABA Kimberly Plateau, Arnhem Land, Cape York and Atherton Plateau [in part] sub-regions and has a species composition more similar to the Northern Desert than to the more mesic phytogeographic area along the eastern coast of Australia. The Northern Desert phytogeographic region overlaps with some parts of the Northern, Eastern and Western Desert ABA sub-regions. The arid zone in our classification split into the Eremaean (including the southern parts of the Eastern Desert and western Desert ABA sub-regions) as well as South-Western region, which is considered one of the world diversity hotspots.

These results are similar to the proposed ABA phytogeographical regions and sub-regions that have been in use for over 120 years see [6,7,8,9,10,11,12,13]. However, because they are based on a rigorous quantitative analysis of a large data set, our results should be used to revise the ABA area taxonomy and area boundaries as well as to extend sub-regions within the Eremaean and Euronotian regions, so that all areas abut. The current provisional area taxonomy within the ABA has few abutting sub-regions (see Figure 3b); our results can re-define these existing areas to create a more accurate area taxonomy for Australia's phytogeographical regions and sub-regions.

Spatial comparison of the new phytogeographical regions to biomes

Our results strongly reflect the northern tropical summer and southern temperate winter rainfall gradients. Precipitation is a more significant environmental correlate in the northern half of the continent whereas high levels of solar radiation and cool temperatures are more important below the Tropic of Capricorn. However, annual precipitation is a predominant correlate of the coastal Queensland region where a tropical/sub-tropical transition zone, the Eastern Queensland phytogeographical regions, is created.

The north - south split between the Eremaean and Northern Desert Region roughly coincides with the Tropic of Capricorn and the summer-winter rainfall line (see Appendix S1 panel a). However, this split is not evident in the previously published biomes or bioregions of Australia (see Appendix S1 panels a-b-d) [10,4,44,49]. These biome descriptions, which are defined by both climate and biota, identify a large arid Eremaean region that is not split north to south into two regions as was found in our analysis. The Eremaean "zone is crossed obliquely by the junction between the summer and winter rainfall systems but floristically the junction is not so strongly marked due to the presence of small ranges of low mountains, which appear to have acted as refugia" [10]. Our evidence suggests that the division line between Eremaean and Northern Desert regions might be related to the effect of the Tropic of Capricorn, which may have resulted from the palaeoclimatic shifts (warmer-cooler-warmer) during the last 65 Ma [58]. It was mentioned in a compilation of Australian phytogeography that "floristic composition from north to south is probably as closely related to temperature gradient and possibly also day length as to available rainfall" [10]. We also observed a west-east climatic division within the Eremaean and Northern Desert regions. Our analysis identifies the Eastern Queensland as a separate phytogeographical region. This region can be described climatically as an inter-zone defined by the summer-winter rainfall variation as previously noted in Burbidge's biomes in Australia [10].

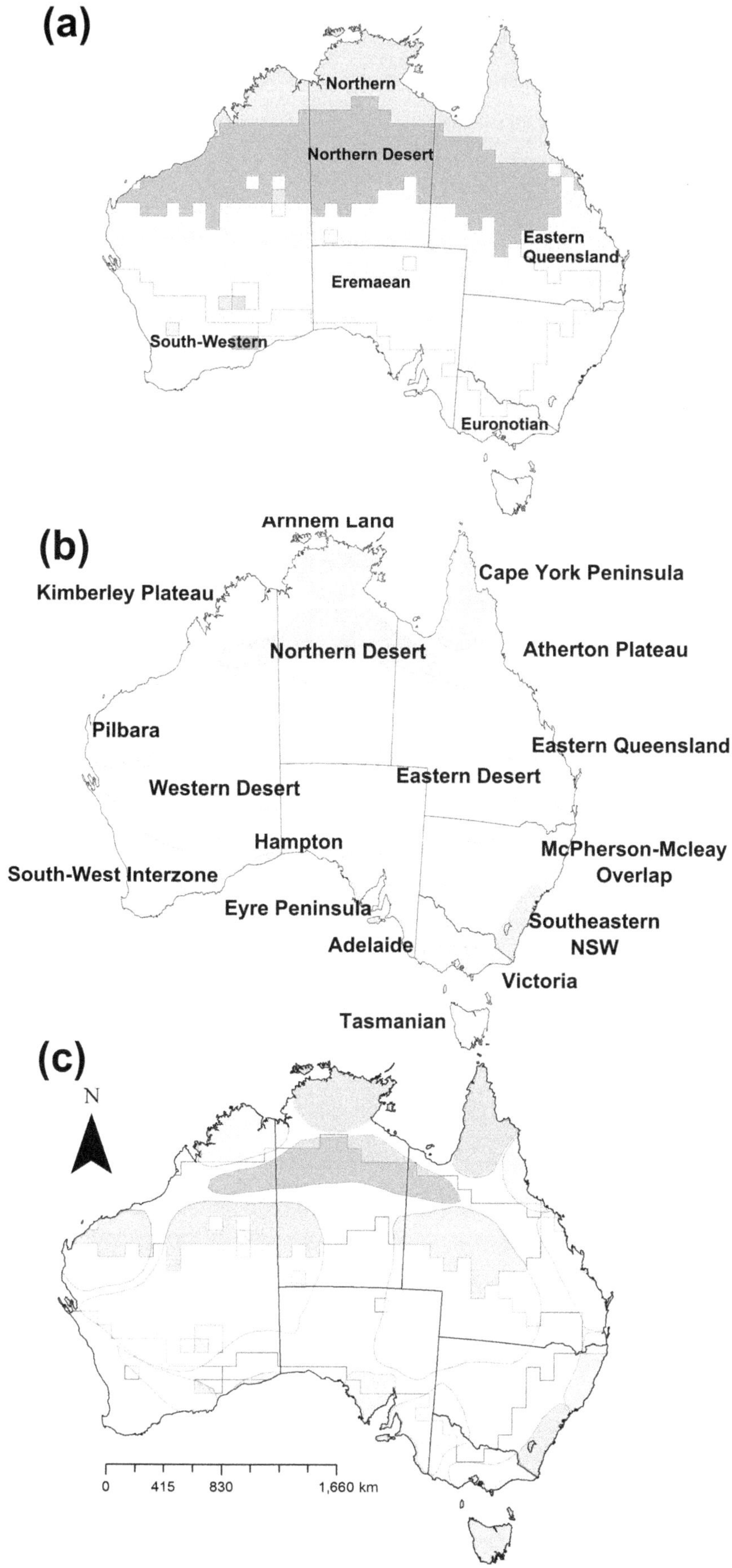
(a)
Northern
Northern Desert
Eastern
Queensland
Eremaean
South-Western
Euronotian
(b)
Arnhem Land
Kimberley Plateau
Cape York Peninsula
Northern Desert
Atherton Plateau
Pilbara
Eastern Queensland
Eastern Desert
Western Desert
Hampton
McPherson-Mcleay
Overlap
South-West Interzone
Eyre Peninsula
Southeastern
NSW
Adelaide
Victoria
Tasmanian
(c)
N
0
415
830
1,660 km

Figure 3. Spatial agreement between our six phytogeographical regions of Australian flora (a) and the terrestrial phytogeographical sub-regions of Australia (ABA) (b) [17]. Shown is the degree of spatial agreement of the ABA and our classification (c) and the percentage of overlap among each of our phytogeographical regions and the ABA sub-regions (d). Equivalent sub-regions from the ABA are noted below and as shown in Fig 2. The Northern Desert Region (red and blue: Northern Desert); Eremaean (red: Western Desert, blue: Eastern Desert); South-Western (blue & orange: Southwest Interzone); Euronotian (red: Eyre Peninsula and Adelaide [in part], blue: Victoria, Southeastern NSW, McPherson - Macleay Overlap [in part]); Eastern Queensland (blue: Atherton Tableland [in part], light blue: Eastern Queensland); Northern Region (red: Kimberley Plateau, orange: Arnhem Land, blue: Cape York Peninsula and Atherton Tableland in part).

Spatial comparison of the new phytogeographical regions to other classifications

The comparison of our regions and sub-regions against geology [45], soils [46], and vegetation types [47] uncovered few congruent patterns (Appendix S1). The results align with the current distribution of major vegetation groups of Australia as cited by the National Vegetation Information Systems (NVIS) [47]. Geology and soils are treated as artificial units (e.g., Formations, Ferrosols etc.), rather than types of rock and soil (e.g., sandstones, sandy loams) and therefore are unlikely to overlap. However, general climatic maps correlate with our results. The six proposed floristic regions (see Appendix S1 panel c) closely agree with Köppen's macro-climatic map of Australia (Appendix S1 panel e; http://www.bom.gov.au/climate/environ/other/kpn_group.shtml) [27]. Regarding Köppen's classification, the tropical zone (see Appendix S1, dark green in panel 3e) maps precisely with the Northern region, the subtropical zone (see Appendix S1, light green in panel e) matches well with our Eastern-Queensland region and the temperate climate group (see Appendix S1, blue in panel e) fit well with our Euronotian region. The main inconsistency is with the desert and grassland groups (see Appendix S1 orange and yellow in panel e) where a split into grassland that covers semi-arid areas is conspicuous, although these grassland areas roughly agree with the eastern subregions of the Northern Desert and Eremaean regions.

Utilization of phytogeographical classifications

Our results support some previous biotic [57] and climatic classifications of Australia [10] but also disagree in some cases [59], and thus add new information to the biogeographical literature. For example, our results suggest for the first time that the flora of arid Australia (Eremaean Region of the ABA) can be divided into distinct phytogeographic regions, first along a north to south gradient and then along an east to west gradient, in contrast to some proposed biogeographic faunal patterns [48]. We show that a unified method for quantifying species turnover can be used to successfully partition a continent into geographically meaningful regions using a broad sample of plant groups. This analysis also demonstrates that biogeographical regionalisation does not have to be convoluted and complex. With fewer factors involved, patterns are easier to explain. For example, the strong evidence of the relationship of sub-regions of species turnover with climatic variables suggests that species assemblages across Australia have responded to changes in weather systems across the continent.

The phytogeographical regions presented here are defined using species turnover and thus relate to taxonomic diversity in the groups studied. Here, the taxonomic groups contain a combination of recent (*Acacia*) and older clades (the bryophyte groups). It is probable that the recently diverged clades are driving the patterns identified because they comprise a large proportion of the species sampled. However, the older bryophytic and pteridophytic clades do not have the same broad continental distributions of the younger clades studied here, which may reflect recent distributional patterns that might not be shared with these older clades. If dominated by the distributions of recently derived species, our results likely will match modern climatic zones, while older species might reflect geological features, tectonic patterns or older palaeo-climatic zones.

Given this, we highlight the importance of generating regionalizations based on large, multi-taxon datasets. Furthermore, basing floristic regions only on species turnover misses out on the full depth of phylogenetic information available. Future studies should compare these results with patterns of spatial similarity generated using measures of phylogenetic turnover [52], to obtain a better picture of the historical relationships among areas within Australia. Understanding the adaptive changes in morphology and physiology that accompanied biome shifts will enable a broad understanding of the adaptive history of organisms and its potential for adaptation in the face of human induced climate change [56].

Table 5. Area taxonomy overlaps between new areas and existing regions and sub-regions from the recently published Australian Bioregionalization Atlas (ABA) [17].

New Areas (this study)	ABA Area Taxonomy Regions	ABA Area Taxonomy Sub-regions
Northern Desert Region	Eremaean	Northern Desert
Eremaean	Eremaean	Western and Eastern Deserts
South-Western	Southwest Australia	Southwest Interzone
Euronotian		McPherson - Macleay Overlap (in part), Southeastern NSW, Victoria, Adelaide (in part), Eyre Peninsula (in part).
Northern Region	Euronotian	Kimberly Plateau, Arnhem Land, Cape York Peninsula, Atherton Tableland (in part)
Eastern Queensland	Euronotian	Atherton Tableland (in part), Eastern Queensland

Note that the new areas abut, while the ABA sub-regions are occasionally separated by undescribed areas (see gaps between regions in Figure 3b).

Supporting Information

Appendix S1 Comparison of our six phytogeographical regions of Australian flora (c) against major biogeographical classifications of Australia. Burbidges biomes [10] (a), Crisp et al biomes [49] (b), IBRA regions [4] (d) and Köppen's macro-climatic map of Australia (e). There is permission to re-print maps on panels A and B, and labels in panels D and E indicate the original publisher (official permission not required because is public material).

Acknowledgments

We would like to thank the two Commonwealth Scientific and Industrial Research Organisation (CSIRO) internal reviewers for their valuable comments. We would like to thank Bernd Gruber for supporting the spatial overlap analyses. BDM thanks CSIRO (Australia) for a Distinguished Visiting Scientist Award.

Author Contributions

Conceived and designed the experiments: CEGO MCE SL JTM. Analyzed the data: CEGO SL JTM NJK. Contributed reagents/materials/analysis tools: CEGO MCE SL AHT NJK ANSL CCC MC NSN BDM JTM. Wrote the paper: CEGO JTM SL MCE.

References

1. Holt BG, Lessard JP, Borregaard MK, Fritz SA, Araújo MB, et al. (2013) An update of Wallace's Zoogeographic regions of the world. Science 339: 74–77.
2. Udvardy MDF (1975) *A classification of the Biogeographical Provinces of the World.* IUCN Occassional Paper (No 18). International Union for Conservation of Nature and Natural Resources, Morges, Switzerland, 49 pp.
3. Integrated Marine and Coastal Regionalisation of Australia (2006) IMCRA – Version 4.0- Department of the Environment and Heritage, Australian Government.
4. Interim Biogeographic Regionalisation for Australia (1995) IBRA -Version 6.1, Canberra, Australia.
5. Pepper M, Doughty P, Keogh S (2013) Geodiversity and endemism in the iconic Australian Pilabara region: a review of landscapes evolution and biotic response in an ancient refugium. Journal of Biogeography doi:10.1111/jbi.12080.
6. Ebach M (2012) A history of biogeographical regionalization in Australia. Zootaxa 3392: 1–34.
7. Mueller F von (1858) Botanical report on the North-Australian expedition, under the command of A.C. Gregory, Esq. Preceding of the Royal Linnaean Society 2: 137–163.
8. Tate R (1889) On the influence of physiological changes in the distribution of life in Australia, Report of the First Meeting of the Australian Association for the advancement of science. pp. 312–326.
9. Nicholls GE (1933) The composition and biogeographical relations of the fauna of western Australia. Reports of the Australian Association for the Advancement of science 21: 93–138.
10. Burbidge N (1960) The phytogeography of the Australian region. Australian Journal of Botany 8: 75–211.
11. Doing H (1970) Botanical Geography and chorology in Australia. Mededelingen van de botanische tuinen en het Belmonte Arboretum der Landbouwhogeschool te Wageningen 13: 81–89.
12. Barlow BA (1984) Proposal for delineation of botanical regions in Australia. Brunonia 7: 195–202.
13. Beard JS (1981) The history of the phytogeographic region concept in Australia. In: Keast A Ecological biogeography of Australia. Dr W Junk Publishers, The Hague, pp. 335–375.
14. Crisp MD, Linder HP, Weston PH (1995) Cladistic biogeography of plants in Australia and New Guinea: congruent pattern reveals two endemic tropical tracks. Systematic Biology 44: 457–473.
15. Ladiges PY, Humpries CJ, Martinelli LW (1991) Austral Biogeography. CSIRO, Melbourne, 227 pp.
16. Ladiges PY, Kellermann J, Nelson G, Humphries CJ, Udovicic F (2005) Historical biogeography of Austral Rhamnaceae, tribe Pomaderae. Journal of Biogeography 32: 1909–1919.
17. Ebach MC, Gill AC, Kwan A, Ahyong ST, Murphy DJ, et al. (2013) Towards an Australian Bioregionalization Atlas: A Provisional Area Taxonomy of Australia's Biogeographical Regions. Zootaxa 3619: 315–342.
18. González-Orozco CE, Laffan SW, Knerr N, Miller JM (2013) A biogeographical regionalization of Australian Acacia species. Journal of Biogeography 40: 2156–2166.
19. González-Orozco CE, Thornhill AH, Knerr N, Laffan SW, Miller JM (2013) Biogeographical regions and phytogeography of the Eucalypts. Diversity and Distributions 20: 46–58.
20. Tuomisto H (2010) A diversity of beta diversities: straightening up a concept gone awry. Part 1. Defining beta diversity as a function of alpha and gamma diversity. Ecography 33, 2–22.
21. Gaston KJ, Davies RG, Orme CD, Olson VA, Thomas GH, et al. (2007) Spatial turnover in the global avifauna. Proc. R. Soc. B 274: 1567–1574.
22. Baselga A, Gomez-Rodriguez C, Lobo JM (2012) Historical legacies in the world amphibian diversity revealed by the turnover and nestedness components of beta diversity. PloS ONE 7: e32341.
23. Kreft H, Jetz W (2010) A framework for delineating biogeographical regions based on species distributions. Journal of Biogeography 37: 2029–2053.
24. Buckley LB, Jetz W (2008) Linking global turnover of species and environments. PNAS 105: 17836–17841.
25. Linder HP, Klerk HM, Born J, Burgess ND, Fjeldsa J, et al. (2012) The partitioning of Africa: statistically defined biogeographical regions in sub-Saharan Africa. Journal of Biogeography 39: 1189–1205.
26. Moore CWE (1959) Interaction of species and soils in relation to the distribution of Eucalypts. Ecology 40: 734–735.
27. Stern H, Hoedt G, Ernst J (1985?)Objective classification of Australian climates. Australian Bureau of Meteorology.
28. Ferrier S, Manion G, Elith J, Richardson K (2007) Using generalized dissimilarity modelling to analyse and predict patterns of beta diversity in regional biodiversity assessment. Diversity and Distributions 13: 252–264.
29. Linder HP, de Klerk HM, Born J, Burgess ND, Fjeldså J, et al. (2012) The partitioning of Africa: statistically defined biogeographical regions in sub-Saharan Africa. Journal of Biogeography 39: 1189–1205.
30. Conran JG (1995) Family distributions in the Liliiflorae and their biogeographical implications. Journal of Biogeography 22: 1023–1034.
31. Australia's Virtual Herbarium (2010) AVH: Council of Heads of Australasian Herbaria -CHAH- (September 2010). Australian Plant Census (http://www.anbg.gov.au/chah/apc/index.html).
32. CHAH -Council of Heads of Australasian Herbaria. (2010a). Australian Plant Census (http://www.anbg.gov.au/chah/apc/index.html).
33. ESRI (2005). ArcGIS 9.2. Environmental Systems Research Institute, Redlands, CA.
34. Laffan SW, Lubarsky E, Rosauer DF (2010) Biodiverse, a tool for the spatial analysis of biological and related diversity. Ecography 33: 643–647.
35. Garcillán PP, Ezcurra E, Riemann H (2003) Distribution and species richness of woody dryland legumes in Baja California, Mexico. Journal of Vegetation Science 14: 475–486.
36. Brooker MIH, Slee AV, Connors JR (2006) EUCLID Eucalypts of southern Australia (third edition); CD; CSIRO Publishing: Collingwood, Australia.
37. Hnatiuk RJ, Maslin BR (1988) Phytogeography of Acacia in Australia in relation to climate and species-richness. Australian Journal of Botany 36: 361–383.
38. R Development Core Team (2005) R: a language and environment for statistical computing. R foundation for Statistical Computing, Vienna. Available at: http://cran.r-project.org/.
39. Houlder DJ (2000) ANUCLIM user's guide. Version 5.1. Centre for Resource and Environmental Studies, Australian National University, Canberra.
40. Hutchinson MF, Houlder DJ, Nix HA, McMahon JP (2000) BIOCLIM Users Guide Version 6.1. Centre for Resource and Environmental Studies, Australian National University, Canberra.
41. Johnston RM, Barry SJ, Bleys E, Bui EN, Moran CJ, et al. (2003) ASRIS: the database. Australian Journal of Soil Research 41: 1021–1036.
42. Laffan SW (2002) Using process models to improve spatial analysis. Int. Journal Geographical Information Science 16: 245–257.
43. Kulheim C, Yeoh SH, Wallis IR, Laffan SW, Moran GF, et al. (2011) The molecular basis of quantitative variation in foliar secondary metabolites in Eucalyptus globules. New Phytologist 191: 1041–1053.
44. Crisp MD, Cook L, Steane D (2004) Radiation of the Australian flora: what can comparisons of molecular phylogenies across multiple taxa tell us about the evolution of diversity in present-day communities? Phil. Trans. R. Soc. Lond. B 359: 1551–1571.
45. Clarke GL (1980) The geology of Australia. Geology 4: 1–11.
46. Beadle NCW (1962) Soil phosphate and the delimitation of plant communities in eastern Australia II. Ecology 42: 281–288.
47. DWER (2007) Australia's Native Vegetation: a summary of Australia's Major Vegetation groups, 2007. NVIS_3. Australian Government: Canberra. http://www.environment.gov.au/erin/nvis/.
48. Hedley C (1894) The faunal regions of Australia. Reports of the Australian Association for the advancement of Science 5: 444–446.
49. Barlow BA (1984) Proposal for delineation of botanical regions in Australia. Brunonia 7: 195–201.
50. Parenti LR, Malte EC (2009) Comparative Biogeography: discovering and classifying biogeographical patterns of a dynamic earth, University of California Press, Berkeley.
51. Wulff EV (1950). An introduction the historical plant geography. Chronica Botanica Company, Waltham, Mass. pp. 223.

52. Rosauer DF, Ferrier S, Williams KJ, Manion G, Keogh JS, et al. (2013) Phylogenetic generalised dissimilarity modelling: a new approach to analysing and predicting spatial turnover in the phylogenetic composition of communities. Ecography 37: 21–32.
53. González-Orozco CE, Laffan SW, Miller JT (2011) Spatial distribution of species richness and endemism of the genus *Acacia* in Australia. Australian Journal of Botany 59: 600–608.
54. Stevenson L, González-Orozco CE, Knerr N, Cargill DC, Miller JM (2012) Species richness and endemism of Australian bryophytes. Journal of Bryology 34: 101–107.
55. Dapporto L, Ramazzotti M, Fattorini S, Talavera G, Vila R, et al. (2013) recluster: an unbiased clustering procedure for beta-diversity turnover. Ecography 36: 1070–1075.
56. Nogués-Bravo D, Araújo MB, Martinez-Rica JP, Errea MP (2007) Exposure of global mountain systems to climate change during the 21st century. Global Environmental Change 17: 420–428.
57. Cracraft J (1991) Patterns of diversification within continental biotas: hierarchical congruence among the areas of endemism of Australian vertebrates. Australian Systematic Botany 4: 211–227.
58. Zachos J, Pagani M, Sloan L, Thomas E, Billups K (2001) Trends, rhythms, and aberrations in the global climate 65 Ma to present. Science 292: 686–693.
59. Mackey BG, Berry SI, Brown T (2008) Reconciling approaches to biogeographical regionalization: a systematic and generic framework examined with a cases study of the Australian continent. Journal of Biogeography 35: 213–229.

6

Lake Sediment Records on Climate Change and Human Activities in the Xingyun Lake Catchment, SW China

Wenxiang Zhang[1], Qingzhong Ming[1], Zhengtao Shi[1], Guangjie Chen[1], Jie Niu[1], Guoliang Lei[2], Fengqin Chang[1], Hucai Zhang[1]*

1 Key Laboratory of the Plateau Surface Process and Environment Changes of Yunnan Province, Key Laboratory of Plateau Lake Ecology and Global Change, Yunnan Normal University, Kunming, China, **2** Key Laboratory of Humid Subtropical Eco-geographical Process, Ministry of education, Fuzhou, China

Abstract

Sediments from Xinyun Lake in central Yunnan, southwest China, provide a record of environmental history since the Holocene. With the application of multi-proxy indicators (total organic carbon (TOC), total nitrogen (TN), $\delta^{13}C$ and $\delta^{15}N$ isotopes, C/N ratio, grain size, magnetic susceptibility (MS) and $CaCO_3$ content), as well as accelerator mass spectrometry (AMS) ^{14}C datings, four major climatic stages during the Holocene have been identified in Xingyun's catchment. A marked increase in lacustrine palaeoproductivity occurred from 11.06 to 9.98 cal. ka BP, which likely resulted from an enhanced Asian southwest monsoon and warm-humid climate. Between 9.98 and 5.93 cal. ka BP, a gradually increased lake level might have reached the optimum water depth, causing a marked decline in coverage by aquatic plants and lake productivity of the lake. This was caused by strong Asian southwest monsoon, and coincided with the global Holocene Optimum. During the period of 5.60–1.35 cal. ka BP, it resulted in a warm and dry climate at this stage, which is comparable to the aridification of India during the mid- and late Holocene. The intensifying human activity and land-use in the lake catchment since the early Tang Dynasty (~1.35 cal. ka BP) were associated with the ancient Dian culture within Xingyun's catchment. The extensive deforestation and development of agriculture in the lake catchment caused heavy soil loss. Our study clearly shows that long-term human activities and land-use change have strongly impacted the evolution of the lake environment and therefore modulated the sediment records of the regional climate in central Yunnan for more than one thousand years.

Editor: Navnith K.P. Kumaran, Agharkar Research Institute, India

Funding: This research was supported by the National Natural Science Foundation of China (Grant No. 41201204, U1133601, U0933604 and 41101189) and the Key Science Research Foundation of the Education Department of Yunnan Province (Grant No. 2012Z014). The funders had no role in study design, data collection and analysis, decision to publish, or preparation of the manuscript.

Competing Interests: The authors have declared that no competing interests exist.

* Email: hucaizhang@yahoo.com

Introduction

The Yunnan Plateau, southwest China, is located in the confluence zone of the Asian monsoon, and the climate is mainly controlled by a system comprising the Asian southwest monsoon, westerly winds and local climatic influences of the Qinghai-Tibet Plateau. Since the Cenozoic, a large number of structurally-controlled lake basins formed following the uplift of the Qinghai-Tibet Plateau [1]. Therefore, it is a key area in which to study the prehistoric Asian monsoon patterns over different time scales.

The climatic transition from the Pleistocene to Holocene changed the activity characteristics and life style of ancient people [2]. Agriculture provided the material foundation for birth of civilization and promoted an increase in population and intensity of human activity [3], [4], which has had a lasting and profound impact on the environment. China is the origin and development centers of rain-fed millet [5] and cultivated rice [6]. Early wheat remains in China emerged mainly in the Tarim Basin [7], the Hexi Corridor [8], [9], the Tianshui Basin [10], Shandong and Henan [11], dated at roughly in late Holocene, and the cultivated rice spread southwardly to Southeast Asia through Guangdong and Yunnan Provinces [12] at the same time.

Lake sediments play an increasingly important role in the research of global change and regional environmental evolution because they can demonstrate continuity and environmental and seasonal sensitivity, therefore providing high resolution and typically abundant environmental and climatic information [13], [14]. Whilst both natural climate change and human-influenced environmental changes are likely to be recorded in Holocene sediments of Yunnan [1], it is critical to be able to evaluate the influence of human activities on the reliability of sedimentary proxies in inferring the past climate change [15], [16]. Few researchers have investigated climate change and potentially associated human-influenced environmental change based on lake-catchment sediments, at millennial time scales.

Based on high-resolution geochemical proxy analysis of sediment core from Xingyun Lake, we aim to establish the processes and stages of the environmental change of Xingyun's catchment, and identify the initial time and manner in which human activities influenced the receiving environment and additionally probe how Xingyun Lake responded to the interaction between climate and humans.

Study area

Xingyun Lake is located in the central Yunnan Plateau (Figure 1), about 80 km south of Kunming and 20 km east of Yuxi. It is a semi-closed shallow lake, situated at 1740 m above sea-level and is surrounded by mountains. The lake has a water surface area of 34.7 km^2 and catchment area of 325 km^2. The mean water depth is approximately 7 m. The lake joins Fuxian Lake to the north via the narrow channel of the Gehe River [17]. Mean annual temperatures recorded at regional weather stations are in the range of 12–16°C, and the mean annual precipitation is approximately 976 mm, with more than 85% falling between May and October [18]. The climate in the catchment is dominated by the Asian southwest monsoon. The native vegetation comprises mainly broadleaved deciduous forests [16]. The small catchment area of Xingyun Lake suggests that the lake sediments will likely be reliable recorders of environmental change of the catchment, given the limited external influence thereon.

Materials and Methods

In June 2011, a sediment core 429 cm in length was recovered using a piston corer at the western part of the deep basin of Xingyun Lake in water 7.5 m deep (24°19′48.78″ N; 102°45′ 47.04″ E, Figure 1). No specific permissions were required for the described study, which complied with all relevant regulations. The field studies did not involve endangered or protected species and specific location can be seen below. The core was sampled in the field at 2.0 cm intervals, and sub-samples were sealed in plastic bags for transport to the laboratory.

The experimental analysis of sedimentary proxies was carried out at the Key Laboratory of Plateau Surface Process and Environment Changes of Yunnan Province. Magnetic susceptibility was measured with a Bartington MS2 magnetic susceptibility meter and mass-specific magnetic susceptibility (χ_{lf}) was also calculated. The $CaCO_3$ content of the samples was measured using the calcimeter method of Bascomb [20]. This involved measuring the amount of CO_2 produced after adding the HCl, and stoichiometrically calculated this into $CaCO_3$ content, and error of <1% was achieved during this analysis. Grain size was measured with a Mastersizer-2000 laser diffraction particle size analyzer (Malvern Instruments Ltd., UK) after treatment with H_2O_2 and HCl to remove organic matter and carbonate [21]. From this analysis, the relative standard deviation of parallel analyses for individual samples obtained was <1%. In the interest of obtaining accurate data, all of above proxies were measured three times, under the same conditions, treatment and analytical methods.

Total organic carbon (TOC), total nitrogen (TN), $\delta^{13}C$ and $\delta^{15}N$ isotope ratios of the samples were measured at 4.0 cm intervals at the Key Laboratory of China Geological Survey of Nanjing Center, Geological institution of the Ministry of Land and Resources, using an elemental analyzer (Flash EA1112 HT, Thermo) and mass spectrometer (MAT253). Samples were decarbonizes with 0.5 mol/L HCl and then rinsed with deionized water until the filtrate was neutral, and then freeze dried. Afterwards, samples were weighed into tinfoil capsules and combusted at 1080°C in excess oxygen. The resulting material was flushed into the MAT253 with a He-carrier flow for analysis [22]. Isotope ratios are reported in δ-notation, where $\delta = (R_s/R_{st} - 1) \times 1000$. R_s and R_{st} are the isotope ratios of the sample and the standard (PDB for carbon, AIR for nitrogen), respectively. Standard sample (lake sediment of the national soil standard reference material, GSS-9) with known $\delta^{13}C$ and $\delta^{15}N$ were measured daily to monitor analytical accuracy. The analytical precision was better than the mean: ±0.1‰ for $\delta^{13}C$, ±0.2 ‰ for $\delta^{15}N$, 0.1 $mg \cdot g^{-1}$ for TOC and 0.05 $mg \cdot g^{-1}$ for TN.

Results

1 Age model and chronology

Eight bulk sediment and plant macrofossil samples were collected from the organic-rich horizons of the core, and accelerated mass spectrometry (AMS) ^{14}C dates were measured at the AMS Laboratory of Beijing University, China (Table 1 and Figure 2). Organic carbon was extracted from each sample and dated following the method described by Nakamura [23]. The conventional ages were converted to calibrate with IntCal13 calibration data [24]. Based on linear-fitting analysis, the uppermost sediments in the core yield an age of 1.2 ka. According to ^{210}Pb dating of the core [16], the anomalously old age can be considered to result from "carbon reservoir effects" on radiocarbon dating of the Xingyun Lake sediments. To produce the age model for the core, we subtracted the reservoir age of 1.2 ka from all the ages, assuming that it is constant throughout the core (Table 1). The age models and the records of environmental proxies show a continuous sedimentation record, extending back to the last ~11 ka BP. The sedimentation rate of the upper layer (0–167 cm) is 1.25 mm/a and 1.24 mm/a, and is 0.17 mm/a and 0.27 mm/a in the lower layers (167–243 cm and 243–351 cm), respectively. The sedimentation rate in the lowest part of the core (351–429 cm) is 0.72 mm/a (Table 1). The results of sedimentation rate are in coincidence with geochronology of Xingyun Lake and other lakes in Yunnan Province [13], [25].

2 Proxy indices

The χ_{lf} value of Xingyun Lake sediments varies from 0.9 to 30.1×10^{-8} m^3/kg with an average value of 10.9×10^{-8} m^3/kg, and demonstrates a substantial increase above 167 cm in the core (Figure 3-A). An average value of the χ_{lf} from 167 cm to the top is 24.24×10^{-8} m^3/kg. The $CaCO_3$ content of Xingyun Lake samples is in the range of 0–61.45% (Figure 3-B), and is hence variable and fluctuates substantially. In the 243-167 cm interval, $CaCO_3$ content peaks at 61.5% (213 cm), and is very low (average of 0.32%) from 109 cm to the top of the core. The medium diameter (Md) is in the range of 2.88–15.11 μm, with an average of 6.63 μm. The sediment is therefore composed principally of fine grain sizes, with the finest material located between 167 cm and the top of the sediment core (Figure 3-D).

Values of TOC vary between 0.73% and 13.20% with a mean value of 6.40% (Figure 3-E), and gradually increase between 429 cm and 351 cm. The TOC content increases to a peak value from 351 cm to 243 cm although there are obvious fluctuations. The lowest value of TOC (approximately 0–3%), from 167 cm to the top of the sediment core, indicates a decrease in the abundance of plants in the catchment. The TN content is generally low, with an average of 0.33% (Figure 3-F), and is well correlated with TOC ($R = 0.88$, Figure 4-A). The C/N ratio of the sediments varied between 6.88 and 49.07, with an average of 18.71 (Figure 3-G). The C/N ratio exceeds a value of 20 at the 429-161 cm interval of the sediment core, with the maximum value between 237 cm and 161 cm. From 109 cm to the top of the core, the C/N ratio is ≤ 10. The C/N ratio displays a positive correlation with the TOC ($R = 0.56$, Figure 4-B), and a very weak correlation with the TN content ($R = 0.11$). This suggests that the TOC more strongly influences the C/N ratio.

Values of $\delta^{13}C$ vary from −24.27‰ to −7.74‰ with an average value of −18.37‰ (Figure 3-H). The highest values of $\delta^{13}C$ were found in the 243-167 cm interval of the sediment core.

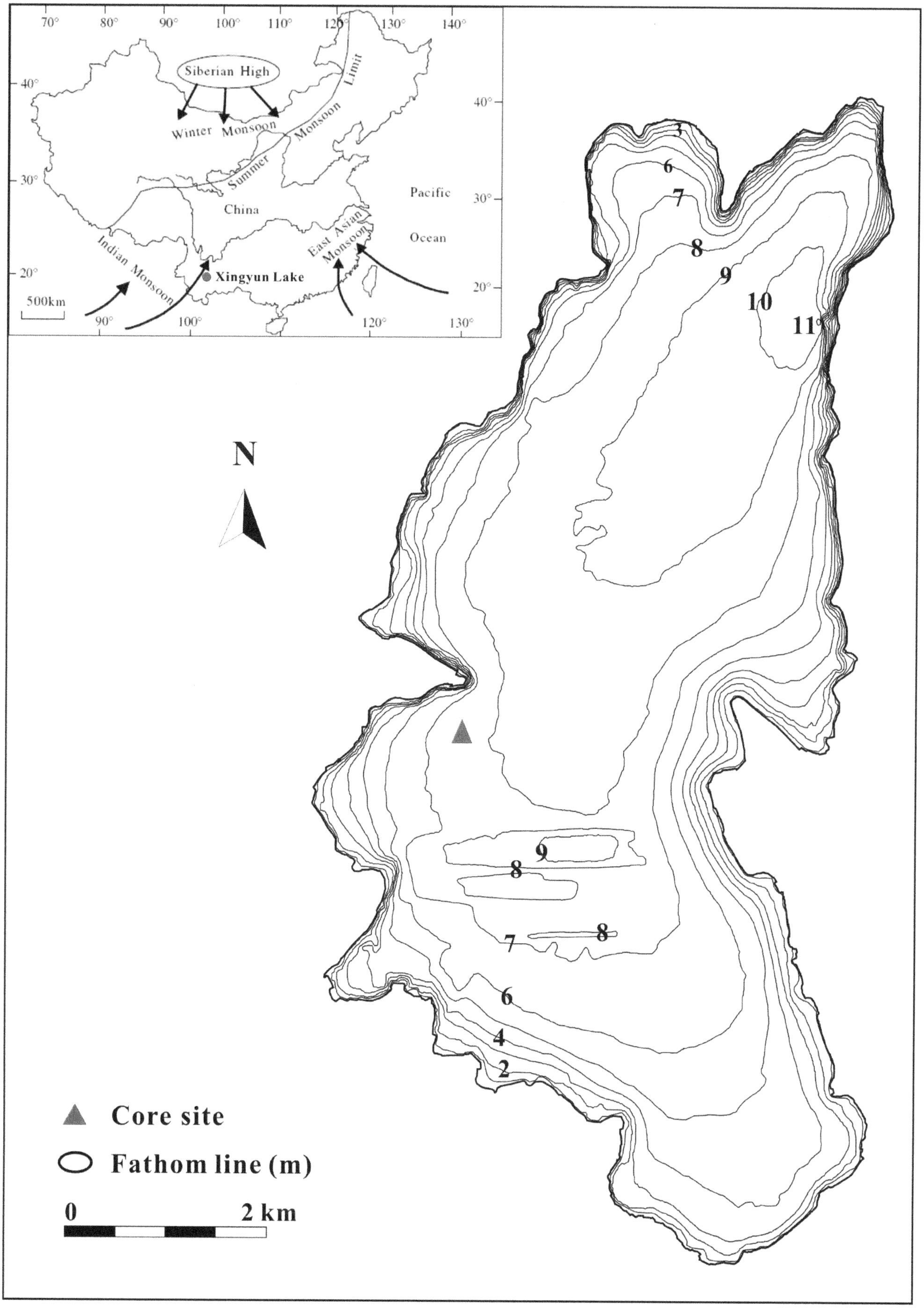
Siberian High
Winter Monsoon
Summer Monsoon Limit
China
Pacific Ocean
Indian Monsoon
East Asian Monsoon
Xingyun Lake
500km
N
Core site
Fathom line (m)
0 2 km

Figure 1. The monsoonal system in China and the location of Xingyun Lake (Based on Wu [19]).

There is a demonstrable positive correlation between $\delta^{13}C$ and the C/N ratio (R = 0.82, Figure 4-C). Values of $\delta^{15}N$ vary between 0.07‰ and 5.55‰ with an average value of 3.11 (Figure 3-C), and the 351-243 cm interval of the sediment core is the most ^{15}N depleted (minimum of 0.07‰). The depth profile of $\delta^{15}N$ is similar to that of TN content, and there is a good inverse correlation between $\delta^{15}N$ and TN content (R = −0.79, Figure 4-D).

Discussion

1 Climatic significance of the proxy indices

Geochemical proxy indices. Sources of TOC in lacustrine sediments have been suggested to be the biomass of aquatic algae, aquatic macrophytes and terrestrial plants in the lake catchment. The contribution of each of these different organic matter types to the TOC is affected by the regional climates, catchment

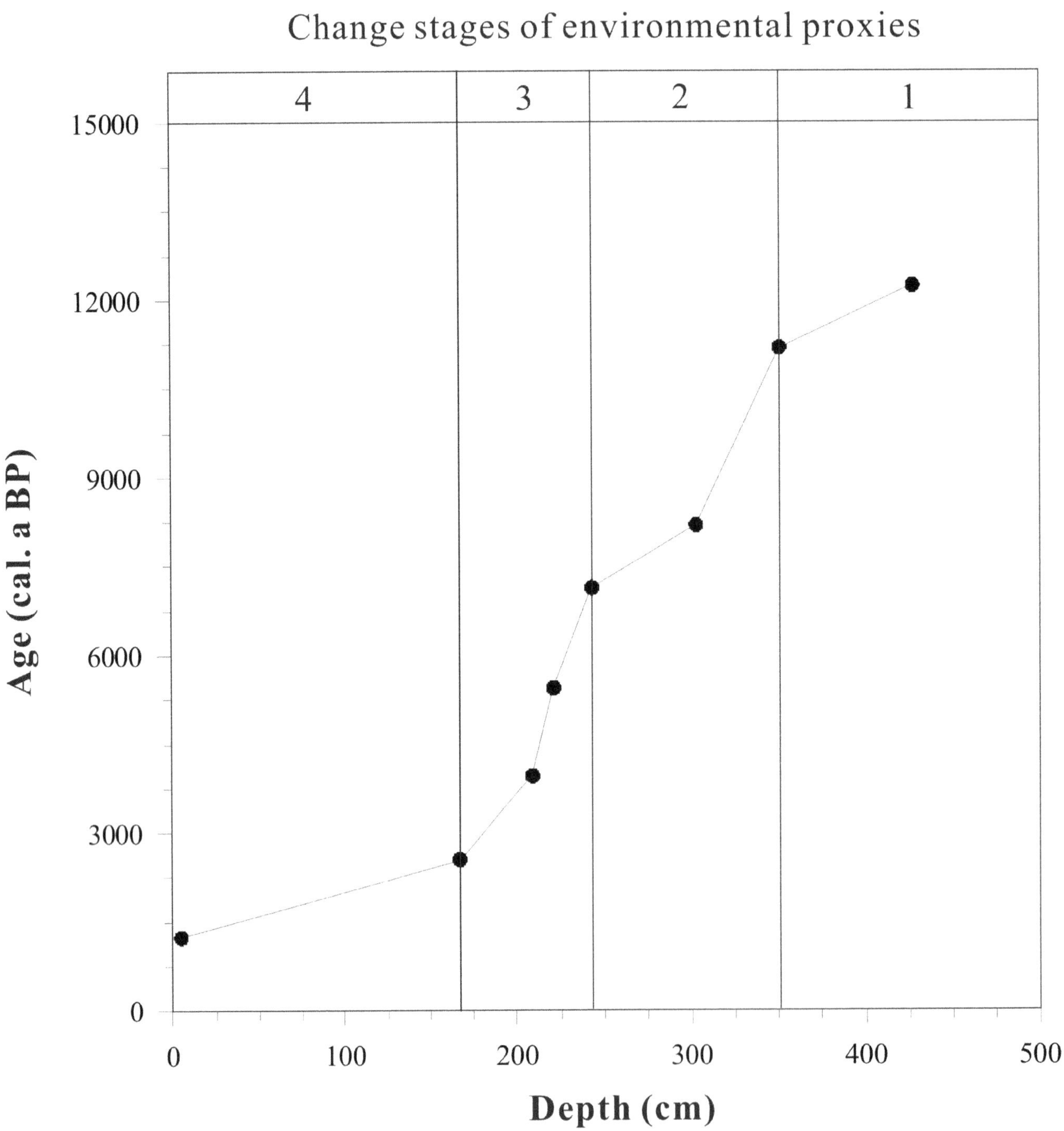

Figure 2. Age-depth curve and sedimentation rate for Xingyun lake core. Radiocarbon dates are listed in Table 1.

Table 1. Radiocarbon dating in Xingyun Lake sediment core and age model.

Laboratory number	Depth (cm)	Dating material	$\delta^{13}C$	AMS ^{14}C age (^{14}C yr BP)	Calibrated ^{14}C (2σ, cal a BP)	Median (cal a BP)	Modelled age (cal a BP)	Sedimentation rate (mm/a)
PA06849	5	Organic matter	−31.42	1335±40	1310–1175	1240	40	1.25
PA06850	167	Organic matter	−26.20	2485±40	2725–2365	2550	1350	1.24
PA06851	209	Organic matter	−19.92	3625±40	4080–3840	3960	2760	0.30
PA06852	221	Organic matter	−24.02	4645±40	5570–5305	5440	4240	0.08
PA06853	243	Organic matter	−26.46	6210±40	7250–7005	7130	5930	0.13
PA06854	303	Organic matter	−22.21	7375±40	8325–8050	8190	6990	0.57
PA06855	351	Organic matter	−28.20	9760±45	11250–11110	11180	9980	0.16
PA06856	427	Organic matter	−30.84	10360±45	12405–12050	12230	11030	0.72

environment and human activity. For example, there would be little terrestrial plant contribution to the lake sediments if deforestation of wooded areas occurred [26], [27]. The TN content of lake sediments is typically indicative of lake's trophic level, and is closely related to the algae production [28].

The C/N ratio is regarded as an effective indicator of the provenance of organic matter in sediments [29], [30]. When sedimentary organic matter primarily originates from endogenous materials, it has a C/N ratio between 4 and 10 because of protein-rich and cellulous-poor aquatic organisms. Exogenous materials tend to result in C/N ratios greater than 20 owing to the protein-poor and cellulous-rich nature of higher terrestrial plants [31], [32]. Lamb [33] suggested that terrestrial plant matter always result in C/N ratios of between 16 and 20 in lacustrine sediments.

Variations in $\delta^{13}C$ are closely related to the sources of organic matter in the lake sediments [34]; C3 and C4 land plants have $\delta^{13}C$ values that range from −37‰ to −24‰ and −19‰ to −9‰ with an average value of −27‰ and −14‰, respectively [35]. Crassulacean acid metabolism (CAM) plants have a broad range of $\delta^{13}C$ values and the $\delta^{13}C$ composition of aquatic macrophytes is complex and broad of range. The $\delta^{13}C$ values of aquatic plants vary between −20‰ and −12‰. These plants take up carbon from HCO_3^- in lake water for photosynthesis, yielding higher $\delta^{13}C$ values than for emergent plants (−30‰ to −24‰).

The $\delta^{15}N$ values of lake sediments can be influenced by the concentration of dissolved nitrate, nitrogen-fixing processes, bacterial decomposition, kinetic isotopic effects and climate change [36]. The dominant factor influencing $\delta^{15}N$ depends on the local physicochemical processes [37]. In general, large amounts of terrigenous organic matter entering the lake would cause a low $\delta^{15}N$ value during warm-humid period and vice versa [32], [38]. Moreover, human activities can increase the $\delta^{15}N$ value in lacustrine sediment significantly [39].

Other proxy indices. In southwest China, especially in Yunnan Province, rainfall is a grain-size determining factor. In short time-scale and high resolution studies (years to decades), it is evident that larger sediment grain-sizes become more migration during high rainfall periods in a wet climate [40]. Sediment grain size can, therefore be used to study the changes in humidity and can be distinguished from environmental indicators linked to the grain size of aeolian sediments. Moreover, the development of agricultural practices is expected to compromise the soil surface structure, and therefore increase the content of finer sediments being transported and entering lake [41].

Magnetic susceptibility is an efficient and sensitive proxy with which to study the lake environment [42]. The research results indicate that ferromagnetic material, concentrated in surface soil and detritus of the lake catchment [43], is one of the main sources of magnetic minerals in the lake sediments. $CaCO_3$ in the lake sediments is mainly composed of authigenic and allochthonic carbonates. The two main factors that induce carbonate precipitation are biological, and physico-chemical, such as temperature variations, evaporation and release of CO_2 [44].

2 Environmental evolution of Xingyun Lake

From the environmental proxies and depositional rates determined for the Xingyun lake sediments, the environmental evolution, including monsoonal and human influences, can be divided into four main stages (Figures 3 and 5), as follows:

Stage 1:11.06-9.98 cal. ka BP (429-351 cm). This period was characterized by a gradual increase in TOC, TN content, $CaCO_3$, χ_{lf} and fluctuating Md, indicating a period of high productivity, enhanced hydraulic conditions and increasing precipitation in the lake catchment. The C/N ratio, $\delta^{13}C$ and

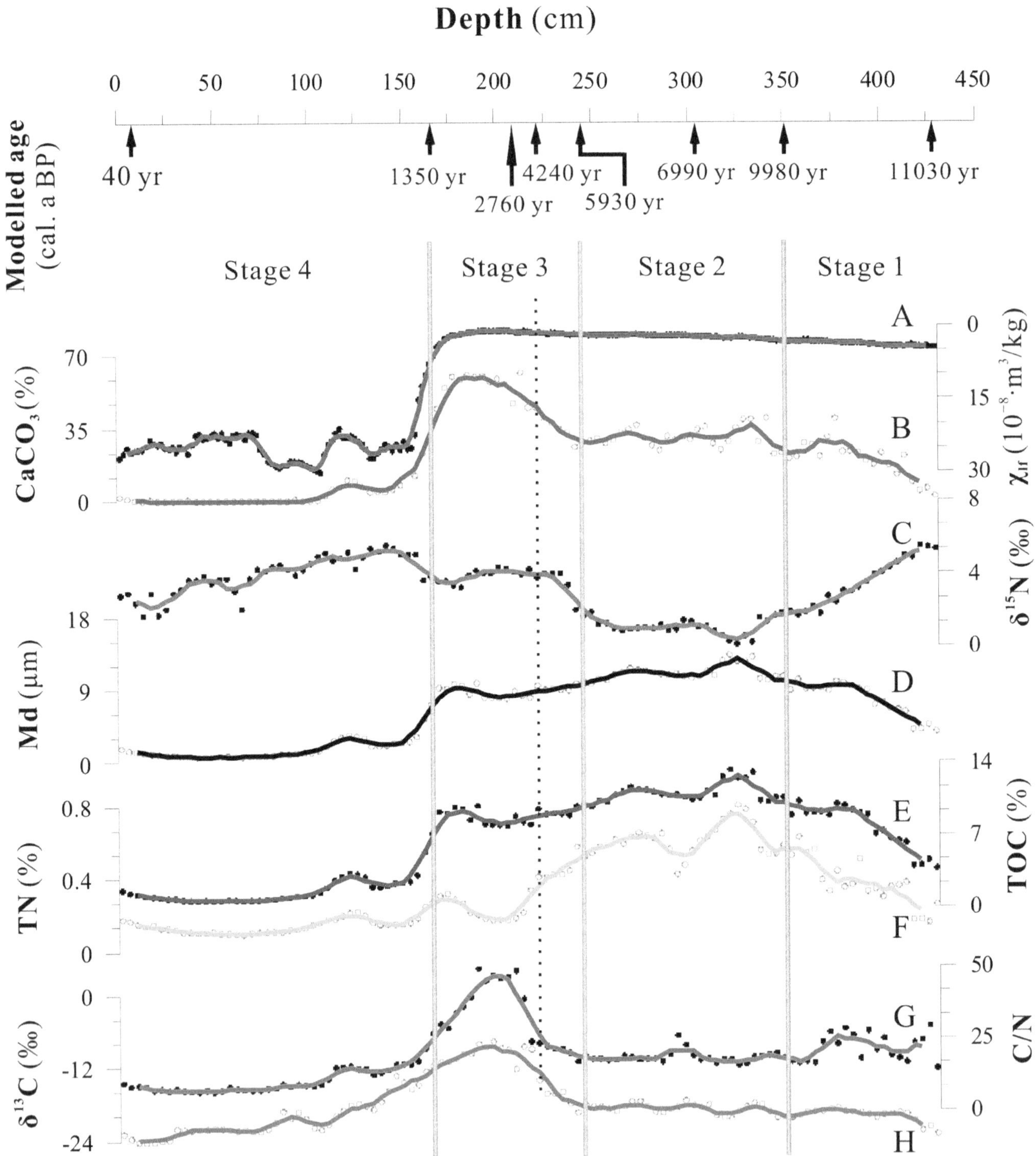

Figure 3. Multi-proxies results from Xingyun lake sediments: (A) magnetic susceptibility χ_{lf} and $CaCO_3$; (B) grain size and $\delta^{15}N$; (C) TOC and TN; (D) C/N and $\delta^{13}C$. ^{14}C AMS control points for Xingyun core are denoted on upper axis, and each curve is the 5 data running average value of the proxies. The gray bars highlight the boundary between Stage 1(429-351 cm, 11060-9980 yr BP, Younger Dryas and the early Holocene), Stage 2 (351-243 cm, 9,980-5930 yr BP, strongest summer monsoon in the Holocene), Stage 3 (243-167 cm, 5930-1350 yr BP, weaker summer monsoon) and Stage 4 (167-0 cm, 1350 yr BP to present, intensified human activities).

relatively low $\delta^{15}N$ value indicate that organic matter in the lacustrine sediments was mainly derived from C4 land plants and that the climate was warm and humid. The change in environmental proxies are considered consistent with an intense influence of the Asian southwest monsoon and overall the climate shifted from cold-wet to warm-humid during this early Holocene period [45].

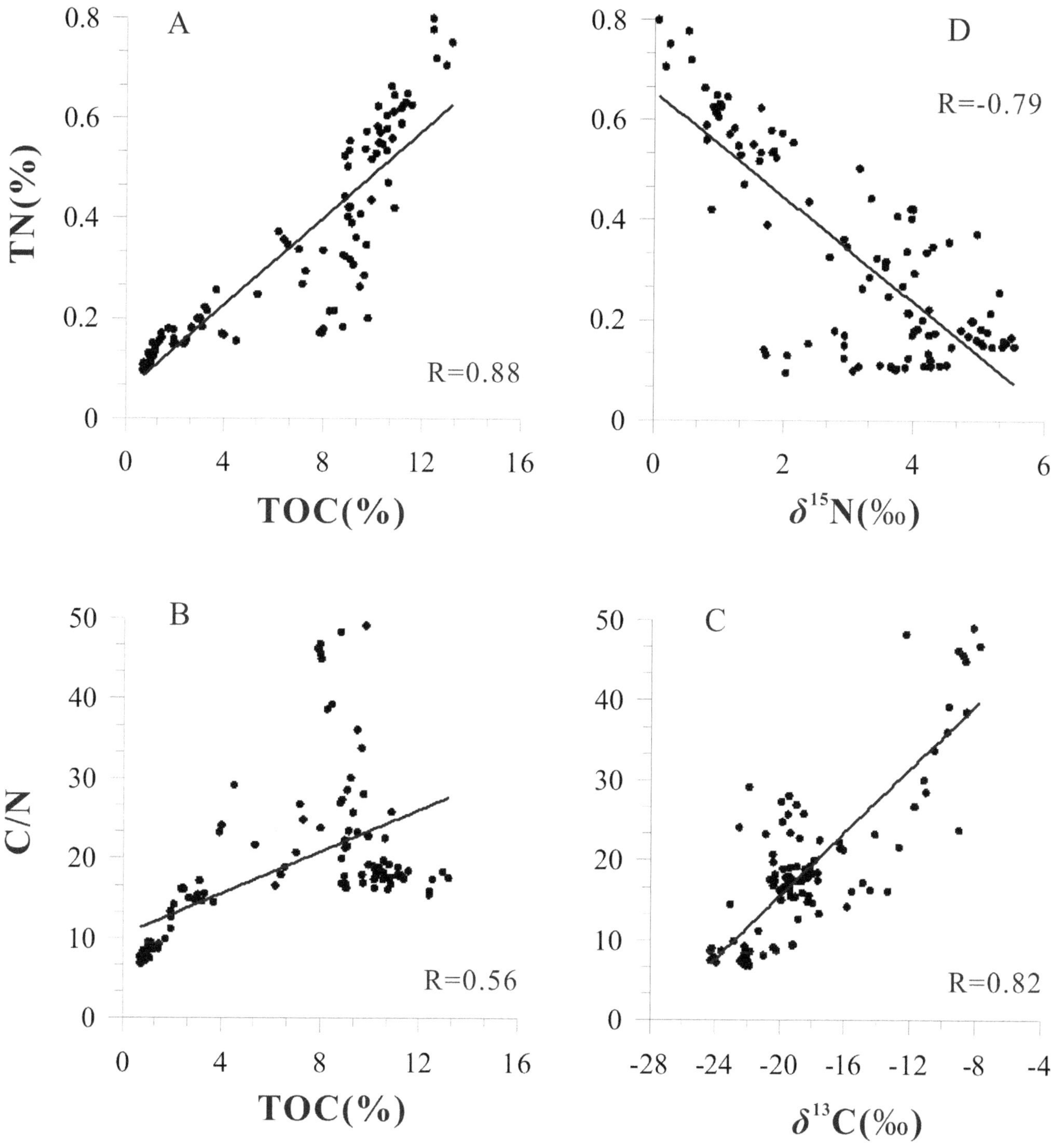

Figure 4. Correlations between TOC contents, TN contents, C/N ratios, $\delta^{13}C$ and $\delta^{15}N$ (n = 108). The correlations were all highly significant (P<0.01).

Stage 2:9.98-5.93 cal. ka BP (351-243 cm). Compared with the early Holocene, this period experienced abundant rainfall and high effective moisture of the lake catchment, as represented by the high values of Md, $CaCO_3$ and χ_{lf}. The high TOC and TN content, stable C/N and $\delta^{13}C$ values suggest that the organic matter was mainly derived from land and aquatic plants, and that vegetation in the lake catchment was flourishing. The low $\delta^{15}N$ value is consistent with a climate characterized by warm and wet conditions strongly influenced by the southwest Asian monsoon. During this period, the warmest stage identified lies in the 8.98-6.10 cal. ka BP interval, which coincides with the global Holocene Optimum [46–50].

Stage 3:5.93-1.35 cal. ka BP (243-167 cm). Unit 1 5.93-4.36 cal. ka BP (243-217 cm) Between 5.93 and 4.36 cal. ka BP, the high TOC, TN content, C/N ratio, the positive $\delta^{13}C$ value and gradually increasing $\delta^{15}N$ value indicate significant climatic fluctuation and a decrease in plant productivity in the lake catchment. This is consistent with the relatively high $CaCO_3$

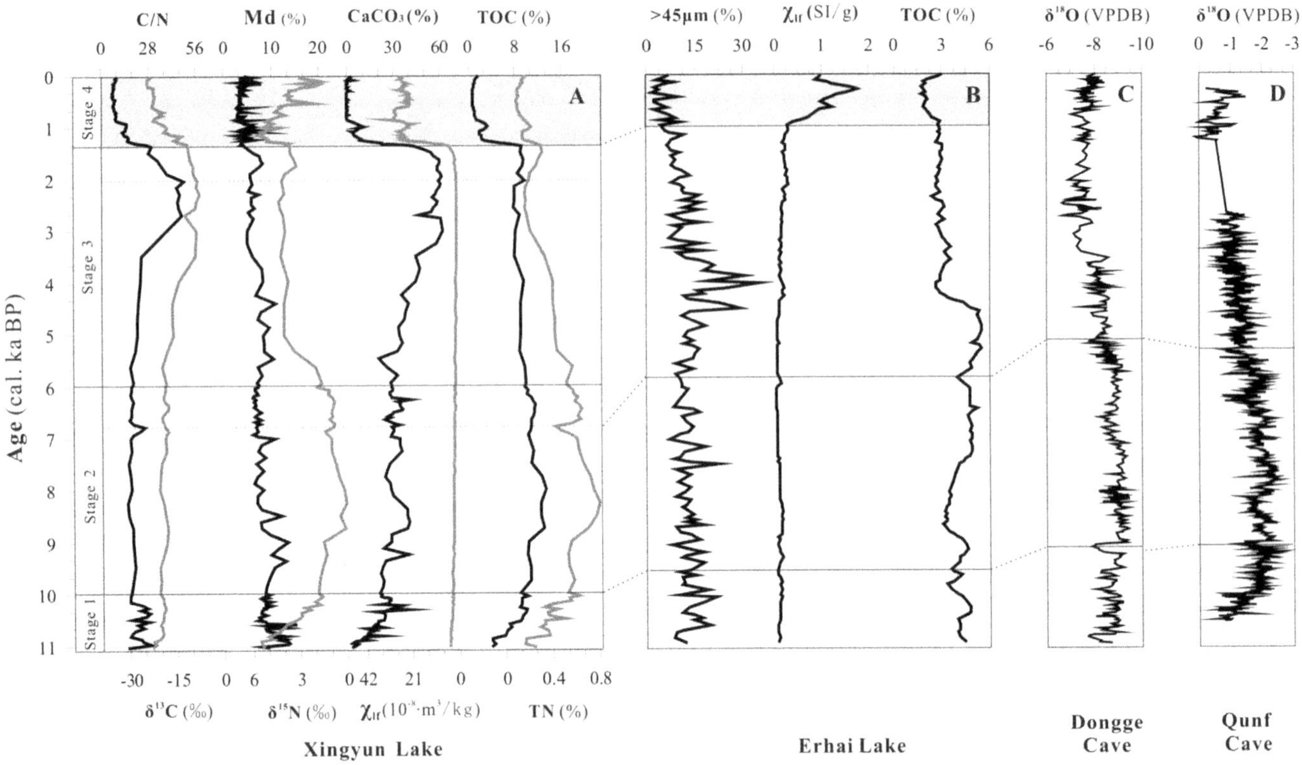

Figure 5. Comparison of the climatic records from Xingyun Lake with the climatic proxies of Erhai Lake and the $\delta^{18}O$ of Dongge Cave and Qunf Cave in the Holocene. The period of intensive human activities and the farming agriculture of Xingyun and Erhai Lake was denoted by the gray area. (A) The climatic proxies of Xingyun Lake indicating strength of Asian summer monsoon and record the strengthened human activities in Yunnan plateau; (B)The climatic proxies (grain size >45 μm, χ_{lf} and TOC content) of Erhai Lake [1]; (C) The stalagmite $\delta^{18}O$ record from Dongge Cave, southwest China [45]; (D) The stalagmite $\delta^{18}O$ record from Qunf Cave, Oman [51].

content, overall fine grain size and lower χ_{lf}. The environmental proxies therefore denote the beginning of a temperature decrease, consistent with a climatic cold event reported from other localities in southwest China in the 6-4 ka BP interval [13]. However, there is no immediate prospect of human activities records of Xingyun Lake sediment in this period.

Unit 2 5.60-1.35 cal. ka BP (217-167 cm) The transition from Unit 1 to 2 is marked by a strong shift in environmental proxies. The highest $CaCO_3$ content and the lowest χ_{lf} show that the temperature was comparatively warm, and the high TOC, C/N ratio, low TN content, positive $\delta^{13}C$ and high $\delta^{15}N$ values all suggest that organic matter entering the lake at this time was mainly derived from C4 terrestrial plants and planktonic productivity was low. The environmental proxies, and relatively slow rate of deposition, are further consistent an ecological change, decreased precipitation and intensification of evaporation within the lake catchment. The climate appeared to warm and dry at this stage, comparable to the warm and dry period of the mid- to late Holocene [51] and aridification of India [52]. Cultural relics unearthed from the Lijiashan ruins in this period depicted a deer falling prey to two leopards [53], indicating that the climate of the Xingyun catchment was warmer than presently. Meanwhile, the massive people immigrated into Yunnan Province, which coincided with the Spring and Autumn and the Warring States Period (~2.0 ka BP) of Chinese history, and it had further exacerbated the ecological environment of the lake catchment.

Stage 4:1.35 cal. ka BP to present (167-0 cm). The consistently low content of $CaCO_3$, TOC and TN reflect the low biomass of the Xingyun catchment during this stage. The fine grain-size, high rate of deposition and high χ_{lf} suggest a heavy loss of soil from the catchment. The C/N ratio, $\delta^{13}C$ and $\delta^{15}N$ values are consistent with high contributions of organic matter from aquatic plant life to the lake sediments, and negligible terrestrial organic matter input. This suggests an abrupt change in the plant community, and also that intensified human activity and land-use in the catchment began from 1.35 cal. ka BP. This is consistent with the period of Dian culture established within the Xingyun catchment [54]. At this time extensive deforestation and development of agriculture in the lake catchment caused heavy loss of soil and fine-grained sediment was easily transported into the lake. In addition, the intensive agriculture would likely have led to organic matter decomposition and release of bound soil particles - as demonstrated by the increase in magnetic minerals present in the lake sediments. During the early Tang Dynasty (600 AD), local population increased in the Xingyun area, which was related to a high rate of immigration into Yunnan. This area was developed into a social and economic center of stockbreeding, irrigation, agriculture and trade [55]. Increased human activity, such as deforestation, reclaiming of land, agriculture and stockbreeding, resulted in the reduced vegetative coverage of the landscape. This inevitably led to increased erosion of surface soil and destruction of the vegetation ecology, as demonstrated by the high χ_{lf}, low TOC and TN content, and increased $\delta^{15}N$ value as a result of human activity.

3 Comparison with Erhai Lake, Qunf and Dongge Cave

To better understand the regional environment change of Xingyun lake catchment, the records of the lake environment are compared to those with high temporal resolution records from different monsoonal regions. Erhai lake is lied on the area of southwest monsoon, which is about 300 km distant from Xingyun lake. Qunf cave (54°18 E, 17°10 N) is located in Oman experienced typical Indian monsoon climate [54], and Dongge cave (108°5 E, 25°17 N) is situated in southwest China, with local climate influenced by both India monsoon and East Asian monsoon [47].

As shown in Figure 5, the changing trend of the environmental proxies of Xingyun Lake was similar with the TOC content of Erhai Lake and the $\delta^{18}O$ value of Qunf and Dongge stalagmite. They exhibit a similar feature of the Holocene summer monsoon variation: intensive summer monsoon in the early Holocene, a decreasing trend during the middle Holocene, and a relatively weak summer monsoon in the late Holocene. Asian monsoon intensity was found to be directly controlled by mid-July solar insolation at 30°N over the entire Holocene. The general consistency of monsoon records of Xingyun Lake with those from other continental records across Asia indicates an in-phase relationship between the Asian summer monsoon on orbital timescales over the Holocene.

The Xingyun Lake records (TOC, and $\delta^{15}N$ isotope, grain-size, $CaCO_3$) show a distinct weakened monsoon event during 8.3-8.1 cal. ka BP, and this 8.2-ka BP weak monsoon event is also clearly exhibited in the $\delta^{18}O$ records in caves of Qunf and Dongge, but to a less degree in the records of Erhai Lake. Meanwhile, the Holocene Optimum of Qunf cave was terminated in 7300 a BP, which was earlier than the lake of Xingyun Lake, Erhai and Dongge cave [1], [45], [51]. This may be due to the fact that insolation-driven Intertropical Convergence Zone migrated southward and India summer monsoon began to decrease, which resulted in less precipitation and gradual influence to north direction after 7.3 cal. ka BP.

During the late Holocene, the records of TOC, TN and $CaCO_3$ in Xingyun Lake, TOC content of Erhai Lake, the $\delta^{18}O$ value of Qunf and Dongge cave indicate the weak summer monsoon. The abrupt changes of χ_{lf} and fine-grained of Xingyun and Erhai Lake sediment suggest the intensive human activity and land-use since 1.35 ka BP, which wasn't recorded in Qunf and Dongge cave. It suggests that this symbiotic human-monsoon relationship may have existed at Yunnan Plateau, southwest China. However, human activity had less influence on cave system compared to that on the lake catchment which was important for agriculture activities, and there stalagmite records can provide more faithful information on climate change.

Conclusions

Based on the application of biochemical (TOC, TN content, $\delta^{13}C$ and $\delta^{15}N$ isotope ratios, C/N ratios) and other environmental proxies (grain-size, χ_{lf}, $CaCO_3$), as well as AMS ^{14}C dating and historical documents, we have identified four climatic stages in the Holocene as recorded in lacustrine sediments on Xingyun Lake. Furthermore, we have identified the impact of human activities during the past ~2000 years on the environmental records. The conclusions of this study can be summarized as follows:

(1) A $\delta^{15}N$ isotope record from a 429-cm sediment core from Xingyun Lake, dating back to 11.06 cal. ka BP, provides the first complete Holocene nitrogen isotope record for this area. The record reveals a long history of climate changes and human influence in the lake catchment in the latter part of this history. The Holocene climatic evolution is characterized of a shift from being dry-cold to humid-warm and finally to dry-warm. There is also evidence that the lake expansion resulted from an intensification of the Asian southwest monsoon during the early Holocene.

(2) Between 9.98 and 5.93 cal. ka BP, the gradually expanded lake may have reached the optimum water depth, causing a marked decline in the coverage of aquatic plants and low planktonic productivity in the lake, again strongly influenced by the Asian southwest monsoon. This coincided with the global Holocene Optimum. Between 5.60 and 1.35 cal. ka BP the temperature was comparatively warm and precipitation decreased, thus comparable to the aridification of India in the mid- and late Holocene. Meanwhile, the massive people immigrated into Yunnan Province, which coincided with the Spring and Autumn and the Warring States Period (~2.0 ka BP) of Chinese history, and it had further exacerbated the environment of the lake catchment.

(3) Human activity and land-use intensified since the early Tang Dynasty (~1.35 cal. ka BP) are corresponded well with the development of the local Dian culture. The extensive deforestation within the catchment could lead to heavy soil loss as revealed in lake sediments.

Acknowledgments

We thank the editor and reviewers for their constructive comments and suggestions for improving our paper. We also thank Lunqing Yang, Kunwu Yang, Jie Chen, Qinglei Li and Xiao Bai for the field work and their assistance during the laboratory work. Thanks are extended to Dr. David Smith for his assistance in checking the English of this paper.

Author Contributions

Conceived and designed the experiments: WXZ HCZ QZM ZTS. Performed the experiments: WXZ JN GLL FQC. Analyzed the data: WXZ GJC. Contributed reagents/materials/analysis tools: JN GLL. Wrote the paper: WXZ.

References

1. Shen J, Yang LY, Yang XD, Matsumoto R, Tong GB, et al. (2005) Lake sediment records on climate change and human activities since the Holocene in Erhai catchment, Yunnan Province, China. Science in China (Series D) 48: 353–363.
2. Bellwood (2005) First Farmers: The Origin of Agricultural Societies. London: Blackwell Publishing.
3. Ruddiman WF, Guo Z, Zhou X, Wu H. (2008) Early rice farming and anomalous methane trends. Quat Sci Rev 27: 1291–1295.
4. Li XQ, Sun N, Dodson J, Zhou X (2012) Human activity and its impact on the landscape at the Xishanping site in the western Loess Plateau during 4800–4300 cal yr BP based on the fossil charcoal record. J Archaeol Sci 39: 3141–3147.
5. Lu HY, Zhang J, Liu KB, Wu N, Li Y, et al. (2009) Earliest domestication of common millet (Panicum milliaceum) in East Asia extended to 10,000 years ago. Proc Natl Acad Sci USA 106: 7367–7372.
6. Crawford GW (2006) East Asian plant domestication. In: Stark M T, ed. Archaeology of Asia. Malden: Blackwell Publishing.
7. Zhao KL, Li XQ, Zhou XY, Dodson J, Ji M (2013) Impact of agriculture on an oasis landscape during the late Holocene: Palynological evidence from the Xintala site in Xinjiang, NW China. Quaternary International 311: 81–86.
8. Flad R, Li SC, Wu XH, Zhao ZJ (2010) Early wheat in China: Results from new studies at Donghuishan in the Hexi Corridor. The Holocene 20: 955–965.
9. Dodson J, Li XQ, Zhou XY, Zhao KL, Sun N, et al. (2013) Origin and spread of wheat in China. Quat Sci Rev 72: 108–111.
10. Li X Q, Dodson J, Zhou XY (2007) Early cultivated wheat and broadening of agriculture in Neolithic China. The Holocene 17: 555–560.
11. Thornton CP, Schurr TG (2004) Genes, language, and culture: An example from the Tarim Basin. Oxford Journal of Archaeology 23: 83–106.
12. Yan W M (1997) The new progress on the rice origin in China. Archaeology 9: 71–76 (in Chinese).

13. Hodell DA, Brenner M, Kanfoush SL, Curtis JH, Stoner JS, et al. (1999) Paleoclimate of southwestern China for the past 50,000 yr. inferred from lake sediment records. Quaternary Research 52: 369–380.
14. Zhang HC, Peng JL, Ma YZ, Chen GJ, Feng ZD, et al. (2004) Late Quaternary palaeolake levels in Tengger Desert, NW China. Palaeogeoraphy, Palaeoclimatology, Palaeoecology 211: 45–58.
15. Yang DP, Brugam R (1997) Human disturbance and trophic status changes in crystal lake, McHenry country, Illinois, USA. Journal of Paleoliminology 17: 369–376.
16. Zhang HL, Li SJ, Feng QL, Zhang ST (2010) Environmental change and human activities during the 20th century reconstructed from the sediment of Xingyun Lake, Yunnan Province, China. Quaternary International 212: 14–20.
17. Yu G, Xue B, Liu J (2001) Lake Records from China and the Palaeoclimate Dynamics. Beijing: China Meteorological Press (in Chinese).
18. Shi ZT, Ming QZ and Zhang HC (2005) A study review on the modern processes and environmental evolution of the typical lakes in Yunnan. Yunnan Geographic Environment Research 17: 24–26. (in Chinese).
19. Wu WX, Liu TS (2004) Possible role of the "Holocene Event 3" on the collapse of Neolithic Cultures around the Central Plain of China. Quaternary International 117: 153–166.
20. Bascomb CL (1961) A Calcimeter for routine use on soil samples. Chemistry and Industry 45: 1826–1827.
21. Konert M, Vandenberghe JEF (1997) Comparison of laser grain size analysis with pipette and sieve analysis: A solution for the un-derestimation of the clay fraction. Sedimentology 44: 523–535.
22. Li X, Liu W, Xu L (2012) Carbon isotopes in surface-sediment carbonates of modern Lake Qinghai (Qinghai–Tibet Plateau): Implications for lake evolution in arid areas. Chemical Geology 300, 88–96.
23. Nakamura T, Niu E, Oda H, Ikeda A, Minami M, et al. (2000) The HVEE tandetron AMS system at Nagoya University. Nuclear Instruments and Methods in Physics Research Section B: Beam Interactions with Materials and Atoms 172: 52–57.
24. Reimer PJ, Bard E, Bayliss A, Beck JW, Blackwell PG, et al. (2013) IntCal13 and Marine13 radiocarbon age calibration curves 0–50,000 years cal BP. Radiocarbon 55(4): 1869–1887.
25. Zhang ZK, Yang XD, Shen J, Li SF, Zhu YX, et al. (2001) Climatic variations recorded by the sediments from Erhai Lake, Yunnan Province, southwest China during the past 8000 a. Chinese Science Bulletin 46: 80–82.
26. Chang FQ, Zhang HC, Chen Y, Yang MS, Niu J, et al. (2008) Changes during the Late Pleistocene of Paleolake Qarhan in the Qaidam Basin. Journal of China University of Geosciences 19: 1–8.
27. Song L, Qiang M, Lang L, Liu X, Wang Q, et al. (2012). Changes in palaeoproductivity of Genggahai Lake over the past 16 ka in the Gonghe Basin, northeastern Qinghai-Tibetan Plateau. Chinese Science Bulletin 57(20), 2595–2605.
28. Lücke A, Schleser GH, Zolitschka B, Negendank JF (2003) A Lateglacial and Holocene organic carbon isotope record of lacustrine palaeoproductivity and climatic change derived from varved lake sediments of Lake Holzmaar, Germany. Quaternary Science Reviews, 22(5): 569–580.
29. Meyers PA (2003) Applications of organic geochemistry to paleolimnological reconstructions: A summary of examples from the Laurentian Great Lakes. Organic Geochem 34: 261–289.
30. Routh J, Meyers PA, Hjorth T, Baskaran M, Hallberg R (2007) Sedimentary geochemical record of recent environmental changes around Lake Middle Marviken, Sweden. J Paleolimn 37: 529–545.
31. Krishnamurthy RV, Bhattacharya SK, Kusumgar S (1986) Palaeoclimatic changes deduced from 13C/12C and C/N ratios of Karewa lake sediments, India. Nature 323: 150–152.
32. Meyers PA, Lallier VE (1999) Lacustrine sedimentary organic matter records of Late Quaternary paleoclimates. Journal of Paleolimnology 21: 345–372.
33. Lamb AL, Leng MJ, Mohammed MU, Lamb HF (2004) Holocene climate and vegetation change in the Main Ethiopian Rift Valley, inferred from the composition (C/N and $\delta^{13}C$) of lacustrine organic matter. Quaternary Science Reviews 23(7): 881–891.
34. Brown R (1991) Isotopes and Climates. London: Elsevier.
35. Smith BN, Epstein S (1971) Two categories of $^{13}C/^{12}C$ ratio for higher plants. Plant Physiol 1971, 47: 380–384.
36. Xu H, Sheng E, Lan J, Liu B, Yu K et al. (2014) Decadal/multi-decadal temperature discrepancies along the eastern margin of the Tibetan Plateau. Quaternary Science Reviews 89: 85–93.
37. Hodell DA, Schelske CL (1998) Production, sedimentation, and isotopic composition of organic matter in Lake Ontario. Limnology and Oceanography 1998, 43(2): 200–214.
38. Watanabe T, Naraoka H, Nishimura M, Kawai T. Biological and environmental changes in Lake Baikal during the late Quaternary inferred from carbon, nitrogen and sulfur isotopes. Earth and Planetary Science Letters 2004, 222: 285–299.
39. Wu J, Shen J (2010) Paleoenviromental and paleoclimatic changes in lake Xingkai inferred from stable carbon and nitrogen isotopes of bulk organic matter since 28 kaBP. Acta Sedimentologica Sinica 28(2): 365–372. (in Chinese).
40. Chen JA, Wan GJ, Zhang F (2004) Environmental records of lacustrine sediments in different time scales: Sediment grain size as an example. Science in China (Series D) 47: 954–960.
41. Vannière B, Bossuet G, Walter-Simonnet AV, Gauthier E, Barral P, et al. (2003) Land use change, soil erosion and alluvial dynamic in the lower Doubs Valley over the 1st millenium AD (Neublans, Jura, France). Journal of Archaeological Science 30: 1283–1299.
42. Hu SY, Deng CL, Appel E, Verosub KL (2002) Environmental magnetic studies of lacustrine sediments. Chinese Science Bulletin 47: 613–616.
43. Prins MA, Postma G, Weltje G (2000) Controls on the terrigenous sediment supply o the Arabian Sea during the late Quaternary: The Makran continental slope. Marine Geology 169: 351–371.
44. Chen JA, Wan GJ, Wang FS (2002) Research of the Carbon Environment Records in the Lake Modern Sediments. Science in China (Series D) 32: 73–80.
45. Wang YJ, Cheng H, Edwards RL, He YQ, Kong XG, et al. (2005) The Holocene Asian monsoon: links to solar changes and North Atlantic climate. Science 308: 854–857.
46. Sirocko F, Sarnthein M, Erlenkeuser H, Lange H, Arnold M, et al. (1993) Century-scale events in monsoonal climate over the past 24,000 year. Nature 323: 48–50.
47. Sirocko F, Garbe-Schönberg D, Mcintyre A, Molfino B (1996) Teleconnections between the subtropical monsoons and high- latitude climates during the last deglaciation. Science 272: 526–529.
48. Thompson LG, Yao TD, Davis ME, K. Henderson A, Thompson EM et al. (1997) Tropical Climate Instability: The Last Glacial Cycle from a Qinghai-Tibetan Ice Core. Science 276: 1821–1825.
49. Wang YJ, Cheng H, Edwards RL, An ZS, Wu JY, et al. (2001) A high-resolution absolute-dated late Pleistocene monsoon record from Hulu Cave, China. Science 294: 2345–2348.
50. IPCC (Intergovernmental Panel on Climate Change) 2007. Climate Change 2007: The Physical Sciences Basis. New York: Cambridge University Press.
51. Fleitmann D, Burns SJ, Mudelsee M, Neff U, Kramers J, et al. (2003) Holocene forcing of the Indian monsoon recorded in a stalagmite from southern Oman. Science 300: 1737–1739.
52. Ponton C, L. Giosan TI, Eglinton DQ, Fuller JE, Johnson P, et al. (2012) Holocene aridification of India. Geophysical Research Letters 39(3), L03704.
53. Li XC, Zhang XN, Han RF, Sun SY (2008) Analysis and study of the Metalware from the Lijiashan Cemetery in Jiangchuan, Yunnan. Archaeology 8: 76–90. (in Chinese).
54. Higham C (2011) The origins of the bronze age of southwest Asia. Journal of world prehistory 24: 227–274.
55. Yu XX, Yu XQ (1987) Study on the ancient Yunnan agricultural development and the evolution of agricultural area. A collection of Chinese history 2: 121–126. (in Chinese).

7

An Abrupt Centennial-Scale Drought Event and Mid-Holocene Climate Change Patterns in Monsoon Marginal Zones of East Asia

Yu Li*, Nai'ang Wang, Chengqi Zhang

College of Earth and Environmental Sciences, Center for Hydrologic Cycle and Water Resources in Arid Region, Lanzhou University, Lanzhou, China

Abstract

The mid-latitudes of East Asia are characterized by the interaction between the Asian summer monsoon and the westerly winds. Understanding long-term climate change in the marginal regions of the Asian monsoon is critical for understanding the millennial-scale interactions between the Asian monsoon and the westerly winds. Abrupt climate events are always associated with changes in large-scale circulation patterns; therefore, investigations into abrupt climate changes provide clues for responses of circulation patterns to extreme climate events. In this paper, we examined the time scale and mid-Holocene climatic background of an abrupt dry mid-Holocene event in the Shiyang River drainage basin in the northwest margin of the Asian monsoon. Mid-Holocene lacustrine records were collected from the middle reaches and the terminal lake of the basin. Using radiocarbon and OSL ages, a centennial-scale drought event, which is characterized by a sand layer in lacustrine sediments both from the middle and lower reaches of the basin, was absolutely dated between 8.0–7.0 cal kyr BP. Grain size data suggest an abrupt decline in lake level and a dry environment in the middle reaches of the basin during the dry interval. Previous studies have shown mid-Holocene drought events in other places of monsoon marginal zones; however, their chronologies are not strong enough to study the mechanism. According to the absolutely dated records, we proposed a new hypothesis that the mid-Holocene dry interval can be related to the weakening Asian summer monsoon and the relatively arid environment in arid Central Asia. Furthermore, abrupt dry climatic events are directly linked to the basin-wide effective moisture change in semi-arid and arid regions. Effective moisture is affected by basin-wide precipitation, evapotranspiration, lake surface evaporation and other geographical settings. As a result, the time scales of the dry interval could vary according to locations due to different geographical features.

Editor: Navnith K.P. Kumaran, Agharkar Research Institute, India

Funding: This research was supported by the National Natural Science Foundation of China (Grant No. 41371009 and 41001116) and the Fundamental Research Fund for the Central Universities (Grant No. lzujbky-2013-127 and lzujbky-2013-129). The funders had no role in study design, data collection and analysis, decision to publish, or preparation of the manuscript.

Competing Interests: The authors have declared that no competing interests exist.

* E-mail: liyu@lzu.edu.cn

Introduction

The Earth's climate zones are classified according to their average temperature and rainfall accumulation, and, in general, form latitudinal, east-west oriented bands on the Earth's surface [1,2]. The mid-latitude climate is affected by two different air-masses, which are cold air masses from the poles and warm air masses from the tropics. In East and Central Asia, the mid-latitude climate is characterized by the interaction between the Asian summer monsoon from the Tropical Ocean and the westerly winds from the mid-to-high latitudes [3,4]. Furthermore, the interaction between the monsoon and the westerly winds is influenced by the Qinghai-Tibet Plateau, which is a natural obstacle for blocking the interaction between the air masses from different latitudes [5,6]. Various climatic zones are distributed in the mid-latitude regions of East and Central Asia [5].

During the Holocene epoch, there are many uncertainties relating to the Holocene climate change in the mid-latitude regions of East and Central Asia, and the climate change processes may differ in different parts of the region [7–11]. Li (1990) suggested the climate in northwest China could be divided into monsoon and westerly wind patterns in terms of the millennium-scale climate change [12]. Herzschuh (2006) synthesized 75 records for the Asian continent and found that the Holocene climate patterns differed between Central Asia and some parts of the Asian monsoon domain [13]. Chen et al. (2008) analyzed the absolutely dated Holocene records for the Asian monsoon domain and arid Central Asia and found that Holocene millennial-scale climate patterns are different for the two regions [14]. Chen et al. (2008) also concluded the millennium-scale difference is mainly due to different evolution histories of the Asian monsoon and westerly winds [14]. Since the Asian summer monsoon reached their highest level during the early Holocene (ca. 9000 cal yr BP), while the westerly winds were relatively dry during the early Holocene on the millennial-scale [13,14]. According to previous studies, the millennial-scale interaction between the Asian summer monsoon and the westerly winds could be a reason for asynchronous Holocene climate changes in East and Central Asia.

Monsoon marginal zones in East Asia are affected both by the Asian monsoon and the westerly winds [5,6]. Abrupt climatic changes during the mid-Holocene records have been reported in different areas of monsoon marginal zones in East and Central

Asia: Mongolia [15,16], the Alashan Plateau [17], central Inner Mongolia [18], eastern Inner Mongolia [19], the Loess Plateau [20]. The mid-Holocene dry interval was also reported in southern China [21]. The time scales and mechanisms of abrupt climate change events in monsoon marginal zones are crucial for understanding the millennial-scale interaction between the Asian monsoon and the westerly winds. Two kinds of hypotheses have been proposed regarding the mid-Holocene abrupt climate events: one is relating the abrupt dry mid-Holocene interval to the increased evaporation [18,22]; the other links the interval to the weakened Asian summer monsoon [17,19]. However, restricted by the ages of the records, the mechanisms about the abrupt mid-Holocene intervals are still in dispute. Absolutely dated mid-Holocene climate records are needed for investigations into abrupt mid-Holocene climatic changes. The Shiyang River drainage basin is located in the northwest margin of the Asian monsoon (Figure 1, Figure 2). Previous studies have reported long-term Holocene environmental changes by the palynology and geochemical proxies [17,23–27]. In relatively weak chronological frames, it is difficult to study the mechanisms of the abrupt mid-Holocene climate events [25–27]. Chen et al. (2003, 2006) have linked the mid-Holocene abrupt changes to the weakening Asian summer monsoon; however, this hypothesis cannot be supported by the gradual change of the mid-Holocene Asian monsoon [17,23,28,29]. The time scales and mechanisms of the abrupt climatic events during the mid-Holocene are still unclear in this area, which is a key area for studying the millennial-scale interaction between the Asian summer monsoon and the westerly winds. In this study, we used chronological and grain size data from QTH01 and QTH02 sections in the terminal lake. Radiocarbon and OSL dates of the sand layers were compared to study the chronologies, while grain size data provide more details about the environments during the abrupt climate change. In addition, mid-Holocene lacustrine sediments (JDT section) from the middle reaches of the basin were also studied for a basin-wide comparison. According to chronologies of the mid-Holocene sediments from different parts of the basin, we proposed a new hypothesis regarding the mid-Holocene climate change patterns in monsoon marginal zones.

Geographical Setting

The Shiyang River basin is located at the east part of the Hexi Corridor of Gansu Province in northwest China. It is on the northern slope of Qilian Mountain at longitude of 101°41′–104°16′E and latitude of 36°29′–39°27′N (Figure 2). The river originates from the southern part of the mountain and ends at the terminal lake, Zhuye Lake. The whole river basin has an area of 41,600 km^2. The river basin spans over three climatic zones from south to north. The south cold semi-arid to semi-humid zone at the highland of Qilian Mountain (altitude 2000–5000 m) has an annual precipitation of 300–600 mm and pan evaporation of 700–1200 mm. The vegetation zones can be divided into the alpine cushion zone, alpine meadow zone, alpine shrub zone, forest zone, and piedmont meadow steppe zone. The middle cool arid zone at the flatland of Hexi Corridor (altitude 1500–2000 m) has an annual precipitation of 150–300 mm and pan evaporation of 1200–2000 mm. The vegetation zone is a desert steppe zone. The north temperate arid zone (altitude 1300–1500 m) has an annual precipitation of less than 150 mm and pan evaporation of more than 2000 mm. Desert vegetation is widely distributed in this area [30,31]. Around the terminal lake and near the edge of the Tengger desert, the annual rainfall is about 50 mm and pan evaporation is 2000–2600 mm. Based on the geographical divisions in China, the Shiyang River drainage basin is in the transition zone of monsoonal and arid regions, and the modern climate of the region is affected by the Asian monsoon and the westerly winds [5]. According to modern climate research, the westerly wind comprises a prevailing westerly wind and a westerly jet, both of which make a large contribution to water vapor transportation in the basin. In the alpine areas of the Shiyang River basin, the summer precipitation is also contributed by the Asian summer monsoon [4]. The terminal lake, Zhuye Lake, has been dried up since the 1950s, a result of the diversion of water from the Shiyang River for irrigation and other purposes. Based

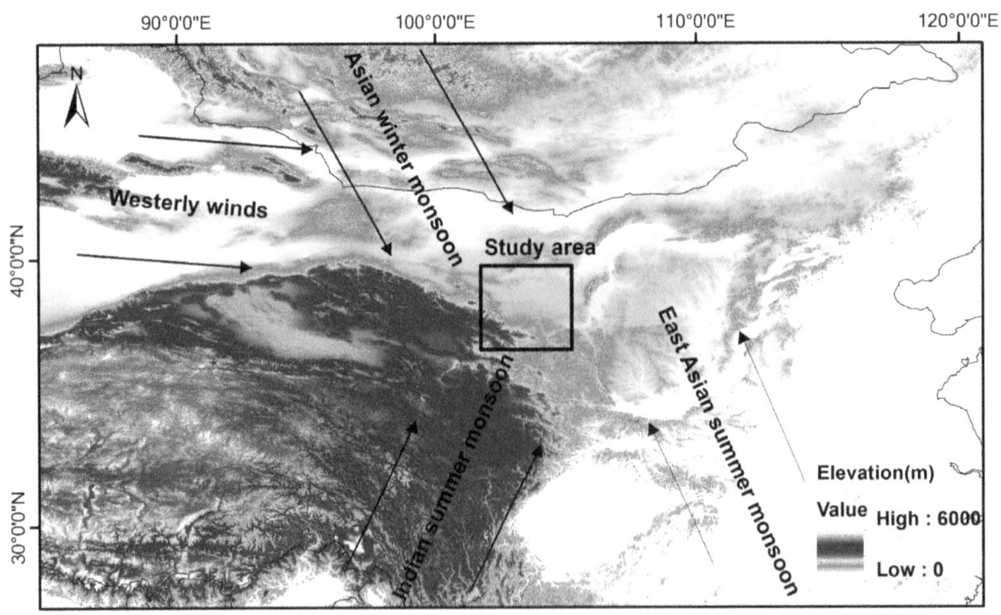

Figure 1. Map showing the study area and the arrows indicate the climate systems affecting China, including the East Asian summer monsoon, the Indian summer monsoon, the Asian winter monsoon, and the westerly winds.

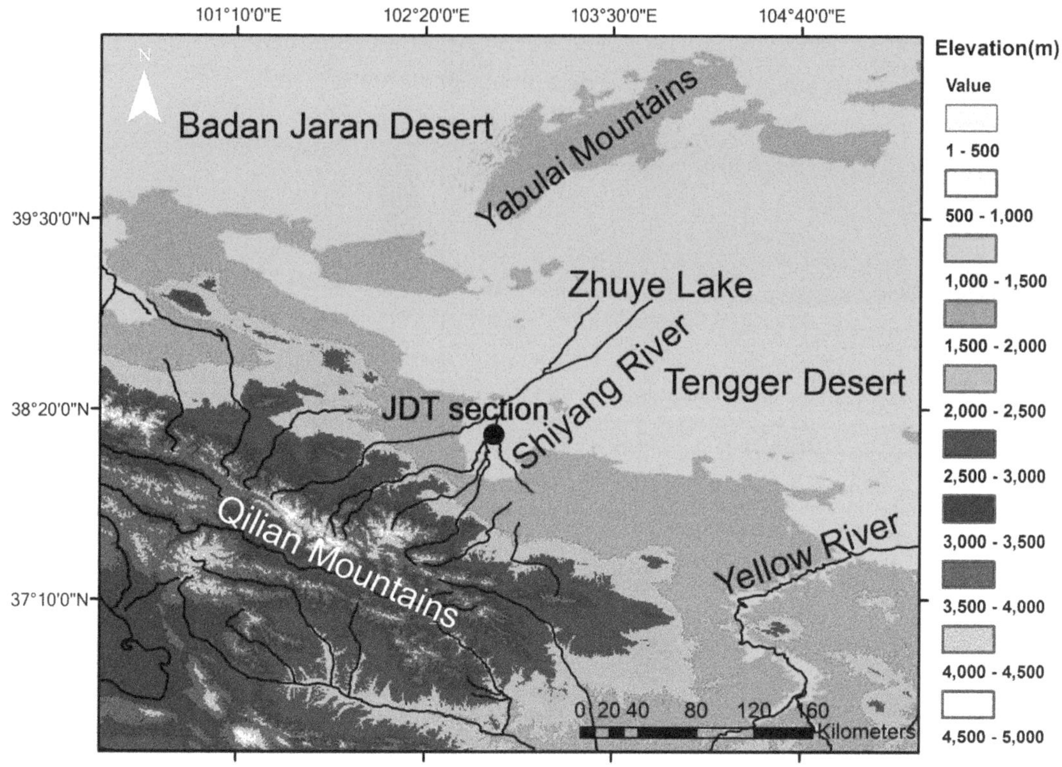

Figure 2. Map showing the topography around Zhuye Lake and the Shiyang River drainage basin. The black circles indicate the JDT section.

on the geomorphologic studies, the lake level is relatively high during the historical period, the Marine Isotopic Stage 3 (MIS3) and the early Holocene periods [32–34]. At present, the dry lakes that exist in several depressions on the Tengger desert margins only fill with water in years with sufficient precipitation.

Materials and Methods

In the terminal lake, the QTH01 and QTH02 sections were excavated in the central part of Zhuye Lake. No specific permissions were required for these excavation works. The field studies did not involve endangered or protected species and the specific location can be seen below. The two sections are at an elevation of 1309 m above sea level. The geographic coordinates are 39°03′00′′N, 103°40′08′′E (Figure 3). Stratigraphically (Figure 4), QTH01 at 0–165 cm are aeolian sediments, composed of yellow and brownish clay, sandy clay and sand deposits, influenced by agricultural activities on the top of the section; 165–230 cm are alluvial sediments, composed of light red silty clay with many brown spots; 230–315 cm are lacustrine deposits formed in shallow lake composed of grey silt and sandy silt, and striped peats with plant fragments and mollusk shells are embedded; 315–450 cm are typical lacustrine deposits with grey silt, sandy silt and carbonate; 450–495 cm is a grey sand layer with some cracked mollusc shells; 495–603 cm is a lacustrine deposit with grey, carbonate enriched silt; 603–692 cm is grey or yellow sand and well sorted. QTH02 shows a similar stratigraphy (Figure 4). QTH01 was sampled at 2 cm intervals at the lake sediment layers and at 5 cm intervals otherwise, resulting in 292 samples for analyses of grain size. QTH02 was sampled at 2 cm intervals, yielding 368 samples for analysis of grain size. The JDT section (38°10′46″N, 102°45′53″E) is located at an elevation of 1460 m asl (Figure 2). The JDT section is 580 cm thick, and a total of 58 samples are obtained and for analyzing grain size. Stratigraphically (Figure 3), JDT at 0–120 cm are aeolian deposits, composed of yellow or brown silty clay; 120–230 cm are a kind of lacustrine deposits, composed of grey carbonate enriched silty clay or silt with plant fragments; 230–300 cm is a light yellow sand layer with many brown spots; 300–330 cm is silt peat with plant fragments; 330–400 cm is grey or brown silty clay or silt with roots and carbonate; 400–580 cm is yellow or grey gravel sand with roots.

In the QTH01 and QTH02 sections, fourteen radiocarbon dates are obtained. Six of them are dated in Peking University Dating Laboratory, others are dated in Lanzhou University Dating Laboratory. Of the 14 radiocarbon dates (Figure 4, Table 1), 11 are from QTH01, the remainder from QTH02. Dates were calibrated using CALIB 6.11 and are reported in calibrated years. For correcting the hard water effects, six optically stimulated luminescence (OSL) samples are dated to provide chronological control at QTH01 section (Figure 4, Table 2) [27]. Quartz (38–63 μm) OSL measurements were performed in the Luminescence Dating Laboratory of the Qinghai Institute of Salt Lakes, Chinese Academic Sciences, using an automated Risø TL/OSL-20 reader. In arid/semiarid regions of China, previous studies showed that radiocarbon dating of lake sediments is likely affected by the hard water effects [35]. For estimating the reliability of ^{14}C dating, we compare the calibrated radiocarbon ages with OSL ages adjusted on AD 1950, at QTH01 section (Figure 4, Table 1, Table 2). The OSL and ^{14}C ages are in good agreement, suggesting that the established chronology is robust. Beside the comparison between the OSL dates and the radiocarbon dates, as shown in Table 1, the age difference of the radiocarbon date for organic matter and the date for shells at the same depth (3.15 m) is only ~30 years. Based on these results,

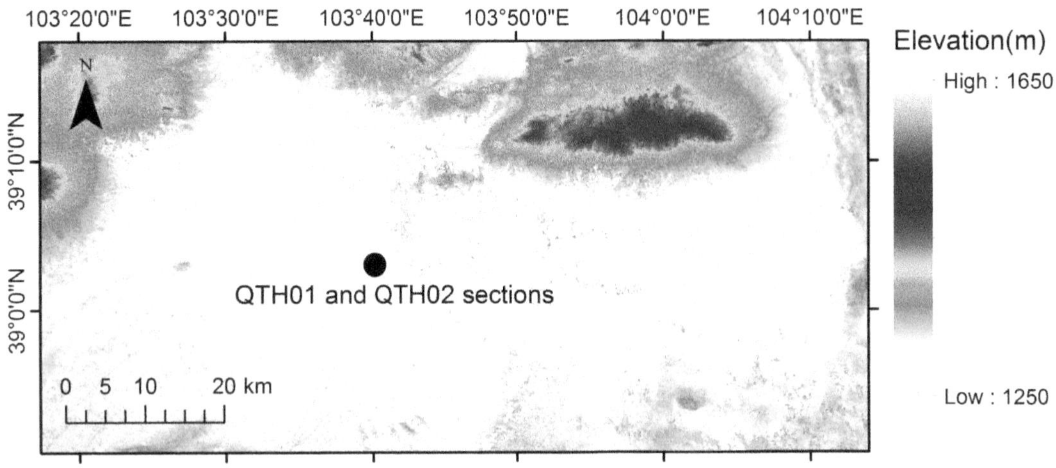

Figure 3. Map showing latitudes, longitudes and elevations of the Zhuye Lake basin. The black circle indicates the location of the QTH01 and QTH02 sections. (The elevation data are based on the ASTER-GDEM dataset, with 30 m resolution).

we concluded the hard water effects were slight in the two sections. Most stratigraphic units are correlated between the two sections. After analyzing lithology and grain size in the two sections, the radiocarbon and OSL dates can be compared from the two sections (Figure 4). The chronology of the JDT section is based on six radiocarbon dates of organic matter and

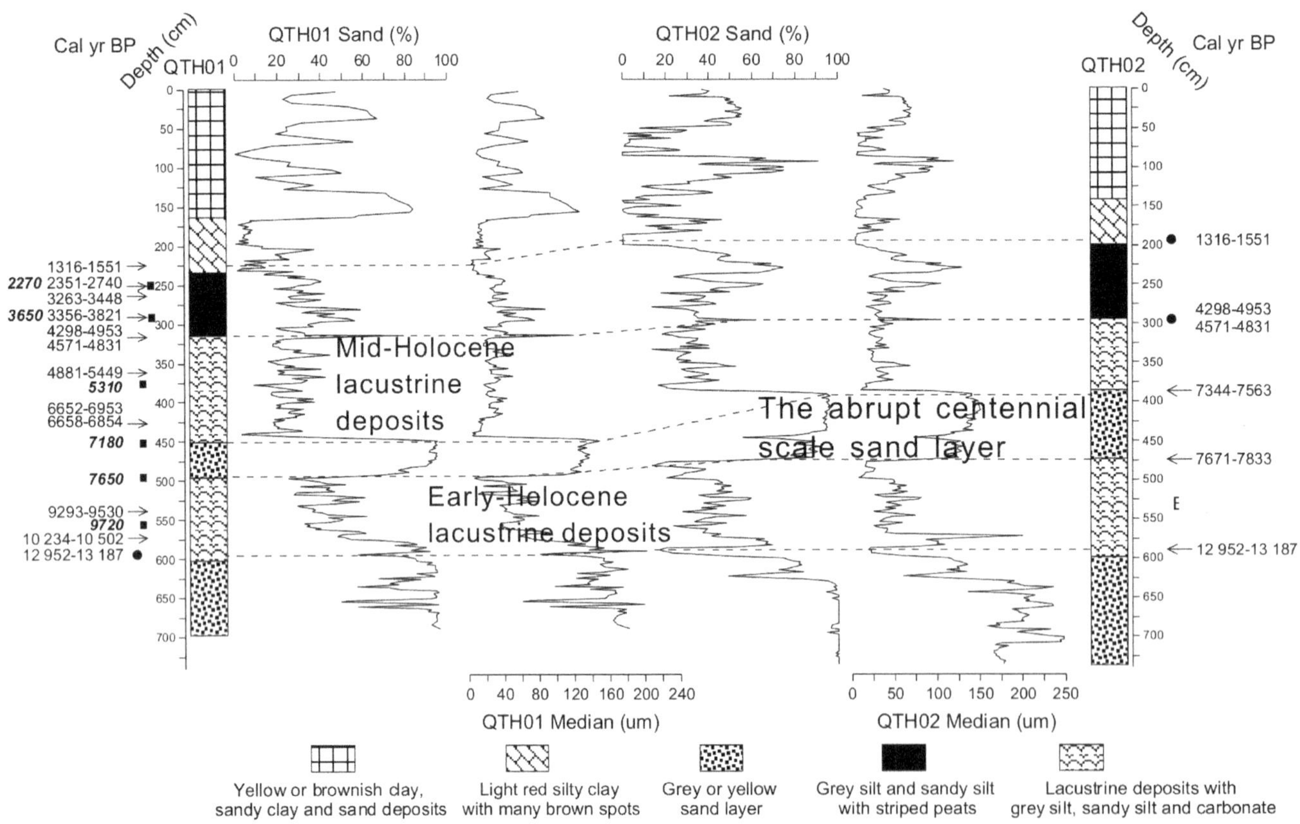

Figure 4. Lithology and ages in the QTH01 and QTH02 sections. Subsurface correlation of the QTH01 and QTH02 sections, based on grain size data (% sand and Median (μm)) plotted against depth. Of the 14 radiocarbon dates (Table 1), 11 are from QTH01, the remainder from QTH02. All the OSL dates are from QTH01 (Table 2). The arrowheads show the radiocarbon dates, and the black rectangles show the OSL dates. The OSL dates are highlighted using italics. The broken lines connect the extrapolated ages between two sections and the extrapolated ages are marked with black circles. During the mid-Holocene three lithologic phases are showed beside the lithologic columns, including the mid-Holocene lacustrine deposits, the abrupt centennial-scale dry interval and the early-Holocene lacustrine deposits.

terrestrial branches analyzed in Lanzhou University Dating Laboratory (Table 1, Figure 5). In the JDT section, at the depth of 4.20 m, the dating material is terrestrial branches, which are less influenced by the hard water effects. Based on a comparison between ages from terrestrial branches and organic matter, the hard water effect is also slight in the middle reaches of the Shiyang River drainage basin. Grain size distribution was determined with a Malvern Mastersizer 2000 particle analyzer that automatically yields the percentages of clay-, silt- and sand-size fractions, as well as median, mean and mode sample diameters. 0.2–0.4 g of sediment was pretreated by heating in 10 ml of 10% H_2O_2 to remove organics, heated in 10 ml 10% HCl to remove carbonate that otherwise would bond different mineral fractions, then shaken in Na-hexametaphosphate to disaggregate the sediment for 1 h prior to analysis. The median grain size and the grain size frequency curves are all based on the sample diameters. Sand percentages are based on sediment fractions that are higher than 63 μm.

Chronological and Grain Size Data for the Abrupt Dry Mid-Holocene Interval

1 Ages of the mid-Holocene sand layers in the QTH01, QTH02 and JDT sections

In the QTH01 section, the mid-Holocene sand layer is dated by OSL dating method. The upper boundary is around 7180 years BP, and the lower boundary is around 7650 years BP (Table 2). In the QTH02 section, the mid-Holocene sand layer is dated by radiocarbon dating method. The upper boundary is around 7461 cal yr BP, and the lower boundary is around 7739 cal yr BP (Figure 6, Table 2). In the JDT section, linear interpolated by the mid-Holocene radiocarbon dates, the upper boundary of the sand layer is 7130 cal yr BP, and the lower boundary is around 7529 cal yr BP (Figure 6, Table 1). Totally, based on the radiocarbon and OSL dates of the three sections in the Shiyang River drainage basin, the abrupt dry mid-Holocene interval

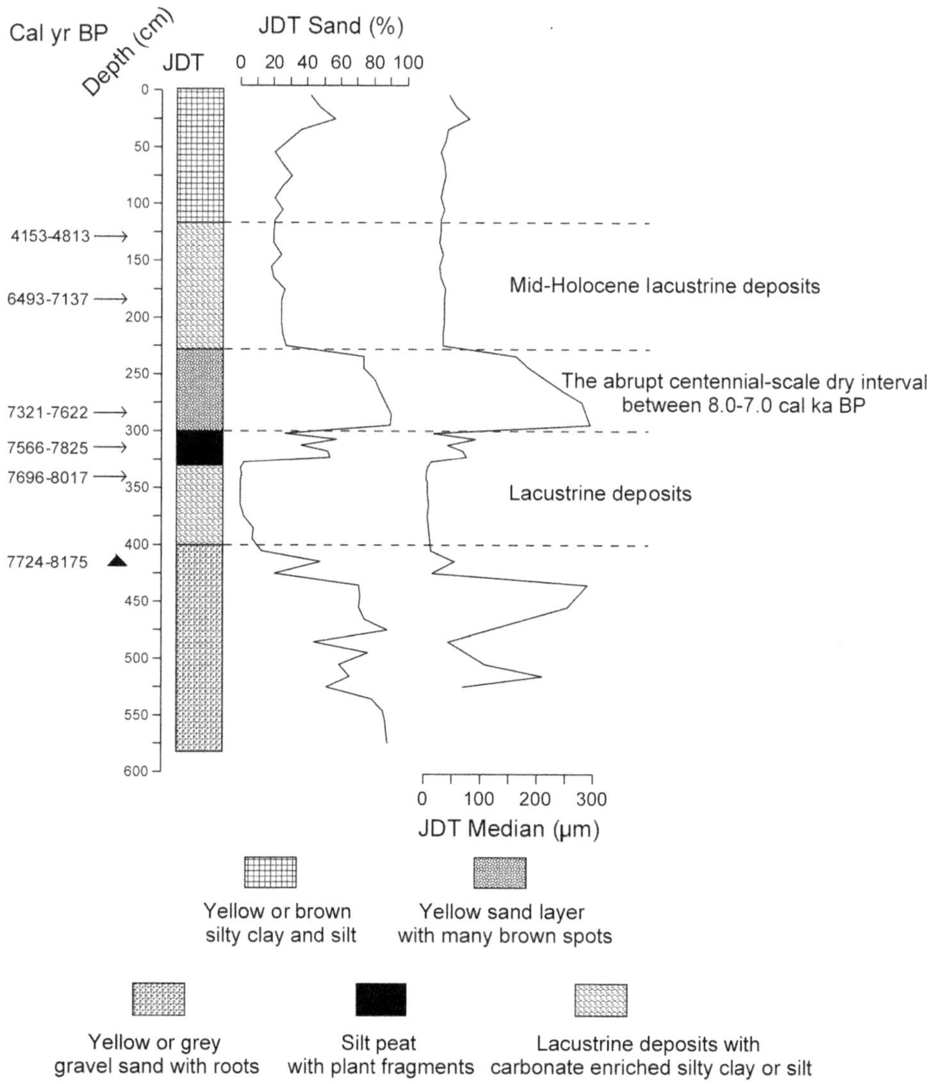

Figure 5. Lithology and ages in the JDT section. Of the 6 radiocarbon dates (Table 1), 5 are dated by bulk samples, and the remainder is dated by terrestrial branches. The arrowheads show the bulk samples' dates, and the black triangle shows the date of the terrestrial branches. During the mid-Holocene, three lithologic phases are showed beside the lithologic column, including the mid-Holocene lacustrine deposits, the abrupt centennial-scale dry interval and the lacustrine deposits below the dry interval.

Table 1. AMS and conventional radiocarbon ages in the QTH01, QTH02 and JDT sections.

Laboratory number	Depth in section (m)	Dating material	^{14}C age (yr BP)	Calibrated ^{14}c age (2σ)(cal yr BP)
	QTH01			
LUG96-44	2.25	Organic matter	1550±60	1316–1551
LUG96-45	2.5	Organic matter	2470±90	2351–2740
BA05223	2.62	shells	3140±40	3263–3448
LUG96-46	2.9	Organic matter	3300±90	3356–3821
LUG96-47	3.15	Organic matter	4130±110	4298–4953
BA05224	3.15	shells	4160±40	4571–4831
LUG96-48	3.6	Organic matter	4530±80	4881–5449
LUG96-49	4.25	Inorganic matter	5960±65	6652–6953
BA05225	4.25	shells	5920±40	6658–6854
LUG02-25	5.37	Organic matter	8412±62	9293–9530
LUG02-23	5.72	Organic matter	9183±60	10234–10502
	QTH02			
BA05222	3.88	shells	6550±40	7344–7563
BA05221	4.75	shells	6910±40	7671–7833
BA05218	5.91	shells	11175±50	12952–13187
	JDT			
LUG96-53	1.33	Organic matter	3980±96	4153–4813
LUG96-51	1.87	Organic matter	5930±100	6493–7137
LUG96-54	2.84	Organic matter	6600±90	7321–7622
LUG96-50	3.15	Organic matter	6820±70	7566–7825
LUG96-55	3.4	Organic matter	7060±85	7696–8017
LUG96-52	4.2	Terrestrial branches	7130±110	7724–8175

appears to be a centennial-scale dry interval between 8.0–7.0 cal kyr BP.

2 Grain size data of the mid-Holocene sand layers

It is described as the sediment sorting principle that the grain size of lake sediments becomes finer and finer from the shore to the center, and sediment belts of different grain size fractions levels can be distinguished, as a result, the grain size of lake sediments is in close relation to water level and the energy of the inflows [26,36]. According to this theory, grain size change in the QTH01 and QTH02 sections of Zhuye Lake can be related to the lake level change. The general mid-Holocene grain size change can be shown by sand percentages (>63 μm) and median grain size (Figure 4). Mid-Holocene sediments can be divided into three phases according to the grain size data from QTH01 and QTH02 sections (Figure 4, Figure 7): the mid-Holocene lacustrine deposits, the abrupt centennial-scale sand layer and the early-Holocene lacustrine deposits. In the QTH01 section, the mid-Holocene lacustrine deposits exist at the depth of 315–450 cm (~7.2–~4.7cal kyr BP). During this phase, the average sand percentage is 25.73% (6.18–68.43%), and the average median grain size is 37.24 μm (0.38–107.89 μm) (Figure 4). The average grain size frequency distribution curve can be divided into 3 parts: 0.03–0.89 μm, 0.89–447.74 μm and 447.74–1782.50 μm. There are also three peaks (0.18 μm, 22.44 μm and 1124.68 μm, relatively) corresponding to the three parts (Figure 7). The abrupt centennial-

Table 2. OSL dates for six samples in the QTH01 section [27]. All the OSL dates were adjusted to AD 1950 to compare with calibrated ^{14}C dates.

Depth (cm)	U (ppm)	Th (ppm)	K (%)	Water content (%)	Dose rate (Gy/kyr)	De (Gy)	Age (kyr)
250	11.19±0.35	2.65±0.16	0.58±0.03	50±5	2.02±0.16	4.70±0.10	2.27±0.19
290	6.27±0.27	4.34±0.20	0.97±0.04	45±5	1.76±0.13	6.50±0.20	3.65±0.30
375	4.08±0.21	1.70±0.14	0.32±0.02	58±5	0.76±0.07	4.10±0.30	5.31±0.60
455	4.47±0.23	2.93±0.18	1.30±0.05	44±5	1.60±0.12	11.80±0.50	7.18±0.63
495	8.10±0.28	2.82±0.17	0.97±0.03	57±5	1.49±0.12	11.50±0.30	7.65±0.67
560	10.65±0.35	5.48±0.25	1.17±0.04	51±5	2.21±0.18	21.60±0.50	9.72±0.81

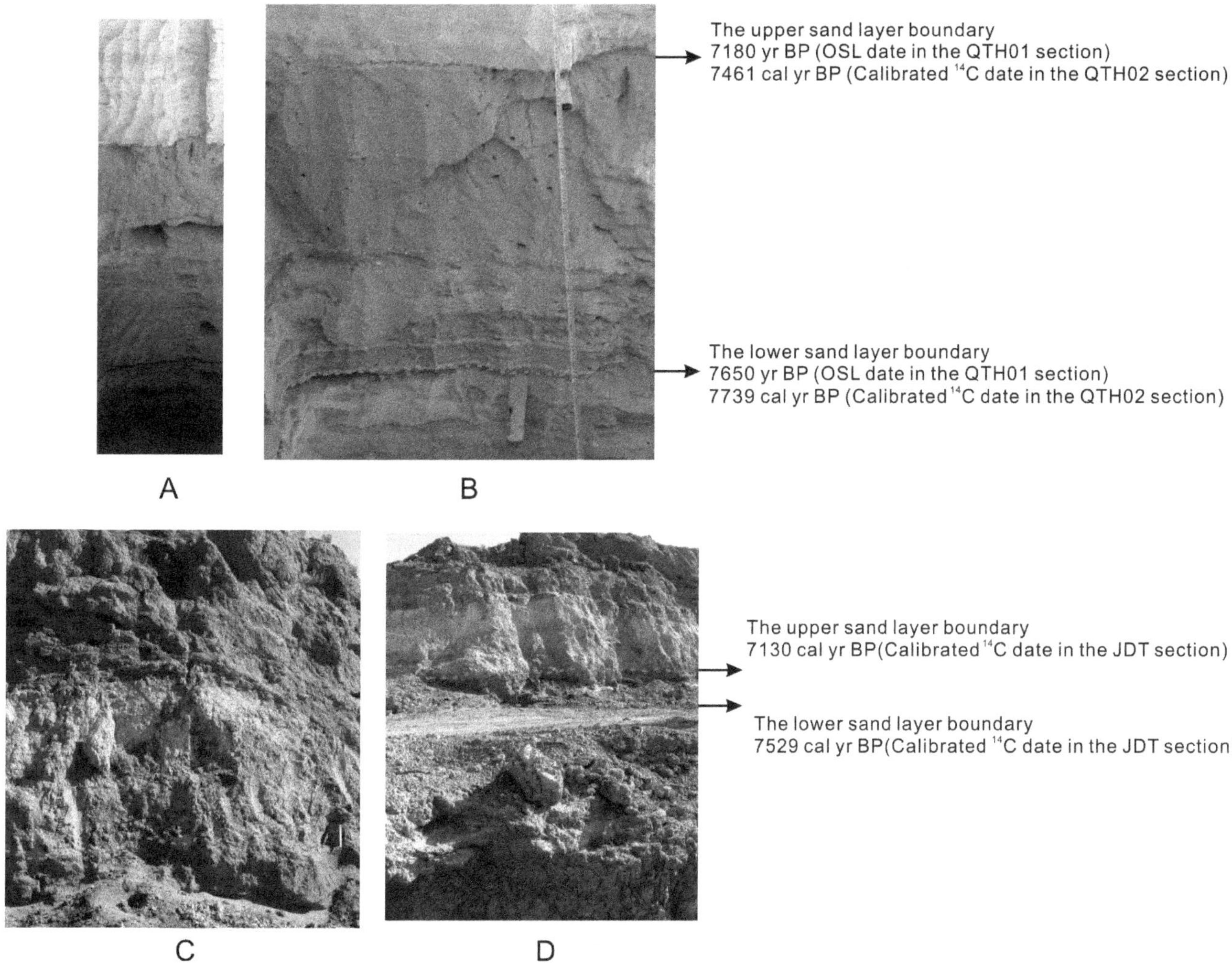

Figure 6. Pictures showing the mid-Holocene sand layers and the dates in the QTH02 and JDT sections. The red broken lines indicate the boundaries of the mid-Holocene sand layers in the QTH02 and JDT sections. A and B show the sand layer in the QTH02 section. C and D show the sand layer in the JDT section.

scale sand layer appears between at the depth of 450–495 cm (~7.6–~7.2cal kyr BP). During this phase, the average sand percentage is 88.24% (65.95–95.48%), and the average median grain size is 128.01 μm (105.30–146.74 μm). The average grain size frequency distribution curve can be divided into 2 parts: 0.05–35.57 μm and 35.57–447.74 μm. Two peaks for the two parts are 12.62 μm and 126.19 μm, respectively. The early-Holocene lacustrine deposits appear at the depth of 495–600 cm (~13.0–~7.6 cal kyr BP). During this phase, the average sand percentage content is 50.51% (12.26–91.50%), and the average median grain size is 77.66 μm (2.51–230.94 μm). The average grain size frequency distribution curve can be divided into 4 parts: 0.04–0.89 μm, 0.89–28.25 μm, 28.25–447.74 μm and 447.74–1782.50 μm.

In the QTH02 section, the mid-Holocene lacustrine deposits appear at the depth of 297-385 cm (~7.4–~4.7 cal kyr BP). During this phase, the average sand percentage is 31.24% (13.52–52.75%), and the average median grain size is 30.73 μm (11.28–68.69 μm) (Figure 4). The average grain size frequency distribution curve can be divided into 3 parts: 0.36–1.00 μm, 1.00–399.05 μm and 399.05–1782.50 μm (Figure 7). The abrupt centennial-scale sand layer appears at the depth of 385–475 cm (~7.7–~7.4 cal kyr BP). The average sand percentage is 87.68% (47.18–96.22%), and the average median grain size is 127.86 μm (55.41–151.23 μm) during this phase. The average grain size frequency distribution curve can be divided into 3 parts: 0.36–31.70 μm, 31.70–400.00 μm and 400.00–1782.50 μm, and there are three peaks (14.16 μm, 126.19 μm and 1002.37 μm relatively) corresponding to the three parts. The early-Holocene lacustrine deposits appear between at the depth of 475–599 cm (~13.0–~7.7 cal kyr BP). During this phase, the average sand percentage is 41.45% (13.55–96.16%), and the average median grain size is 49.74 μm (7.4–200.81 μm). The average grain size frequency distribution curve can be divided into 4 parts, 0.36–1.12 μm, 1.12–31.70 μm, 31.70–316.98 μm and 316.98–1415.89 μm.

In the middle reaches of the basin, the JDT section is located on a terrace of the Hongshui River, which is a tributary of the Shiyang River. The sedimentary facies and grain size of the JDT section are mostly controlled by the environmental conditions in the middle reaches of the basin. Located beside the Tengger Desert, lacustrine sediments and peat are formed during the humid period; on the contrary, sediments are dominated by sand and aeolian sediments during the arid period. Based on the grain size data (Figure 5), the mid-Holocene sediments can be divided

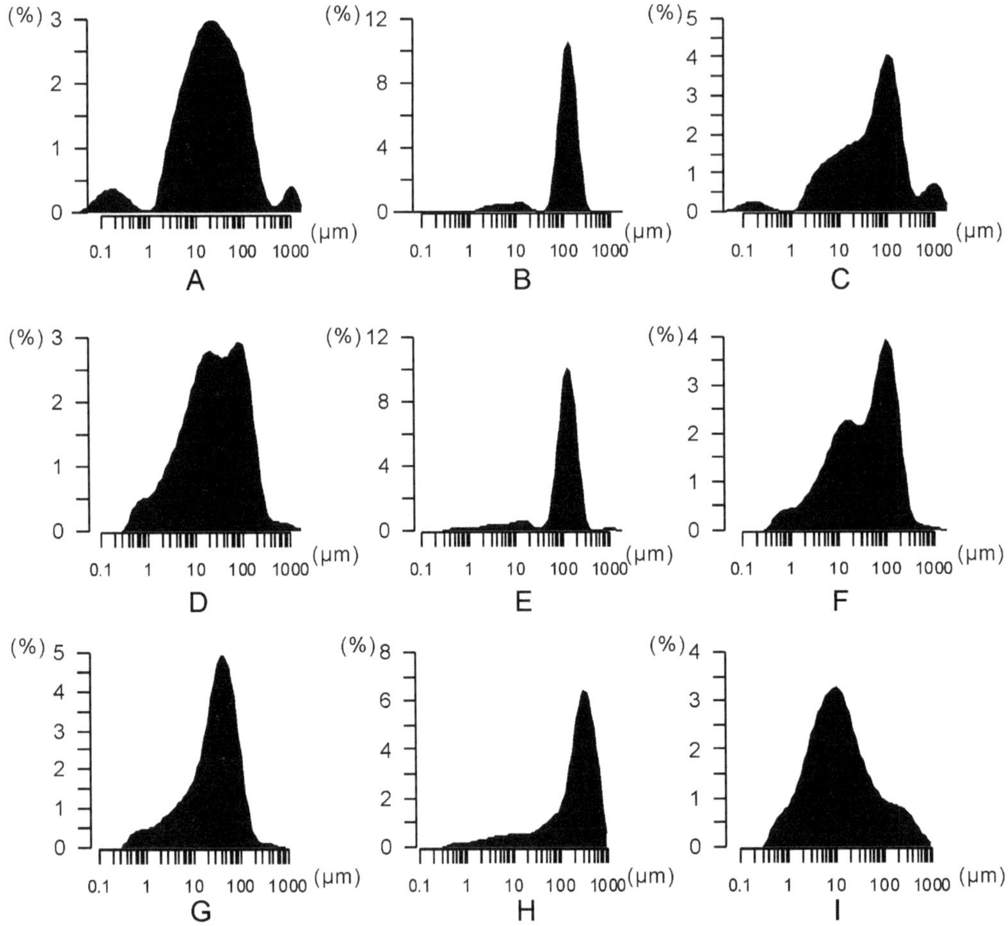

Figure 7. Average grain size frequency distribution curves in the QTH01 and QTH02 and JDT sections. A, B and C represent the three mid-Holocene phases, including the mid-Holocene lacustrine deposits, the abrupt centennial-scale dry interval and the early-Holocene lacustrine deposits in the QTH01 section. D, E and F represent the three mid-Holocene phases, including the mid-Holocene lacustrine deposits, the abrupt centennial-scale dry interval and the early-Holocene lacustrine deposits in the QTH02 section. G, H and I represent the three mid-Holocene phases, including the mid-Holocene lacustrine deposits, the abrupt centennial-scale dry interval and the lacustrine deposits below the dry interval in the JDT section.

into three phases: the mid-Holocene lacustrine deposits, the abrupt centennial-scale sand layer and the lacustrine deposits below the sand layer (Figure 5, Figure 7). In the JDT section, the mid-Holocene lacustrine deposits appear between at the depth of 120–230 cm (~7.1–~4.4 cal kyr BP). During this phase, the average sand percentage is 23.23% (18.67–27.16%), and the average median grain size is 33.14 μm (27.26–37.90 μm) (Figure 5). The average grain size frequency distribution curve can be divided into 3 parts, 0.36–1.12 μm, 1.12–200.00 μm and 200.00–563.68 μm (Figure 7). The abrupt centennial-scale sand layer appears at the depth of 230–300 cm (~7.5–~7.1 cal kyr BP). During this phase, the average sand percentage is 83.00% (73.99–90.54%), and the average median grain size is 264.88 μm (164.09–372.61 μm). The average grain size frequency distribution curve can be divided into 2 parts: 0.40–20.00 μm and 20.00–893.37 μm. The lacustrine deposits below the sand layer appear at the depth of 300–400 cm (~7.9–~7.5 cal kyr BP). During this phase, the average sand percentage is 16.41% (0.13–58.17%), and the average median grain size is 264.88 μm (164.09–372.61 μm). The average grain size frequency distribution curve can be divided into 3 parts: 0.36–1.26 μm, 1.26–89.34 μm and 89.34–893.37 μm.

In the QTH01, QTH02 and JDT sections, grain size data indicate that sediments from the mid-Holocene sand layers are much coarser than those of the mid-Holocene lacustrine deposits from different parts of the basin. In the terminal lake, mid-Holocene environmental changes implied by QTH01 and QTH02 sections are closely related to lake level changes, while the abrupt sand layer can show an abrupt decline in lake level. In the JDT section, the coarser deposits from the mid-Holocene sand layer could indicate the arid environment and the expansion of the desert in the middle reaches of the basin. The average grain size frequency distribution curves of the sand layers from the QTH01, QTH02 and JDT sections are similar to each other (Figure 7), showing the deposits are well sorted with a symmetrical distribution and mainly distributed between 100 and 300 μm. This kind of grain size frequency distribution is similar to that of the desert surface samples in the surrounding regions, northern China [37]. This could indicate that the deposits of the mid-Holocene sand layers are mainly from the surrounding deserts. In the three sections, grain size of the mid-Holocene lacustrine deposits can be generally divided into 3 or more than 3 parts and poorly sorted. The relatively fine components, below 40 μm, generally account for a large fraction, which can be the suspended load [37,38]. At the same time, lacustrine deposits from arid regions can be transported by many dynamic processes, such as wind transportation, saltation and suspension transportations in water [37]. The poorly sorted grain size data and the characteristics of grain size frequency distribution show multiple transport powers in the region.

Discussion

Based on the radiocarbon and OSL dates of the mid-Holocene sand layers from QTH01, QTH02 and JDT sections in the Shiyang River drainage basin, the abrupt centennial-scale dry interval is between 8.0 and 7.0 cal kyr BP, and the abrupt sand layer is embedded between the early and middle Holocene lacustrine sediments. The sedimentary sequences show that the relatively humid early and middle Holocene was interrupted by an abrupt dry event in the basin. In the west of Zhuye Lake, Chen et al. (2003) reported a dry middle Holocene interval (~7.0–~5.0 cal kyr BP) [17]. In the middle reaches of the Shiyang River, Zhang et al. (2000) also found a mid-Holocene sand layer (~6.3–~5.7 cal kyr BP) [39]. However, previous studies failed to establish an absolutely dated mid-Holocene climate change record due to a relatively loose age control. Based on a comparison between OSL and radiocarbon ages in the terminal lake and the mid-Holocene record in the middle reaches of the basin, this work provide an accurate chronological framework for an abrupt dry mid-Holocene interval that happened between 8.0 and 7.0 cal kyr BP, while the grain size data indicate a low lake level in the terminal lake and an arid environment in the middle reaches during the abrupt dry interval. The 8.2 kyr BP event (the major cooling episode 8200 years before present) has been widely regarded as the most remarkable cooling episode during the early Holocene [40]. Although the radiocarbon and the OSL dating methods have some uncertainties on the centennial-scale, the abrupt dry event recovered from the Shiyang River Basin was dated by different dating methods and found at different sites of the drainage basin. All the ages at the boundaries of the sand layers indicate that the dry interval is not beyond 8.0 cal kyr BP. Furthermore, early Holocene records in surrounding areas did not show an obvious cool/dry signal of the 8.2 kyr BP event [24,27]. As a result, the abrupt dry interval between 8.0–7.0 cal kyr BP cannot be linked to the 8.2 kyr BP event directly. This drainage basin is located in the northwest margin of the Asian summer monsoon, where the modern climate is affected both by the Asian summer monsoon and the westerly winds. A comparative study between this record and mid-Holocene climatic records from the Asian monsoon and westerly winds domains is critical for understanding the mechanism of the abrupt dry interval.

1 Mid-Holocene Asian summer monsoon evolution

In the east and south Asian continent, the high-resolution and precisely dated speleothem records provide a reliable Holocene Asian monsoon evolution history, which are widely accepted as an appropriate proxy to reconstruct the Holocene climate change [28,29]. Although Clemens et al. (2010) argued the cave $\delta^{18}O$ (‰) cannot be interpreted as reflecting the Asian summer monsoon alone on the orbital scale [41]. During the Holocene epoch, a comparison between speleothem and lake records in the Qinghai-Tibet Plateau and East Asia shows relatively consistent results regarding the Asian summer monsoon evolution [28,29,35,42–44]. In this study, we chose speleothem records of Qunf Cave (17°10′ N) [29], Dongge Cave (25°20′ N) [28,45], Lianhua Cave (29°29′ N) [46], Heshang Cave (30°27′ N) [47], Sanbao Cave (31°40′ N) [48] and Jiuxian Cave (33°34′ N) [49] from low to mid latitudes in East Asia. In addition, Qinghai Lake is the largest saline lake in China and located in the northern Qinghai-Tibet Plateau, where the location is sensitive to advances and retreats of the Asian summer monsoon and a good place for the Holocene Asian summer monsoon reconstruction [43,44]. The Qinghai Lake (36°50′ N) record is also involved in this study. As shown by the $\delta^{18}O$ (‰) records in Figure 8, the Asian summer monsoon histories are relatively consistent from south to north and the monsoon intensity reaches the highest level during the early-Holocene (ca. 9000 years ago), and then decreases gradually over the next several thousand years. This pattern closely follows the changes of summer insolation at low latitudes (Figure 8) [50]. In Qinghai Lake, the pollen concentration and $\delta^{18}O$ (‰) of ostracode shells also show a similar trend, compared with the speleothem records. According to a synthesis of the Asian summer monsoon, the mid-Holocene summer monsoon evolution is relatively stable and showing a trend of gradually decreasing.

2 Mid-Holocene lake evolution in arid Central Asia

In arid Central Asia, Chen et al. (2008) synthesized 11 lake records to evaluate spatial and temporal patterns of moisture changes during the Holocene [14]. The results show that the Holocene moisture change is relatively synchronous and showing a coherent trend: the early-Holocene is relatively dry and the climate becomes wet during the mid-Holocene. In this study, we chose a Holocene moisture evolution trend and a Holocene moisture index of arid Central Asia synthesized by Chen et al. (2008) [14]. Other three lakes, Bosten Lake [14,51], Telmen Lake [14,52] and Wulungu Lake [53] were selected for a comparative study. Figure 9 shows a relatively dry early-Holocene and a wet mid-Holocene in arid Central Asia. Between 8.0 and 7.0 cal kyr BP, it is a transitional period between dry and humid phases. As shown in Figure 9, the Holocene summer insolation at 30°N crosses with the Holocene moisture evolution trend of arid Central Asia at 8.0–7.0 cal kyr BP. This period (8.0–7.0 cal kyr BP) is characterized by the gradually decreasing Asian summer monsoon and the relatively arid climate in arid Central Asia. In monsoon marginal zones, the climate is both affected by the Asian summer monsoon and the westerly winds. Therefore, the marginal regions are easily affected by the decreasing monsoon and the arid climate in arid Central Asia, both of which triggered the unstable environment conditions in these regions at 8.0–7.0 cal kyr BP.

3 Dry mid-Holocene intervals in monsoon marginal zones

As has been introduced in the Introduction part, there are many dry mid-Holocene records found in the marginal regions of the Asian monsoon. Most of them are recovered from lake records. However, the time scales of the dry events are inconsistent while the explanations regarding the dry mid-Holocene intervals are inconsistent. Generally speaking, the dry mid-Holocene lake records are embedded between early and middle Holocene humid lake records. The early-to-mid Holocene humid environment was interrupted by an abrupt dry interval, which can be seen as a mid-Holocene climate change pattern; although the time scales of the dry mid-Holocene intervals are different. The dating uncertainty can play a very important role for the different time scales. However, in addition to dating errors, how to explain lake records in arid and semi-arid regions is an important to the issue of different time scales. Lake evolution and proxies in arid areas are actually affected by the effective moisture change in the entire basin, which is controlled by the precipitation, evapotranspiration, lake surface evaporation, etc. Li and Morrill (2010) have confirmed the effects of evapotranspiration and lake surface evaporation to lake level changes in East Asia [9]. Various

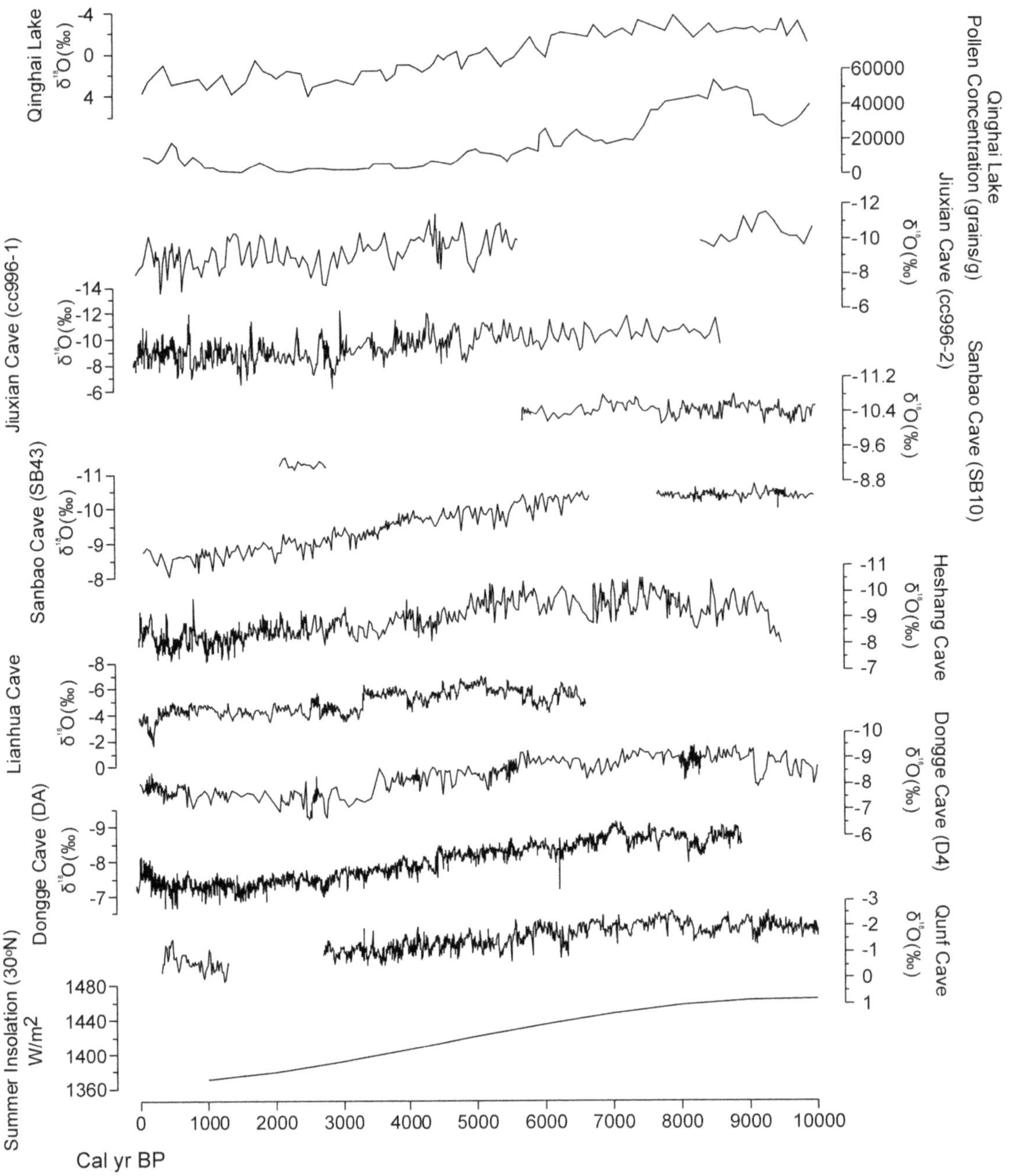

Figure 8. The absolutely dated Holocene speleothem records and the Qinghai Lake record showing the Holocene Asian summer monsoon evolution. The records (from bottom to top) are arranged by latitudes from low to high, which are Summer Insolation at 30°N [50], $\delta^{18}O$(‰) for Qunf Cave (17°10′ N)[29], $\delta^{18}O$(‰) of DA for Dongge Cave (25°20′ N)[28], $\delta^{18}O$(‰) of D4 for Dongge Cave (25°20′ N)[45], $\delta^{18}O$(‰) for Lianhua Cave (29°29′ N)[46], $\delta^{18}O$(‰) for Heshang Cave (30°27′ N)[47], $\delta^{18}O$(‰) of SB43 for Sanbao Cave (31°40′ N)[48], $\delta^{18}O$(‰) of SB10 for Sanbao Cave (31°40′ N)[48], $\delta^{18}O$(‰) of cc996-1 for Jiuxian Cave (33°34′ N), $\delta^{18}O$(‰) of cc996-2 for Jiuxian Cave (33°34′ N)[49], pollen concentration (grains/g) for Qinghai Lake (36°50′ N)[43], $\delta^{18}O$(‰) of ostracode shells for Qinghai Lake (36°50′ N)[44].

drainage basins are located in monsoon marginal zones and the geographical settings (e.g. topography, geomorphology, soil, vegetation, and hydrology) for different basins vary according to their locations. During the same period, the effective moisture changes could differ from different drainage basins due to the different geographical settings. Therefore, it is normal to find the mid-Holocene dry intervals are asynchronous. In this study, the correlation between the abrupt dry interval in the Shiyang River drainage basin and the long term evolutions of the Asian monsoon and westerly winds provide a clue for detecting the mechanism of the dry mid-Holocene events. The mid-Holocene climate change pattern that the early-to-mid Holocene humid period was interrupted by a centennial-scale abrupt dry interval between 8.0–7.0 cal kyr BP could influence other parts of monsoon marginal zones; however, this mid-Holocene climate pattern could have different time scales due to various effective moisture change patterns, which are related to the geographical settings of East Asia.

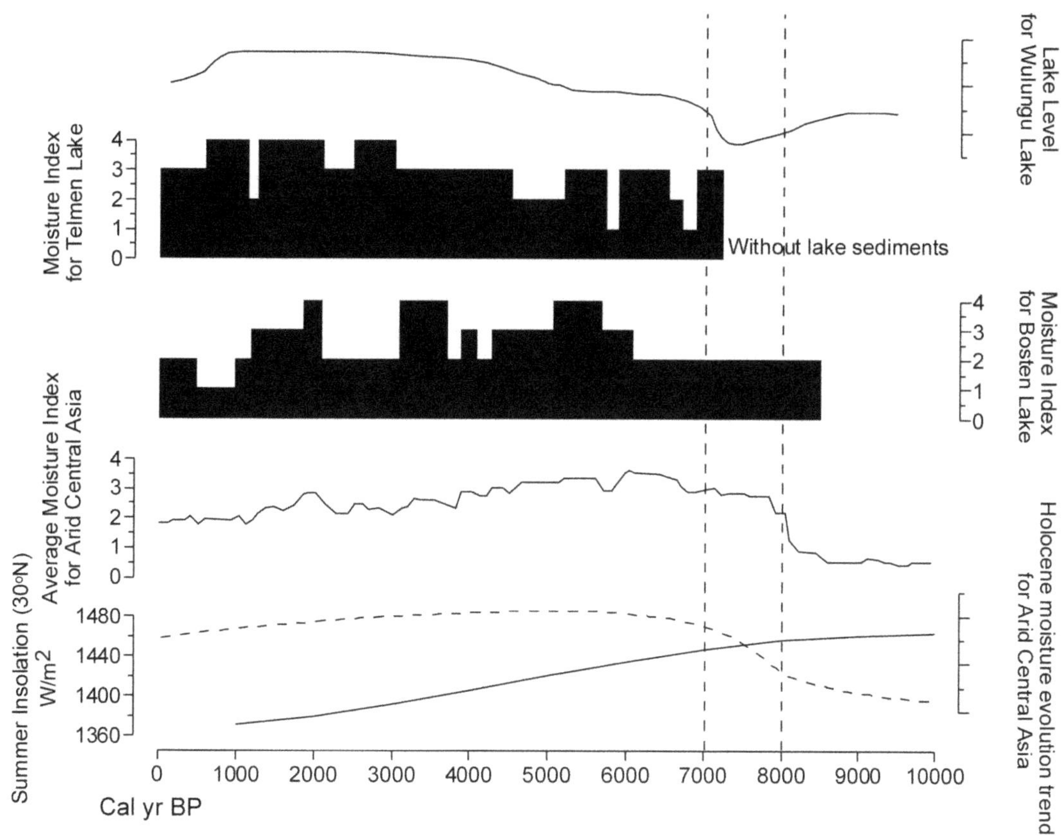

Figure 9. Holocene lake records and moisture indices in arid Central Asia. The two broken lines show the interval between 8.0–7.0 cal kyr BP. From bottom to top, Summer Insolation at 30°N [50], the Holocene moisture evolution trend for arid Central Asia [14], the average moisture index for arid Central Asia [14], the moisture index for Bosten Lake [14,51], the moisture index for Telmen Lake [14,52] and the lake level change for Wulungu Lake [53].

Conclusions

Mid-Holocene climate records were recovered from the terminal lake and middle reaches of the Shiyang River drainage basin. A centennial-scale sand layer was found from QTH01, QTH02 and JDT sections. According to grain size data, the sand layer embedded between lacustrine sediments is related to an abrupt decline in lake level and arid environment in the basin. A comparison between radiocarbon and OSL ages shows the abrupt dry mid-Holocene interval happened between 8.0 and 7.0 cal kyr BP. The mid-Holocene climate record was compared with the long-term Asian monsoon evolution and the moisture history in arid Central Asia. Based on the comparative study, the abrupt centennial-scale dry interval can be related to the long term evolutions of the Asian summer monsoon and the westerly winds. Along with the Holocene insolation change for the low-latitude regions of the Northern Hemisphere, the Asian summer monsoon shows a weakening tendency since the early-Holocene (ca. 9000 years ago). In arid Central Asia, the lake records show a climate transition around 8.0–7.0 cal kyr BP, which indicates a climatic shift from arid to humid. In monsoon marginal zones, the abrupt centennial-scale dry mid-Holocene interval can be influenced by the weakening Asian summer monsoon and relatively arid westerly winds between 8.0–7.0 cal kyr BP. Dry mid-Holocene lake records are widely distributed in monsoon marginal zones. The humid early-to-middle Holocene interrupted by an abrupt dry interval can be seen as a mid-Holocene climate change pattern. The mid-Holocene climate change pattern also can be affected by the long term evolutions of the Asian summer monsoon and the westerly winds. Dating uncertainties can take an important role for different time scales of those dry mid-Holocene intervals. In addition, lake evolution is mainly controlled by the effective moisture change in the basin, while the effective moisture is related to the basin-wide geographical settings. In semi-arid and arid regions, the effective moisture changes vary according to their locations due to various geographical features. The different effective moisture history is another reason for the different time scales of the dry mid-Holocene intervals.

Acknowledgments

We thank the editor and reviewers for their constructive comments and suggestions for improving our paper. We also thank Dr. Hao Long, who was of great assistance with the OSL dating. Thanks are extended to Dr. Carrie Morrill for her comments and Mr. Allan Grey for his assistance in checking the English of this paper.

Author Contributions

Conceived and designed the experiments: YL. Performed the experiments: CZ. Analyzed the data: NW. Contributed reagents/materials/analysis tools: NW. Wrote the paper: YL.

References

1. Ahrens CD (1985) Meteorology Today (2nd ed). St. Paul: West Publishing Company.
2. Lin C (1993) The Atmosphere and Climate Change. Dubuque: Kendall/Hunt Publishing Company.
3. Zhang JC, Lin ZG (1992) Climate of China. New York: Wiley.
4. Wang K, Jiang H, Zhao H (2005) Atmospheric water vapor transport from westerly and monsoon over the Northwest China. Advances in Water Science 16: 432–438. (in Chinese)
5. Zhao SQ (1983) A new scheme for comprehensive physical regionalization in China. Acta Geogr Sin 38: 1–10. (in Chinese)
6. Peel MC, Finlayson BL, McMahon TA (2007) Updated world map of the Köppen–Geiger climate classification. Earth Syst Sci 11: 1633–1644.
7. An Z, Porter SC, Kutzbach JE, Wu X, Wang S, et al. (2000) Asynchronous Holocene optimum of the East Asian monsoon. Quat Sci Rev 19: 743–762.
8. He Y, Wilfred H, Zhang Z, Zhang D, Yao T, et al. (2004) Asynchronous Holocene climatic change across China. Quat Res: 61, 52–63.
9. Li Y, Morrill C (2010) Multiple factors causing Holocene lake–level change in monsoonal and arid central Asia as identified by model experiments. Clim Dynam 35: 1119–1132.
10. Wang Y, Liu X, Herzschuh U (2010) Asynchronous evolution of the Indian and East Asian Summer Monsoon indicated by Holocene moisture patterns in monsoonal central Asia. Earth–Sci Rev 103: 135–153.
11. Li Y, Wang N, Chen H, Li Z, Zhou X, et al. (2012) Tracking millennial–scale climate change by analysis of the modern summer precipitation in the marginal regions of the Asian monsoon. J Asian Earth Sci 58: 78–87.
12. Li J (1990) The patterns of environmental changes since late Pleistocene in northwestern China. Quat Sci 3: 197–204. (in Chinese)
13. Herzschuh U (2006) Palaeo–moisture evolution in monsoonal Central Asia during the last 50, 000 years. Quat Sci Rev 25: 163–178.
14. Chen FH, Yu ZC, Yang ML, Ito E, Wang SM, et al. (2008) Holocene moisture evolution in arid central Asia and its out–of–phase relationship with Asian monsoon history. Quat Sci Rev 27: 351–364.
15. An C, Chen F, Barton L (2008) Holocene environmental changes in Mongolia: A review. Global Planet Change 63: 283–289.
16. Wang W, Ma Y, Feng Z, Narantsetseg T, Liu K, et al. (2011) A prolonged dry mid–Holocene climate revealed by pollen and diatom records from Lake Ugii Nuur in central Mongolia. Quat Int 229: 74–83.
17. Chen FH, Wu W, Holmes J, Madsen DB, Zhu Y, et al. (2003) A mid–Holocene drought interval as evidenced by lake desiccation in the Alashan Plateau, Inner Mongolia, China. Chin Sci Bull 48: 1–10.
18. Chen CT, Lan HC, Lou JY, Chen YC (2003) The dry Holocene Megathermal in Inner Mongolia. Palaeogeogr Palaeoclimatol Palaeoecol 193: 181–200.
19. Jiang WY, Liu TS (2007) Timing and spatial distribution of mid–Holocene drying over northern China: Response to a southeastward retreat of the East Asian Monsoon. J Geophys Res Atmos 112: 1–8.
20. Guo Z, Petit–Maire N, Kropelin S (2000) Holocene non–orbital climatic change events in present–day arid areas of northern Africa and China. Global Planet Change 26: 97–103.
21. Zhou W, Yu X, Timothy AJ, Burrb G, Xiao JY (2004) High–resolution evidence from southern China of an Early–Holocene optimum and a Mid–Holocene dry event during the past 18,000 years. Quat Res 62: 39–48.
22. An C, Feng Z, Barton L (2006) Dry or humid? Mid–Holocene humidity changes in arid and semi–arid China. Quat Sci Rev 25: 351–361.
23. Chen FH, Cheng B, Zhao Y, Zhu Y, Madsen DB (2006) Holocene environmental change inferred from a high–resolution pollen record, Lake Zhuyeze, arid China. The Holocene 16: 675–684.
24. Zhao Y, Yu Z, Chen FH, Li J (2008) Holocene vegetation and climate change from a lake sediment record in the Tengger Sandy Desert, northwest China. J Arid Environ 72: 2054–2064.
25. Li Y, Wang N, Cheng H, Long H, Zhao Q (2009) Holocene environmental change in the marginal area of the Asian monsoon: a record from Zhuye Lake, NW China. Boreas 38: 349–361.
26. Li Y, Wang N, Morrill C, Cheng H, Long H, et al. (2009) Environmental change implied by the relationship between pollen assemblages and grain size in N.W. Chinese lake sediments since the Late Glacial, Rev Palaeobot Palynol 154: 54–64.
27. Long H, Lai Z, Wang N, Li Y (2010) Holocene climate variations from Zhuyeze terminal lake records in East Asian monsoon margin in arid northern China. Quat Res 74: 46–56.
28. Wang Y, Cheng H, Edwards RL, He Y, Kong X, et al. (2005) The Holocene Asian Monsoon: Links to Solar Changes and North Atlantic Climate. Science 308: 854–857.
29. Fleitmann D, Burns SJ, Mudelsee M, Neff U, Kramers J, et al. (2003) Holocene forcing of the Indian monsoon recorded in a stalagmite from Southern Oman. Science 300: 1737–1739.
30. Chen LH, Qu YG (1992) Water–land resources and reasonable development and utilization in the Hexi region. Beijing: Science Press. (in Chinese)
31. Huang DX (1997) Gansu Vegetation. Lanzhou: Gansu Science and Technology Press. (in Chinese)
32. Pachur HJ, Wünnemann B, Zhang H (1995) Lake Evolution in the Tengger Desert, Northwestern China, during the last 40,000 Years. Quat Res 44: 171–180.
33. Wünnemann B, Pachur HJ, Zhang HC (1998) Climatic and environmental changes in the deserts of Inner Mongolia, China, since the Late Pleistocene. In: Alsharhan AS, Glennie KW, Whittle GL, editors. Balkema, Rotterdaman: Quatary Deserts and Climatic Changes. pp. 381–394.
34. Zhang HC, Peng JL, Ma Y, Chen GJ, Feng ZD, et al. (2004) Late Quatary palaeolake–levels in Tengger Desert, NW China. Palaeogeogr Palaeoclimatol Palaeoecol 211: 45–58.
35. Morrill C, Overpeck JT, Cole JE, Liu KB, Shen CM (2006) Holocene variations in the Asian monsoon inferred from the geochemistry of lake sediments in central Tibet. Quat Res 65: 232–243.
36. Lerman A (1978) Lake: Chemistry, Geology, Physics. Berlin: Springer–Verlag.
37. Sun D, Bloemendal J, Rea DK, Vandenberghe J, Jiang F, et al. (2002) Grain–size distribution function of polymodal sediments in hydraulic and aeolian environments, and numerical partitioning of the sedimentary components. Sediment Geol 152: 263–277.
38. Middleton GV (1976) Hydraulic interpretation of sand size distributions. Journal of Geology 84: 405–426.
39. Zhang HC, Ma YZ, Wünnemann B, Pachur HJ (2000) A Holocene climatic record from arid northwestern China. Palaeogeogr Palaeoclimatol Palaeoecol 162: 389–401
40. Alley RB (2000) Ice–core evidence of abrupt climate change. PNAS 97: 1331–34.
41. Clemens SC, Prell WL, Sun Y (2010) Orbital–scale timing and mechanisms driving Late Pleistocene Indo–Asian summer monsoons: Reinterpreting cave speleothem $\delta^{18}O$. Paleoceanography 25: PA4207.
42. Gasse F, Arnold M, Fontes JC, Fort M, Gibert E, et al. (1991) A 13,000 year climate record from western Tibet. Nature 353: 742–745.
43. Shen J, Liu X, Wang S, Matsumoto R (2005) Palaeoclimatic changes in the Qinghai Lake area during the last 18000 years. Quat Int 136: 131–140.
44. Liu XQ, Shen J, Wang SM, Wang YB, Liu WG (2007) Southwest monsoon changes indicated by oxygen isotope of ostracode shells from sediments in Qinghai Lake since the late Glacial. Chin Sci Bull 52: 539–544.
45. Dykoski CA, Edwards RL, Cheng H, Yuan D, Cai Y, et al. (2005) A high–resolution, absolute–dated Holocene and deglacial Asian monsoon record from Dongge Cave, China. Earth Planet Sci Lett 233: 71–86.
46. Cosford J, Qing HR, Eglington B, Mattey D, Yuan DX, et al. (2008) East Asian monsoon variability since the Mid–Holocene recorded in a high–resolution, absolute–dated aragonite speleothem from eastern China. Earth Planet Sci Lett 275: 296–307.
47. Hu C, Henderson GM, Huang J, Xie S, Sun Y, et al. (2008) Quantification of Holocene Asian monsoon rainfall from spatially separated cave records. Earth Planet Sci Lett 266: 221–232.
48. Dong J, Wang Y, Cheng H, Hardt B, Edwards RL, et al. (2010) A high–resolution stalagmite record of the Holocene East Asian monsoon from Mt Shennongjia, central China. The Holocene 20, 257–264.
49. Cai Y, Tan L, Cheng H, An Z, Edwards RL, et al. (2010) The variation of summer monsoon precipitation in central China since the last deglaciation. Earth Planet Sci Lett 291: 21–31.
50. Berger A, Loutre MF (1991) Insolation values for the climate of the last 10,000,000 years. Quat Sci Rev 10: 297–317.
51. Wünnemann B, Mischke S, Chen FH (2006) A Holocene sedimentary record from Bosten Lake, China. Palaeogeogr Palaeoclimatol Palaeoecol 234: 223–238.
52. Fowell SJ, Hansen BCS, Peck JA, Khosbayar P, Ganbold E (2003) Mid to late–Holocene climate evolution of the Lake Telmen basin, North Central Mongolia, based on palynological data. Quat Res 59: 353–363.
53. Liu XQ, Herzschuh U, Shen J, Jiang Q, Xiao X (2008) Holocene environmental and climatic changes inferred from Wulungu Lake in northern Xinjiang, China. Quat Res 70: 412–425.

Genetic, Ecological and Morphological Divergence between Populations of the Endangered Mexican Sheartail Hummingbird (*Doricha eliza*)

Yuyini Licona-Vera, Juan Francisco Ornelas*

Departamento de Biología Evolutiva, Instituto de Ecología, AC, Xalapa, Veracruz, Mexico

Abstract

The Mexican Sheartail (*Doricha eliza*), an endangered hummingbird, is endemic to Mexico where two populations have a disjunct distribution. One population is distributed along the northern tip of the Yucatan Peninsula whereas the other is mostly restricted to central Veracruz. Despite their disjunct distribution, previous work has failed to detect morphological or behavioral differences between these populations. Here we use variation in morphology, mtDNA and nuDNA sequences to determine the degree of morphological and molecular divergence between populations, their divergence time, and historical demography. We use species distribution modeling and niche divergence tests to infer the relative roles of vicariance and dispersal in driving divergence in the genus. Our Bayesian and maximum likelihood phylogenetic analyses revealed that *Doricha eliza* populations form a monophyletic clade and support their sister relationship with *D. enicura*. We found marked genetic differentiation, with reciprocal monophyly of haplotypes and highly restricted gene flow, supporting a history of isolation over the last 120,000 years. Genetic divergence between populations is consistent with the lack of overlap in environmental space and slight morphological differences between males. Our findings indicate that the divergence of the Veracruz and Yucatan populations is best explained by a combination of a short period of isolation exacerbated by subsequent divergence in climate conditions, and that rather than vicariance, the two isolated ranges of *D. eliza* are the product of recent colonization and divergence in isolation.

Editor: João Pinto, Instituto de Higiene e Medicina Tropical, Portugal

Funding: This project was funded by competitive grants (grant numbers 61710 and 155686) from Consejo Nacional de Ciencia y Tecnología (CONACyT, http://www.conacyt.mx) and research funds from INECOL (20030/10563) awarded to Juan Francisco Ornelas. Yuyini Licona-Vera was supported by a Master of Science scholarship (no. 262561). The funders had no role in study design, data collection and analysis, decision to publish, or preparation of the manuscript.

Competing Interests: The authors have declared that no competing interests exist.

* Email: francisco.ornelas@inecol.mx

Introduction

The Mexican Sheartail Hummingbird (*Doricha eliza*) is an endemic to Mexico, and globally is a near threatened species according to the IUCN Red List [1]. It is locally endangered with population declines owing to habitat loss and degradation [2], and is thus facing risk of extinction in the wild. These hummingbirds of the monophyletic assemblage Mellisugini [3], known as bees, and were originally included in the genus *Trochilus* [4]. They have since moved into different genera (*Calliphlox*, *Calothorax*, *Myrtis*, *Thaumastura*, *Rhodopis*, and *Doricha*; [4–5]). Although the recent use of mitochondrial DNA sequences placed *D. eliza* in the bees group [6], additional taxa likely to be nested within the Mellisugini monophyletic assemblage, including the putative sister species, the Slender Sheartail (*D. enicura*), and multiple loci are needed to fully resolve the phylogenetic position of the Mexican Sheartail within the Mellisugini [6–7].

Sheartails are small hummingbirds with long, arched bills and are common in semi-open scrubby areas [8]. Males with glittering, rose-pink gorgets display rocking pendulum flights (shuttle displays) to females, along with high climbs and steep dives (**Figure S1** and **Video S1**; personal observation, [8–11]). The breeding range of the Mexican Sheartail is divided into two widely separated geographical areas, one in central Veracruz and the other mainly on the northern fringe of the Yucatan Peninsula [8,12–14]. In 2002, the Veracruz population was estimated at about 2500 individuals, and the Yucatan population at no more than 6000–10,000 individuals [12]. Both the Veracruz and Yucatan populations are declining, locally threatened [1,15], and subject to different threats. The Veracruz population is facing severe habitat degradation as a result of livestock grazing, sugarcane cultivation and residential development [1,13–15], while the Yucatan population is under pressure mainly from the development of its coastal dune habitat for tourism [1,14]. To our knowledge, no information has been published that documents the evolutionary divergence of the Veracruz and Yucatan populations of the Mexican Sheartail. There are breeding and feeding records suggesting that these separated populations are allopatric all year round [8,10,12–15]. During the breeding season (March-August), the Yucatan population is exclusively found in a narrow coastal strip mainly in the ecotone between mangroves and tropical deciduous forest [9,14], but also breeds in gardens and urban areas [1,14]. The Veracruz population occurs in undisturbed, dry deciduous forest and heavily disturbed agricultural landscapes c. 25 km inland [1,13]. There is no historical evidence that the two populations interbreed, and the question of when the two populations diverged is still open. Despite the distance between the Veracruz and Yucatan populations (c. 780 km) and the differences in the habitats occupied, there are no apparent morphological or behavioral differences [13]. However, geographic distance as a driver of the divergence between the two populations of Mexican Sheartail in isolation, has not been investigated. This question is particularly important because each population is facing different threats and in a different environment, requiring locally adapted conservation schemes.

In this study, we ask the following questions: (1) what is the phylogenetic position of *Doricha eliza* within the monophyletic assemblage of Mellisugini? (2) What is the level of genetic differentiation between the Veracruz and Yucatan populations? (3) Are disjunct populations currently connected by gene flow? (4) When did the Veracruz and Yucatan populations split? And (5) was the divergence between the two disjunct populations caused by vicariance or dispersal? To answer these questions, we conducted Bayesian and maximum likelihood phylogenetic analyses of mitochondrial and nuclear DNA markers and time estimates of intraspecific genetic divergence. We also used morphological data, genetic diversity and historical demographic indices, modeling ancestral distribution, and use of niche divergence tests to infer the history of the Veracruz and Yucatan populations and the relative roles of dispersal and vicariance in driving divergence in the genus.

Materials and Methods

Ethics Statement

We obtained the collecting permit to conduct this work from Mexico's Secretaría de Medio Ambiente y Recursos Naturales, Instituto Nacional de Ecología, Dirección General de Vida Silvestre (permit number: INE SGPA/DGVS/07701/11) for the field study described. This collecting permit specifically allowed for the collection of tail feathers from the birds. Manipulation of birds in the field was minimal. Birds were captured with mist nets, measured, and their two outermost tail feathers were removed for genetic analyses before the birds were released. All procedures with birds were carried out in accordance with the Guidelines for the Use of Wild Birds in Research proposed by the North American Ornithological Council and the ethics of experimental procedures were revised and authorized by the Animal Care and Use Committee under the Graduate Studies Committee (Maestría en Biodiversidad y Sistemática; No. INECOL/SP/CAP/2012/103) of the Instituto de Ecología, A.C. (INECOL). While the field studies involve an endangered and protected species, no specific permits are required for field studies such as the one described here.

Sample Collection

Feather samples were collected from a total of 25 *D. eliza* during the 2011 and 2012 breeding seasons. Ten hummingbirds were collected in central Veracruz at the following locations: Xalapa, El Lencero, Miradores and Chavarrillo. Feather samples were collected from 15 individuals of the Yucatan population at Rio Lagartos and Chicxulub (**Table S1**). We sequenced the mitochondrial nicotinamide adenine dinucleotide dehydrogenase subunit 2 gene (ND2) and the complete ATP synthase 6 and ATP synthase 8 coding region (ATPase), and the nuclear 20454 locus from tail feathers of 25 *D. eliza*, and sequenced or downloaded sequence data from GenBank for the sister species *D. enicura* and for the outgroups, the bee hummingbirds *Calothorax lucifer*, *C. pulcher*, *Selasphorus rufus*, *S. sasin*, *S. platycercus*, *S. calliope*, *Atthis heloisa*, *Archilochus colubris* and *Tilmatura dupontii*, and the emerald *Amazilia cyanocephala* (**Table S2**). We also obtained ND2 sequences from GenBank for an additional 17 species of the bee hummingbird group (*Archilochus alexandri*, *Calliphlox amethystina*, *C. bryantae*, *C. mitchellii*, *Calypte anna*, *C. costae*, *Chaetocercus bombus*, *Ch. mulsant*, *Doricha enicura*, *Eulidia yarrellii*, *Microstilbon burmeisteri*, *Myrmia micrura*, *Myrtis fanny*, *Rhodopis vesper*, *Selasphorus flammula* and *Thaumastura cora*), 11 representative taxa of the mountain gems group and 12 species of the emeralds group to be used for sequence alignment and as outgroups (**Table S2**).

DNA Isolation, Amplification and Sequencing Protocols

Total genomic DNA was extracted using the Qiagen DNeasy blood and tissue extraction kit (Qiagen, Inc., Valencia, CA, USA), following the manufacturer's protocol. Using polymerase chain reaction (PCR), we amplified fragments from three mitochondrial DNA (mtDNA) coding genes: ND2 (350 bp, primers pair L5216 and H5578 [16]); ATPase 6–8 (727 bp, primer pair L8929 and H9947 [17]); and 20454 (502 bp, primer pair 20454F and 20454R [18]). PCR reactions (20 μL total volume) for genes contained 0.72×buffer, 0.58 Mm of each dNTP, 0.4 μg/μL BSA, 0.04 U *Taq* polymerase (Promega, Madison, WI, USA), 4.0 mM $MgCl_2$, and 0.29 μM of each primer. PCR reactions were performed in a 2720 thermal cycler (Applied Biosystems, Carlsbad, CA, USA) or in an Eppendorf Mastercycler thermal cycler (Eppendorf AG, Hamburg, Germany). For amplification of the ND2, cycling parameters consisted of initial denaturation at 94°C for 3 min, followed by 40 cycles at 94°C for 45 sec, annealing at 47–48°C for 45 sec, 72°C for 30 sec, and a final step at 72°C for 5 min. The protocol for amplifying ATPase 68 was an initial denaturation at 95°C for 2 min, followed by 40 cycles at 92°C for 40 sec, annealing at 47–50°C for 1 min, 73°C for 2 min, and a final step at 73°C for 3 min. Amplification of the 20454 locus included initial denaturation at 94°C for 1.30 min, followed by 40 cycles at 94°C for 30 sec, annealing at 50–52°C for 30 sec, 72°C for 45 sec, and a final step at 72°C for 10 min. PCR products were purified with QIAquick (Qiagen Inc.) and sequenced in both directions to check the validity of sequence data using the Big Dye Terminator Cycle Sequencing kit (Applied Biosystems). The products were read on a 310 automated DNA sequencer (Applied Biosystems) at the INECOL's sequencing facility. Finally, sequences were assembled using Sequencher v4.9 (Gene Codes Corp., Ann Arbor, MI, USA) and then manually aligned using SE-AL v2.0a11 (http://tree.bio.ed.ac.uk/software/seal). All sequences are deposited under the following GenBank accession numbers: KJ710519–KJ710624 (**Table S2**).

Individual haplotypes from 20454 sequences were statistically inferred using PHASE v2.1 [19–20] with the following parameters: 100,000 iterations, a thinning interval of 10, and a burn-in of 1000. PHASE uses a Bayesian statistical method to determine the most probable pair of alleles or haplotypes. Heterozygous sites in nuclear sequences were identified when two different nucleotides were present at the same position in the electropherograms of both strands. Three runs were conducted to check the consistency of results obtained by examining the allele frequencies and coalescent goodness-of-fit measures estimated for each run, and only highly supported haplotype pairs (probability 0.70–0.90) were maintained.

Phylogenetic Analyses of mtDNA and nuDNA

Phylogenetic relationships among mtDNA sequences of *D. eliza* were reconstructed using Bayesian inference (BI) and maximum likelihood (ML) methods. The BI analyses were run in MrBayes v3.12 [21] and the ML analyses in RAxML v7.4.4 [22] using the CIPRES Science Gateway [23]. Phylogenetic analyses were performed using three data sets. We first ran analyses with a ND2 data set for *Doricha eliza* and available sequences for North

American and South American members of the Mellisugini clade [7] retrieved from GenBank. The second set of analyses was run with the combined mtDNA data set (ND2 and ATPase) for *D. eliza* and outgroups, and the third with the nuclear 20454 locus data set. All DNA markers used and their accession numbers are listed in **Table S2**. We used jModeltest v1.1 [24] to choose the model of molecular evolution that best fit our sequence data under the Akaike information criterion (AIC; [25]), GTR+I+G (base frequencies: A = 0.3401, C = 0.3710, G = 0. 0760, T = 0.2183; gamma distribution shape parameter = 1.1860) for ND2; GTR+I (base frequencies: A = 0.2304, C = 0.0958, G = 0.3565, T = 0.3173) for ND2+ATPase; and HKY+G (base frequencies: A = 0.2480, C = 0.2034, G = 0.2310, T = 0.3177; gamma distribution shape parameter = 0.0170) for the nuclear 20454 locus. For each data set, two parallel Markov chain Monte Carlo (MCMC) analyses were executed simultaneously, and each was run for 10 million generations, sampling every 1000 generations. A majority consensus tree was obtained (50% majority-rule), showing nodes with a posterior probability (PP) of 0.6 or more. Bayesian PP values were calculated from the sampled trees remaining after 10% burn-in samples were discarded [21] to only include trees after stationarity/convergence was reached as checked in Tracer v1.5 [31]. Nodes with PP≥95 were considered to be strongly supported [26]. The consensus tree was later visualized in FigTree v1.2.3 (http://tree.bio.ed.ac.uk/software/figtree/). The BI analyses included two additional sets of analyses using the combined data set (ND2+ATPase+20454): the first used a single model for the entire combined loci data set (the 'unpartitioned' analyses), and the second set employed partition-specific DNA evolution models of each gene. For each data set, two parallel Markov chain Monte Carlo (MCMC) analyses were executed simultaneously, and each was run for 50 million generations, sampling every 1000 generations. We computed Bayes factors with the harmonic means [27] to determine whether applying partition-specific models significantly improved the explanation of the data.

ML analyses were performed using default values and the same evolution models as in the Bayesian analyses. Node support for the ML tree was estimated with 1000 bootstrap replicates, and nodes were considered highly supported when bootstrap values were ≥ 70% [28].

Species Tree and Divergence Time Estimation

To estimate relationships between populations of *D. eliza*, we used ND2, ATPase and 20454 sequences for all *D. eliza* samples under the multispecies coalescent method of *BEAST [29–30] implemented in BEAST v1.7.4 [31]. This method models the lineage sorting process between units for groups of individuals not connected by gene flow above, at, or below the species level to obtain a species tree [32]. *Doricha enicura*, *Calothorax lucifer* and *C. pulcher* were the outgroups. We employed a relaxed molecular clock model with branch rates drawn independently from a lognormal distribution and the Yule process as a tree prior under a continuous population size model. The models of molecular evolution that best fit our sequence data under the Akaike information criterion (AIC; [25]) were HKY (base frequencies: A = 0.2247, C = 0.1467, G = 0.3356, T = 0.2930) for ND2; HKY+G (base frequencies: A = 0.2325, C = 0.1003, G = 0.3430, T = 0.3242; gamma distribution shape parameter = 0.0160) for ATPase; and HKY+G (base frequencies: A = 0.2447, C = 0.1990, G = 0.2310, T = 0.3254; gamma distribution shape parameter = 0.0160) for the nuclear 20454 locus. We performed three independent runs of 10 million generations each, sampling every 1000 generations, and discarding the first 1 million generations of every replicate as burn-in. Replicate results were combined in LogCombiner v1.7.4 (http://beast.bio.ed.ac.uk/LogCombiner) and the convergence of runs was confirmed by effective sample sizes (ESS) >200 for all parameters and by visual inspection of traces within and between replicates using Tracer v1.5 [31]. The resulting posterior sample of trees was summarized in a Maximum Clade Credibility (MCC) tree using TreeAnnotator v1.7.4 (http://beast.bio.ed.ac.uk/TreeAnnotator). Nucleotide substitution models selected with jModeltest v1.1 [24] were incorporated, and we used a relaxed clock model with an uncorrelated lognormal distribution. To calibrate the tree we used the mean rates of 2.9×10^{-2} substitutions/site/lineage/million years (s/s/l/My) for ND2, 2.2×10^{-2} s/s/l/My for ATPase, and 1×10^{-3} s/s/l/My for 20454 based on rates obtained for Hawaiian honeycreepers [33]. We prefer the rates suggested by Lerner et al. [33] because these are likely more appropriate for the lower taxonomic level of *Doricha* species [34–35] than the low substitution rates obtained for major bird orders [36].

BEAST v1.7.4 [31] was used to estimate the time of the most recent common ancestor (TMRCA) of clades in *D. eliza*. We ran two individual analyses to estimate TMRCA, one with all the ND2 sequences of bee hummingbirds available from previous studies [6–7,37–38] and retrieved from GenBank, and the second with the mtDNA sequences (ND2 and ATPase) generated in our study for *D. eliza* and the other hummingbird species listed above. All sequences used and those retrieved from GenBank are listed with their accession numbers in **Table S2**. The best-fit model of evolution, GTR+I+G for the ND2 and GTR+I for the ND2+ATPase data set, was estimated from the data sets using jModeltest and an uncorrelated lognormal relaxed model selected in BEAST as the clock model. A coalescent model assuming constant population size was used to model the tree prior. The coalescent tree prior used in these analyses appears to fit better when mixed data sets are predominantly intraspecific data [39]. To calibrate the root in both analyses, we used the divergence time between the bee and emerald hummingbird groups (normal prior, mean 13.97 Ma, SD 3.0; [6]) as a secondary calibration. To calibrate the tree, we used the average divergence time for the basal split between North and South American bee hummingbirds (normal prior, mean 6.1, SD 1.0, range of 7.74-4.45 Ma; [40]). Twenty-seven species from the bee hummingbird group, and representatives of the mountain gems (11 taxa) and emeralds (13 taxa) were included as outgroups in the analysis using the ND2 data set, and a fewer outgroups were used in the analysis using the ND2+ATPase data set separately (see Sample Collection; **Table S2**). All of the samples of *D. eliza* were used in both analyses, rather than just the unique haplotypes, to avoid overestimating evolutionary time scales [41]. For each of the analyses, we performed three independent runs of 10 million generations, sampling every 1000 steps, and discarding the first 10% of trees as burn-in. We combined the log and trees files from each independent run using LogCombiner, then viewed the combined log file in Tracer to ensure that ESS values for all priors and the posterior distribution were >200, and then annotated the trees using TreeAnnotator summarized as a maximum clade credibility tree with mean divergence times and 95% highest posterior density (HPD) intervals of age estimates, visualized in FigTree.

Genetic Structure and Genetic Diversity

To infer genealogical relationships among haplotypes, the ND2, ATPase and phased 20454 sequence data sets were separately analyzed using the statistical parsimony algorithm, implemented in TCS v2.1 [42] with the 95% connection limit. The genetic structure of mtDNA sequence data was further explored through pairwise comparisons of F_{ST} values and analysis of molecular variance (AMOVA [43]). The AMOVA was run grouping individuals into two groups according to the observed divergence in the BI analysis (see Results), and using the Jukes and Cantor model, and 16,000 permutations to determine the significance of the AMOVA using Arlequin v3.1 [44]. Lastly, we calculated corrected genetic distances [45] for mtDNA data sets between populations of *D. eliza* and other species within the Mellisugini clade (*D. enicura*, *Calothorax pulcher*, *C. lucifer*) using DnaSP v5.1 [46], and assessed genetic variation within populations by calculating the haplotype diversity (*Hd*) and nucleotide diversity (π) using Arlequin [44].

Historical Demography

The demographic histories of the Veracruz and Yucatan populations of *D. eliza* were inferred by means of neutrality tests and mismatch distributions carried out in Arlequin v3.1 [44]. Tajima's *D* [47] and Fu's *Fs* [48] were calculated to test whether populations evolved under neutrality, and mismatch distributions [49] were calculated using the sudden expansion model [50] with 1000 bootstrap replicates. The validity of the sudden expansion assumption was determined using the sum of squares differences (SSD), which is higher in stable, nonexpanding populations [51]. To validate the estimated demographic and geographic expansion tests 16,000 permutations were used in Arlequin. We also used Bayesian skyline plots (BSP; [52]) to assess changes in effective population size (*Ne*) over time in BEAST. This analysis was performed for each genetic group separately and for the two groups combined (concatenated). Concatenated analysis has been proposed to satisfy the assumption of lineages interbreeding in scenarios where divergence is recent and there is low genetic structure [53]. The time axis was scaled using the mean rates of 2.9×10^{-2} substitutions/site/lineage/million years (s/s/l/My) for ND2 and 2.2×10^{-2} s/s/l/My for ATPase based on rates for Hawaiian honeycreepers [33].

We used the 'isolation-with-migration' coalescent model as implemented in the program IMa [54–55] to estimate the time of divergence (*t*) between the Veracruz and Yucatan populations of *D. eliza*, the effective number of migrants per generation ($m_{V\ to\ Y}$ and $m_{Y\ to\ V}$), and the effective population size of the ancestral (q_A) and descendant populations (q_V and q_Y). We used mitochondrial and nuclear phased haplotypes to produce maximum-likelihood estimates and confidence intervals for splitting times, effective population sizes, and gene flow [55]. Every locus was tested for evidence of recombination using IMgc [56]. This program removes either sites or haplotypes to obtain the longest region to pass the four-gamete test [57]. Three independent runs of 25 million generations were performed under Hasegawa-Kishino-Yano (HKY) model for mitochondrial loci and the Infinite Sites (IS) model for nuclear locus. Each run used identical conditions, but different starting seed values, and a burn-in period of 3 million steps with parameter values empirically determined in the preliminary runs to verify the convergence of independent analyses. To improve the mixing of the Markov chains (to facilitate convergence), we ran multiple heated chains and kept monitoring the autocorrelation and estimates of ESS [55]. Using estimated ESS values in IMa [55], we considered stationarity to have been reached when the ESS value for each independent run was >50. The rates of 2.9×10^{-2} substitutions/site/lineage/million years (s/s/l/My) for ND2, 2.2×10^{-2} s/s/l/My for ATPase and 1×10^{-3} s/s/l/My for 20454 obtained for Hawaiian honeycreepers [33] were provided in the IMa input file and the mean rates for all genes were used to estimate the effective population sizes of each genetic group. We used a 2.5 year generation time assuming that the sexual maturity of Mexican Sheartail begins approximately 2 years after hatching and assuming an annual survivorship of 0.35, as estimated for other bee hummingbirds [58–59], to convert the effective population size estimates. Migration rates per generation were converted to population migration rates per generation using estimates of the effective population size. The approximate average generation time (T) is calculated according to $T = a+[s/(1-s)]$ [60–61], where *a* is the time to maturity and *s* is the adult annual survival rate. Based on this, the estimate for T was 3.04 years. To convert the time since divergence parameter of IMa to years, *t*, we divided the time parameter (*B*) by the mutation rate per year (*U*) converted to per locus rate by multiplying by the fragment length in base pairs, and calculated for the rates described above.

Species Distribution Models

We constructed a species distribution model (SDM [62]) to predict where populations of *D. eliza* resided during the Last Glacial Maximum (LGM, 21,000-18,000 years ago) and Last Interglacial (LIG, 120,000–140,000 years ago). We assembled a data set of 121 unique records (51 for Veracruz and 76 for the Yucatan) from georeferenced museum (Atlas Aves de México, [63]) specimens obtained through http://vertnet.org and the Global Biodiversity Information Facility (GBIF, http://data.gbif.org/species/browse/taxon), and analyzed the data with the maximum entropy algorithm in MaxEnt [64–65]. Present climate layers (temperature and precipitation variables, BIO1–BIO19) were drawn from the WorldClim database (*c.* 1 km^2; [66]). Using ArcView v3.2 (ESRI, Redlands, CA, USA), we first extracted GIS data from the 19 WorldClim layers at *D. eliza*'s occurrence points, and then ran a correlation analysis to eliminate correlated environmental variables using the program PAST v2.12 [67]. When the correlation coefficient was higher than 0.80 the variables were considered highly correlated, and for each pair of correlated variables we selected the one that was more temporally inclusive. After removing the highly correlated variables, six variables were used in the analysis (BIO1 [Annual Mean Temperature], BIO2 [Mean Diurnal Range], BIO3 [Isothermality], BIO4 [Temperature Seasonality], BIO12 [Annual Precipitation], and BIO14 [Precipitation of Driest Month]). MaxEnt was set to randomly use 70% of the values for training and 30% of values for testing the model. We constructed the species distribution models using MaxEnt because it provides robust performance with small sample sizes (restricted distribution) of presence only data [64]. Model performance was evaluated using the area under the receiver operating characteristic curve (AUC; [68]). The model for the present was also projected to past climate scenarios, and past climate layers were drawn from WorldClim for two LGM past climate scenarios developed by the Paleoclimate Modelling Intercomparison Project Phase II [69]: the Community Climate System Model (CCSM; [70]) and the Model for Interdisciplinary Research on Climate (MIROC; [71]), and for the LIG [72]. Both CCSM and MIROC climate models simulate climate conditions as they are calculated to have been during the LGM, with a stronger temperature decrease assumed in CCSM than in MIROC [73]. Climate suitability was displayed in ArcView v3.2. (ESRI, Redlands, CA, USA).

Niche Divergence Tests

We employed a multivariate method [74] to test for niche divergence/conservatism. Briefly, we tested for niche divergence using climate data extracted from occurrence points and used the six uncorrelated BIO variables described above to generate ENMs, and then drew minimum convex polygons around occurrence points of each lineage using the Hawth's Tools package in ArcMap v9 [74]. We defined the background characteristics of each group using 1000 random points inside each polygon, and then conducted a principal components analysis (PCA) using these data. The first three PC (niche) axes explained a high percentage of the variance (89%) and were thus used in further analyses. Niche divergence or conservatism was evaluated on each niche axis by comparing the observed difference between the means for each lineage on that axis to the mean difference in their background environments on the same axis [74]. A null distribution of background divergence was created by recalculating the score of background divergence over 1000 jackknife replicates with 75% replacement. Significance for rejecting the null was evaluated at the 95% level. These analyses were conducted using Stata v10 (StataCorp, College Station LP, Texas, USA).

Morphological Variation

To examine differences in morphological variation between the mist-netted Veracruz and Yucatan adult hummingbirds used in the genetic analyses, six measures were taken using a dial calliper with a precision of 0.1 mm and a wing ruler: total body length (BL; the distance from the tip of its bill to the tip of longest tail feather); exposed culmen (EC; from the base of the bill to the tip of the upper mandible); bill width at the base (BB; by the location of the nostrils); and wing chord (WC; the distance from the carpal joint to the tip of the longest unflattened primary) for both males and females, and tail length (TL; from the uropygial gland to the tip of the longest rectrix) for females, and the length of the outermost rectrices (r5), from the base of the uropygial gland to the tip of the longest rectrix (left and right) for males. All measurements were taken by YLV. Measurements for two juvenile males from Yucatan were discarded from the analysis. To examine morphological differences between populations, for males and females we conducted a one-way non-parametric Kruskal-Wallis test using genetic group as fixed factors and morphological measures as dependent variables. These analyses were performed using SPSS v17 for Mac (SPSS, Armonk, NY, USA).

Results

Phylogenetic Analysis of mtDNA and nuDNA

Interspecific phylogenetic relationships among mtDNA and nuDNA sequences of *D. eliza* and other species in the bee hummingbird group were reconstructed using Bayesian inference (BI) and Maximum Likelihood (ML). The BI and ML analyses yielded the same general topologies with minor differences in the position of some terminal branches. Only BI trees are shown. Both the BI and ML trees of the ND2 sequence data set confirmed that *D. eliza* and *D. enicura* form a highly supported monophyletic clade (PP = 0.95, bootstrap = 79%; **Figure S2**). A highly supported relationship between *Doricha* and *Calothorax* species was retrieved (P = 0.94, bootstrap = 75%), yet the relationships within these genera of sheartails and those between other members of the Mellisugini are not fully resolved (**Figure S2**). The level of polymorphism found in the Mellisugini of the nuclear 20454 locus was low and several haplotypes were shared among species. Most interspecific relationships in the Mellisugini were not resolved when only using 20454 (**Figure S3**). In contrast, the interspecific phylogenetic relationships among species in the Mellisugini were more fully resolved when using the combined ND2+ATPase+20454 data set (**Figure 1**). The relationship between *D. eliza* and *D. enicura* is retrieved with high support in both the BI and ML analyses (PP = 0.99, bootstrap = 98%), and monophyly of sheartails (between *Doricha* and *Calothorax* species) is also retrieved with high support (PP = 1.0, bootstrap = 99%). Individuals of *D. eliza* are retrieved as a monophyletic group (PP = 0.99, bootstrap = 95%), with a split separating the Veracruz and Yucatan populations (**Figure 1**). The BI inference using the combined data set of both mtDNA and nuDNA (ND2+ATPase+20454), with unpartitioned and partitioned DNA evolution models of each gene yielded the same relationships. The Bayes factor indicated that the BI tree obtained with the unpartitioned data was more informative (harmonic mean loglikelihood, unpartititoned = –4616.68, partitioned by each gene = –4547.25, logB10 = 69.43, 2logB10 = 138.68), and this difference was very strong (the PP values for the tree obtained with partitioned data are shown in **Figure 1**).

Species Tree and Divergence Time Estimation

Relationships between *D. eliza* populations estimated in *BEAST (**Figure 2**) strongly supported common ancestry for the Veracruz and Yucatan populations (PP = 1.0). Results from *BEAST (ND2+ATPase+20454) suggest that the divergence between the Veracruz and Yucatan clades occurred at c. 120,000 years ago (95% HPD 240,000-31,000 ka), and the divergence between *Doricha* species (TMRCA) in BEAST was estimated to be 1.03 Ma (95% HPD 1.608-0.309 Ma) and 1.46 Ma (95% HPD 2.104-0.852 Ma) between *Calothorax* species (**Figure 2**). Divergence time between *D. eliza* populations was estimated to be 541,000 years ago (95% HPD 902,000-224,000 ka, PP = 0.99) when using the ND2 data set of Mellisugini representatives and 222,000 years ago (95% HPD 352,000-107,000 ka, PP = 1.0) when using the ND2+ATPase data set. Estimates of the TMRCA for *Doricha* species and for the sheartails (*Doricha* and *Calothorax*) indicate that the splits occurred at 1.04 Ma (95% HPD 1.674-0.486 Ma, PP = 0.99) and 2.06 Ma (95% HPD 3.224-1.038 Ma, PP = 1.0) when using the ND2 data set, and at 0.7 Ma (95% HPD 1.791-0.761 Ma, PP = 1.0) and 1.21 Ma (95% HPD 1.719-0.761 Ma, PP = 1.0) when using the ND2+ATPase data set.

Genetic Structure and Genetic Diversity

Sequencing two mtDNA markers in 25 individuals of *D. eliza* (**Tables S1 and S2**) produced 10 haplotypes for ND2, and 8 haplotypes for ATPase, resulting in 10 haplotypes for the concatenated sequence (1077 bp). Phylogenetic analysis of the ND2, ATPase, and concatenated ND2+ATPase haplotypes revealed genetic divergence between the Veracruz and Yucatan populations with no shared haplotypes (**Figure 2**). Haplogroups are connected by more than one step for the ND2, one step for the ATPase, and by four steps for the concatenated ND2+ATPase. Haplotype diversity (h) and nucleotide diversity (π) were moderate for both the Veracruz ($h = 0.666 \pm 0.163$, $\pi = 0.0007 \pm 0.0006$) and Yucatan ($h = 0.561 \pm 0.143$, $\pi = 0.0006 \pm 0.005$) populations. When samples were combined, overall h was 0.796, and overall π was 0.0020 indicating relatively high levels of genetic diversity in the Mexican Sheartail hummingbird. The AMOVA results revealed strong population structure ($F_{CT} = 0.79$, df = 1,24, $P < 0.05$) when samples were grouped by geographic area. Mitochondrial divergence between the Veracruz and Yucatan populations was low ($Dxy = 0.35\%$), whereas genetic divergence between populations of *D. eliza* and the other members of the sheartails ranged from 0.46% to 3.47% (**Table 1**).

The sequencing of the nuDNA locus 20454 produced 8 haplotypes for the phased sequences. One haplotype was shared between the Veracruz and Yucatan populations (**Figure 2**). However, five haplotypes were found only in the Veracruz population and the other two in the Yucatan population, indicating some genetic structure in this locus.

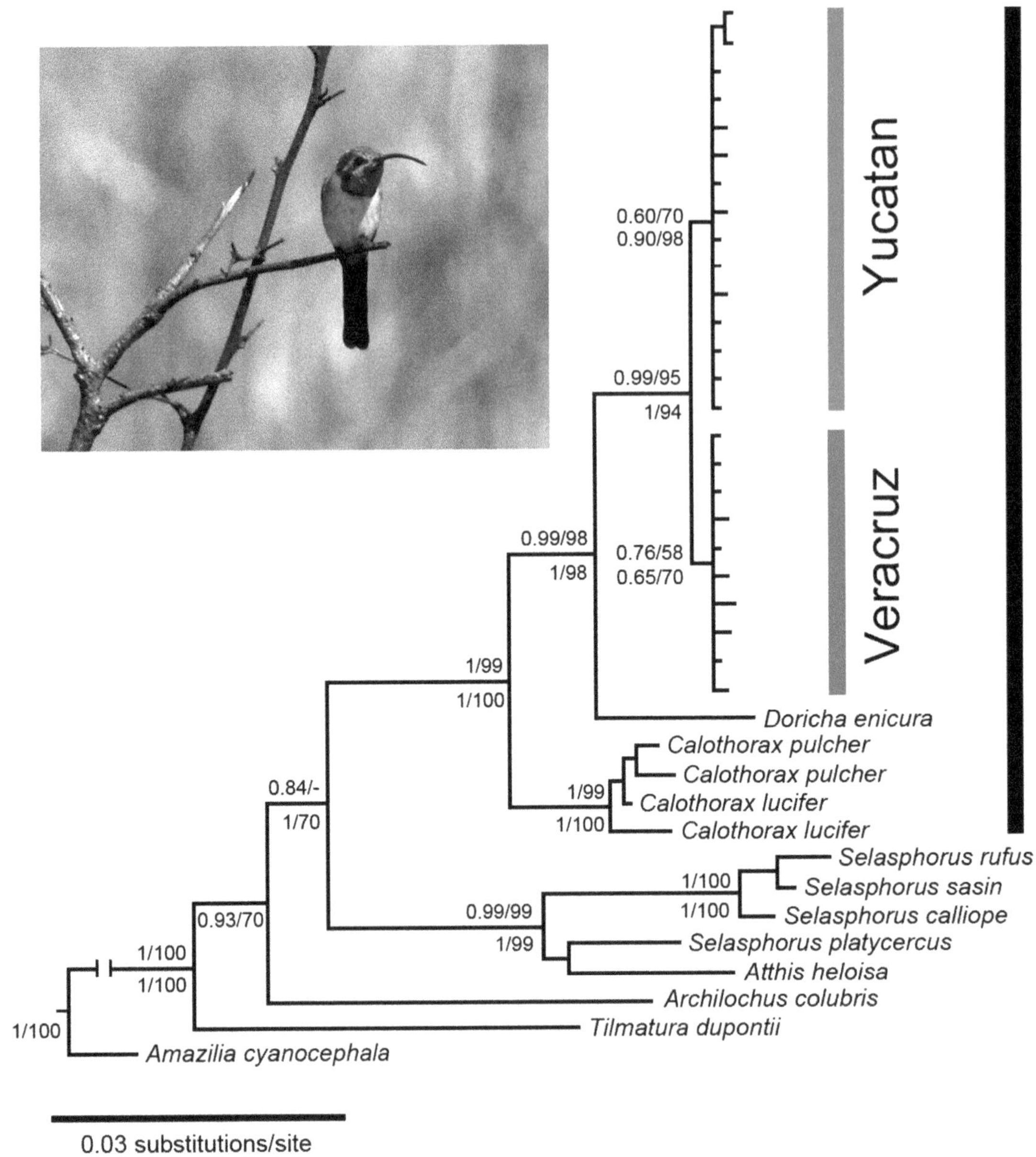

Figure 1. Bayesian posterior probabilities and bootstrap support for MrBayes and Maximum Likelihood analyses. Illustration of tree topology based on ND2+ATPase+20454 concatenated sequences of *Doricha eliza* and outgroups. Values above branches denote posterior probabilities (left) and bootstrap values (right) and those below branches denote the same values for phylogenetic analyses based on the ND2+ ATPase data set (20454 excluded).

Historical Demography

We conducted demographic analyses for the Veracruz and Yucatan populations and for all populations of *D. eliza* using the concatenated mtDNA data set. Neutrality tests revealed low and negative values in all cases, except that the Tajima's D value for the whole population was not significant (**Table 2**). In the mismatch distribution (**Figure 3**), sudden demographic expansion (SSD values) was not rejected for all cases (**Table 2**). The BSP of N_e over time showed no evidence for population expansion; BSP for the Veracruz and Yucatan lineages were flat over time and there was an increase in population size around the LGM (c. 21,000 years ago) when the Veracruz and Yucatan populations were pooled (**Figure 3**).

IMa results are summarized in **Table 3**. Results are reported as highest point estimates and 90% highest probability density (HPD). Based on the mutation rates obtained for Hawaiian honeycreepers, the ancestral population size (N_A) was estimated to be 5,380 (90% HPD, 819–10,600) and the sizes of the two descendant populations were $N_{VERACRUZ} = 1{,}830$ (90% HPD, 1,120–2,520) and $N_{YUCATAN} = 1{,}410$ (90% HPD, 993–1,830). Migration rates between genetic groups ($m_{YUCATAN \rightarrow VERACRUZ}$ and $m_{VERACRUZ \rightarrow YUCATAN}$) were 1.28 (90% HPD, 0.316–2.790) and 1.02 (90% HPD, 0.299–2.120), respectively, and the divergence time (t) between genetic groups was estimated to be 22,100 years ago (90% HPD, 27,000-17,400 ka).

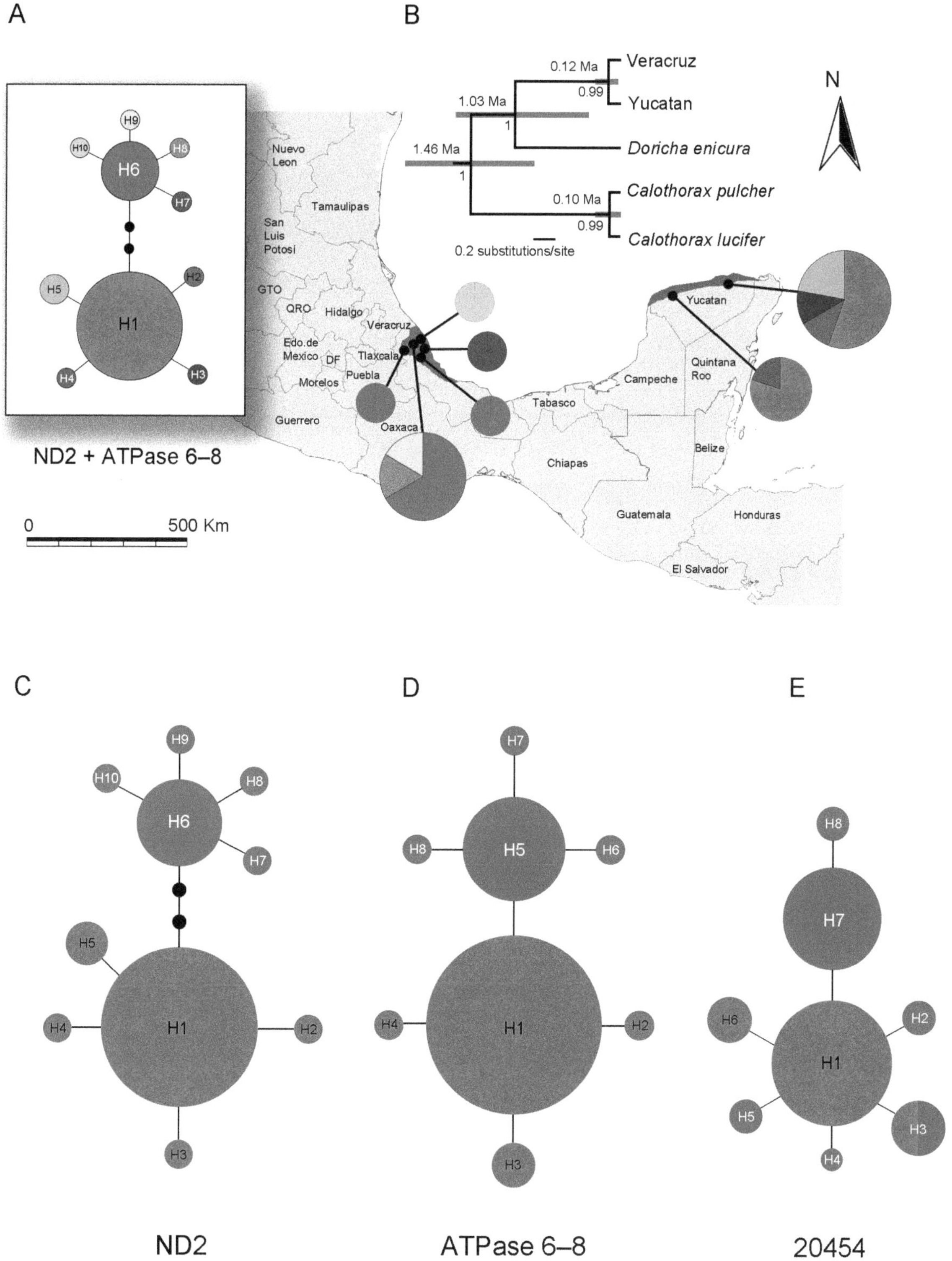

Figure 2. Genetic divergence of *Doricha eliza* populations in Veracruz and the Yucatan Peninsula, Mexico. (A) Haplotype network for ND2+ATPase 6–8 concatenated sequences overlaid on a relief map showing the geographical distribution of *D. eliza*. (B) Species tree and time divergence estimates (95% HPD) in years from the *BEAST analysis based on both mitochondrial (ND2+ATPase) and nuclear DNA (20454 locus). Numbers below branches denote Bayesian posterior probabilities (PP). (C) Haplotype network for ND2. (D) Haplotype network for ATPase. (E) Haplotype network for 20454. Haplotypes are represented by circles, their size proportional to their frequency in the population. Each branch represents a single nucleotide change, with additional mutations indicated by black dots along branches. The color-coding of haplotypes is the same in all figures, blue colors for Veracruz and rose-pink colors for the Yucatan.

Table 1. Percent genetic distances between populations corrected for intra-population polymorphism (%*Dxy*).

X	*Y*	*%Dxy*
Veracruz	Yucatan	0.35
Veracruz	*Doricha enicura*	2.62
Veracruz	*Calothorax lucifer*	3.43
Veracruz	*Calothorax pulcher*	3.47
Yucatan	*Doricha enicura*	2.72
Yucatan	*Calothorax lucifer*	3.33
Yucatan	*Calothorax pulcher*	3.37
Doricha enicura	*Calothorax lucifer*	3.48
Doricha enicura	*Calothorax pulcher*	3.62
Calothorax lucifer	*Calothorax pulcher*	0.46

Data shown for differences between the Veracruz and Yucatan populations of *D. eliza* and between species of *Doricha* and *Calothorax* based on concatenated mtDNA (1077 bp).

Species Distribution Models

The current distribution predicted by MAXENT (**Figure 4**) closely matched the known range of *D. eliza* (**Figure 2**), and the models performed well (all AUC values >0.948). The ENM for the current climate variables using both the Veracruz and Yucatan records predicted well the distribution of the species well and over-predicted the distribution of the Veracruz population (**Figure 4**). When models were projected onto past climatic layers based on two LGM climate scenarios (MIROC and CCSM), predictions suggest that suitable habitat for both *D. eliza* populations expanded in Veracruz and the Yucatan Peninsula with a large geographical disjunction. Lastly, models projected onto LIG climatic layers revealed a different scenario to the predicted ENM for the present (**Figure 4**). Predictions suggest that there was almost no suitable habitat for *D. eliza* in Veracruz, and that potentially suitable habitat for *D. eliza* was restricted to a smaller area in the tip of the Yucatan Peninsula, small areas in the arid central valleys of Oaxaca (low probability), and Guatemala.

Niche Divergence Tests

The PCA of environmental data that together three niche axes explained 88.7% of the variation in *D. eliza* (Veracruz and Yucatan records). The first niche axis (38.8% of variation) was associated with isothermality (BIO3) and precipitation of driest month (BIO14). The second niche axis (34.2%) was associated with annual precipitation (BIO12) and annual mean temperature (BIO1), and the third axis (15.7%) was associated with mean annual range (BIO2) and temperature seasonality (BIO4). Tests of niche divergence and conservatism on these three niche axes showed evidence for niche conservatism on niche axis 1 and niche divergence on niche axis 2 and 3 (**Table 4**).

Morphological Variation

Morphological analysis detected no significant differences in the mean values of most traits between the Veracruz and Yucatan populations (Kruskal-Wallis tests, $P>0.05$; **Table S3** and **Figures S4 and S5**), yet males from the Yucatan population had significantly smaller left outermost rectrices (r5; mean = 34.3 mm, SD = 0.05) than those from Veracruz (mean = 37.3 mm, SD = 0.11; Kruskal-Wallis test, $H = 7.5$, $P < 0.01$; **Figure S4**).

Discussion

Mellisugini Phylogeny and the Molecular Placement of the Mexican Sheartail

A molecular phylogeny combining the available ND2 sequences to resolve the relationships within the Mellisugini (bees) clade was not available until now. Here we used a ND2 data set from four taxa assigned to the genera *Doricha* and *Calothorax*, as well as 22 samples of all other genera within Mellisugini [3] to determine the place of *D. eliza* within this phylogeny. We included our ND2 (350 bp) sequences and the available ND2 (1041 bp) sequences were downloaded from GenBank, which led to a high percentage of missing characters. Although some studies found that incomplete data could bias the ML and BI analysis [75], other studies have argued that missing data does not affect the accuracy of phylogenies in either the ML or BI analysis, and that phylogenetic

Table 2. Results of demographic analyses of *Doricha eliza*.

Group	**N**	**H**	*D*	*Fs*	**SSD**
Veracruz	10	5	−1.6670*	−2.847***	0.0311*
Yucatan	15	5	−1.5181*	−2.676***	0.0170*
Doricha eliza	25	10	−0.8057	−3.260**	0.0337*

N = number of individuals, H = number of haplotypes, *D* = Tajima's *D*, *Fs* = Fu's *Fs*, SSD = differences in the sum of squares or mismatch distribution.
*$P<0.05$,
**$P<0.01$,
***$P<0.001$.

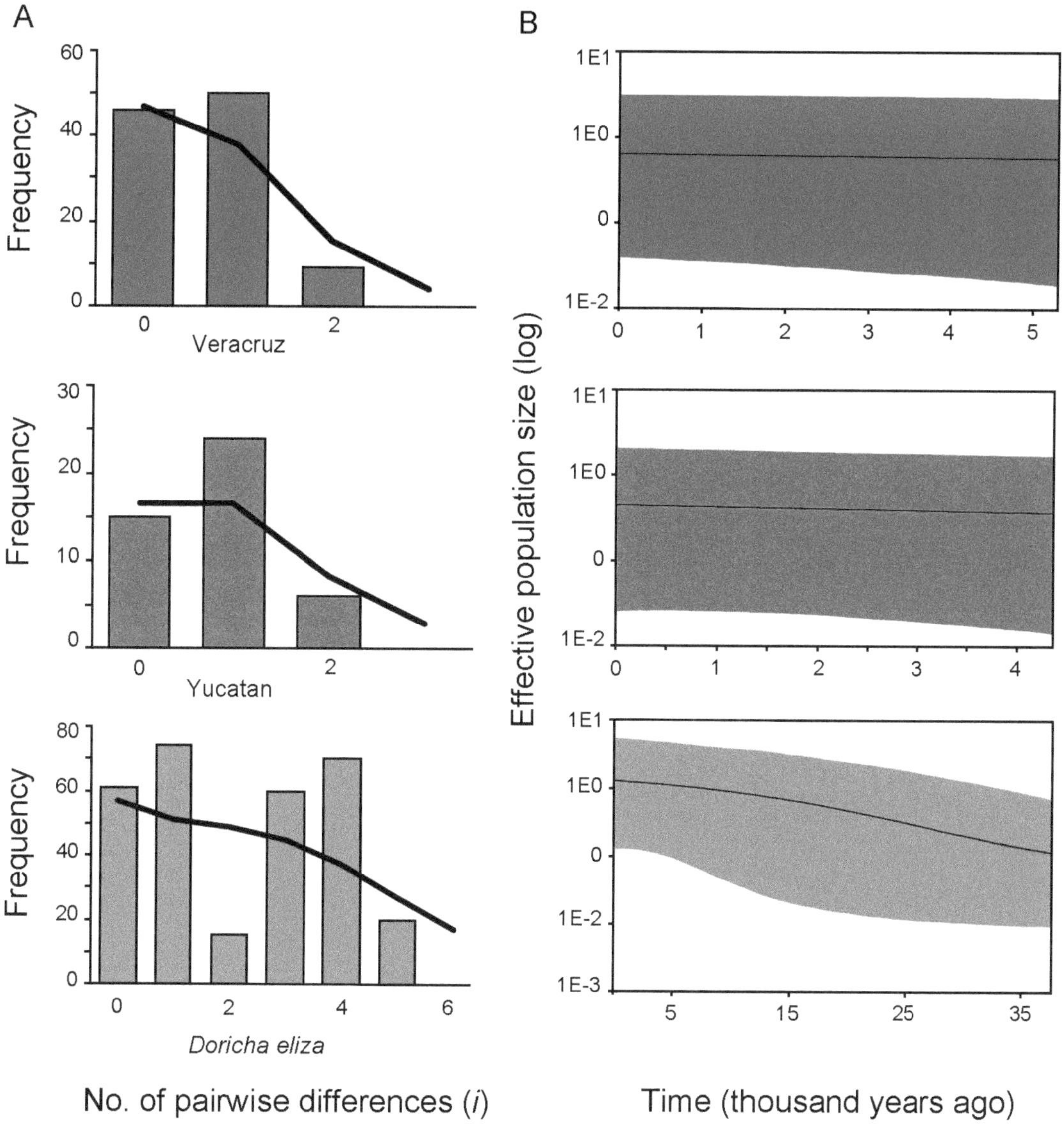

Figure 3. Mistmatch distributions (A) and Bayesian skyline plots (B) showing historical demographic trends of Veracruz, the Yucatan and *Doricha eliza* populations using mitochondrial sequences. Histograms correspond to observed frequencies of pairwise nucleotide differences, and lines represent the expected frequencies under a sudden expansion model. The *y* axis of the skyline plots is the product between effective population size and the generation time and the *y* axis is time in thousands of years. A mutation rate of 2.9×10^{-2} substitutions/site/lineage/million years (s/s/l/My) for ND2, 2.2×10^{-2} s/s/l/My for ATPase, and 1×10^{-3} s/s/l/My for 20454 based on rates obtained for Hawaiian honeycreepers [33]. Solid lines represent median estimates and shaded areas represent 95% confidence intervals. The color-coding, as in Fig. 2, is blue for Veracruz, rose-pink color for the Yucatan, and orange for all populations of *D. eliza*.

accuracy is typically improved with the addition of characters even if much of the information for those characters is missing [76–77]. Most suspected members of the Mellisugini [3,78], *Archilochus, Atthis, Calothorax, Calliphlox, Calypte, Chaetocercus, Doricha, Eulidia, Microstilbon, Myrmia, Myrtis, Rhodopis, Selasphorus* (incl. *Stellula*), *Thaumastura* and *Tilmatura*, are recovered within two groups (sheartails and "Selasphorus" + woodstars). A previous study [6] using mtDNA sequences confirmed the inclusion of *Doricha, Calothorax, Atthis* and *Tilmatura* in the Mellisugini as suggested by McGuire et al. [3], with *Tilmatura dupontii* as the only representative of the woodstars and sister to all other bees in that study, whereas our phylogenetic analyses place *T. dupontii* closer to South American woodstars.

According to our phylogenetic analyses of the combined data set (ND2+ATPase+20454), the Mexican Sheartail hummingbird (*D. eliza*) is strongly supported as the sister group to *D. enicura*, and together they appear as the sister to *Calothorax* species forming the group of sheartails with strong support. The relationship between sheartails and woodstars, however, received moderate support, and *Archilochus colubris* and *A. alexandri* cluster with the woodstars. In a recent study surveying Mellisugini relationships using nuclear and mtDNA sequences [78], *Archilochus* species appeared in a clade with *Calliphlox evelynae* and *Mellisuga minima* but a sister relationship between this clade and sheartails (*Calothorax lucifer* and *Doricha eliza*) was not supported. More data is necessary, including that of *Mellisuga helenae, Chaetocercus heliodor, Ch. astreans, Ch. berlepschi* and *Ch. jourdanii*, to verify this position and to corroborate the monophyly of woodstars. Based on the BI and ML analyses of the combined data set (ND2+ATPase+20454), we propose that *D.*

Table 3. Results of isolation-with-migration model (IMa) for the splits between the Veracruz (V) and Yucatan (Y) populations of *Doricha eliza*.

	Model parameter estimates					
	q_V	q_Y	q_A	t	$m_{V\ to\ Y}$	$m_{Y\ to\ V}$
Veracruz vs. the Yucatan						
Mean	2.213	1.712	6.538	1.917	1.155	1.192
HPD95Lo	1.360	1.206	0.994	1.505	0.465	0.495
HPD95Hi	3.056	2.206	12.883	2.341	1.823	1.911
	Demographic parameter estimates					
	N_V	N_Y	N_A	*t*	$Nm_{Y\ to\ V}$	$Nm_{V\ to\ Y}$
Veracruz vs. the Yucatan						
Mean	1,830	1,410	5,380	22,100	1.280	1.020
HPD95Lo	1,120	993	819	17,400	0.316	0.299
HPD95Hi	2,520	1,830	10,600	27,000	2.790	2.120

Model parameters indicate estimates without using the molecular rate of evolution for six parameters (IMa output values). Demographic rates represent parameters scaled to rates of molecular evolution; *q* parameters in thousands of effective population size (*Ne*), *m* in genes per generation of effective migration rate (*Nm*), *t* parameter in thousands of years.

eliza is sister to *D. enicura* and both form a monophyletic clade with *Calothorax* species.

Divergence Date Estimates

An important question that is implicit to our study is how much time after isolation (or colonization) is required for genetic and morphological variation to arise in natural populations. For Mexican Sheartails, the monophyly of *D. eliza* is indicative of a single isolation or relatively recent colonization event from the Yucatan to Veracruz, perhaps in the last 120,000 years. The star-shaped haplotype network recovered in the ND2, ATPase, 20454 data sets and in the combined ND2+ATPase data set, and the lack of shared mtDNA haplotypes between the Veracruz and Yucatan populations also suggest a recent isolation or colonization followed by haplotype differentiation *in situ*. In contrast to the mtDNA pattern, one of the nuDNA locus 20454 low-frequency haplotypes is shared between populations, suggesting that the nuclear genome also became differentiated after a short history of isolation or colonization. The quasi star-shaped haplotype networks with some low frequency singletons separated from high frequency central haplotypes by a single mutational step, the moderate levels of differentiation between populations, and a mismatch distribution of pairwise differences among haplotypes indicating a sudden increase in expansion from a single population are all expected for a species that rapidly expanded from a single refugium with high levels of gene flow [51,79–80]. The modeled paleodistribution suggests that suitable LGM habitat for the Mexican Sheartail would have expanded under both the MIROC and CCSM scenarios, but suitable habitat conditions were not predicted in Veracruz during the LIG. While populations may have expanded during the LGM, the disjunction persisted and, therefore, our genetic results along with those of paleodistribution modeling correspond to the hypothesis of a relatively recent colonization event from the Yucatan to Veracruz.

Colonization

Mellisugini are a recently diverged lineage [3,6–7], and are part of a radiation that includes the evolution of several species of Nearctic-Neotropical migrants [37,81]. Despite the observed genetic differentiation between the two populations of *D. eliza*, the question of how their isolation occurred remains unanswered. It is likely that the Veracruz population represents a relatively recent colonization event, though it is difficulty to directly observe immigration events in nature [82]. Colonization has been more important than large-scale vicariance in determining the phylogenetic structure of hummingbird faunas, particularly the insular Mellisugini species assemblage of the West Indies [83], owing to their high dispersal ability, their capacity to adapt to novel environments [82–83], and the fact that migratory behavior can evolve rapidly in response to selection [37]. The Mellisugini are highly opportunistic generalists that, seasonally and altitudinally, cover large distances to track floral resources [5,84–85]. These migratory habits confer a natural vagility and may have predisposed them to fly long distances and tolerate a wide range of ecological regimes [84]. Although migratory behavior might have increased the colonization success of Mellisugini in the West Indies and remote geographic areas with a seasonal climate, vagrancy does not appear to predict the colonization of oceanic islands or remote areas [83], and it is not known whether migratory Mellisugini species are more prone to vagrancy than sedentary hummingbird species such as the Mexican Sheartail. An alternative explanation is that ancestral colonizers arrived naturally from Yucatan to Veracruz, a direction potentially assisted by the prevailing east-to-west trade winds and hurricanes. Our estimates of historical gene flow indicating a general trend of unidirectional gene flow between populations correspond to a Yucatan-to-Veracruz direction of historical migration.

Genetic and Morphological Differentiation between Disjunct Populations

Our results reveal moderate mtDNA divergence between the Veracruz and Yucatan populations of *D. eliza* but reciprocal monophyly of haplotypes, supporting the hypothesis of a short history of isolation. Moderate levels of haplotype and nucleotide diversity of populations suggest relatively small population sizes and founder effects. Significant genetic differentiation between

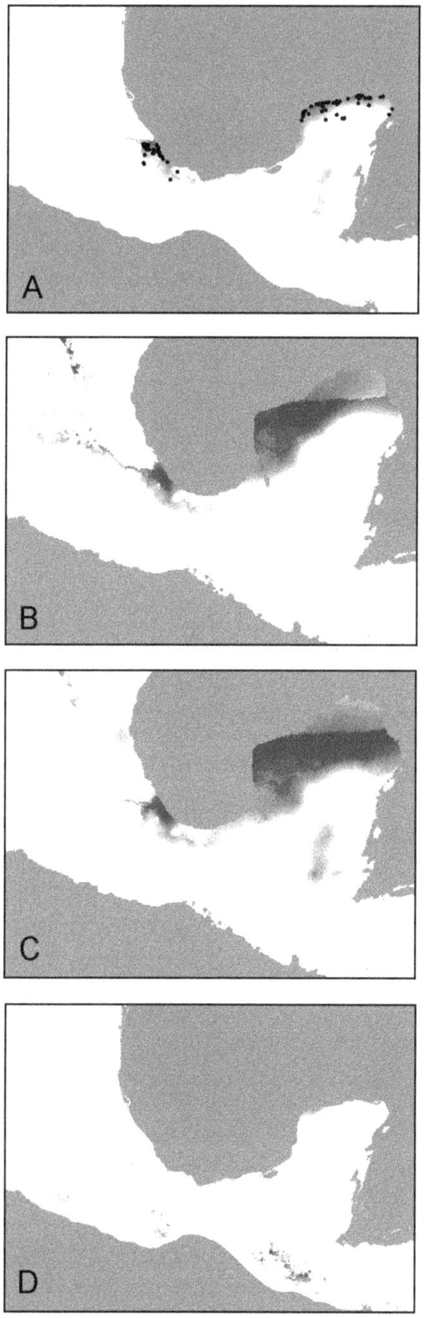

Figure 4. Distributional records and species distribution models for *Doricha eliza* at present (A), the Last Glacial Maximum (LGM, 21 ka) (B, CCSM model; C, MIROC model), and the Last Interglacial (LIG, 140–120 ka) (D) climate conditions. The darkest colors indicate the highest predicted probability of occurrence.

populations and limited gene flow resulting from barriers to dispersal have been found for montane hummingbird species in particular, such as Speckled Hummingbird (*Adelomyia melanogenys*) [86], Wedge-tailed Sabrewing (*Campylopterus curvipennis*) [87], Azure-crowned Hummingbird (*Amazilia cyanocephala*) [38,88], Broad-tailed Hummingbird (*Selasphorus platycercus*) [37] and Amethyst-throated Hummingbird (*Lampornis amethystinus*) [J. F. Ornelas, C. González, B. Hernández-Baños and J. García-Moreno, unpublished data]. Limited differentiation has been found in other species with a lowland distribution, such as the Rufous-tailed Hummingbird (*Amazilia tzacatl*) [89] and the Long-billed Hermit (*Phaethornis longirostris*) [90]. The Escudo Hummingbird (*A. t. handleyi*), endemic to the Caribbean island Escudo de Veraguas in western Panama, initially described as a distinct species on the basis of its considerably larger size and darker plumage, is slightly differentiated (ND2; 2 substitutions; uncorrected distance 0.2–0.5%) from the mainland *A. tzacatl* c. 10 km away [89].

Overall, population differentiation in the Mexican Sheartail seems primarily enhanced by isolation, which is reasonable for populations separated by a long distance. The Veracruz colonization hypothesis is consistent with the lower migration rate of the Veracruz population to Yucatan than was found for the opposite direction, and with the results of the tests of niche conservatism that suggest that the Veracruz colonization with gene flow was facilitated by niche similarity (PC1). Consequently, following geographic isolation, the populations of *D. eliza* separated by the Gulf of Mexico would have been exposed and eventually adapted to the different environmental conditions. Populations of the Mexican Sheartail separated by 780 km (and by the Gulf) are distributed in a unique environmental space, implying that the different environmental conditions in the Yucatan Peninsula and in Veracruz would have reduced gene flow, as shown by the IMa results, and this would have reinforced the divergence of the two mtDNA haplogroups following the initial spatial separation. This scenario is supported by our tests of niche divergence and conservatism that compared the amount of climatic divergence to the null expectation of background climatic divergence and that showed evidence for niche divergence between the *D. eliza* records of Veracruz and the Yucatan on two axes of environmental space related to annual precipitation and mean diurnal temperature range (PC2 and PC3). These findings support the hypothesis that climatic niche dissimilarity between *D. eliza* populations separated by the Gulf seems to have reduced gene flow. Our analyses of *D. eliza*, combining a phylogeographic and species distribution modeling approach, suggest that the observed patterns of genetic variation and divergence between the Veracruz and Yucatan populations are best explained by a combination of isolation exacerbated by subsequent climate differentiation between regions. Although the latter may be true for species that disperse poorly or that are reluctant to cross areas of less hospitable habitat for physiological reasons, niche divergence for other species with poor dispersal may mean enhanced opportunities for isolation and reduced gene flow, thereby increasing the likelihood of speciation.

Our morphological analysis confirmed that *D. eliza* hummingbirds from Veracruz are similar in most trait mean values to individuals from the Yucatan population (see also [12]). Patterns of limited population differentiation in size trait values were surprising given the large geographic separation between the two populations and habitat differences. Studies of hummingbirds, such as *A. melanogenys* [86], *C. curvipennis* [87], *A. cyanocephala* [38], and a member of the Mellisugini, *S. platycercus* [37], found significant size differences between populations in different habitats, yet separated by shorter distances. In all these cases, the genetic break at the potential barriers corresponds to differences in morphology and to the lack of overlap in environmental space between lineages on both sides of the barrier. One possible explanation for this pattern is that vicariance and ecological divergence have both played an important role in the strong morphological differentiation between populations that are physically separated [38,86–87]. While this hypothesis may hold true for these hummingbird species, for which divergence times between populations were estimated to have occurred c. 700,000 years ago, the hypothesis alone is insufficient to explain the limited

Table 4. Loadings of the environmental variables for each PC axis and tests of niche divergence and conservatism.

	Niche Axes		
	PC1	**PC2**	**PC3**
BIO1 Annual Mean Temperature	−0.2471	−0.5043	0.1829
BIO2 Mean Diurnal Range	0.4634	−0.0849	0.7049
BIO3 Isothermality	0.5661	−0.3322	0.0686
BIO4 Temperature Seasonality	−0.4005	0.2751	0.6713
BIO12 Annual Precipitation	0.0082	0.6121	0.0565
BIO14 Precipitation of Driest Month	0.4932	0.4217	−0.1052
Percent variance explained	38.8	34.2	15.7
Observed differences	0.184*	**3.637***	**0.613***
Null distribution	(1.801–1.820)	(1.973–1.989)	(0.334–0.3353)

Observed differences in climatic niche of *Doricha eliza* lineages (Veracruz and the Yucatan) on each PC axis. Bold values indicate significant niche divergence of the differences between their environmental backgrounds compared to the middle 95th percentile of a null distribution (in parentheses).
*Significance level, $P<0.05$.

morphological variation between the Veracruz and Yucatan populations of *D. eliza*. Interestingly, the time of divergence between *D. eliza* populations was estimated at 120,000 years ago, supporting the hypothesis of a short period of isolation and limited morphological differentiation. However, male individuals from the Yucatan population had smaller values for the outermost rectrices than did males from Veracruz. It remains to be tested with larger sample sizes whether these differences represent significant levels of variation affecting the males' acrobatic displays ([10], **Figure S1** and **Video S1**), and thus increased sexual selection in the smaller population of Veracruz.

Conservation and Management Considerations

Our results reveal that the Veracruz and Yucatan populations of *D. eliza* are genetically differentiated, and that the outermost rectrices of male hummingbirds from Veracruz are longer than those of the males from Yucatan. The Mexican Sheartail Hummingbird is globally near threatened and both the Veracruz and Yucatan populations are locally endangered with population declines in Veracruz resulting from severe habitat degradation caused by livestock grazing, sugarcane cultivation and residential development, while the Yucatan population is under pressure mainly from the development of its coastal dune habitat for tourism [1,13–15]. Here we have identified that the disjunct populations of *D. eliza* constitute distinct genetic lineages, and that the importance of these populations as reservoirs of endemic genetic diversity require different management approaches and merit targeted conservation efforts to preserve the unique genetic pools of both populations and their habitats.

Supporting Information

Figure S1 Stills from video recording, showing moments of a rocking pendulum flight displayed by a *Doricha eliza* male to a female at the nest. (A) Photograph shows a male *D. eliza* from the Veracruz population. Photograph by Gerardo Sánchez Vigil. (B) Photograph shows a female *D. eliza* from the Veracruz population. Photograph by Yuyini Licona Vera. (C–J) The male begins the courtship display doing a pendulum flight (from left to right) in front of the female. During the display, the male extends his throat feathers and fully displays tail rectrices, while approaching the female repeatedly. The entire time, the female at the nest follows the male's movements (red arrows). This pendulum flight is done repeatedly around the female (from right to left and from left to right) and is finished with an upward flight (not shown in the video). The video is available as supplementary material – Video S1.

Figure S2 Bayesian posterior probabilities and bootstrap support for MrBayes and Maximum Likelihood analyses. Illustration of tree topology based on ND2 sequences for North American and South American members of the Mellisugini clade. Values above branches denote posterior probabilities (PP) and those below branches denote bootstrap values.

Figure S3 Bayesian posterior probabilities and bootstrap support for MrBayes and Maximum Likelihood analyses. Illustration of tree topology based on the nuDNA locus 20454 unphased sequences from *D. eliza* and outgroups. Values above branches denote posterior probabilities (PP) and those below branches denote bootstrap values.

Figure S4 Morphological differences between the Veracruz and Yucatan populations of *D. eliza* males. Data are means and 95% confidence intervals for total body length (A), exposed culmen (B), base bill-width (C), wing chord (D), length of right outermost rectrix (E), and length of left outermost rectrix (F). Measurements are in mm.

Figure S5 Morphological differences between the Veracruz and Yucatan populations of *D. eliza* females. Data are means and 95% confidence intervals for total body length (A), exposed culmen (B), base bill-width (C), wing chord (D), and tail length (E). Measurements are in mm.

Table S1 Code for identification (ID), sex, state and locality of origin, geographic coordinates and elevation of sampled individuals of *Doricha eliza*.

Table S2 Species names, sequence data and GenBank accession numbers for *Doricha eliza* (25) and outgroups (51) used in this study.

Table S3 Code for identification (ID), sex, morphological data and GenBank accession numbers for *Doricha eliza* individuals used in this study.

Video S1 Rocking pendulum flight as displayed by male Mexican Sheartails. The video was recorded at Miradores, Veracruz during the breeding season (07 September 2012).

Acknowledgments

We extend our sincere gratitude to C. Bárcenas, I. Chávez-Domínguez, R. Díaz-Valenzuela, C. González, R.A. Jiménez, A. Malpica, G. Mejía, A. Montes de Oca, A.E. Ortiz Rodríguez, F. Rodríguez-Gómez, E. Ruíz-González, E. Ruiz-Sánchez, M.J. Pérez Crespo, Pronatura Veracruz and A. Vásquez for field and laboratory assistance, and to G. Sánchez Vigil for permission to use his photographs in Figures 1 and 2. A. Navarro-Sigüenza, C. Gutiérrez-Rodríguez, E. Ruiz-Sánchez, and three anonymous reviewers provided useful comments on previous versions of the manuscript. Our fieldwork was conducted with the assistance of Irving Chávez and the permission of the Mexican Government, Secretaría de Medio Ambiente y Recursos Naturales, Subsecretaría de Gestión para la Protección Ambiental, Dirección General de Vida Silvestre (permit number SGPA/DGVS/07701/11). This work constitutes partial fulfillment of YLV's degree requirements in Biodiversity and Systematics at INECOL.

Author Contributions

Conceived and designed the experiments: YLV JFO. Performed the experiments: YLV JFO. Analyzed the data: YLV JFO. Contributed reagents/materials/analysis tools: YLV JFO. Wrote the paper: YLV JFO.

References

1. IUCN (2013) The IUCN Red List of Threatened Species. Version 3.1. Available: http://www.iucnredlist.org. Accessed 2013 Nov. 23.
2. SEMARNAT (2010) Norma Oficial Mexicana NOM-059-ECOL-2010, Protección ambiental-Especies nativas de México de flora y fauna silvestres-categorías de riesgo y especificaciones para su inclusión, exclusión o cambio-lista de especies en riesgo. Diario Oficial de la Federación 30 December 2010. México, DF, 1–77.
3. McGuire JA, Witt CC, Remsen JV Jr, Dudley R, Altshuler DL (2009) A higher-level taxonomy for hummingbirds. J Ornithol 150: 155–165.
4. Ridgway R (1911) The Birds of North and Middle America. Bull US Nat Mus 50: 650–651.
5. Schuchmann KL (1999) Family Trochilidae (Hummingbirds). In: del Hoyo J, Elliott A and Sargatal J, editors. Handbook of the Birds of the World, Vol. 5, Barn-owls to hummingbirds. Barcelona, Spain: Lynx Editions, 468–535.
6. Ornelas JF, González C, Espinosa de los Monteros A, Rodríguez-Gómez F, García-Feria LM (2014) In and out of Mesoamerica: temporal divergence of *Amazilia* hummingbirds pre-dates the orthodox account of the completion of the Isthmus of Panama. J Biogeogr 41: 168–181.
7. McGuire JA, Witt CC, Altshuler DL, Van Remsen J Jr (2007) Phylogenetic systematics and biogeography of hummingbirds: Bayesian and maximum likelihood analyses of partitioned data and selection of an appropriate partitioning strategy. Syst Biol 56: 837–856.
8. Howell SNG, Webb S (1995) A guide to the birds of Mexico and northern Central America. Oxford, UK: Oxford University Press.
9. Scott PE (1994) Lucifer Hummingbird (*Calothorax lucifer*). In: Poole A and Gill F, editors. The Birds of North America, no. 134, Philadelphia, PA: Academy of Natural Sciences.
10. Díaz-Valenzuela R, Lara-Rodríguez NZ, Ortiz-Pulido R, González-García F, Ramírez Bautista A (2011) Some aspects of the reproductive biology of the Mexican Sheartail (*Doricha eliza*) in central Veracruz. Condor 113: 177–182.
11. Clark CJ, Feo TJ, Van Dongen WFD (2013) Sounds and courtship displays of the Peruvian Sheartail, Chilean Woodstar, Oasis Hummingbird, and a hybrid male Peruvian Sheartail × Chilean Woodstar. Condor 115: 558–575.
12. Ortiz-Pulido R, Peterson AT, Robbins MB, Díaz R, Navarro-Sigüenza AG, et al. (2002) The Mexican Sheartail (*Doricha eliza*): morphology, behavior, distribution, and endangered status. Wilson Bull 114: 153–160.
13. Ornelas JF, Licona-Vera Y (2014) *Doricha eliza* (Lesson & DeLattre, 1839). In: Hernández Baz F and Rodríguez Vargas DU, editors. Libro Rojo de la Fauna del Estado de Veracruz, Xalapa, Veracruz: Gobierno del Estado de Veracruz, 148–150.
14. Santamaría-Rivero W, MacKinnon B, Leyequién E (2013) Registros de anidación del colibrí tijereta mexicano (*Doricha eliza*) en el estado de Yucatán, México. Huitzil 14: 139–145.
15. Ortiz-Pulido R, Flores Ceballos E, Ortiz Pulido R (1998) Descripción del nido de *Doricha eliza* y ampliación de su rango. Orn Neotrop 9: 223–224.
16. Sorenson MD, Ast JC, Dimcheff DE, Yuri T, Mindell DP (1999) Primers for a PCR-based approach to mitochondrial genome sequencing in birds and other vertebrates. Mol Phylogenet Evol 12: 105–114.
17. Eberhard JR, Bermingham E (2004) Phylogeny and biogeography of the *Amazona ochrocephala* (Aves: Psittacidae) complex. Auk 121: 318–332.
18. Backström N, Fagerberg S, Ellegren H (2008) Genomics of natural bird populations: a gene-based set of reference markers evenly spread across the avian genome. Mol Ecol 17: 964–980.
19. Stephens M, Donelly P (2003) A comparison of Bayesian methods for haplotype reconstruction from population genotype data. Am J Human Genet 73: 1162–1169.
20. Stephens M, Smith NJ, Donelly P (2001) A new statistical method for haplotype reconstruction from population data. Am J Human Genet 68: 978–989.
21. Ronquist F, Huelsenbeck JP (2003) MrBayes 3: Bayesian phylogenetic inference under mixed models. Bioinformatics 19: 1572–1574.
22. Stamatakis A (2006) RAxML-VI-HPC: maximum likelihood-based phylogenetic analyses with thousands of taxa and mixed models. Bioinformatics 22: 2268–2690.
23. Miller MA, Pfeiffer W, Schwartz T (2010) Creating the CIPRES Science Gateway for inference of large phylogenetic trees. Proceedings of the Gateway Computing Environments Workshop (GCE), 14 Nov. 2010, New Orleans, LA, 1–8.
24. Posada D (2008) jModelTest: phylogenetic model averaging. Mol Biol Evol 25: 1253–1256.
25. Alfaro ME, Huelsenbeck JP (2006) Comparative performance of Bayesian and AIC-based measures of phylogenetic model uncertainty. Syst Biol 55: 89–96.
26. Felsenstein J (1985) Confidence limits on phylogenies: an approach using the bootstrap. Evolution 39: 783–791.
27. Nylander JAA, Ronquist F, Huelsenbeck JP, Nieves-Aldrey JL (2004) Bayesian phylogenetic analysis of combined data. Syst Biol 53: 47–67.
28. Hillis DM, Bull JJ (1993) An empirical test of bootstrapping as a method for assessing confidence in phylogenetic analyses. Syst Biol 42: 182–192.
29. Heled J, Drummond AJ (2010) Bayesian inference of species trees from multilocus data. Mol Biol Evol 27: 570–580.
30. Degnan JH, Rosenberg NA (2009) Gene tree discordance, phylogenetic inference and the multispecies coalescent. Trends Ecol Evol 24: 332–340.
31. Drummond AJ, Rambaut A (2007) BEAST: Bayesian evolutionary analysis by sampling trees. BMC Evol Biol 7: 214.
32. Heled J, Drummond AJ (2010) Bayesian inference of species trees from multilocus data. Mol Biol Evol 27: 570–580.
33. Lerner HRL, Meyer M, James HF, Hofreiter M, Fleischer RC (2011) Multilocus resolution of phylogeny and timescale in the extant adaptive radiation of Hawaiian honeycreepers. Current Biol 21: 1838–1844.
34. Hosner PA, Nyári AS, Moyle RG (2013) Water barriers and intra-island isolation contribute to diversification in the insular *Aethopyga* sunbirds (Aves: Nectariniidae). J Biogeogr 40: 1094–1106.
35. Voelker G, Bowie RCK, Klicka J (2013) Gene trees, species trees and Earth history combine to shed light on the evolution of migration in a model avian system. Mol Ecol 22: 3333–3344.
36. Pacheco MA, Battistuzzi FU, Lentino M, Aguilar RF, Kumar S, et al. (2011) Evolution of modern birds revealed by mitogenomics: timing the radiation and origin of major orders. Mol Biol Evol 28: 1927–1942.
37. Malpica A, Ornelas JF (2014) Postglacial northward expansion and genetic differentiation between migratory and sedentary populations of the broad-tailed hummingbird (*Selasphorus platycercus*). Mol Ecol 23: 435–452.
38. Rodríguez-Gómez F, Gutiérrez-Rodríguez C, Ornelas JF (2013) Genetic, phenotypic and ecological divergence with gene flow at the Isthmus of Tehuantepec: the case of the azure-crowned hummingbird (*Amazilia cyanocephala*). J Biogeogr 40: 1360–1373.
39. Ho SYW, Lanfear R, Bromham L, Phillips MJ, Soubrier J, et al. (2011) Time-dependent rates of molecular evolution. Mol Ecol 20: 3087–3101.
40. Smith BT, Klicka J (2010) The profound influence of the Late Pliocene Panamanian uplift on the exchange, diversification, and distribution of New World birds. Ecography 33: 333–342.
41. Marino IAM, Pujolar JM, Zane L (2011) Reconciling deep calibration and demographic history: Bayesian inference of post glacial colonization patterns in *Carcinus aestuarii* (Nardo, 1847) and *C. maenas* (Linnaeus, 1758). PLoS One 6: e28567.
42. Clement M, Posada D, Crandall KA (2000) TCS: a computer program to estimate genealogies. Mol Ecol 9: 1657–1659.

43. Excoffier L, Smouse PE, Quattro JM (1992) Analysis of molecular variance inferred from metric distances among DNA haplotypes: application to human mitochondrial DNA restriction data. Genetics 131: 479–491.
44. Excoffier L, Laval G, Schneider S (2005) Arlequin ver. 3.0: an integrated software package for population genetics data analysis. Evol Bioinformatics 1: 47–50.
45. Nei M (1987) Molecular Evolutionary Genetics. New York, NY: Columbia University Press.
46. Librado P, Rozas J (2009) DnaSP v5: a software for comprehensive analysis of DNA polymorphism data. Bioinformatics 25: 1451–1452.
47. Tajima F (1989) Statistical-method for testing the neutral mutation hypothesis by DNA polymorphism. Genetics 123: 585–595.
48. Fu YX (1997) Statistical neutrality of mutations against population growth, hitchhiking and background selection. Genetics 147: 915–925.
49. Harpending RC (1994) Signature of ancient population growth in a low-resolution mitochondrial DNA mismatch distribution. Human Biol 66: 591–600.
50. Schneider S, Excoffier L (1999) Estimation of demographic parameters from the distribution of pairwise differences when the mutation rates vary among sites: Application to human mitochondrial DNA. Genetics 152: 1079–1089.
51. Rogers AR, Harpending H (1992) Population growth makes waves in the distribution of pairwise differences. Mol Biol Evol 9: 552–569.
52. Drummond AJ, Rambaut A, Shapiro B, Pybus OG (2005) Bayesian coalescent inference of past population dynamics from molecular sequences. Mol Biol Evol 22: 1185–1192.
53. Ho WYS, Shapiro B (2011) Skyline-plot methods for estimating demographic history from nucleotide sequences. Mol Ecol Res 11: 423–434.
54. Hey J, Nielsen R (2004) Multilocus methods for estimating population sizes, migration rates and divergence time, with applications to the divergence of *Drosophila pseudo-obscura* and *D. persimilis*. Genetics 167: 747–760.
55. Hey J, Nielsen R (2007) Integration within the Felsenstein equation for improved Markov chain Monte Carlo methods in population genetics. Proc Natl Acad Sci USA 104: 2785–2790.
56. Woerner AE, Cox MP, Hammer MF (2007) Recombination-filtered genomic datasets by information maximization. Bioinformatics 23: 1851–1853.
57. Storchová R, Reif J, Nachman MW (2009) Female heterogamety and speciation: reduced introgression of the Z chromosome between two species of nightingales. Evolution 64: 456–471.
58. Calder WA, Calder LL (1992) Broad-tailed Hummingbird. In: Poole A, Stettenhein P and Gill F, editors. The Birds of North America. Philadelphia, PA: Academy of Natural Sciences of Philadelphia, 1–16.
59. Ruiz-Gutiérrez V, Doherty PF, Santana E, Contreras Martínez S, Schondube J, et al. (2012) Survival of resident Neotropical birds: considerations for sampling and analysis based on 20 years of bird-banding efforts in Mexico. Auk 129: 500–509.
60. Lande R, Engen S, Sæther BE (2003) Stochastic population dynamics in ecology and conservation. Oxford, UK: Oxford University Press.
61. Spellman GM, Klicka J (2006) Testing hypotheses of Pleistocene population history using coalescent simulations: phylogeography of the pygmy nuthatch (*Sitta pygmaea*). Proc R Soc Lond B 273: 3057–3063.
62. Elith JH, Phillips SJ, Hastie T, Dudík M, Chee YE, et al. (2011) A statistical explanation of MaxEnt for ecologists. Divers Dist 17: 43–57.
63. Navarro AG, Peterson AT, Gordillo-Martínez A (2003) Museums working together: the atlas of the birds of Mexico. In: Collar N, Fisher C and Feare C, editors. Why museums matter: avian archives in an age of extinction. Bull Brit Ornithologists' Club Suppl 123A.
64. Phillips SJ, Anderson RP, Schapire RE (2006) Maximum entropy modeling of species geographic distributions. Ecol Modell 190: 231–259.
65. Phillips SJ, Dudík M (2008) Modeling of species distributions with MaxEnt: new extensions and a comprehensive evaluation. Ecography 31: 161–175.
66. Hijmans RJ, Cameron SE, Parra JL, Jones PG, Jarvis A (2005) Very high resolution interpolated climate surfaces for global land areas. Int J Climatol 25: 1965–1978.
67. Hammer O (2011) PAST (Paleontological Statistics). Natural History Museum, University of Oslo. Available: http://folk.uio.no/ohammer/past/index.html.
68. Mertz CE (1978) Basic principles in ROC analysis. Sem Nuclear Med 8: 283–298.
69. Braconnot P, Otto-Bliesner B, Harrison S, Joussaume S, Peterchmitt JY, et al. (2007) Results of PMIP2 coupled simulations of the Mid-Holocene and Last Glacial Maximum – Part 2: feedbacks with emphasis on the location of the ITCZ and mid- and high latitudes heat budget. Climate Past 3: 279–296.
70. Collins WD, Bitz CM, Blackmon ML, Bonan GB, Bretherton CS, et al. (2004) The community climate system model: CCSM3. J Climate 19: 2122–2143.
71. Hasumi H, Emori S (2004) K-1 coupled GCM (MIROC) description, K-1 Tech. Rep. 1, Climate Syst Res, Kashiwa, Japan.
72. Otto-Bliesner BL, Marshall SJ, Overpeck JT, Miller GH, Hu A (2006) Simulating Arctic climate warmth and icefield retreat in the Last Interglaciation. Science 311: 1751–1753.
73. Otto-Bliesner BL, Hewitt CD, Marchitto TM, Brady E, Abe-Ouchi A, et al. (2007) Last Glacial Maximum ocean thermohaline circulation: PMIP2 model intercomparisons and data constraints. Geophys Res Lett 34: L12707.
74. McCormack JE, Zellmer AJ, Knowles LL (2010) Does niche divergence accompany allopatric divergence in *Aphelocoma* jays as predicted under ecological speciation?: insights from tests with niche models. Evolution 64: 1231–1244.
75. Lemmon AR, Brown JM, Stanger-Hall K, Lemmon EM (2009) The effect of ambiguous data on phylogenetic estimates obtained by maximum likelihood and Bayesian inference. Syst Biol 58: 130–145.
76. Wiens JJ, Morrill MC (2011) Missing data in phylogenetic analysis: reconciling results from simulations and empirical data. Syst Biol 60: 719–731.
77. Wiens JJ, Tiu J (2012) Highly incomplete taxa can rescue phylogenetic analyses from the negative effects of limited taxon sampling. PLoS One 7: e42925.
78. McGuire JA, Witt CC, Remsen JV Jr, Corl A, Rabosky DL, et al. (2014) Molecular phylogenetics and the diversification of hummingbirds. Current Biol 24: 910–916.
79. Slatkin M, Hudson RR (1991) Pairwise comparisons of mitochondrial DNA sequences in stable and exponentially growing populations. Genetics 129: 555–562.
80. Hutchinson DW, Templeton AR (1999) Correlation of pairwise genetic and geographic distance measures: inferring the relative influences of gene flow and drift on the distribution of genetic variability. Evolution 53: 1898–1914.
81. Berns CM, Adams DC (2013) Becoming different but staying alike: patterns of sexual size and shape dimorphism in bills of hummingbirds. Evol Biol 40: 246–260.
82. Lees AC, Gilroy JJ (2014) Vagrancy fails to predict colonization of oceanic islands. Global Ecol Biogeogr 23: 405–413.
83. Bleiweiss R (1998) Origin of hummingbird faunas. Biol J Linn Soc 65: 77–97.
84. Des Granges JL (1979) Organization of a tropical nectar feeding bird guild in a variable environment. Living Bird 17: 199–236.
85. Ornelas JF, Arizmendi MC (1995) Altitudinal migration: implications for the conservation of the Neotropical migrant avifauna of western Mexico. In: Wilson M and Sader S, editors. Conservation of Neotropical migratory birds in Mexico. Maine, USA: Maine Agriculture and Forestry Experimental Station, 98–112.
86. Chaves JA, Pollinger JP, Smith TB, LeBuhn G (2007) The role of geography and ecology in shaping the phylogeography of the speckled hummingbird (*Adelomyia melanogenys*) in Ecuador. Mol Phylogenet Evol 43: 795–807.
87. González C, Ornelas JF, Gutiérrez-Rodríguez C (2011) Selection and geographic isolation influence hummingbird speciation: genetic, acoustic and morphological divergence in the wedge-tailed sabrewing (*Campylopterus curvipennis*). BMC Evol Biol 11: 38.
88. Rodríguez-Gómez F, Ornelas JF (2014) Genetic divergence of the Mesoamerican Azure-crowned hummingbird (*Amazilia cyanocephala*, Trochilidae) across the Motagua-Polochic-Jocotán fault system. J Zool Syst Evol Res 52: 142–153.
89. Miller MJ, Lelevier MJ, Bermingham E, Klicka JT, Escalante P, et al. (2011) Phylogeography of the rufous-tailed hummingbird (*Amazilia tzacatl*). Condor 113: 806–816.
90. Arbeláez-Cortés E, Navarro-Sigüenza AG (2013) Molecular evidence of the taxonomic status of western Mexican populations of *Phaethornis longirostris* (Aves: Trochilidae). Zootaxa 3716: 81–97.

9

The Earliest Giant *Osprioneides* Borings from the Sandbian (Late Ordovician) of Estonia

Olev Vinn[1]*, Mark A. Wilson[2], Mari-Ann Mõtus[3]

1 Department of Geology, University of Tartu, Tartu, Estonia, **2** Department of Geology, The College of Wooster, Wooster, Ohio, United States of America, **3** Institute of Geology, Tallinn University of Technology, Tallinn, Estonia

Abstract

The earliest *Osprioneides kampto* borings were found in bryozoan colonies of Sandbian age from northern Estonia (Baltica). The Ordovician was a time of great increase in the quantities of hard substrate removed by single trace makers. Increased predation pressure was most likely the driving force behind the infaunalization of larger invertebrates such as the *Osprioneides* trace makers in the Ordovician. It is possible that the *Osprioneides* borer originated in Baltica or in other paleocontinents outside of North America.

Editor: Andrew A. Farke, Raymond M. Alf Museum of Paleontology, United States of America

Funding: O.V. is indebted to the Sepkoski Grant program (Paleontological Society), Estonian Science Foundation grant ETF9064, Estonian Research Council grant IUT20-34 and a target-financed project from the Estonian Ministry of Education and Science (SF0180051s08; Ordovician and Silurian climate changes, as documented from the biotic changes and depositional environments in the Baltoscandian Palaeobasin) for financial support. M-A.M was also supported by a target-financed project from the Estonian Ministry of Education and Science (SF0140020s08). The funders had no role in study design, data collection and analysis, decision to publish, or preparation of the manuscript.

Competing Interests: The authors have declared that no competing interests exist.

* E-mail: olev.vinn@ut.ee

Introduction

The oldest macroborings in the world are the small simple holes of *Trypanites* reported in Early Cambrian archaeocyathid reefs in Labrador [1,2]. The next oldest macroborings are found in carbonate hardgrounds of Early Ordovician age [3,4,5,6]. There was a great increase in bioerosion intensity and diversity in the Ordovician, now termed the Ordovician Bioerosion Revolution [7]. In the Middle and Late Ordovician, shells and hardgrounds are often thoroughly riddled with holes, most of them attributable to *Trypanites* and *Palaeosabella* [8]. In addition, Ordovician bioerosion trace fossils include bivalve borings (*Petroxestes*), bryozoan etchings (*Ropalonaria*), sponge borings (*Cicatricula*), *Sanctum* (a cavernous domichnium excavated in bryozoan zoaria by an unknown borer) and *Gastrochaenolites* [8,9]. Bioerosion was very common in the Middle Paleozoic, especially in the Devonian [10]. Later in the Mesozoic bioerosion intensity and diversity further increased [6,9,11,12], and deep, large borings became especially common [13].

The bioerosion trace fossils of Ordovician of North America are relatively well studied [14,15,16]. In contrast, there is a limited number of works devoted to the study of bioerosional trace fossils in the Ordovician of Baltica. The earliest large boring occurs in the Early to Middle Ordovician hardgrounds and could belong to *Gastrochaenolites* [4,17]. Abundant *Trypanites* borings are known from brachiopods of the Arenigian [18] and Sandbian [19]. Wyse Jackson and Key [20] published a study on borings in trepostome bryozoans from the Ordovician of Estonia. They identified two ichnogenera, *Trypanites* and *Sanctum,* in bryozoans of Middle and Upper Ordovician strata of northern Estonia.

The aims of this paper are to: 1) determine whether the shafts in large Sandbian bryozoans belong to previously known or a new bioerosional ichnotaxon for the Ordovician; 2) determine the systematic affinity of the trace fossil; 3) discuss the ecology of the trace makers; 4) discuss the paleobiogeographic distribution of the trace fossil; and 4) discuss the occurrence of large borings during the Ordovician Bioerosion Revolution.

Geological Background and Locality

During the Ordovician, the Baltica paleocontinent migrated from the temperate to the subtropical realm [21,22]. The climatic change resulted in an increase of carbonate production and sedimentation rate on the shelf during the Middle and Late Ordovician. In the Upper Ordovician the first carbonate buildups are recorded, emphasizing a striking change in the overall character of the paleobasin [23].

The total thickness of the Ordovician in Estonia varies from 70 to 180 m [23]. The Ordovician limestones of Estonia form a wide belt from the Narva River in the northeast to Hiiumaa Island in the northwest [23]. In the Middle Ordovician and early Late Ordovician, the slowly subsiding western part of the East-European Platform was covered by a shallow, epicontinental sea with little bathymetric differentiation and an extremely low sedimentation rate. Along the extent of the ramp a series of grey calcareous - argillaceous sediments accumulated (argillaceous limestones and marls), with a trend of increasing clay and decreasing bioclasts in the offshore direction [22].

The material studied here was collected from the Hirmuse Creek (Fig. 1) and Alliku Ditches (Fig. 1) of Sandbian age (Haljala

Stage) (Fig. 2). Hirmuse Creek is located in Maidla parish of Ida-Viru County. Clayey and skeletal limestones with interlayers of marls are exposed on the creek bed and in its banks. The fossil assemblage includes algae (*Mastopora*), brachiopods (*Clinambon, Leptaena, Platystrophia, Apartorthis, Porambonites, Pseudolingula*), conulariids, gastropods (*Lesueurilla, Holopea, Bucanella, Pachydictya, Megalomphala, Cymbularia*), ichnofossils (*Amphorichnus, Arachnostega*), sponges, receptaculitids (*Tettragonis*), rugosans (*Lambelasma*), bryozoans, and asaphid trilobites. The Alliku Ditches are located in Harju County near the village of Alliku. Clayey limestones with interlayers of marls are exposed here. The fauna includes algae, brachiopods, bryozoans (*Aluverina, Annunziopora, Batostoma, Ceramoporella, Coeloclema, Constellaria, Corynotrypa, Crepipora, Dekayella, Diazipora, Diplotrypa, Enallopora, Esthoniophora, Graptodictya, Hallopora, Hemiphragma, Homotrypa, Homotrypella, Kukersella, Lioclemella, Mesotrypa, Monotrypa, Nematopora, Nematotrypa, Oanduella, Moyerella, Pachydictya, Phylloporina, Prasopora, Proavella, Pseudohornera, Rhinidictya, Stictoporella*), echinoderms, gastropods, ostracods, rugosans, receptaculitids and trilobites according to Rõõmusoks [24].

No permits were required for the described study, which complied with all relevant Estonian regulations, as our study did not involve collecting protected fossil species. Three described bryozoan specimens with the *Osprioneides* borings are deposited at the Institute of Geology, Tallinn University of Technology (GIT), Ehitajate tee 5, Tallinn, Estonia, with specimen numbers GIT-398-729, GIT 665-18 and GIT 665-19.

Results

Numerous unbranched, single-entrance, large deep borings with oval cross sections were found in three large trepostome bryozoan colonies (Figs. 3, 4, 5, 6). The borings are vertical to subparallel to the bryozoan surfaces and have a tapered to rounded terminus. Several borings have lost their roofs due to erosion. The boring apertures' minor axis is 2.7 to 7.0 mm (M = 5.05, sd = 1.34, N = 12) and major axis is 7.0 to 15.0 mm (M = 10.37, sd = 2.60, N = 12) long. The axial ratio (major axis/minor axis) of the borings ranges from 1.60 to 2.59 (M = 2.08, sd = 0.29, N = 12). Three completely preserved borings are 25 mm (aperture 12×6 mm), 28 mm (aperture 9×4.5 mm), and 32 mm (aperture 13×6 mm) deep. Two unroofed borings have depths of 35 mm and 50 mm. The borings are abundant in the studied samples (Figs. 3, 4, 5, 6). They occasionally truncate each other, which somewhat resembles a branching pattern. There are no linings or septa inside the borings. The growth lamellae of the bryozoans show no reactions around the borings. Small *Trypanites* borings occur inside the large boring with oval cross section. The apertures of the large borings occur on both the upper and lower surfaces of the bryozoans (the upper and lower surface of bryozoans was determined by looking at skeletal growth).

Discussion

Taxonomic Affinity of the Borings and the Possible Trace Maker

The borings in these bryozoans resemble somewhat *Petroxestes* known from Late Ordovician bryozoans and hardgrounds of North America [14]. Both are of unusually large size for Ordovician borings, and both have oval-shaped apertures. However, in *Petroxestes* the aperture width is much greater than the boring's depth. In contrast, the depth of the borings in bryozoans is much greater than their apertural width. Unlike *Petroxestes*, the Sandbian borings examined here have a tapering terminus and somewhat sinuous course. The axial ratio of *Petroxestes* borings aperture (major axis/minor axis) is also much greater than observed in these borings.

The other similar large Palaeozoic boring is *Osprioneides*, which is known from the Silurian of Baltica, Britain and North America [13]. We assign borings in the bryozoans studied here to *Osprioneides kampto* because of their similar general morphology. They have a single entrance, an oval cross section, and significant depth similar to *Osprioneides kampto*. Their straight, curved to somewhat sinuous shape also resembles that of *Osprioneides*. Both *Osprioneides* and these borings in bryozoans have a tapered to rounded terminus.

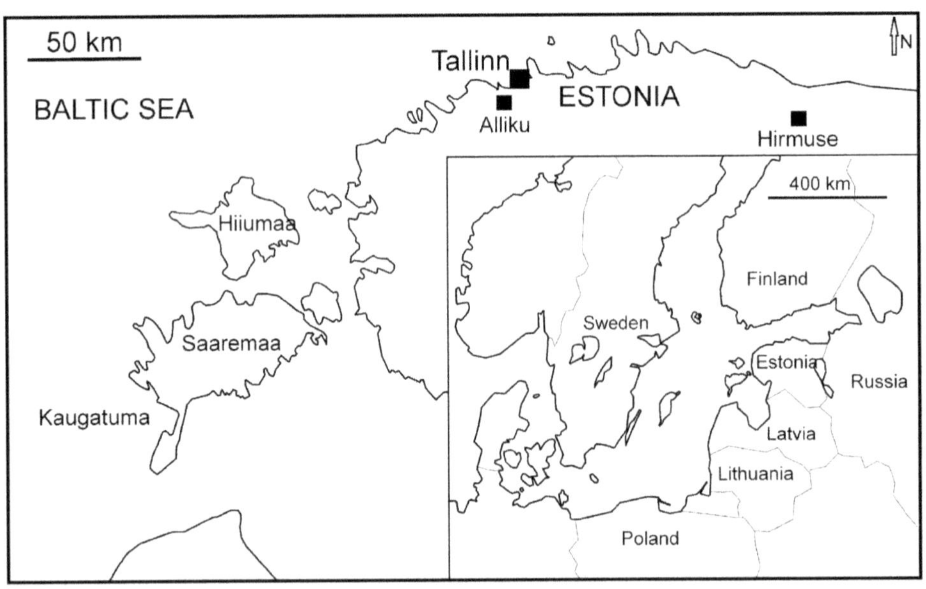

Figure 1. Locality map. Location of Hirmuse Creek and Alliku Ditches in North Estonia.

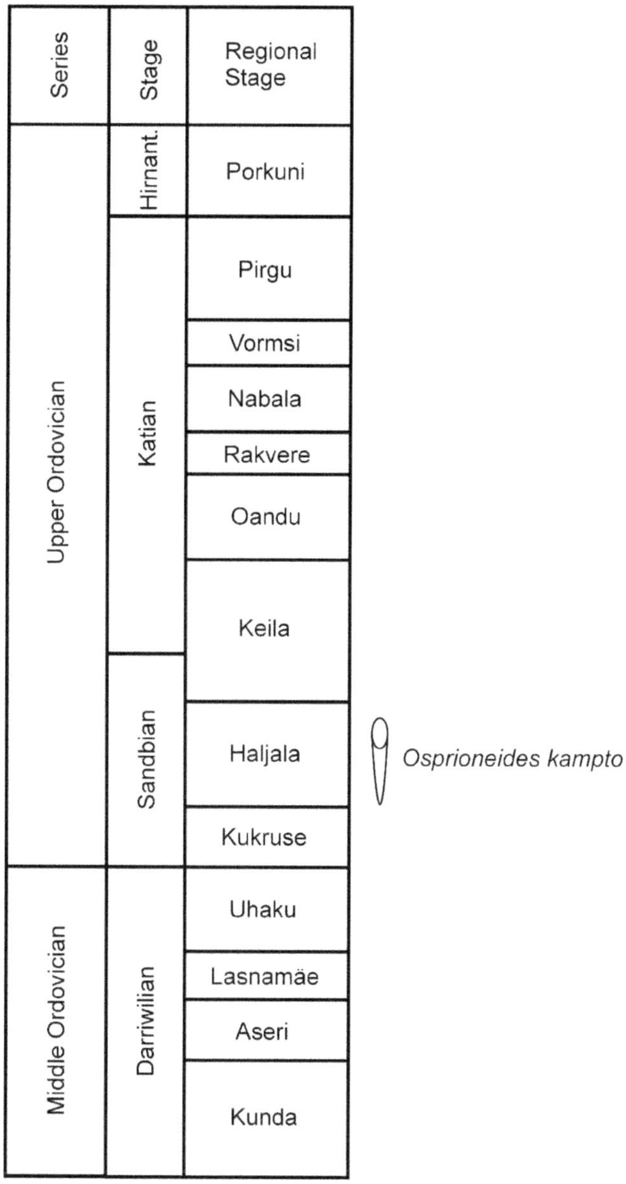

Series
Stage
Regional Stage
Upper Ordovician
Middle Ordovician
Hirnant.
Katian
Sandbian
Darriwilian
Porkuni
Pirgu
Vormsi
Nabala
Rakvere
Oandu
Keila
Haljala
Kukruse
Uhaku
Lasnamäe
Aseri
Kunda
Osprioneides kampto

Figure 2. Stratigraphic scheme. The Middle and Upper Ordovician in Estonia. Location of *Osprioneides kampto* borings. Modified after Hints et al. (2008).

Most likely the *Osprioneides* trace maker was a soft-bodied animal similar to polychaete worms that used chemical means of boring as suggested by Beuck et al. [13]. This is supported by the slightly curved to sinuous course of several borings and their variable length. The presence of a tapered terminus in *Osprioneides* means bivalves were very unlikely to have been the trace makers.

Paleoecology and Taphonomy

Osprioneides borings were made post mortem because the growth lamellae of the bryozoan do not deflect around the borings. There are also no signs of skeletal repair by the bryozoans. Several *Osprioneides* borings truncate other *Osprioneides* borings that were likely abandoned by the trace maker by that time. Similarly, empty *Osprioneides* borings were colonized by *Trypanites* trace makers. This indicates that the *Osprioneides* borings may have appeared relatively

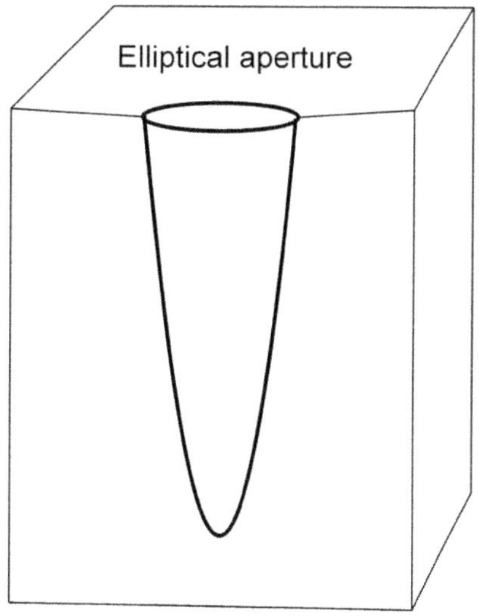

Figure 3. ***Osprioneides kampto.*** Schematic line drawing showing a straight boring.

early in the ecological succession. Overturning of the bryozoan zoaria can explain the occurrence of *Osprioneides* borings apertures on both upper and lower surfaces. There is no sign of encrustation on the walls of the studied *Osprioneides* borings, suggesting relatively rapid burial of the host bryozoans shortly after the *Osprioneides* colonization.

It is likely that *Osprioneides* trace makers were suspension feeders similar to the *Trypanites* animals due to their stationary life mode [25]. Bryozoan skeletons may have offered them protection against predators and a higher tier for suspension feeding. Previously known host substrates of *Osprioneides* comprise stromatoporoids and tabulate corals. This new occurrence of *Osprioneides* borings in large bryozoans shows that the trace maker possibly selected its substrate only by size of skeleton because the traces are not found in smaller fossils. However, they are not found in any Ordovician hardgrounds that provide more area than do the bryozoan colonies. Wyse Jackson and Key [20] suggest that large bryozoan colonies were exploited by borers because they would have been easy to bore into.

Ordovician Bioerosion Revolution

Morphological diversification was not the only result of the Ordovician Bioerosion Revolution. Most of the large bioerosional traces of the Paleozoic had their earliest appearances in the Ordovician [7,8]. The earliest known large borings are those of *Gastrochaenolites* from the Early Ordovician of Baltica [4,17]. Later, during the Late Ordovician, large *Petroxestes* borings appeared in North America. At the same time the *Osprioneides* borings described here appeared in Baltica. Thus the Ordovician was also the time of great increase in quantities of hard substrate removed by single trace makers. The biological affinities of Ordovician *Gastrochaenolites* are not known [8], but it may have been a soft-bodied animal. The Late Ordovician *Petroxestes* was almost certainly produced by the facultatively boring bivalve *Corallidomus scobina* [26]. Boring polychaetes were the likely *Osprioneides* trace makers, which is suggested by the somewhat sinuous shape of some borings. This indicates that more than one group of animal was involved in the appearance of large bioerosional traces during the Ordovician Bioerosion Revolution. Increased predation pressure [27] was most likely the driving force behind the infaunalization of larger

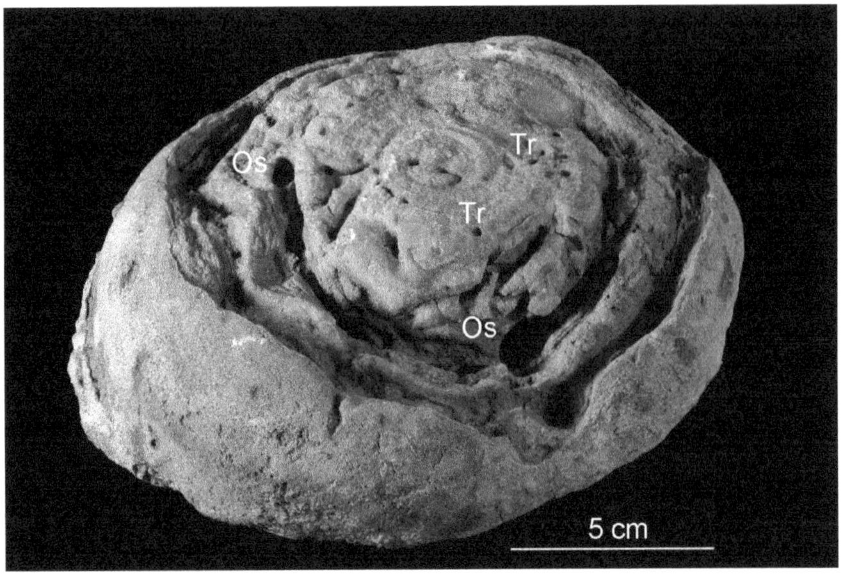

Figure 4. ***Osprioneides kampto*** **borings (Os).** A bryozoan from Hirmuse Creek, Sandbian, Upper Ordovician, Estonia. Tr – *Trypanites* borings. GIT 398–729.

invertebrates such as the *Osprioneides* trace makers in the Ordovician. On the other hand in echinoids, for example, infaunalization was presumably the result of colonization of unoccupied niche space [28].

Paleobiogeography

Osprioneides is a relatively rare fossil compared to the abundance of *Trypanites* in the Silurian of Baltica [29]. In the Silurian, *Osprioneides* borings also occur outside of Baltica. They are known from the Llandovery of North America and Ludlow of the Welsh Borderlands [30]. *Osprioneides* is presumably absent in the Ordovician of North America because Ordovician bioerosional trace fossils of North America are relatively well studied [15,16]. Thus, it is possible that the *Osprioneides* trace maker originated in Baltica or elsewhere and migrated to North America in the Silurian. This may well be connected to the decreased distance between Baltica and Laurentia (the closing of the Iapetus Ocean) and the loss of provinciality of faunas in the Silurian.

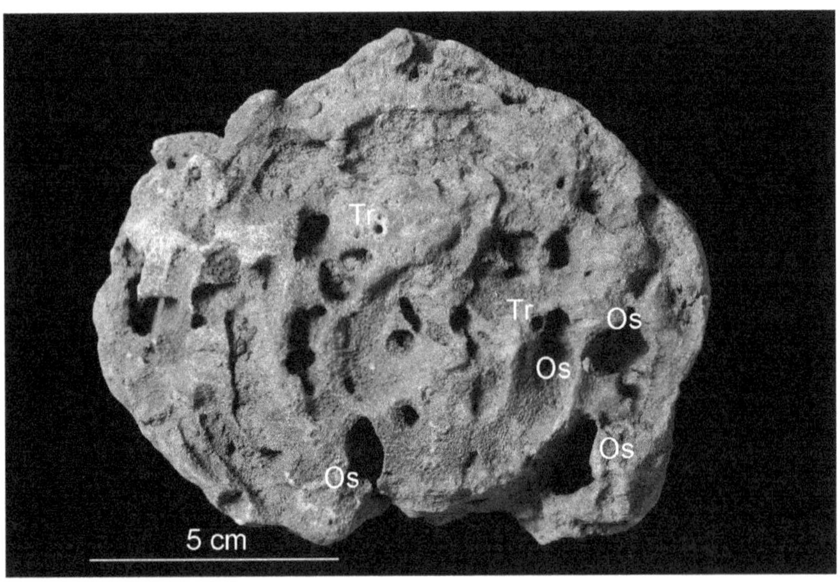

Figure 5. ***Osprioneides kampto*** **borings (Os).** A bryozoan from Hirmuse Creek, Sandbian, Upper Ordovician, Estonia. Tr – *Trypanites* borings. GIT 665-18.

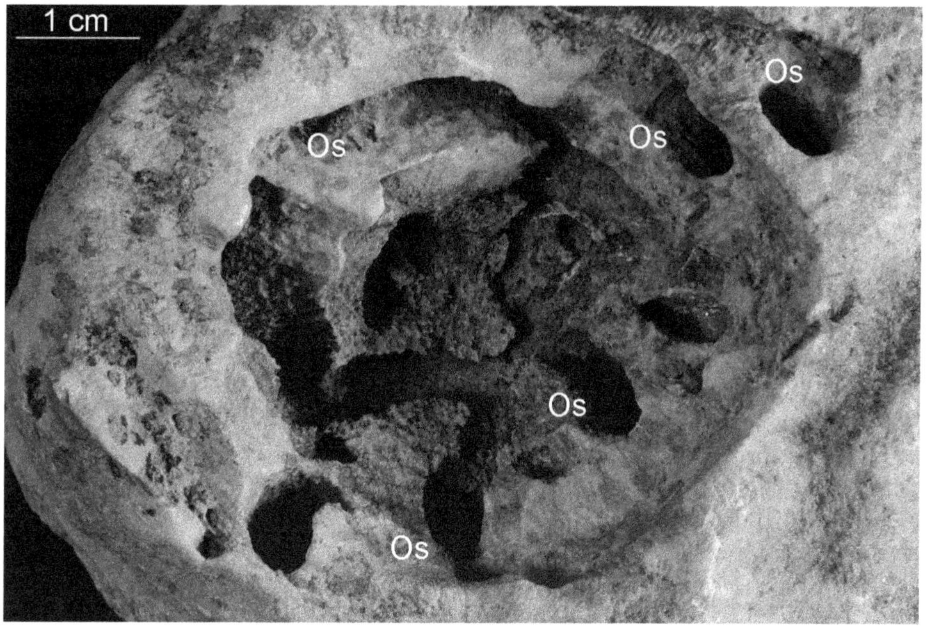

Figure 6. ***Osprioneides kampto*** **borings (Os).** A bryozoan from Hirmuse Creek, Sandbian, Upper Ordovician, Estonia. Tr – *Trypanites* borings. GIT 665-19.

Acknowledgments

We are grateful to Ursula Toom for finding the specimens among the old collections of the Institute of Geology, Tallinn University of Technology. Ursula Toom, Prof. Dimitri Kaljo, Dr. Linda Hints, Dr. Helje Pärnaste from the Institute of Geology, Tallinn University of Technology and Dr. Mare Isakar from the Geological Museum of the University of Tartu Natural History Museum are thanked for identifications of the associated fossils. We are grateful to G. Baranov from Institute of Geology at TUT for technical help with images. We are grateful to Dr. Harry Mutvei and an anonymous reviewer for the constructive reviews.

Author Contributions

Conceived and designed the experiments: OV MAW MAM. Performed the experiments: OV MAW MAM. Analyzed the data: OV MAW MAM. Contributed reagents/materials/analysis tools: OV MAW MAM. Contributed to the writing of the manuscript: OV MAW MAM.

References

1. James NP, Kobluk DR, Pemberton SG (1977) The oldest macroborers: Lower Cambrian of Labrador. Science 197: 980–983.
2. Kobluk DR, James NP, Pemberton SG (1978) Initial diversification of macroboring ichnofossils and exploitation of the macroboring niche in the lower Paleozoic. Paleobiology 4: 163–170.
3. Palmer TJ, Plewes CR (1993) Borings and bioerosion in the fossil record: Geology Today 9: 138–142.
4. Ekdale AA, Bromley RG (2001) Bioerosional innovation for living in carbonate hardgrounds in the Early Ordovician of Sweden. Lethaia 34: 1–12.
5. Dronov AV, Mikuláš R, Logvinova M (2002) Trace fossils and ichnofabrics across the Volkhov depositional sequence (Ordovician, Arenigian of St. Petersburg Region, Russia). J Czech Geol Soc 47: 133–146.
6. Taylor PD, Wilson MA (2003) Palaeoecology and evolution of marine hard substrate communities. Earth Sci Rev 62: 1–103.
7. Wilson MA, Palmer TJ (2006) Patterns and processes in the Ordovician Bioerosion Revolution. Ichnos 13: 109–112.
8. Wilson MA (2007) Macroborings and the evolution of bioerosion. In Miller III, W (ed.). Trace fossils: concepts, problems, prospects. Amsterdam: Elsevier. 356–367.
9. Bromley RG (2004) A stratigraphy of marine bioerosion, p.455–481. In McIlroy, D. (ed.) The application of ichnology to palaeoenvironmental and stratigraphical analysis. Geological Society of London Special Publications, 228p.
10. Zatoń M, Zhuravlev AV, Rakociński M, Filipiak P, Borszcz T, et al. (2014) Microconchid-dominated cobbles from the Upper Devonian of Russia: Opportunism and dominance in a restricted environment following the Frasnian–Famennian biotic crisis. Palaeogeogr Palaeoclimat Palaeoecol 401: 142–153.
11. Zatoń M, Machocka S, Wilson MA, Marynowski L, Taylor PD (2011) Origin and paleoecology of Middle Jurassic hiatus concretions from Poland. Facies 57: 275–300.
12. Zatoń M, Wilson MA, Zavar E (2011) Diverse sclerozoan assemblages encrusting large bivalve shells from the Callovian (Middle Jurassic) of southern Poland. Palaeogeogr Palaeoclimat Palaeoecol 307: 232–244.
13. Beuck L, Wisshak M, Munnecke A, Freiwald A (2008) A giant boring in a Silurian stromatoporoid analysed by computer tomography. Acta Palaeont Pol 53: 149–160.
14. Wilson MA, Palmer TJ (1988) Nomenclature of a bivalve boring from the Upper Ordovician of the Midwestern United States. J Paleont 62: 306–308.
15. Erickson JM, Bouchard TD (2003) Description and interpretation of *Sanctum laurentiensis*, new ichnogenus and ichnospecies, a domichnium mined into Late Ordovician (Cincinnatian) ramose bryozoan colonies. J Paleont 77: 1002–1010.
16. Tapanila L, Copper P (2002) Endolithic trace fossils in the Ordovician-Silurian corals and stromatoporoids, Anticosti Island, eastern Canada. Acta Geol Hisp 37: 15–20.
17. Vinn O, Wilson MA (2010a) Early large borings from a hardground of Floian-Dapingian age (Early and Middle Ordovician) in northeastern Estonia (Baltica). Carnets Géol CG2010_L04.
18. Vinn O (2004) The earliest known *Trypanites* borings in the shells of articulate brachiopods from the Arenig (Ordovician) of Baltica. Proc Est Acad Sci Geol 53: 257–266.
19. Vinn O (2005) The distribution of worm borings in brachiopod shells from the Caradoc Oil Shale of Estonia. Carnets Géol CG2005_A03.
20. Wyse Jackson PN, Key MM Jr (2007) Borings in trepostome bryozoans from the Ordovician of Estonia: two ichnogenera produced by a single maker, a case of host morphology control. Lethaia 40: 237–252.
21. Torsvik TH, Smethurst MA, van der Voo R, Trench A, Abrahamsen N, et al. (1992) Baltica. A synopsis of Vendian–Permian palaeomagnetic data and their palaeotectonic implications. Earth Sci Rev 33: 133–152.
22. Nestor H, Einasto R (1997) Ordovician and Silurian carbonate sedimentation basin. 192–204. In A. Raukas and A. Teedumäe (eds.), Geology and mineral resources of Estonia. Estonian Academy Publishers, Tallinn, 436 pp.
23. Mõtus MA, Hints O (2007) Excursion Guidebook. In 10th International Symposium on Fossil Cnidaria and Porifera. Excursion B2: Lower Paleozoic geology and corals of Estonia. August 18–22, 2007. Institute of Geology at Tallinn University of Technology, 66 p.
24. Rõõmusoks A (1970) Stratigraphy of the Viruan series (Middle Ordovician) in northern Estonia. University of Tartu, Tallinn, 346 pp.

25. Nield EW (1984) The boring of Silurian stromatoporoids – towards an understanding of larval behaviour in the Trypanites organism. Palaeogeogr Palaeoclimat Palaeoecol 48: 229–243.
26. Pojeta J, Jr, Palmer J (1976) The origin of rock boring in mytilacean pelecypods. Alcheringa 1: 167–179.
27. Huntley JW, Kowalewski M (2007) Strong coupling of predation intensity and diversity in the Phanerozoic fossil record. PNAS 38: 15006–15010.
28. Borszcz T, Zatoń M (2013) The oldest record of predation on echinoids: evidence from the Middle Jurassic of Poland. Lethaia 46: 141–145.
29. Vinn O, Wilson MA (2010) Occurrence of giant borings of *Osprioneides kampto* in the lower Silurian (Sheinwoodian) stromatoporoids of Saaremaa, Estonia. Ichnos 17: 166–171.
30. Newall G (1970) A symbiotic relationship between *Lingula* and the coral *Heliolites* in the Silurian. Geol J Spec Issue 3: 335–344.

Occlusal Enamel Complexity in Middle Miocene to Holocene Equids (Equidae: Perissodactyla) of North America

Nicholas A. Famoso*, Edward Byrd Davis

Department of Geological Sciences and Museum of Natural and Cultural History, University of Oregon, Eugene, Oregon, United States of America

Abstract

Four groups of equids, "Anchitheriinae," Merychippine-grade Equinae, Hipparionini, and Equini, coexisted in the middle Miocene, but only the Equini remains after 16 Myr of evolution and extinction. Each group is distinct in its occlusal enamel pattern. These patterns have been compared qualitatively but rarely quantitatively. The processes influencing the evolution of these occlusal patterns have not been thoroughly investigated with respect to phylogeny, tooth position, and climate through geologic time. We investigated Occlusal Enamel Index, a quantitative method for the analysis of the complexity of occlusal patterns. We used analyses of variance and an analysis of co-variance to test whether equid teeth increase resistive cutting area for food processing during mastication, as expressed in occlusal enamel complexity, in response to increased abrasion in their diet. Results suggest that occlusal enamel complexity was influenced by climate, phylogeny, and tooth position through time. Occlusal enamel complexity in middle Miocene to Modern horses increased as the animals experienced increased tooth abrasion and a cooling climate.

Editor: Alistair Robert Evans, Monash University, Australia

Funding: Funding for this project was provided by the University of Oregon Museum of Natural and Cultural History, Paleontological Society Richard K. Bambach Award, and Geological Society of America Graduate Student Grant. The funders had no role in study design, data collection and analysis, decision to publish, or preparation of the manuscript.

Competing Interests: The authors have declared that no competing interests exist.

* E-mail: nfamoso2@uoregon.edu

Introduction

Horses have long been used as a primary example of evolution through adaptation to a changing environment [1,2,3]. Horse adaptations to changing climates, specifically through dental evolution in response to an increasingly abrasive diet, have been qualitatively analyzed, but rarely investigated quantitatively [4,5,6,7]. Grass phytoliths have often been invoked as a primary driver of ungulate dental evolution [8], but recent work has suggested a much greater role for grit from drier environments and a reduced or even no role for phytoliths [9,10,11,12]. Previous work on equid adaptation to an abrasive diet focused on changes in hypsodonty and enamel microstructure [8,12,13]. Evolution of horse teeth through an increase in hypsodonty, quantified as Hypsodonty Index (HI, the ratio of mesostyle crown height to occlusal length) [14,15,16,17,18], has been documented in the Oligocene through Pleistocene fossil record, primarily for North America [19]. Increased tooth height provides more resistive enamel over an animal's lifetime. These changes have been interpreted as an adaptation to feeding in open habitats as cooling and drying climates changed woodlands to grasslands, requiring horses to adapt to increased rates of tooth wear created by environmental grit and the phytoliths of grasses [2,8,12]. Pfretzschner [13] investigated changes in equid enamel microstructure, concluding that adaptation to increased tooth wear was in place by the rise of "*Merychippus*" at about 19 Ma. The prisms and interprismatic matrix that make up enamel at the microscopic level stiffen enamel and the arrangement of these prisms strengthens it with respect to mechanical stress patterns from grinding against opposing teeth and food [13].

Miocene and later equid teeth are marked by complex, sinuous bands of enamel on their occlusal (chewing) surface (Fig. 1). These bands have taxonomically distinct patterns, with workers suggesting that members of the Equine tribe Hipparionini have more complex enamel bands than members of the tribe Equini [4,5]. Previous workers have observed qualitatively that occlusal enamel increases in complexity over the evolutionary history of horses [5]. This change is suggestive because, in a similar way to increases in hypsodonty, increasing the occlusal enamel complexity of teeth should allow them to last longer simply by distributing lifetime tooth wear over a greater total resistive cutting area. Recent work has begun exploring the relationship between the complexity of ungulate occlusal enamel and abrasiveness of diet using quantitative methods [7,20,21,22]. Here we assess the evolution of enamel complexity in Miocene and later North American equids in terms of occlusal enamel complexity, specifically investigating whether enamel complexity evolves in a pattern consistent with that expected as a response to increasing dietary abrasion. Additionally, we provide the first quantitative test of the relative complexity of hipparionine and equine occlusal enamel bands.

Questions

Given current hypotheses of horse phylogeny and diversification in response to environmental changes and the extremely large available sample size (>2,581 known North American localities

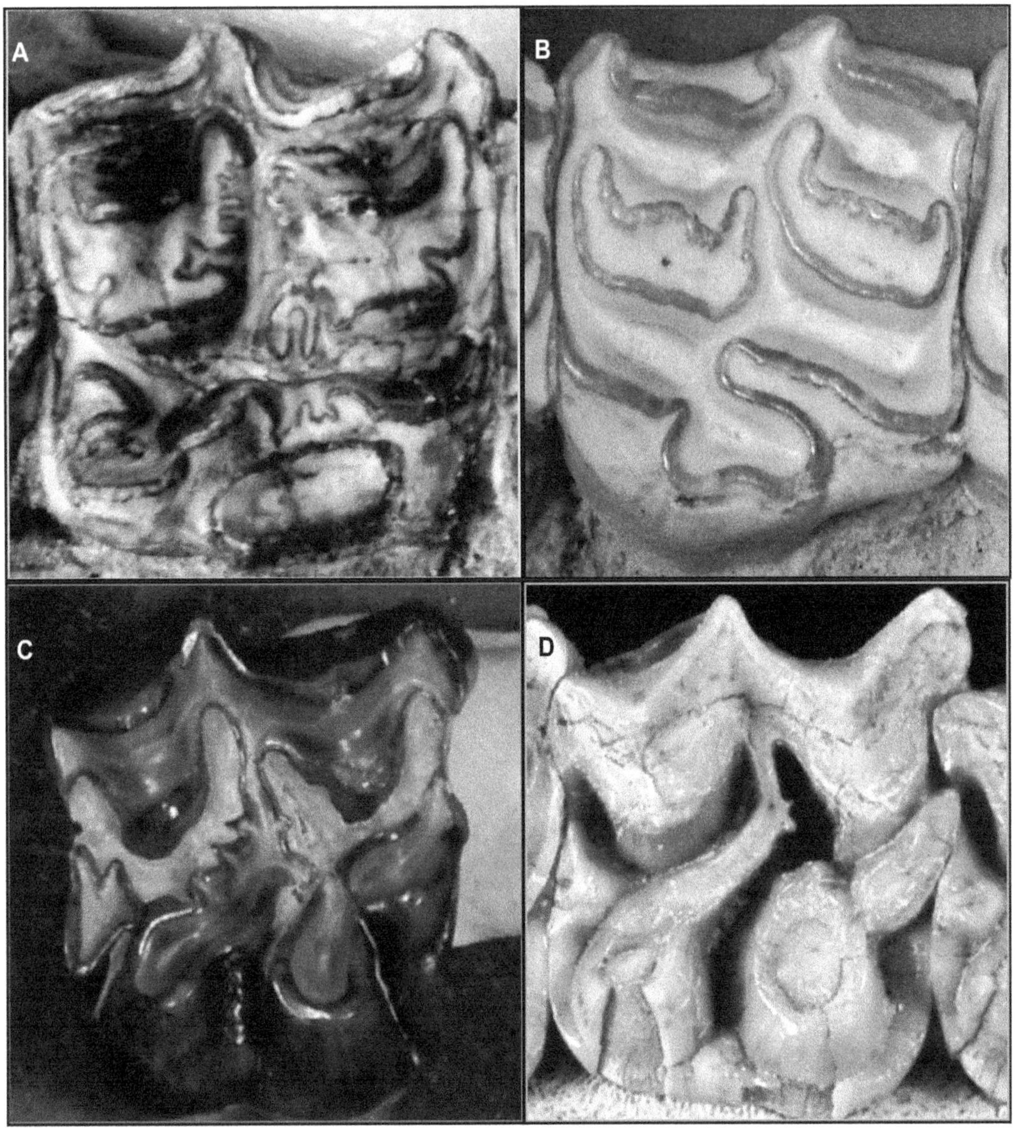

Figure 1. Representative teeth of each tribal-level group in this study. (A) Hipparionini, (B) Equini, (C) "Merychippini," and (D) "Anchitheriini." Each tribe has a distinct enamel pattern; the patterns decrease in complexity from A to D.

with fossil equids), we can use equid occlusal enamel band length and complexity of the occlusal surface to investigate the evolution of morphology in response to an increasingly abrasive diet. These observations lead to a series of questions: Do equids change their enamel complexity from the Miocene through the Recent? If so, does complexity increase over time, as would be expected for increasing adaptation to an abrasive diet? Is there a difference in enamel complexity between equid tribes, especially Hipparionini and Equini? If the evolution of enamel complexity is consistent with dietary adaptation, are there compromises between hypsodonty and enamel complexity? If so, do the two tribes make different compromises?

Hypothesis

We hypothesize that increased abrasion in equid diets produced a selective advantage for teeth with greater resistive cutting area (occlusal enamel complexity).We will test this hypothesis by statistical analysis of enamel complexity derived from images of fossil horse teeth. If the statistical analysis shows a distinct pattern, then equids responded to increased abrasion through an increase in occlusal enamel complexity, providing an increased resistive cutting area for food processing during mastication. If the statistical analysis shows a pattern indistinguishable from random, we will be unable to reject the null hypothesis of no unifying adaptive significance to changes in occlusal enamel complexity or that some other process that we have not tested is controlling occlusal enamel complexity. Occlusal enamel complexity will vary as a consequence of phylogenetic constraint and evolutionary response to changes to ecological role through time. If our hypothesis is correct, the complexity of enamel on the occlusal surface of equid teeth should increase through time, tracking changes in the abrasiveness in diet as climates changed through the Neogene.

It is possible that phylogenetic constraint, inherited developmental or other limits to adaptation, may control the compromises different lineages of horses find between hypsodonty and enamel complexity for their adaptation to tooth abrasion. If so, we would

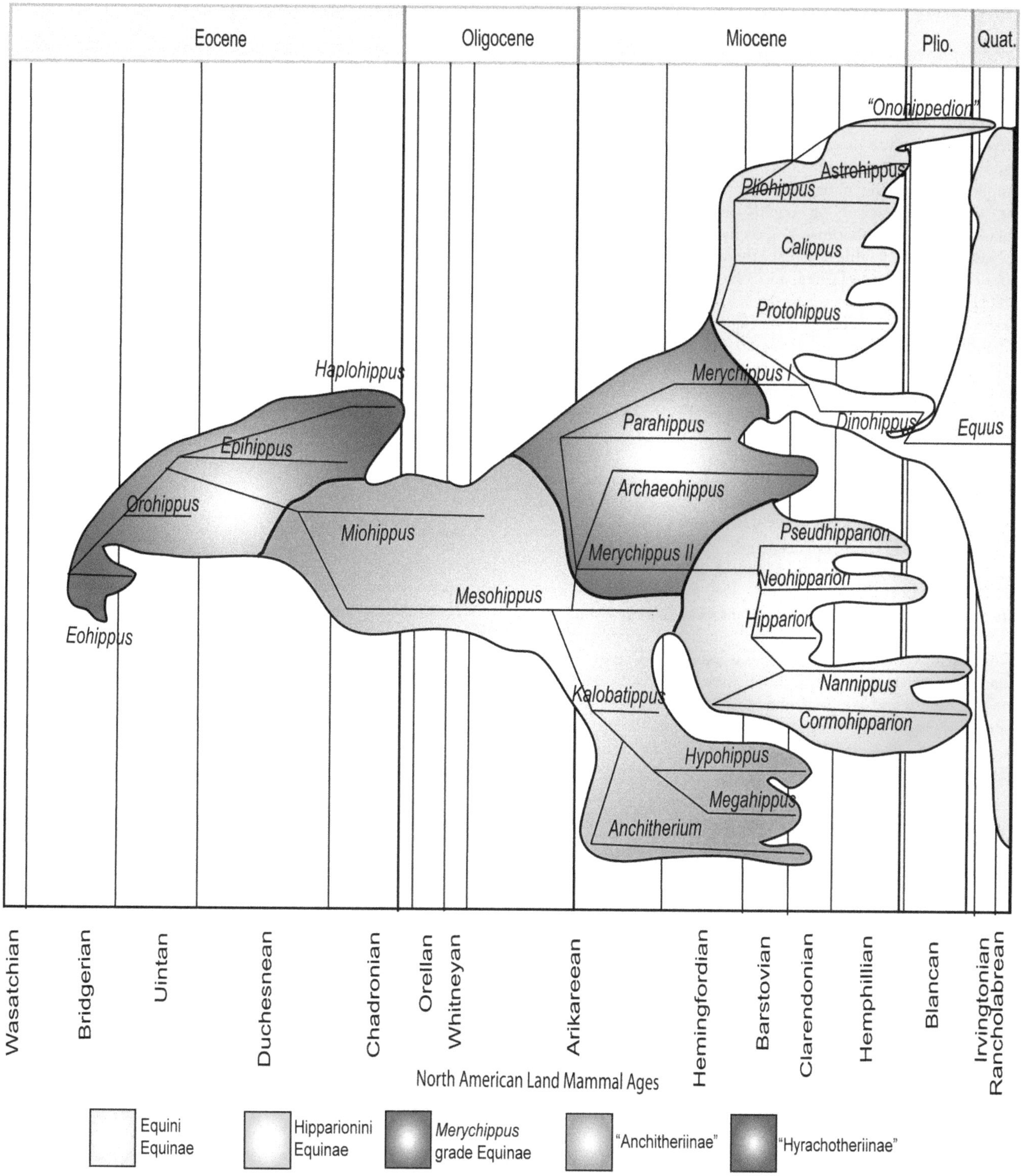

Figure 2. Phylogeny of Equidae used in this study (after MacFadden [25]). North American Land Mammal Ages indicated on the bottom. The size of the colored regions represents relative diversity among the groups. Horizontal lines represent time ranges of each genus or clade. This study begins with the Barstovian to capture the most advanced Equinae with derived enamel prismatic structure.

expect each tribe to have distinct differences in their occlusal enamel complexity in comparison to their hypsodonty. Published qualitative observations of equid tooth morphology and its relationship to diet [7,21,22] suggest to us that Hipparionini should have the most complicated occlusal enamel, followed by Equini, then the "*Merychippus*" grade horses, and finally "Anchitheriinae".

Background

Evolutionary Context

Analyses of evolutionary adaptations must be investigated within the context of phylogeny [23]. Linnean taxonomy is a hierarchical naming system that was originally created in a pre-Darwinian context to describe similarity amongst organisms. Like most natural systems, phylogenetic relationships are more complicated than the initial set of categories defined by man. The current consensus on equid phylogeny includes three subfamilies, "Hyracotheriinae," "Anchitheriinae," and Equinae [5,24,25] (Fig. 2). Within Equinae, there are two sub-clades, the tribes Hipparionini and Equini, and a basal grade mostly assigned to "*Merychippus*." This genus has long been considered a paraphyletic taxon, maintained through convenience to include all basal equines that do not possess apomorphies of either Equini or Hipparionini. Typical "*Merychippus*" have an upper dentition that maintains the plesiomorphic features of the basal "Anchitheriinae," a paraphyletic grade below Equinae (Fig. 1), but also share characters with derived Equinae [5,26,27]. Hipparionini and Equini have distinct tooth morphologies as well (Fig. 1). Members of the tribe Hipparionini are hypsodont, but relatively lower crowned and have more complicated enamel borders than their equin counterparts [4,5,24]. The two tribes of Miocene horses, Hipparionini and Equini, are diagnosed on the basis of differences of the structures formed by the folding of enamel on the occlusal surface of their teeth [4,5,6,24,25]. The shape of the occlusal pattern was shown to be an important character in equin and hipparionin phylogeny [5,24,28]. This qualitative difference leads us to ask whether complexity of occlusal enamel evolved differently because of phylogenetic constraint and/or climatic pressures between Equini and Hipparionini.

Because species are phylogenetically related to differing degrees, they cannot be considered as independent for statistical analysis [23]. To accommodate this dependence, Felsenstein [23] proposed the method of independent contrasts, incorporating the phylogenetic relationships into regression analysis. Independent contrasts has been developed into a broad field of phylogenetic comparative methods [29,30,31], but at this point all of them require phylogenies with branch lengths derived from models of molecular evolution. Ideally, we would use one of these comparative methods for testing our hypothesis of variations in the context of phylogeny, but current methods require known branch lengths and have yet to be adapted to fossil-based morphological phylogenies [32,33,34].

We will accommodate phylogenetic interdependence amongst the fossil horses by using nested variables in a multi-way analysis of variance (ANOVA). In this way, we are able to model phylogeny using the hierarchical taxonomic system as a proxy for phylogeny [7]. Using these nested variables in an ANOVA is not ideal for phylogeny, because it does not completely take the topology of a phylogenetic tree into account, but as a coarse approximation, it functions for this scale of analysis.

Table 1. Results of Tooth Position Wilcoxon Test.

Level	Count	Score Sum	Expected Score	Score Mean	(Mean-Mean0)/Std0
M1	70	8707	10010	124.386	−2.175
M2	72	10655	10296	147.986	0.593
P3	68	9454	9724	139.029	−0.454
P4	75	11939	10725	159.187	1.981

Measures of Complexity

Species and other higher taxonomic groups in horses are primarily diagnosed by qualitative characters; in fact, a majority of equid diagnoses rely upon differences in pattern of occlusal enamel [4,24]. A complicated enamel pattern should have longer occlusal

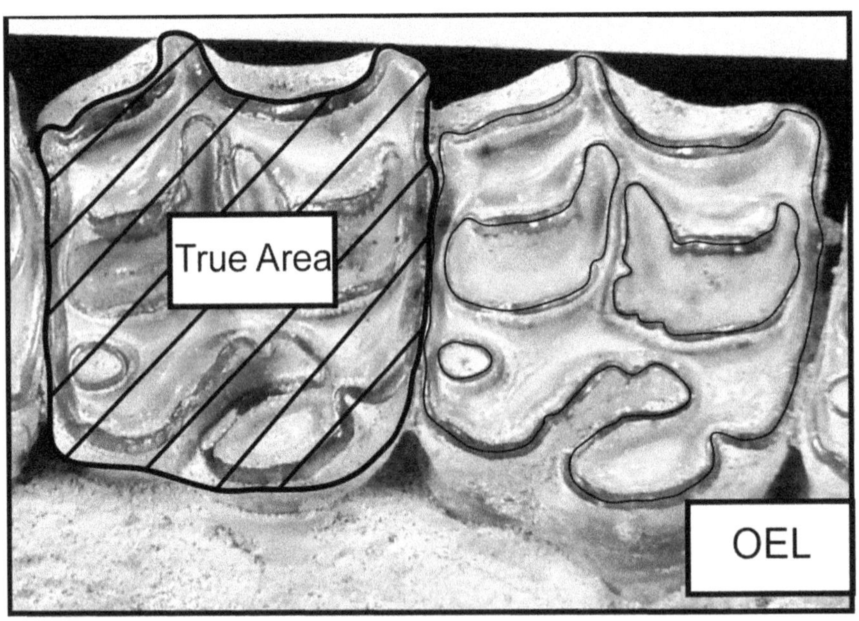

Figure 3. Examples of True Area and Occlusal Enamel Length (OEL) taken on digital image of *Pseudhipparion* sp. (UNSM 125531). True Area is a different measurement than length by width. These measurements are calculated with ImageJ. Figure is based on methodology presented by Famoso et al. [7].

Table 2. Results of Tooth Position ANOVA and Tukey-Kramer Test.

Tooth Position	Group	Mean OEI
P4	A	17.676
M2	AB	16.415
P3	AB	16.357
M1	B	15.857

enamel length thus producing more enamel per unit surface area on the occlusal plane. Famoso et al. [7] introduced a numerical method to quantitatively measure and test the differences in enamel complexity in ungulates, a unit-less value called Occlusal Enamel Index (OEI): $OEI = OEL/\sqrt{(True\ Area)}$ where *OEL* is the total length of enamel bands exposed on the occlusal surface as measured through the center of the enamel band, and *True Area* is the occlusal surface area constructed as a polygon following the outer edge of the occlusal surface, including any cementum that may exist outside of the enamel, where cementum on the lingual side is part of the occlusal surface while that on the buccal is not (Fig. 3). The True Area is not an occlusal length multiplied by width, but is instead representative of the area actually contained within the curved occlusal boundaries of the tooth. We are measuring True Area as a 2D projection, so we do not account for any increases in area that might arise from topography on the occlusal surface of the tooth. Because most equid teeth are on the low-relief end of the mesowear spectrum [34], this projection will have little effect on our current study; however, studies that extend this methodology to high-relief teeth might find improvements from a 3D approach. Analyzing images of teeth in the computer allows us to use the more precise true area instead of the more traditional technique of multiplying the measured length and width of the occlusal surface. True area is a proxy for body size, so OEI removes the effects of absolute scale on complexity; however, the effects of body size are not completely removed, as OEI does not adjust for size-related differences in complexity, i.e., allometry [7].

Becerra et al. [36] have introduced a similar enamel complexity metric, applying it to rodents. The enamel index (EI) is calculated as: $EI = OEL/(True\ Area)$. OEI differs from EI in that the occlusal area is treated differently. OEI produces a unitless metric while EI does not, producing values in units of 1/length, so consistent length scales would have to be used to maintain comparability among analyses. Becerra et al. [36] found evidence to suggest that selective pressures from regional habitats, in particular vegetation, have shaped the morphological characteristics of the dentition of caviomorph rodents in South America.

We use OEI for this study for three reasons: (1) we expect the unitless index to more completely account for isometric changes of enamel length with mass, (2) we want our results to be directly comparable to Famoso et al. [7], and (3) the unitless index is methodologically aligned to the unitless HI commonly used in horse paleoecology.

Two recent studies have analyzed enamel complexity within the Order Artiodactyla, using a slightly different approach that focuses more on visible enamel band orientation. Heywood [21] analyzed molar occlusal surfaces and characterized them on the basis of length, thickness, and shape of the enamel bands, concluding that plant toughness is a primary driver of occlusal enamel form in bovids. Kaiser et al. [22] investigated the arrangement of occlusal enamel bands in the molars of ruminants with respect to diet and phylogeny, finding that larger ruminants or those with higher grass content in their diet have a higher proportion of enamel ridges aligned at low angles to the direction of the chewing stroke.

Previous work on occlusal enamel patterns in equids has been limited to the observation that patterns change through wear stages [5,37]. Famoso and Pagnac [6] suggested that the differences in occlusal enamel patterns through wear correspond to evolutionary relationships in Hipparionini. To date, attempts at quantifying the patterns of evolutionary change in occlusal enamel complexity between and within these equid tribes have been limited by small sample sizes [6,7].

Tooth Position

Beyond the pressures of the environment, differential expression by tooth position is another aspect of enamel band evolution that may be linked to phylogeny. Famoso et al. [7] demonstrated that enamel complexity is expressed significantly differently at each tooth position. Equid P2 and M3 are easily identifiable in isolation: the P2 has a mesially pointed occlusal surface while the M3 is tapered distally. The middle four teeth (P3-M2) are more difficult to identify to position when isolated as they have uniformly square occlusal surfaces. Premolars tend to be larger than molars within a single tooth-row, but size variation within a population overwhelms this difference for isolated teeth. As with many mammals, the majority of identifiable fossil equid material tends to be isolated teeth, as teeth are composed of highly resistant materials (enamel, dentine, and cementum) in comparison to the surrounding cranial bone. Many taxa, including *Protohippus placidus*, *Pliohippus cumminsii*, and *Hipparion gratum*, are only known from isolated teeth [1,5,24]. Because of their relative abundance in each tooth-row, a majority of isolated teeth tend to be the more difficult to distinguish P3 to M2. Including isolated teeth in our analysis would increase geographic and taxonomic diversity, but variation in enamel complexity amongst the tooth positions could overwhelm the signal. Optimizing the sample size in our study design makes it important to identify whether tooth position has a significant effect on OEI for P3 - M3.

Methods

Materials

Our data consists of scaled, oriented digital photographs of the occlusal surface of fossil and modern equid upper dentitions. We

Table 3. Results of the Nested Multi-way ANOVA.

Dependent Variable	NALMA	Tooth Position	Subfamily	Tribe [Subfamily]
OEI	$p<0.0001$	$p<0.0001$	$p<0.0001$	$p<0.0001$
F test value	0.310	0.027	0.080	0.139

Table 4. Results of Wilcoxon test for OEI vs Tribe.

Level	Count	Score Sum	Expected Score	Score Mean	(Mean-Mean0)/Std0
Anchitheriini	36	4820	11574	133.889	−6.246
Merychippini	45	16014.5	14467.5	355.878	1.289
Equini	375	115304	120563	307.476	−2.27
Hipparionini	186	70265	59799	377.769	4.909

measured a total of 800 teeth from a broad selection of North American equids ranging from 16 Ma to recent (Table S1). Photographs were taken with Kodak DC290, Fujifilm Finepix A345, Olympus Stylus Tough, and Canon Digital EOS Rebel SLR cameras. Some data were collected from Famoso et al. [7]. Some images were used with permission from the UCMP online catalog (http://ucmpdb.berkeley.edu/). Specimen numbers and repository information are reported in Table S1, and geographic locations of repositories are indicated in the Institutional Abbreviations section. Each named museum listed in the Institutional Abbreviations section gave us permission to access their collections. Care was taken to select individuals in medial stages of wear (no deciduous premolars and no teeth in extreme stages of wear). Skulls and complete to nearly complete tooth rows were preferred because we can be more confident in taxonomic identification and tooth position. Isolated teeth were also included when more complete tooth-rows were not available for a taxon.

Institutional Abbreviations

UNSM = University of Nebraska State Museum, Lincoln, NE; **UOMNCH** = University of Oregon Museum of Natural and Cultural History, Eugene, OR; **UCMP** = University of California Museum of Paleontology, Berkeley, CA; **MVZ** = University of California Museum of Vertebrate Zoology, Berkeley, CA; **AMNH F:AM** = Frick Collection, American Museum of Natural History, New York, NY; **AMNH FM** = American Museum of Natural History, New York, NY; **UF** = University of Florida Museum of Natural History, Gainesville, FL; **JODA** = John Day Fossil Beds National Monument, Kimberly, OR; **CIT** = California Institute of Technology (Cast at JODA); **UWBM** = University of Washington Burke Museum of Natural History and Culture, Seattle, WA; **SDSM** = South Dakota School of Mines and Technology Museum of Geology, Rapid City, SD; **USNM** = United States National Museum of Natural History, Washington, DC.

Occlusal Enamel Index

Enamel length and True Area of each tooth were measured using the NIH image analysis program ImageJ (http://rsb.info.nih.gov/ij/). Site geology (formation and member), time period (epoch and North American Land Mammal Age [NALMA]), tooth position (if known), physiographic region, political region, and taxonomy (subfamily, tribe, genus, and species) were recorded for each specimen. Data were stored in a Microsoft Excel 2010 spreadsheet (Table S1). OEI was calculated following Famoso et al. [7] (Fig. 3).

Table 5. Results of ANOVA and Tukey-Kramer Test for OEI vs Tribe.

Tribe	n	Group	Mean OEI
Hipparionini	186	A	10.026
"Merychippini"	45	AB	9.903
Equini	375	B	9.602
"Anchitheriini"	36	C	8.350

We used one-way and multi-way analysis of variances (ANOVAs) in JMP Pro 9 to determine whether the relationship between tooth size and enamel length fit our predictions. We used a Shapiro-Wilk W test [38] to test whether OEI values were normally distributed and the Bartlett test of homogeneity [39] to determine whether the variances in OEI among groups were homogeneous. If OEI is normally distributed and the variances are homogeneous among groups, then the data will not violate the assumptions of the ANOVA and a parametric test can be performed. ANOVA is generally robust to violations of both of these assumptions, particularly if the sample sizes amongst groups are similar [40]. Our sample sizes are not similar among all of our groups, so we have supplemented ANOVAs with nonparametric Wilcoxon tests [41] when one or both of these assumptions are violated. When data from all tooth positions were pooled, they did not display a normal distribution. Upon further investigation, we determined that all but one position in the tooth row was normally distributed and excluded the non-normal tooth (M3) from further analysis. As discussed below, we used nested (hierarchical) ANOVAs to account for evolutionary relatedness in our analysis. Nested ANOVAs include levels of independent factors which occur in combination with levels of other independent factors. Because ANOVAs can only provide a test of all factors together, we have included Tukey-Kramer tests where needed to investigate statistically significant groupings [40].

An analysis of tooth position was run on a subset of the data (n = 528 teeth) with known tooth position. This ANOVA allowed us to determine whether there was a tooth position or group of tooth positions with indistinguishable OEI values, allowing us to limit the number of specimens to be measured for the subsequent analyses. The results of this analysis would provide a justification for the selection of a subset of teeth to consistently measure. We ran a multi-way ANOVA with OEI as the dependent variable and tribe, region, NALMA, and tooth position as the independent factors. P2 and M3 were excluded as they have an overall different shape and are statistically different in OEI from the teeth from the middle of the tooth-row [7]. We additionally ran a one-way ANOVA with OEI as the dependent variable and tooth position excluding P2 and M3 as the independent factor. Tukey-Kramer tests [42] were also performed to investigate the origin of significance for independent factors. We also ran a one-way ANOVA with OEI as the dependent variable and tooth position excluding P2 and M3 for the subset of the data that only belonged to the genus *Equus*, the genus with the largest overall sample size. Using just one genus would eliminate any influence from higher

Table 6. Results of Wilcoxon tests for OEI vs NALMA.

Level	Count	Score Sum	Expected Score	Score Mean	(Mean-Mean0)/Std0
Barstovian	147	39029	46525.5	265.503	−3.865
Blancan	67	22358	21205.5	333.701	0.815
Clarendonian	156	46218.5	49374	296.272	−1.594
Hemphillian	96	25101.5	30384	261.474	−3.206
Irv/Rancho	126	59166	39879	469.571	10.517
Recent	40	8155	12660	203.875	−4.03

level evolutionary relationships. A one-way ANOVA with OEI as the dependent variable and tooth position excluding P2 and M3 by tribe (just Equini, just Hipparionini, and just "Anchitheriinae") allowed us to test whether variation in tooth position was consistent at this level of lineage. Tribal affiliations were used as a proxy for phylogenetic relationships, therefore all genera needed a tribal level affiliation to be included in the ANOVAs, but the basal members of the Equinae (members of the "*Merychippus*" grade) do not belong to the Hipparionini or Equini, so we applied the place-holder paraphyletic tribe "Merychippini." Similarly, for all members of the paraphyletic subfamily "Anchitheriinae," the place-holder name "Anchitheriini" was applied.

Running our analyses above the genus level limits the influence of lumping and splitting at the genus and species levels which arise from qualitative analysis of characters found in isolated elements. While working through museum collections, we found several *nomen nudum* manuscript names assigned to specimens. We assigned these specimens to the most appropriate, currently-established genus name and left the species as indeterminate. Even for published species of equids, there are ongoing controversies about the validity of names. Major problem areas include genera and species split from the paraphyletic form genus "*Merychippus*" [5,43,44,45,46] as well as the number and identity of Plio-Pleistocene and recent *Equus* species [5,42,47,48]. There has been controversy as to the validity of the number of genera and species that belong to Hipparionini [5,15,37,43,45,49,50]. Leaving the analysis above the genus level removes any effect taxonomic uncertainty at the generic and specific levels.

Limiting the taxonomy to the Tribe and above also allows a more robust sample size. Equid genera are typically diagnosed through a combination of dental and cranial characters [5,24,51,52]. Most isolated dental specimens can only be identified to genus because of the lack of diagnostic features, so a genus or tribal cutoff for our analysis allows us to access the rich supply of isolated teeth.

Table 7. Results of ANOVAs and Tukey-Kramer Tests for OEI vs NALMA.

Level	Group	Mean OEI
Irv/Rancho	A	10.799
Blancan	B	9.790
Clarendonian	BC	9.504
Hemphillian	BC	9.301
Barstovian	BC	9.300
Recent	C	8.957

It was necessary to combine two of the NALMAs, the Irvingtonian and Rancholabrean, to have sufficient sample size for the analyses used here. This combination is not ideal as it eliminates a portion of the temporal resolution of our study. The Irvingtonian and Rancholabrean are both part of the Pleistocene. The Irvingtonian was not well sampled enough to analyze on its own, and by combining it with the Rancholabrean we were also able to include specimens from the Pleistocene in the temporal bin when their NALMA was not known.

To accurately investigate OEI through hierarchical taxonomic relationships and changing regions through time, it was necessary to use nested terms in our analyses. Nesting tests hypotheses about differences among samples which are placed in hierarchical groups. Nested factors are usually random-effects factors, or a factor with multiple levels but only a random sample of levels is included in the analysis. When applied to an ANOVA, it is considered a modified one-way ANOVA [40] where one variable is the random-effects factor and the other is considered a subsample. Including nested factors accounts for within-group variability.

To make a single overall test of our hypothesis, we constructed a multi-way ANOVA with OEI as the dependent variable and tooth position, nested taxonomy (tribe within subfamily), and time (NALMA) as independent factors (Listed in the Results section as Nested Multi-way Analysis of Variance). In addition, we ran three groups of one-way ANOVAs with Tukey-Kramer tests to test our hypothesis of the influence of climate and phylogeny as on OEI through time. Our one-way ANOVAs use OEI as the dependent variable. Our first group of one-way ANOVAs (in Results as ANOVA 1: OEI vs. Tribe) uses tribe as the independent variable to investigate how OEI differs between lineages. Next, we used NALMA as the independent variable to examine how overall OEI changes through time (ANOVA 2: OEI vs. NALMA). Finally, we used tribe as the independent variable and separated by NALMA to explore whether the different lineages are distinct in OEI at different periods of time (ANOVA 3: OEI vs. Tribe within Each NALMA).

Results

All datasets were tested for the assumptions of ANOVA, Gaussian distribution and equality of variances among groups. For concision, only significant violations of these assumptions are noted.

Tooth Position

The Bartlett test of equal variance for this ANOVA showed significant differences among variances for this subset of the data,

Table 8. Results of Wilcoxon test for OEI vs Tribe within each NALMA.

Hemphillian					
Level	**Count**	**Score Sum**	**Expected Score**	**Score Mean**	**(Mean−Mean0)/ Std0**
Equini	53	1929.5	2570.5	36.406	−4.719
Hipparionini	43	2726.5	2085.5	63.407	4.719

so we supplemented the standard ANOVA with a Wilcoxon test for the comparison (Table 1). The Chi Square approximation of the Wilcoxon was not significant ($p = 0.0757$), in contrast to the one-way ANOVA with OEI as the dependent variable and tooth position (excluding P2 and M3), which was significant at $p = 0.0124$. The Tukey-Kramer test indicates that P3, P4, and M2 are not significantly different from one another and P3, M1, and M2 are not significantly different from one another (Table 2). The P4 and M1 appear to be significantly different from each other, but recall that the Tukey-Kramer test relies upon the pooled variances of the ANOVA. The Bartlett test of equal variances was not significant for the one-way ANOVA with OEI as the dependent variable and tooth position (excluding P2 and M3) for *Equus*. The Tukey test for *Equus* shows a slightly different pattern from the overall data set, P4 and M1 are significantly different from each other, P3 overlaps with P4 and M2, and M2 overlaps with P3 and M1. The Bartlett test of equal variance for the one-way ANOVAs with OEI (dependent variable) and tooth position (excluding P2 and M3) was not significant for Equini, but was significant for Hipparionini. The ANOVA for Equini was not significant, showing no significant differences among tooth positions. The Chi Square approximation of the Hipparionini Wilcoxon test was not significant ($p = 0.3334$). The Hipparionini ANOVA was also not significant ($p = 0.0687$) and the Tukey test showed no significant differences among tooth positions.

Nested Multi-way Analysis of Variance

All independent variables are significant for OEI at the $\alpha = 0.05$ level. Table 3 shows the p values for each variable.

Table 9. Results of ANOVAs and Tukey-Kramer Tests for OEI vs Tribe within each NALMA.

NALMA	Tribe	N	Group	Mean OEI
Barstovian	Hipparionini	35	A	9.927
Barstovian	"Merychippini"	38	A	9.897
Barstovian	Equini	43	B	8.868
Barstovian	"Anchitheriini"	30	B	8.453
Clarendonian	"Merychippini"	5	A	10.31
Clarendonian	Hipparionini	95	A	9.969
Clarendonian	Equini	52	B	8.712
Clarendonian	"Anchitheriini"	4	B	7.771
Hemphillian	Hipparionini	43	A	10.028
Hemphillian	Equini	53	B	8.711
Blancan	Hipparionini	7	A	10.857
Blancan	Equini	60	B	9.666

ANOVA 1: OEI vs. Tribe

The Bartlett test of equal variance for this ANOVA was significant, so we supplemented the standard ANOVA with a Wilcoxon test for the comparison (Table 4). The Chi Square approximation of the Wilcoxon was significant ($p<0.0001$), matching the ANOVA results ($p<0.0001$) (Table 5). Tukey test results indicate that Hipparionini and Equini are separate from one another. "Merychippini" is between the Hipparionini and Equini. The "Anchitheriini" is in its own distinct group.

ANOVA 2: OEI vs. NALMA

The Bartlett test of equal variance for this ANOVA was significant, so we supplemented the standard ANOVA with a Wilcoxon test for the comparison. Results are presented in Table 6. The Chi Square approximation was significant ($p<0.0001$). The Wilcoxon test yields similar results to the standard ANOVA (Table 7), which was also significant ($p<0.0001$). The Irvingtonian/Rancholabrean stands out as a unique period of time with the highest OEI values. The Blancan has the next highest OEI values. The Recent is grouped alone with the lowest OEI values. The Clarendonian, Hemphillian, and Barstovian overlap with the Blancan and the Recent and have OEI values that are intermediate between the two groups.

ANOVA 3: OEI vs. Tribe within Each NALMA

The Bartlett test for the Hemphillian ANOVA was significant, so we supplemented the standard ANOVA with a Wilcoxon test for that interval (Table 8. All tests for NALMAs were significant (Table 9). The Barstovian ($p<0.0001$) had two statistical groupings; one group is the "Merychippini" and Hipparionini, and the other is the Equini and "Anchitheriini." The Clarendonian ($p<0.0001$) had the same two groups. The Hemphillian ($p<0.0001$) and the Blancan ($p = 0.0013$) both have two distinct groups, the Hipparionini and Equini. The groupings of tribes stay the same through time.

Discussion

Tooth position does not significantly affect OEI for the middle four teeth (P3- M2) of the upper tooth row at the tribal level. Our investigation into tooth position indicates that we can safely include isolated molariform teeth in our study without taking tooth position into account if we exclude the P2 and M3. These two teeth have already been shown to be different from the other molariform teeth [7]. We also found that our data were normally distributed when the P2 and M3 were excluded. We found more variation in OEI for the P4 than for the M1, M2, and P3 according to Bartlett's test. We suggest subsequent work should focus on M1, M2, or P3 to take advantage of this lower variance. It is important to note that the Wilcoxon test was not significant for the main body of the data. While the ANOVA was significant, this dataset violated the assumption of equal variance, so the Wilcoxon is the more appropriate test. In the end, all of our analyses of tooth position suggest that the middle four teeth are not significantly different from one another and can be used interchangeably in an analysis at this broad a level. Our investigation into tooth position also explored whether the variation in OEI for the various tooth positions were the same among horse lineages. Within each tribe tooth position is not significant for the four square middle teeth. Tooth position OEI varies significantly between tribes, suggesting that each lineage is adapting differently for each tooth.

The results of our nested multi-way ANOVA indicate that time, tooth position, and nested taxonomy are all significant factors for the length of enamel in horse teeth. Each of the subsequent one-way ANOVAs allowed us to tease apart the details of the multi-way ANOVA result.

Generally, OEI increases from the Miocene NALMAs to the Pleistocene, correlating with the overall cooling climate from the mid-Miocene Climactic Optimum (16 Ma) to recent [53]. This increase in OEI over time is compatible with our hypothesis that, as climate became cooler and dryer and the abrasiveness of the equid diet increased [12], increased OEI was selected for across horse lineages. OEI in the late Miocene is lower than in the Pliocene and the increase continues through to the Pleistocene. In the Holocene, we see a decrease in complexity to levels similar to that of the late Miocene. Increase in OEI though time matches the documented increase in HI through time [35]. OEI and HI are measures of ways in which ungulates increase the amount of enamel available for a lifetime of chewing abrasive foodstuffs, so higher values of either metric could suggest higher abrasiveness in diet [7].

The drop in complexity we observe in our Holocene sample could be influenced by the limitation in taxonomic sampling available for extant Equini. The only animals available for inclusion are influenced by conditions of artificial selection and human management, and are descended from the Old World lineage of horses unlike the New World fossils included in our dataset. These animals do not have the same diet, behavior, or morphology as they would in the wild [54,55], so if enamel complexity is phenotypically plastic and reflects diet during tooth development, as suggested for elephantids and rodents [56,57,58], their simpler enamel may reflect the dietary conditions under domestication. This possibility warrants further investigation but is beyond the scope of our study. More likely is that the domestic and feral horses in our Holocene dataset are descended from animals in distinctly different selective regimes in the Old World; future studies with larger spatial sampling would be needed to test this hypothesis. At this point, we do not feel confident interpreting the drop in OEI from the Pleistocene to the Recent as an evolutionary change, but instead interpret it as suggestive of the biogeography of this trait.

The overall analysis of OEI by tribes (ANOVA 1) strongly supports the hypothesis that the Equini and Hipparionini had distinct evolutionary responses in occlusal enamel evolution. The results of the Tukey-Kramer test very closely reflect the evolutionary relationships of the family. Hipparionini and Equini are sister taxa, and both are in distinct groups from one another. "Merychippini" includes the common ancestor between these two within the subfamily and is grouped with both the Hipparionini and Equini as is expected in light of the phylogeny. "Anchitheriini" is the paraphyletic stem group ancestral to "Merychippini." The "Anchitheriini" is in its own group statistically and has the lowest OEI. Members of "Anchitheriini" are low-crowned, or have low HI [5], and can be interpreted as either browsers or intermediate feeders with a low percentage of abrasive material in their diet. Browse comprises a larger portion of the diet for "Anchitheriini" than any of the other tribes, and if diet is shaping occlusal enamel evolution, this group should have the lowest OEI, as indeed it does. We can use geography to tease apart diet and environmental change. Incorporating independent diet proxies (e.g., stable isotopes from enamel and/or microwear) combined with a regional biogeographic approach in a future study would identify the relative impact of local environmental change versus changing diets in shaping the evolution of OEI.

ANOVAs for tribes by NALMAs present an interesting pattern that enhances our interpretation of occlusal enamel evolution in horses. Ancestry seems to be an important influence on enamel length: the characteristic OEI values for a group are established at its origin and persist through time. When we consider interpretations of diet for each group [25] we find an unexpected pattern: many of the Barstovian equin horses are interpreted to be grazers but have OEI values consistent with contemporaneous browsing taxa. Interestingly, HI values for these equin horses are higher than those of hipparionin horses, while the converse is true for OEI. This supports our qualitative assertion that equin horses have more hypsodont yet less complicated teeth then their hipparionin relatives.

When the four tribes are present, Hipparionini and "Merychippini" are grouped together. Equini and "Anchitheriini" are also grouped. This pattern is only seen in the Barstovian and Clarendonian. Groupings may either represent tribes closely competing for resources or more evidence for the importance of phylogenetic constraint in this character. That is, the sample for "Merychippini" may be dominated by ancestral forms of Hipparionini, producing the observed connection between the tribes. We suspect this may be the case because more of the equin "*Merychippus*" have been split out into their own genera [5,43,44,45,46].

Conversely, it is possible that typical fossil members of Equini were more intermediate feeders and competing with browsing "Anchitheriini" for resources. Notably, in Great Plains Clarendonian faunas, Equini and "Anchitheriini" both compose a small percentage of the relative abundance of horses [6]. The similarity in OEI and relative abundance between these two groups warrants further investigation because previous workers have assigned the equines to grazing niches on the basis of their hypsodonty and isotopic data [25], but their OEI values would suggest that they are browsing along with their contemporaneous anchitherine relatives.

In terms of species richness, Hipparionini were the most successful tribe during the Clarendonian in the Great Plains, but were eventually replaced by Equini at the end of the Blancan. "Anchitheriini" and "Merychippini" go extinct by the Hemphillian, leaving Equini and Hipparionini (Fig. 2). The two tribes are significantly different in the Hemphillian and Blancan. Hipparionini are constrained to the southern latitudes during the Blancan and are extinct by the end of the Blancan [59]. Hipparionini remain in regions closer to the equator where the effects of climate change would not have been as strong [53,60]. In those regions, they continue to have higher OEI than their equin counterparts. The food source for hipparionines may have been restricted to warmer climates as the globe cooled, thus restricting the range of the tribe. The warmer regions may have served as refugia for North American hipparionin horses. We can better understand the drivers of occlusal enamel complexity when we can look across geography because we can compare regional patterns unfolding under slightly different environmental changes. Adding these data would allow us to investigate changes in response to regional climate changes through time.

We would like to apply these methods to other megafauna which have adaptations to increased ingested abrasiveness, such as camels, rhinos, African large primates, and South American notoungulates. A majority of enamel complexity in Equids is found in the hypsodont forms which originate in the Barstovian and are included in this study. However, it would be interesting to extend our methods back deeper into the Anchintheriinae and perhaps include the *Eohippus*-grade equids to see whether they also reflect other metrics of changing ecology. We would also like to test differences within Plio-Pleistocene *Equus* (e.g., caballine and stilt-legged horses), comparing them to Hipparionini genera to see if any equin horses are independently evolving complex enamel patterns similar to hipparionin horses as or after those hipparionins go extinct. This way we could test whether these Equini species converged on vacated niche space left by the extinct hipparionines.

The results of our Occlusal Enamel Index (OEI) study suggest that the complexity of the occlusal enamel of equid teeth is influenced by a combination of evolutionary relatedness, developmental constraint (tooth position), and changing environments over time. Equini seem to have an overall lower OEI than Hipparionini which supports the qualitative hypothesis that Equini have less occlusal enamel than Hipparionini. Our study shows that enamel band shapes are being influenced by climate and evolutionary history. As climate dries through time, we see an overall increase in enamel complexity. Phylogenetic relationships also have an influence on relative enamel complexity between clades (i.e., Equini tends to have less complex enamel than Hipparionini). Our results are consistent with the hypothesis that horses increase their enamel complexity in response to increased tooth abrasion from the Miocene through the Holocene.

Acknowledgments

We would like to thank R. G. Corner, G. Brown, R. Secord, R. Skolnick, S. Tucker, R. Otto, G. Boardman, P. Holroyd, J. Frankel, C. Sidor, B. Eng, J. and B. Orcutt, D. Pagnac, S. Shelton, A. Carr, H. Minkler, J. Samuels, C. Schierup, R. Evander, J. Galkin, J. Meng, S. K. Lyons, C. Ito, P. Wagner, R. Feranec, and E. Scott for access to collections, hospitality, and important discussions. We would also like to thank Q. Jin and S. Frost, for their feedback and suggestions on this project. The members of the University of Oregon Paleontology Group (past and present) have also contributed constructive feedback, advice, and hospitality related to this project. Finally we thank Alistair Robert Evans and Jessica Theodor for their constructive reviews of this manuscript.

Author Contributions

Analyzed the data: NAF. Wrote the paper: NAF EBD. Photographed specimens in museum collections: NAF. Measured digital images: NAF.

References

1. Osborn HF (1918) Equidae of the Oligocene, Miocene, and Pliocene of North America, iconographic type revision: Memoirs of the American Museum of Natural History 2: 1–217.
2. Simpson GG (1951) Horses: the horse family in the modern world and through sixty million years of history: New York, Oxford University Press.
3. Franzen JL (2010) The rise of horses: 55 million years of evolution. The Johns Hopkins University Press, Baltimore.
4. Quinn JH (1955) Miocene Equidae of the Texas Gulf Coastal Plain. University of Texas Publication 5516: 1–102.
5. MacFadden BJ (1998) Equidae In: Janis CM, Scott KM, Jacobs LL (Eds.), Evolution of Tertiary Mammals of North America, v.1: Cambridge University Press, New York, 537–559.
6. Famoso NA, Pagnac D (2011) A comparison of the Clarendonian Equid assemblages from the Mission Pit, South Dakota and Ashfall Fossil Beds, Nebraska. Transactions of the Nebraska Academy of Sciences 32: 98–107.
7. Famoso NA, Feranec RS, Davis EB (2013) Occlusal enamel complexity and its implications for lophodonty, hypsodonty, body mass and diet in extinct and extant ungulates. Palaeogeogr Palaeoclimatol Palaeoecol 387: 211–216. DOI: http://dx.doi.org/10.1016/j.palaeo.2013.07.006.
8. Strömberg CAE (2006) Evolution of hypsodonty in Equids: testing a hypothesis of adaptation. Paleobiology 32: 236–258.
9. Jardine PE, Janis CM, Sahney S, Benton MJ (2012) Grit not grass: Concordant patterns of early origin of hypsodonty in Great Plains ungulates and Glires. Palaeogeogr Palaeoclimatol Palaeoecol 365–366: 1–10. DOI: http://dx.doi.org/10.1016/j.palaeo.2012.09.001.
10. Lucas PW, Omar R, Al-Fadhalah K, Almusallam AS, Henry AG, et al. (2013) Mechanisms and causes of wear in tooth enamel: implications for hominin diets. J R Soc Interface 10: 1742–5662. DOI: http://dx.doi.org/10.1098/rsif.2012.0923.
11. Sanson GD, Kerr SA, Gross KA (2007) Do silica phytoliths really wear mammalian teeth? J Archaeol Sci 34: 526–531. DOI: http://dx.doi.org/10.1016/j.jas.2006.06.009.
12. Damuth J, Janis CM (2011) On the relationship between hypsodonty and feeding ecology in ungulate mammals, and its utility in palaeoecology. Biological Reviews 86: 733–758.
13. Pfretzschner HU (1993) Enamel microstructure in the phylogeny of the Equidae. Journal of Vertebrate Paleontology 13: 342–349.
14. Forsten A (1975) Fossil horses of the Texas Gulf Coastal Plain: A revision. Pierce-Sellards Series, Texas Memorial Museum 22: 1–86.
15. MacFadden BJ (1984) Systematics and phylogeny of Hipparion, Neohipparion, Nannippus, and Cormohipparion (Mammalia, Equidae) from the Miocene and Pliocene of the New World. Bull. Am. Mus. Nat. Hist. 179: 1–195.
16. MacFadden BJ (1988) Fossil horses from "Eohippus" (Hyracotherium) to Equus; 2. Rates of dental evolution revisited. Biol J Linn Soc Lond 35: 37–48.
17. Hulbert RC (1988) Calippus and Protohippus (Mammalia, Perissodactyla, Equidae) from the Miocene (Barstovian-Early Hemphillian) of the gulf coastal plain. Florida Museum of Natural History 32: 221–340.
18. Hulbert RC (1988) Cormohipparion and Hipparion (Mammalia, Perissodactyla, Equidae) from the late Neogene of Florida. Florida Museum of Natural History 33: 229–338.
19. Ungar PS (2010) Mammal teeth: origin, evolution, and diversity. The Johns Hopkins University Press, Baltimore.
20. Rensberger JM, Forsten A, Fortelius M (1984) Functional evolution of the cheek tooth pattern and chewing direction in Tertiary horses. Paleobiology 10: 439–452.
21. Heywood JJN (2010) Functional anatomy of bovid upper molar occlusal surfaces with respect to diet. J Zool 281: 1–11.
22. Kaiser TM, Fickel J, Streich WJ, Hummel J, Clauss M (2010) Enamel ridge alignment in upper molars of ruminants in relation to their diet. J Zool 81: 12–25.
23. Felsenstein J (1984) Phylogenies and the comparative method. Am Nat 125(1): 1–15.
24. MacFadden BJ (1992) Fossil horses: systematics, paleobiology, and evolution of the family Equidae. Cambridge University Press, New York.
25. MacFadden BJ (2005) Fossil Horses–Evidence for Evolution. Science 307: 1728–1730.
26. MacFadden BJ, Hulbert RC (1988) Explosive speciation at the base of the adaptive radiation of Miocene grazing horses. Nature 33(1): 466–468.
27. Hulbert RC, MacFadden BJ (1991) Morphological transformation and cladogenesis at the base of the adaptive radiation of Miocene hypsodont horses. Am. Mus. Novit., 3000: 1–61.
28. Prado JL, Alberdi MT (1996) A cladistics analysis of the horses of the Tribe Equini. Palaeontology 39: 663–680.
29. Martins EP, Diniz-Filho JAF, Housworth EA (2002) Adaptive constraints and the phylogenetic comparative method: a computer simulation test. Evolution 56(1): 1–13.
30. Rohlf FJ (2006) A comment on phylogenetic correction. Evolution 60(7): 1509–1515.
31. Hansen TF, Pienaar J, Orzack SH (2008) A comparative method for studying adaptation to a randomly evolving environment. Evolution 62(8): 1965–1977.
32. Stack JC, Harmon LJ, O'Meara B (2011) RBrownie: an R package for testing hypotheses about rates of evolutionary change. Methods Ecol Evol 2: 660–662.
33. Cayuela L, Granzow-de la Cerda Í, Albuquerque FS, Golicher DJ (2012) Taxonstand: An r package for species names standardization in vegetation databases. Methods Ecol Evol 3: 1078–1083.
34. Slater GJ, Harman LJ (2013) Unifying fossils and phylogenies for comparative analyses of diversification and trait evolution. Methods Ecol Evol 4(8): 699–702.
35. Mihlbachler MC, Rivals F, Solounias N, Semperbon GM (2011) Dietary change and evolution of horses in North America. Science 331: 1178–1181.

36. Becerra F, Vassallo AI, Echeverría AI, Casinos A (2012) Scaling and adaptations of incisors and cheek teeth in Caviomorph rodents (Rodentia, Hystricognathi). Journal of Morphology 273: 1150–1162.
37. Skinner MF, MacFadden BJ (1977) Cormohipparion n. gen. (Mammalia, Equidae) from the North American Miocene (Barstovian-Clarendonian). J Paleontol 51: 912–926.
38. Shapiro SS, Wilk MB (1965) An analysis of variance test for normality (complete samples). Biometrika 52: 591–611.
39. Bartlett MS (1937) Properties of sufficiency and statistical tests. Proc R Soc Lond A 160: 268–282.
40. Zar JH (2010) Biostatistical analysis, 5th Edition. Pearson Prentice-Hall, Upper Saddle River.
41. Wilcoxon F (1945) Individual Comparisons by Ranking Methods. Biometrics 1: 80–83.
42. Kramer CY (1956) Extension of multiple range tests to group means with unequal numbers of replications. Biometrics 12: 307–310.
43. Stirton RA (1940) Phylogeny of North American Equidae. University of California Publications, Bulletin of the Department of Geological Sciences 25: 165–198.
44. MacDonald JR, MacDonald ML, Toohey LM eds (1992) The species, genera, and tribes of the living and extinct horses of the world 1758–1966: Dakoterra 4: 1–429.
45. Kelly TS (1995) New Miocene Horses from the Caliente Formation, Cuyama Valley Badlands, California. Contributions in Science 445: 1–33.
46. Pagnac D (2006) Scaphohippus, a new genus of horse (Mammalia: Equidae) from the Barstow Formation of California. J. Mammal. Evol. 13: 37–61.
47. Azzaroli A, Voorhies MR (1993) The genus *Equus* in North America. The Blancan species. Palaeontographia Italica 80: 175–198.
48. Weinstock J, Willerslev E, Sher A, Tong W, Ho SYW, et al. (2005) Evolution, Systematics, and Phylogeography of Pleistocene Horses in the New World: A Molecular Perspective. PLoS Biol 3(8), e241.
49. Whistler DP (1991) Geologic history of the El Paso Mountains region. San Bernardino County Museum Association Quarterly 38(3): 108–113.
50. Hulbert RC, Whitmore FC (2006) Late Miocene mammals from the Mauvilla Local Fauna, Alabama. Florida Museum of Natural History 46(1): 1–28.
51. Eisenmann V, Alberdi MT, DeGiuli C, Staesche U (1988) Methodology, in: Woodburne MO, Sondarr P (Eds.), Studying fossil horses, 1–71.
52. Woodburne MO (2007) Phyletic diversification of the Cormohipparion occidentale complex (Mammalia, Perissodactyla, Equidae), late Miocene, North America, and the origin of the Old World Hippotherium datum. Bull. Am. Mus. Nat. Hist., 306: 3–138.
53. Zachos JC, Dickens GR, Zeebe RE (2008) An early Cenozoic perspective on greenhouse warming and carbon-cycle dynamics. Nature 451: 279–283.
54. Price EO (1999) Behavioral development in animals undergoing domestication. Appl. Anim. Behav. Sci. 65(3): 245–271.
55. O'Regan HJ, Kitchener AC (2005) The effects of captivity on the morphology of captive, domesticated and feral mammals. Mamm Rev 35: 215–230. DOI: 10.1111/j.1365-2907.2005.00070.x.
56. Renaud S, Auffray JC (2009) Adaptation and plasticity in insular evolution of the house mouse mandible. J Zool Syst Evol Res 48: 138–150. DOI: 10.1111/j.1439-0469.2009.00527.x.
57. Roth VL (1989) Fabricational noise in elephant dentitions. Paleobiology 15: 165–179.
58. Roth VL (1992) Quantitative variation in elephant dentitions: Implications for the delimitation of fossil species. Paleobiology 18: 184–202.
59. Carrasco MA, Kraatz BP, Davis EB, Barnosky AD (2005) Miocene Mammal Mapping Project (MIOMAP). University of California Museum of Paleontology. Available: http://www.ucmp.berkeley.edu/miomap/.
60. Roy T, Bopp L, Gehlen M, Schneider B, Cadule P, et al. (2011) Regional Impacts of Climate Change and Atmospheric CO_2 on Future Ocean Carbon Uptake: A Multimodel Linear Feedback Analysis. J Clim 24: 2300–2318. DOI: 10.1175/2010JCLI3787.1.

11

Millennium-Scale Crossdating and Inter-Annual Climate Sensitivities of Standing California Redwoods

Allyson L. Carroll*, Stephen C. Sillett, Russell D. Kramer

Department of Forestry and Wildland Resources, Humboldt State University, Arcata, California, United States of America

Abstract

Extremely decay-resistant wood and fire-resistant bark allow California's redwoods to accumulate millennia of annual growth rings that can be useful in biological research. Whereas tree rings of *Sequoiadendron giganteum* (SEGI) helped formalize the study of dendrochronology and the principle of crossdating, those of *Sequoia sempervirens* (SESE) have proven much more difficult to decipher, greatly limiting dendroclimatic and other investigations of this species. We overcame these problems by climbing standing trees and coring trunks at multiple heights in 14 old-growth forest locations across California. Overall, we sampled 1,466 series with 483,712 annual rings from 120 trees and were able to crossdate 83% of SESE compared to 99% of SEGI rings. Standard and residual tree-ring chronologies spanning up to 1,685 years for SESE and 1,538 years for SEGI were created for each location to evaluate crossdating and to examine correlations between annual growth and climate. We used monthly values of temperature, precipitation, and drought severity as well as summer cloudiness to quantify potential drivers of inter-annual growth variation over century-long time series at each location. SESE chronologies exhibited a latitudinal gradient of climate sensitivities, contrasting cooler northern rainforests and warmer, drier southern forests. Radial growth increased with decreasing summer cloudiness in northern rainforests and a central SESE location. The strongest dendroclimatic relationship occurred in our southernmost SESE location, where radial growth correlated negatively with dry summer conditions and exhibited responses to historic fires. SEGI chronologies showed negative correlations with June temperature and positive correlations with previous October precipitation. More work is needed to understand quantitative relationships between SEGI radial growth and moisture availability, particularly snowmelt. Tree-ring chronologies developed here for both redwood species have numerous scientific applications, including determination of tree ages, accurate dating of fire-return intervals, archaeology, analyses of stable isotopes, long-term climate reconstructions, and quantifying rates of carbon sequestration.

Editor: Benjamin Poulter, Montana State University, United States of America

Funding: This research was funded by the Save-the-Redwoods League's Redwoods and Climate Change Initiative (RCCI) and the endowment creating the Kenneth L. Fisher Chair in Redwood Forest Ecology at Humboldt State University. Earlier work in Humboldt Redwoods State Park was funded by the National Science Foundation (IOB-0445277). The funders had no role in study design, data collection and analysis, decision to publish, or preparation of the manuscript.

Competing Interests: The authors have declared that no competing interests exist.

* Email: allyson.carroll@gmail.com

Introduction

Coast redwood (*Sequoia sempervirens*, SESE) and giant sequoia (*Sequoiadendron giganteum*, SEGI) drew interest from early dendrochronologists because their decay-resistant heartwood, fire-resistant bark, and consequently great longevity provided access to intact millennium-scale tree-ring records. These are Earth's tallest, heaviest, and among the oldest trees [1,2]. Accurately dated tree-ring series of these species (Figure 1A and 1B) can thus provide a reliable basis for numerous scientific applications, including climate reconstructions and physiological analyses. While SEGI has been a cornerstone of modern dendrochronology, crossdating SESE has proven much more difficult. Accordingly, fundamental aspects of SESE dendrochronology, such as the development of range-wide tree-ring chronologies and assessment of potential drivers of year-to-year ring-width variation, have remained elusive.

During the early 1900s, A.E. Douglass formalized the precise dating of growth rings based on the common patterns of ring widths across a population of trees. The subsequent acceptance of crossdating as a valuable technique was due in part to Douglass's success in developing a 3,200-year tree-ring chronology for SEGI [3]. Generally complacent growth rings (i.e., low degree of inter-annual variation) punctuated with strong marker years (i.e., consistently small or otherwise distinct rings) allowed accurate crossdating of this species [4]. SEGI chronologies have since been used to produce millennium-scale histories of fire [5] and drought [6]. Shortly after his pioneering work with SEGI, Douglass documented complicated and unsuccessful attempts to crossdate SESE [7].

Certain growth characteristics render SESE problematic for crossdating. Frequent discontinuous or missing annual rings (Figure 1C) reflect an absence of wood production at a given cambium location and preclude using simple ring counts to estimate tree age. Missing rings are often associated with a wedging or pinching pattern where multiple rings merge (Figure 1D) [8,9]. Other confounding SESE growth attributes include patterns of spiral compression wood (Figure 1E) [9] and complacent rings with little annual variability [10]. Moreover, annual wood production occurs across an enlarging surface of cambium with increasing age, leading to narrower rings [11] even though whole-trunk wood volume growth continues to increase through old age [12].

Despite these challenges, there has been some progress crossdating SESE. Working on Douglass' range-wide collection, Schulman crossdated sections from limited areas noting that while many samples showed non-climatic growth irregularities, some had "intelligible record[s]" of climate-limited annual rings, especially those from higher on trunks of co-dominant trees [13]. Indeed, the probability of obtaining a maximum ring count along one radius increased with height on main trunks of SESE in a second-growth forest and was higher for co-dominant versus suppressed trees [14]. A crossdated SESE chronology (1750–1985) created from partial cross-sections of 15 old-growth stumps and logs (>6 m above ground) was used to reconstruct fire history at Prairie Creek Redwoods State Park [10]. Unsuccessful attempts to crossdate SESE continued to be reported at sites across the range (e.g., Pt. Reyes National Seashore and Jackson State Demonstration Forest) as work on fire frequencies progressed [15,16].

Climatic conditions generally drive the inter-annual variation in growth rings that underlies crossdating. Spatial synchrony of ring-width variation among sites and species points to broader-scale climate signals. Rather than closely tracking temperature or precipitation, SEGI's rings better record extreme events such as severe drought [6,17,18]. There is minimal dendroclimatic knowledge of SESE, although thin rings corresponded to extreme low precipitation years at sensitive locations [13]. For the Prairie Creek chronology, ring width correlated positively with July temperature and precipitation [19]. These findings underscore the dendroclimatic potential of SESE, which has yet to be assessed across its geographic distribution. Isotopic analyses of ~50 years of carbon and oxygen in tree-ring cellulose revealed correlations with maximum summer temperature, hinting that isotope-based climate reconstructions may be possible in SESE [20]. Any long-term reconstructions, however, will depend upon properly crossdated tree-ring series.

Our goals in this study were to develop tree-ring chronologies for SESE and SEGI throughout their geographic distribution in California and to assess their synchrony and climatic sensitivities. While positions higher on trunks of co-dominant trees may have more interpretable annual rings, access to such locations is limited by both the large stature and protected status of remaining old-growth forests. Use of downed trees is complicated by the unknown final year of growth to anchor crossdating. We employed rope-based tree climbing to overcome these difficulties in 14 forest reserves. Within-tree replication at multiple heights along the trunk facilitated crossdating and provided a basis for comparison among locations and across species. Here we focus on inter-annual variation in radial growth to assess climatic sensitivities. Specifically, we 1) describe crossdating success with an emphasis on SESE, 2) investigate site-to-site synchrony of the tree-ring signals within and between SESE and SEGI and with other tree species, and 3) quantify correlations between ring width and climate variation, including temperature, precipitation, PDSI (Palmer Drought Severity Index), and summer cloudiness during recent time series >100 years.

Methods

Study Area

Both redwood species grow within restricted ranges predominantly in California (Figure 2). SESE occurs along a narrow belt from its northern extent in far southwest Oregon to its southern extent along the Big Sur coast. A Mediterranean climate of wet winters and dry summers typifies this region (Figure S1) with coastal clouds contributing to SESE's annual hydrologic input via fog drip [21] and foliar absorption [22]. Precipitation generally increases with elevation and latitude (Figure 2). Northern rainforests support the largest, oldest, and most structurally complex trees, while drier southern forests are more prone to tree-killing fires leading to a younger age distribution [23]. SEGI occurs in a narrow belt of naturally distinct groves along the western slope of the Sierra Nevada generally between 1,525 and 2,290 m with access to ample soil moisture and most groves located in the southern third of its range [24,25]. Similar to SESE, the climate is Mediterranean but with a greater temperature range and most precipitation falling as snow between December and March [26]. Study locations spanned the native distributions of both species from north to south in high-productivity, old-growth forests exhibiting minimum human disturbance, except the partially logged Whitaker Forest (Table 1). Eight SESE sites spanned 670 km and six SEGI sites spanned 279 km (Figure 2).

Ethics Statement

Research was conducted in Jedediah Smith Redwoods State Park, Prairie Creek Redwoods State Park, Redwood National Park, Humboldt Redwoods State Park, Montgomery Woods State Natural Reserve, Samuel P. Taylor State Park, Big Basin Redwoods State Park, Landels-Hill Big Creek Reserve, Calaveras Big Trees State Park, Whitaker Forest, Kings Canyon National Park, Sequoia National Park, Mountain Home State Demonstration Forest, and Sequoia National Monument (Table 1). We obtained necessary permits from California Department of Parks and Recreation, Redwood National Park, Sequoia and Kings Canyon National Parks, and Sequoia National Monument.

Tree Selection and Core Sampling

A total of 76 SESE and 44 SEGI trees (Table 1) were selected for sampling. The majority of study trees occurred within 16 permanent plots located in undisturbed forests containing some of the largest and most structurally complex trees at each location. In each plot, the tallest tree plus shorter and suppressed trees were climbed and measured. Additional trees from nearby forests were included so that the complete dataset contained the full range of tree sizes, crown structures, and canopy positions occurring in old-growth forests of both species, excluding trees unsafe to climb. Tree heights varied from 18.1 to 115.7 m for SESE and 26.1 to 96.3 m for SEGI. Minimum age calculations ranged from 110 to 2,510 years for SESE and 40 to 3,240 years for SEGI (Sillett et al., unpublished data). We also sampled two recently fallen trees to increase sample size at the southernmost SESE location (LH).

We climbed trees with low impact rope techniques (i.e., no spikes) and used 24″ and 32″ increment borers to collect 1–7 cores around trunk circumference at regular height intervals (about every 10 m). The number of cores collected and positions sampled varied depending on trunk size and evidence of past injuries or anomalous growth. Some cores reached beyond pith to include growth from the opposite side equating to two series (i.e., full or partial radii) per core. The lowest samples were collected above basal buttressing, typically no lower than 5 m above ground level, to avoid irregular growth and creating lower trunk wounds that weep persistently in these species (Sillett, personal observation). All cores were collected between 2005 and 2012.

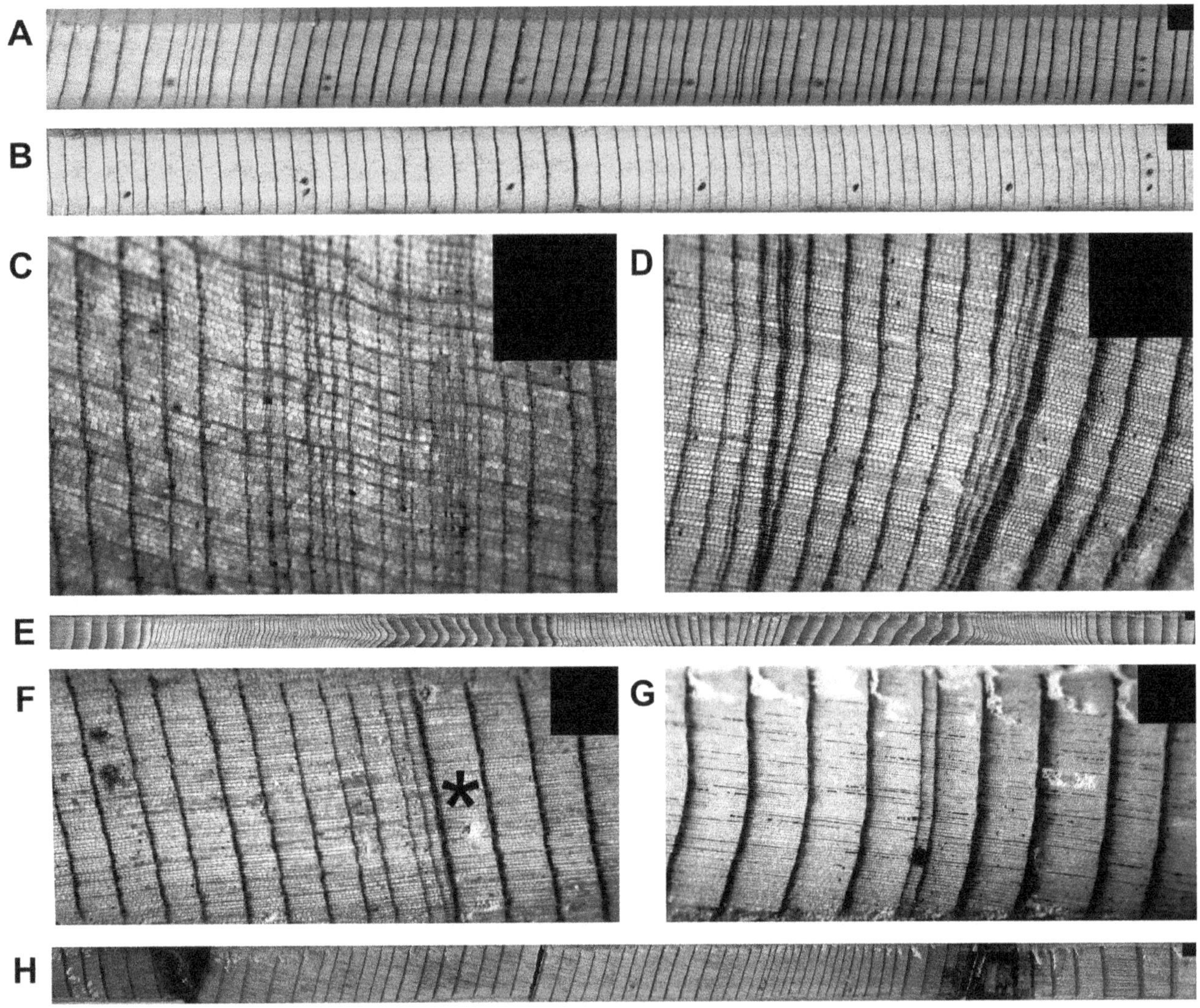

Figure 1. Tree-ring characteristics of *Sequoia sempervirens* and *Sequoiadendron giganteum*. (A) Crossdated SESE annual rings with marks beginning 1900 (three dots far right) and ending 1960 (one dot far left). (B) Crossdated SEGI annual rings with same years and marks as panel A. (C) Area of tight SESE annual rings, including missing rings. (D) SESE ring wedging, where discontinuous rings merge. (E) SESE spiral compression wood. (F) SESE annual rings from northernmost location (JS) showing 1739 event. After large 1738 ring (*), 1739 and 1740 are 1-cell-wide micro-rings merged with 1738 latewood, 1741 and 1742 rings are tight, and 1750 ring is marked with two dots on left. (G) SEGI annual rings from GF showing 1580 event, marked with one dot. Of 12 cores collected from the largest SEGI sampled, this is the only core showing 1580. (H) Two fire scars on core collected 45 m above ground in SESE rainforest (JS). Black boxes in upper right of each panel cover 1 mm^2.

Crossdating Techniques

Cores were glued along shallow grooves in wooden mounts and polished with a sandpaper progression from 220 to 1,500 grit. We inspected each core under a stereo microscope and further polished sections with tight or wedged rings, which often revealed micro-rings along sides of SESE cores. We scanned all cores at ≥1,200 dpi and measured growth rings to the nearest 0.001 mm using WinDendro [27]. Microscope work was the primary source for deciphering growth rings, and WinDendro was used as the measuring tool.

Cores were crossdated by visually identifying marker years and specific ring-width patterns. Narrow rings provided the primary visual markers; however, large growth rings and rings with thick latewood were also used, especially in SESE. Marker years and patterns were recorded using the list method of crossdating [28]. We used COFECHA as a quality control program to confirm crossdating via correlations and to generate skeleton plots [29]. Crossdating occurred independently of previously published chronologies or marker years and was first accomplished at the tree-level and then confirmed at individual locations.

Given SESE's tendency for radial growth irregularities, care was taken in determining final dates. Each tree's chronology was initially built upon the best series (e.g., series with distinct ring boundaries, strong marker years, and limited sections of extremely narrow or wedged rings). Strong correlations in COFECHA ($r \geq 0.32$, $P < 0.01$) verified inclusion of these series in a preliminary chronology. Missing rings were located based on visual crossdating with other cores from the same tree, morphological characteristics of adjacent rings (e.g., wedging, narrow rings, or damaged cells), and confirmation in COFECHA. Series with large sections of difficult rings were floated as unknowns against the master chronology in COFECHA to identify possible dates for earlier years to examine visually. To manage complex series, we classified rings into four categories based on degree of annual resolution

Table 1. Characteristics and tree-ring sampling intensity for two species at 14 locations.

Species	Location	Latitude (°N)	Longitude (°W)	No. trees	No. series	No. rings
SESE	JS, Jedediah Smith Redwoods State Park	41.8	124.1	8	87	34,373
SESE	PC, Prairie Creek Redwoods State Park	41.4	124.0	9	106	37,943
SESE	RNP, Redwood National Park	41.2	124.0	14	143	71,119
SESE	HR, Humboldt Redwoods State Park	40.3	123.9	22	298	49,947
SESE	MW, Montgomery Woods State Natural Reserve	39.2	123.4	5	60	20,025
SESE	SPT, Samuel P. Taylor State Park	38.0	122.7	5	49	10,736
SESE	BB, Big Basin Redwoods State Park	37.2	122.2	5	55	13,128
SESE	LH, Landels-Hill Big Creek Reserve	36.1	121.6	8	66	13,259
SEGI	CBT, Calaveras Big Trees State Park	38.2	120.2	5	63	27,249
SEGI	WF, Whitaker Forest	36.7	118.9	19	266	57,390
SEGI	RMG, Redwood Mountain, Kings Canyon National Park	36.7	118.9	4	43	28,466
SEGI	GF, Giant Forest, Sequoia National Park	36.6	118.7	6	78	51,393
SEGI	MH, Mountain Home State Demonstration Forest	36.2	118.7	5	58	36,011
SEGI	FC, Freeman Creek, Sequoia National Monument	36.1	118.5	5	94	32,673

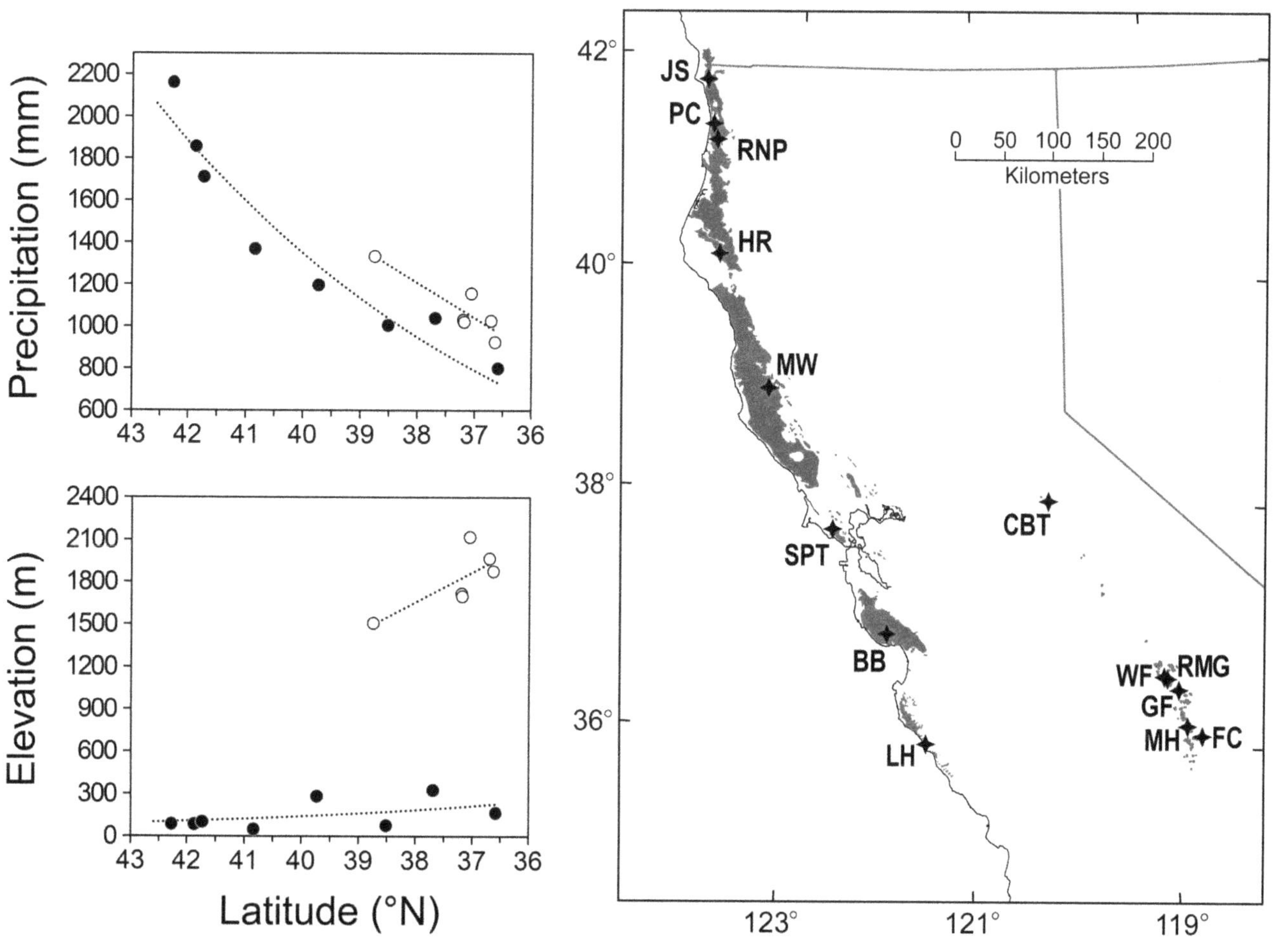

Figure 2. Average annual precipitation, elevation, and geographic distribution of 14 locations for two redwood species. Filled circles are *Sequoia sempervirens* locations and open circles are *Sequoiadendron giganteum* locations. Precipitation values for each location are 114-year averages (1895–2008) at 800-m resolution using PRISM data [36]. Red-shaded areas indicate native ranges of *Sequoia sempervirens* (left) and *Sequoiadendron giganteum* (right), and stars denote sampling locations.

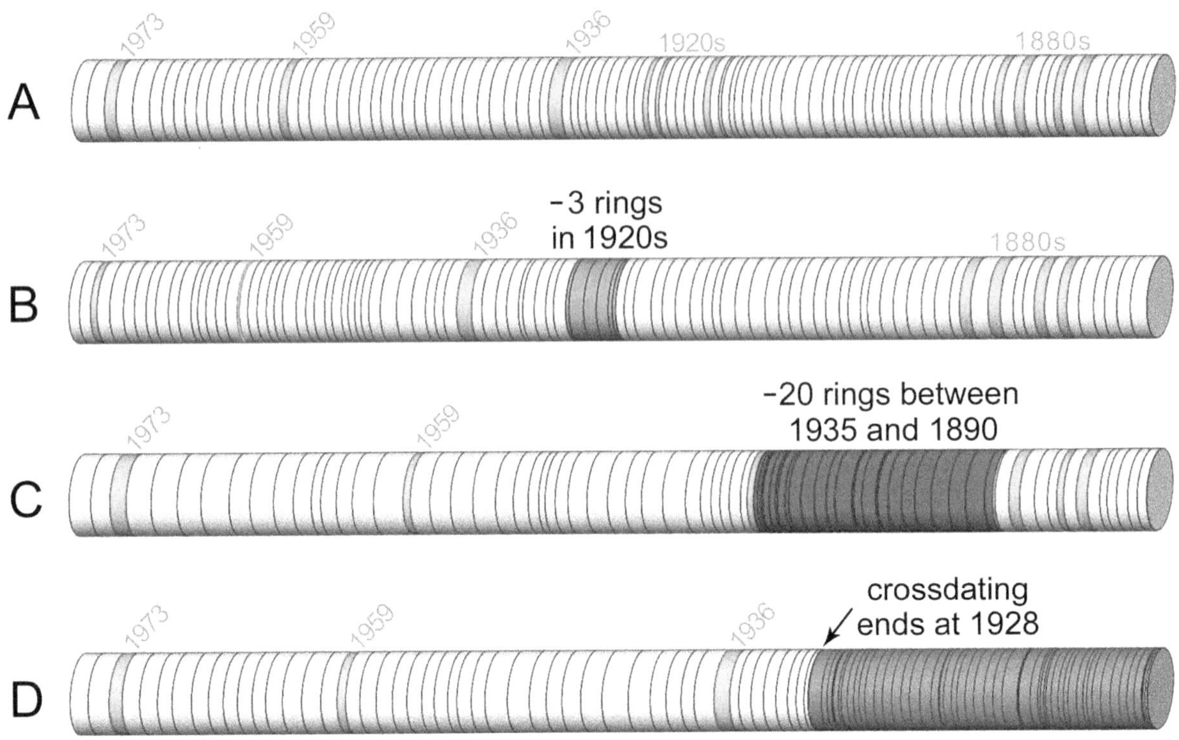

Figure 3. Classification system for crossdating confidence based on annual resolution. Marker years are denoted in blue. Examples are drawn from *Sequoia sempervirens* ring-width data. (A) *High crossdating confidence* or continuous annual resolution with no or few missing rings. (B) *Moderate crossdating confidence* (orange). Although missing rings are placed in their most likely location, alternative positions are possible (e.g., 3 missing rings between 1918–1926). Also assigned to sections of miniscule growth (e.g., missing rings among micro-rings). (C) *Bounded with no annual resolution* (red). Total number of missing rings in a section deduced from surrounding crossdating, but there is no indication of their most likely placement, often due to many missing rings (e.g., 20 missing rings between 1890–1935). (D) *Crossdating cessation* (purple). No annual resolution for a section where crossdating terminated and interior rings are neither resolved nor bound by known markers (e.g., crossdating progressed towards pith and ceased at 1928, leaving an undated section of core with a minimum ring count).

achieved: 1) *high crossdating confidence*, 2) *moderate crossdating confidence*, 3) *bounded with no annual resolution*, and 4) *crossdating cessation* (Figure 3).

Crossdating Statistics

Patterns of crossdating success (*high* plus *moderate confidence*) and missing rings were assessed for each species at the tree level. Differences among species were confirmed using Levene's test for homogeneity of variance and ANOVA with robust Tukey contrasts [30]. The same method was applied using suppression status and spiral compression wood as categorical variables to help explain some of the variation in crossdating success within SESE trees. Suppressed trees were those overtopped by neighbors or heavily shaded on their southern flank. Spiral compression wood was assigned to trees if the pattern was exhibited by at least one core.

Core-level data were used to compare within-tree crossdating variability for each species and to assess the degree of difficulty in crossdating basal versus non-basal cores in SESE. We calculated the standard deviation among cores for each tree and used ANOVA with robust Tukey contrasts to test for significant differences. To test the assumption that basal cores were harder to crossdate than other cores in SESE, a paired Wilcoxon signed-rank test was applied to the mean proportion of crossdated rings of the lowest position versus all higher positions on each tree. All crossdating analyses were performed in R [31].

Creation and Synchrony of Chronologies

We created a standardized chronology at each site to present a template for crossdating and investigation of inter-annual climate sensitivity. First, we created tree-level chronologies, including all series from a given trunk >50 years with *high crossdating confidence*. We used ARSTAN [32] to standardize ring widths around a dimensionless index of 1.0 and applied a 32-year cubic smoothing spline to remove the age-related geometric growth trend and other low-frequency variation. This approach optimized expression of high-frequency variation most useful for crossdating and suitable for inter-annual climate analysis [33]. Tree-level chronologies (standard version) were then combined into site-level chronologies with no further detrending. Standard versions of site chronologies were used for crossdating and identification of marker years, whereas residual versions, which removed first-order autocorrelation and stabilized variance [34], were used for analyses of synchrony and climate sensitivity.

Synchrony of chronologies was assessed by comparing marker years among locations and by correlation analysis. Marker years were categorized using the 10 smallest standardized ring widths per century, and years with three or fewer locations in common were removed to identify the strongest regional markers. We used correlation analysis to compare the common variance among residual chronologies over the well-replicated time period 1750–2008. To evaluate how correlations between these chronologies changed with distance, we computed Euclidean distance between each pair of locations using UTM coordinates and elevations.

Table 2. Statistical characteristics of standard tree-ring chronologies for two species at 14 locations.

Species	Location	Years	1 tree cutoff	Mean sensitivity[a‡]	Interseries correlation[b‡]	Unfiltered auto-correlation[c‡]	Between trees rbar[d]*	EPS[e]*	No. trees common period*
SESE	JS	1311–2009	1370	0.288	0.554	0.773	0.530	0.887	7
SESE	PC	1049–2010	1313	0.289	0.489	0.809	0.403	0.844	8
SESE	RNP	328–2012	1093	0.277	0.529	0.797	0.370	0.804	7
SESE	HR	1071–2011	1326	0.247	0.601	0.762	0.454	0.924	16
SESE	MW	822–2009	1347	0.277	0.634	0.804	0.448	0.803	5
SESE	SPT	1422–2010	1523	0.286	0.580	0.805	0.627	0.834	3
SESE	BB	1415–2011	1600	0.220	0.585	0.737	0.472	0.781	4
SESE	LH	1653–2010	1705	0.323	0.725	0.695	0.472	0.843	6
SEGI	CBT	783–2009	1278	0.182	0.564	0.765	0.459	0.809	5
SEGI	WF	1085–2009	1098	0.182	0.610	0.767	0.455	0.893	10
SEGI	RMG	824–2011	925	0.176	0.566	0.822	0.549	0.829	4
SEGI	GF	474–2011	712	0.187	0.629	0.797	0.509	0.862	6
SEGI	MH	782–2011	830	0.169	0.676	0.774	0.643	0.900	5
SEGI	FC	544–2010	1226	0.159	0.640	0.729	0.453	0.805	5

[a]Measure of year-to-year variability of annual rings [46].
[b]Average correlation of every series to the master (excluding series in question) [46].
[c]Reflects prior year's influence on current year's growth [46].
[d]Correlation between tree-ring chronologies, representing strength of common variability [85].
[e]Degree to which the chronology represents a hypothetical perfect chronology. Function of rbar and sample size, with 0.85 as a recommended cutoff for confidence [44,85].
‡Average tree-level statistics, using series >50 years with high and moderate crossdating confidence. Calculated using COFECHA.
*Common period 1901–2000.

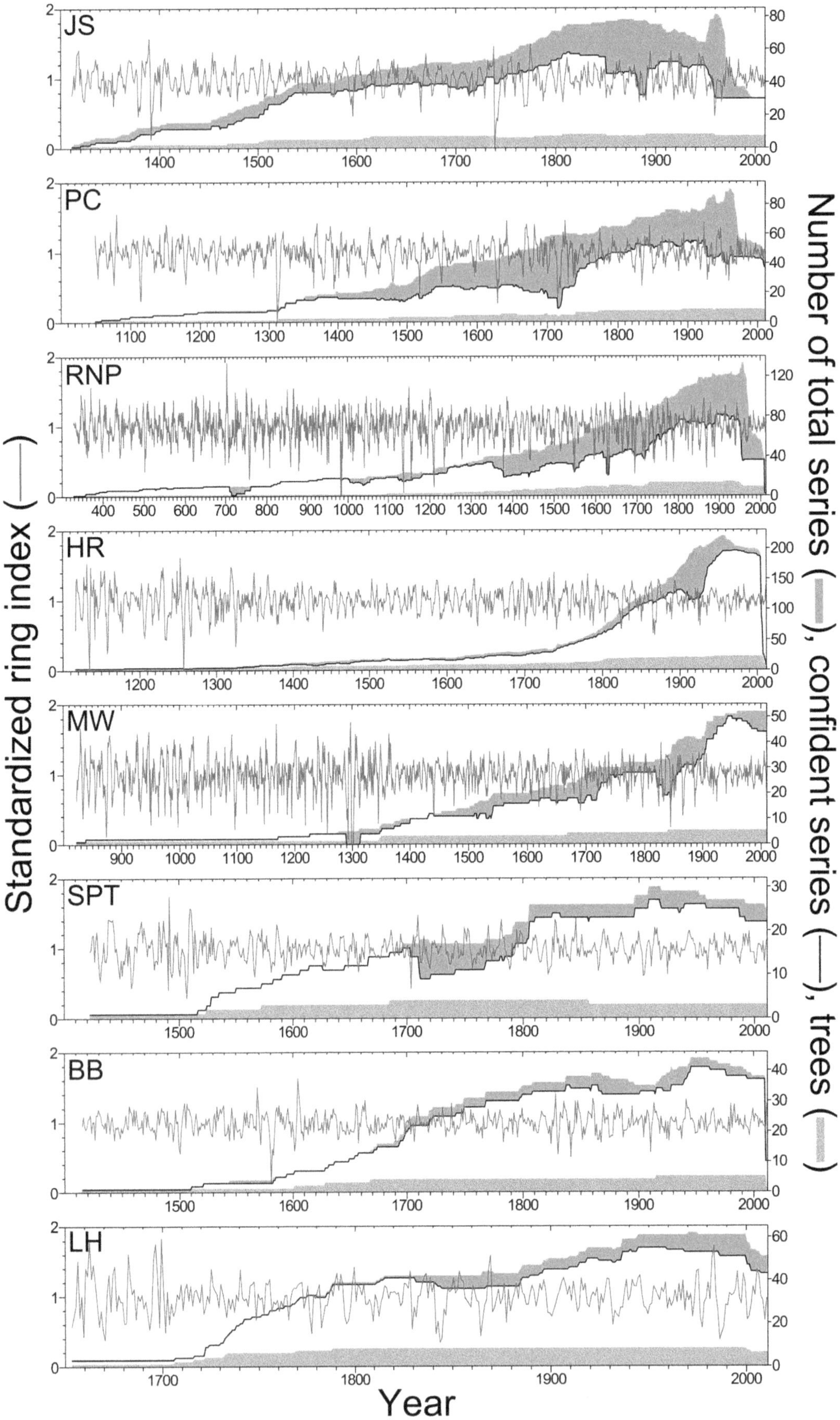

JS
PC
RNP
HR
MW
SPT
BB
LH
Standardized ring index (—)
Number of total series (▬), confident series (—), trees (▬)
Year

Figure 4. Standardized tree-ring chronologies and sample depths for eight *Sequoia sempervirens* locations. Blue lines indicate ring-width indices for each location, using series >50 years in length with *high crossdating confidence* and detrended with a 32-year cubic smoothing spline. Pink shading denotes difference between total number of series sampled and number of series with *high crossdating confidence*. Tree sample sizes indicated in gray.

Regional Assessment

We used correlation analysis to compare inter-annual variability of SESE and SEGI radial growth with that of other tree species. Species chosen on the basis of proximity to study locations, known climate sensitivities, and chronology length included *Pseudotsuga menziesii* (PSME), *Chamaecyparis lawsoniana* (CHLA), *Juniperus occidentalis* (JUOC), *Pinus ponderosa* (PIPO), *Quercus douglasii* (QUDG), and *Pinus jeffreyi* (PIJE). Tree-ring data were accessed via the International Tree Ring Data Bank [35] except the Oregon PSME chronologies (Bryan Black, unpublished data). Raw ring-width series were subjected to the same steps of chronology creation described in the previous section. SESE and SEGI chronologies were collapsed into regional chronologies represented by northern SESE sites (JS, PC, RNP, HR), the southernmost SESE site (LH), and all SEGI sites. The maximum common period of 1760–1980 was used in these analyses.

Climate and Radial Growth

Monthly maximum temperature, minimum temperature, and precipitation data were obtained for each location using the PRISM (Parameter-elevations Regressions on Independent Slopes Model) climate mapping system with 800 m resolution [36]. PRISM uses point measurements, a digital elevation model, knowledge of geospatial climatology, and considers both topography and proximity to the Pacific Ocean [37,38]. Monthly precipitation as snow at 4 km resolution was acquired for each SEGI location from Climate Western North America (ClimateWNA), which relies on PRISM data for a baseline [39].

Palmer Drought Severity Index (PDSI), which incorporates the effects of temperature, precipitation, and evapotranspiration [40], provided a proxy for soil moisture availability. PDSI utilizes a water balance model where positive values reflect wet conditions and negative values reflect dry conditions. Monthly location-specific PDSI as well as California and regional summer (average June–September) PDSI data were used to assess growth sensitivities. Monthly PDSI data at 4 km resolution were obtained from Western Regional Climate Center's WestWide Drought Tracker, which uses both PRISM and North American Land Data Assimilation System Phase 2 data [41]. State and regional PDSI data were obtained from the National Climatic Data Center, using regions CA01 (North Coast California) for all SESE north of the San Francisco Bay, CA04 (Central Coast California) for all SESE south of San Francisco Bay, and CA05 (Sierra Nevada) for all SEGI.

We used reconstructed northern California 1901–2008 airport fog (i.e., cloud base ≤400 m elevation) derived from Arcata and Monterrey airport data (1951–2008) and the inland-coast difference in maximum temperature [42]. A 32-year spline was applied to detrend these data and emphasize inter-annual variation. Average June–September airport fog was used to express summer cloudiness in SESE sites, all of which occurred below 400 m elevation (Figure 2).

Relationships between monthly climate and residual chronologies were examined with combined bootstrapped Pearson's correlation and response function analyses (RFA) for the period 1895–2008 using the bootRES package in R [43]. RFA removed multicollinearity between months before assessing dendroclimatic relationships [43]. Significant relationships revealed by both Pearson's correlations and RFA were emphasized. To account for potential effects of the previous growing season on current radial growth, our analysis window included the beginning of previous through end of current growing season (March–October for SESE; May–October for SEGI). We also analyzed correlations between residual chronologies and both summer cloudiness (1901–2008) and summer PDSI (1895–2008).

Results

Crossdating

We analyzed 76 trees, 864 series, and 250,530 growth rings at eight locations spanning the years 328–2012 for SESE and 44 trees, 602 series, and 233,182 growth rings at six locations spanning the years 474–2012 for SEGI (Table 1). All rings were classified by annual resolution: 1) *high crossdating confidence*, 2) *moderate crossdating confidence*, 3) *bounded with no annual resolution*, and 4) *crossdating cessation* (Figure 3). We crossdated 82.6% of SESE rings (76.4% *high confidence* and 6.2% *moderate confidence*), compared to 99.4% of SEGI rings (98.9% *high confidence* and 0.5% *moderate confidence*). Missing rings represented 4.4% of SESE rings, which was a minimum estimate; the *crossdating cessation* category (7.9% of rings) was not included in the calculation, and this category had higher proportions of tight and likely missing rings. By comparison, SEGI was missing only 0.2% of rings and had only 0.3% of rings in the *crossdating cessation* category. Classification of SESE rings in the *bounded without annual resolution* category (9.4%) often occurred in series with many missing rings. Tight rings were more common in SESE with 8.2% of widths <0.25 mm and 2.4% <0.10 mm, compared to 6.3% and 0.7% for SEGI respectively.

Tree-level proportions of crossdated and missing rings differed between species. The proportion of crossdated rings was 15% lower in SESE than SEGI ($P<0.001$). The range of tree-level crossdating success was also considerably larger in SESE, 3.7–100% (mean = 84.9), compared to 89.7–100% (mean = 99.5) in SEGI. While many trees of both species had all rings annually resolved, 71% of SESE trees compared to 20% of SEGI trees had less than perfect crossdating. Within SESE, both suppressed trees and those exhibiting spiral compression wood had a lower percentage of crossdated rings ($P = 0.030$ and $P = 0.007$, respectively). Suppressed trees averaged 66.6% crossdated rings compared to 87.9% in unsuppressed trees, and trees with spiral compression wood averaged 60.3% crossdated rings compared to 88.1% in trees without this trait. The tree-level proportion of missing rings was higher and varied more in SESE than SEGI ($P<0.001$). For SESE and SEGI respectively, the mean proportion of missing rings per tree was 3.1% versus 0.1% and ranged from 0–12.5% versus 0–0.9%.

Within-tree variation in crossdating differed between species. The standard deviation of percent crossdated rings among cores from the same tree was greater ($P<0.001$) in SESE (range 0–48.3, mean 14.3) than SEGI (range 0–13.4, mean 1.0). Higher variation in SESE was attributable, in part, to poorer crossdating of lowest trunk cores, which crossdated an average of 10.6% less often than cores higher up trunks ($P<0.001$).

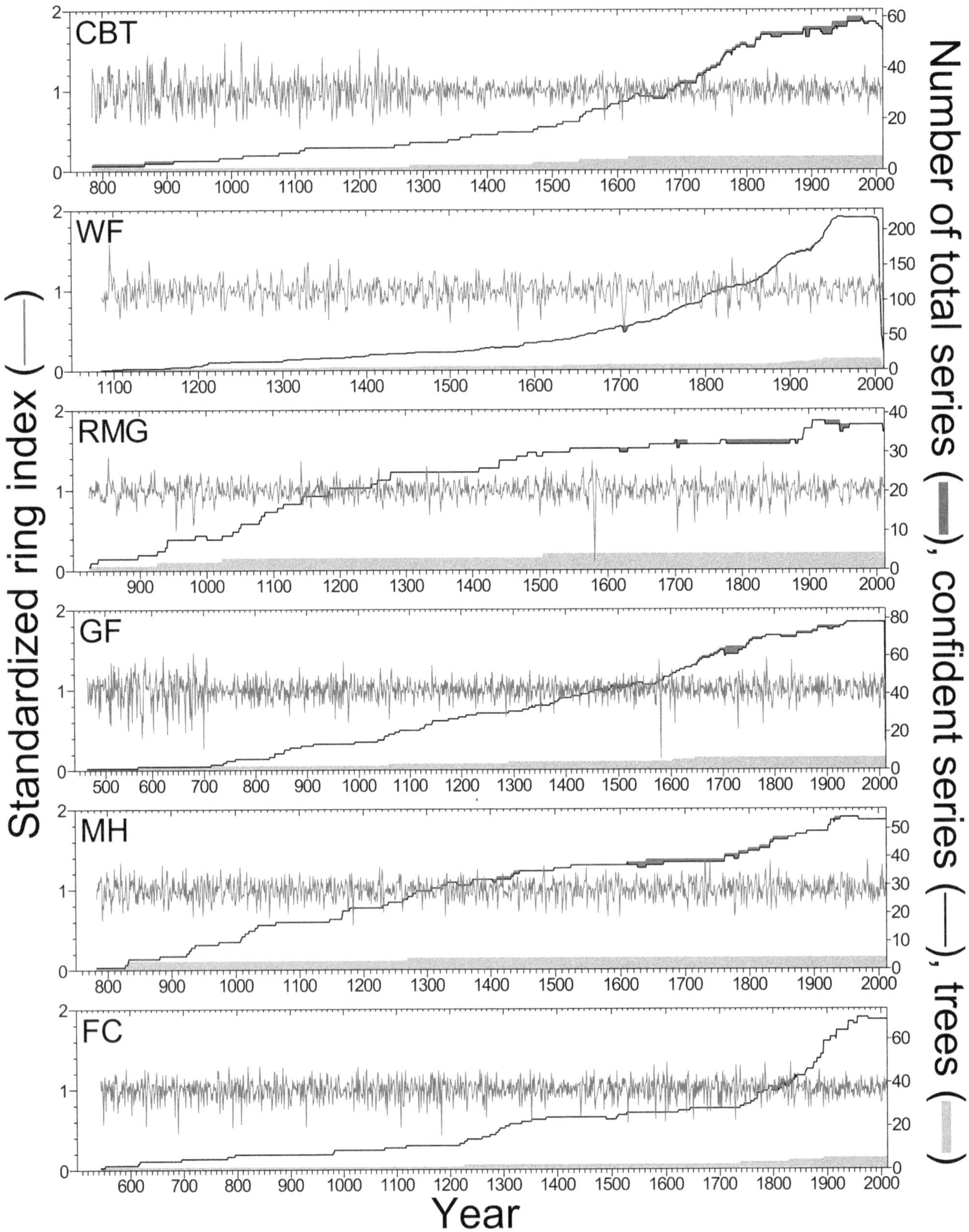

Figure 5. Standardized tree-ring chronologies and sample depths for six *Sequoiadendron giganteum* locations. Blue lines indicate ring-width indices for each location, using series >50 years in length with *high crossdating confidence* and detrended with a 32-year cubic smoothing spline. Red shading denotes difference between total number of series sampled and number of series with *high crossdating confidence*. Tree sample sizes indicated in gray.

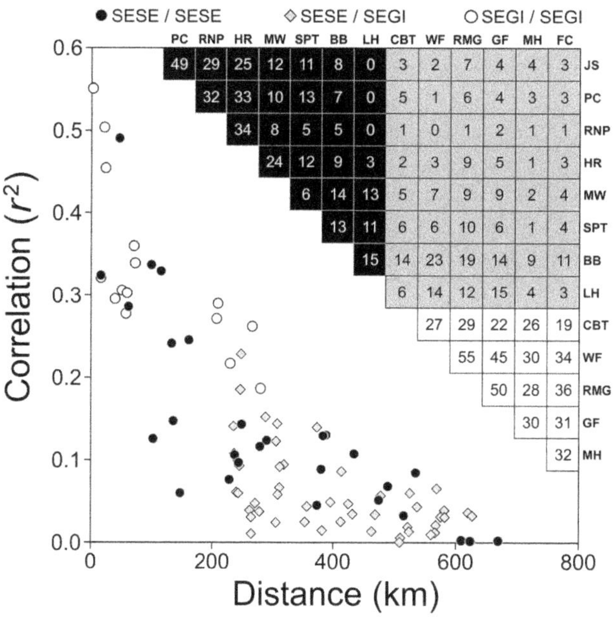

Figure 6. Correlations among redwood tree-ring chronologies and relationships with Euclidean distances between locations. Values in cells are % variance explained ($100 \times r^2$) for pairwise correlations over a 259-year common period (1750–2008). Locations arranged by latitude from north (left and top) to south (right and bottom) within species.

Chronology Characteristics and Synchrony

We developed crossdated tree-ring chronologies for all 14 locations using series with *high crossdating confidence* detrended with a 32-year spline (Table S1). SESE chronologies spanned 1,685 years (328–2012) and were longer for middle and northern latitude locations where older trees persist (Figure 4). SEGI chronologies spanned 1,538 years (474–2011) with all chronologies reaching to at least 1085 (Figure 5). Several statistics described crossdating strength and chronology quality, highlighting the need for more tree replication in SESE locations (Figure 4, Table 2). Years with a greater difference between possible and actual series indicated less successful crossdating, often associated with missing or extremely narrow rings. Crossdating success among series was uniformly high for all SEGI chronologies (Figure 5). The dip in replication for the most recent years was caused by selecting only *highly confident* series >50 years long, and high ring-width variation apparent early in the chronologies was attributable to low sample size [44]. Chronologies at each location were established independently except for the years 1289–1313 at MW where placement of six missing rings was based on the HR chronology. Chronologies exhibited reasonable quality and signal strength given the low tree replication (Table 2).

Marker years in both species aided crossdating but in a different manner. For SESE, crossdating was complex with frequent missing, tight, wedged, and abnormal rings (mean tree-level sensitivity = 0.273), which had to be confirmed nearly every decade by marker years. Accordingly, crossdating SESE involved recognizing not only low-growth marker years (e.g., 1824, 1924), but also years with consistently large ring widths (e.g., 1936, 1983), signature ring-width patterns (e.g., small 1883, 1887, and 1889), rings with distinctively thick latewood (e.g., 1923), and growth reductions (i.e., several years of sustained low-growth rings) (e.g., 1865–1869). Of the eight low-growth SESE years listed by Schulman [13] for the time period 1824–1924, five were in the lowest decile of ring widths at MW, about 20 km away. For SEGI, intermittent low-growth marker years approximately every few decades guided crossdating among rings with similar widths (mean tree-level sensitivity = 0.178) and infrequent missing rings. Seventy percent of marker years listed by Hughes and Brown [6] common to the lowest decile of their three SEGI sites (GF, MH, and Camp Six) were among the lowest decile of ring widths for our unified SEGI chronology (sample period of 782–1988 with three-site replication).

Common marker years and patterns of inter-annual variation emerged at the regional scale (Figure S2). Several marker years were common to both species (e.g., 1500, 1598, 1729, 1865, 1924, 1959), while others clustered by species (e.g., 1739, 1742, 1850 for the northern SESE and 1580, 1637, 1733, 1827, 1829, 1841 for SEGI). Southern SESE locations often shared strong marker years more frequently with SEGI than northern SESE locations (e.g., 1580, 1733). The common climate response indicated by shared marker years was reflected in the synchrony of residual chronologies, which was clearly stratified by distance (Figure 6). Northern locations (JS, PC, RNP, HR) correlated strongly with each other ($P<0.0001$), while southern SESE locations (BB, LH) correlated more strongly with SEGI than with northern SESE locations. Indeed, the southernmost SESE chronology (LH) was virtually independent of northern SESE chronologies. All SEGI chronologies correlated significantly ($P<0.0001$) with each other, but the most northern and isolated location (CBT) had the weakest correlations.

Regional Assessment

Common ring-width variation between redwoods and other species reflected spatial synchrony of climate forcing (Table 3). The northern SESE chronology correlated most strongly with the CHLA chronology from the nearby Klamath-Siskiyou Mountains of southwestern Oregon. Of the PSME chronologies, northern SESE co-varied most with the two coastal Oregon chronologies and correlated more with PSME from the Olympic Peninsula of Washington than with PSME from Point Reyes, California. Moreover, the northern SESE chronology correlated more consistently with PIPO, JUOC, PIJE, and QUDC chronologies from northern California than with those from the Sierra Nevada and central California. The southernmost SESE chronology (LH) showed consistently strong correlations with QUDG chronologies throughout California, especially those from the central California Coast Range. The only PSME chronology strongly correlated with the LH chronology was from Point Reyes. The LH chronology also correlated strongly with PIJE chronologies of the southern Sierra Nevada and exhibited weaker correlations with PIPO, JUOC, and PIJE chronologies of northern California. The unified SEGI chronology showed consistent, positive correlations with tree-ring chronologies throughout California as well as the CHLA chronology from southwestern Oregon.

Dendroclimatic Relationships

Consistent relationships between monthly climate and residual tree-ring chronologies were evident in SESE (Figure 7). PDSI during the growing season correlated positively with radial growth at all locations with only one location (RNP) lacking any significant response functions. The positive relationship between PDSI and radial growth extended earlier in the growing season at southern locations. Precipitation correlated positively with radial growth at all locations, and response functions were significant for April at JS, June at PC, July at RNP, April and May at HR and MW, January at SPT, prior October at BB, and four months during winter and spring at LH. Temperatures correlated with radial

Table 3. Correlations between tree-ring indices of SESE and SEGI and six other western North American tree species.

Species	Location	Latitude (°N)	Longitude (°W)	Elevation (m)	State	Northern SESE (*r*)	Southernmost SESE (*r*)	Unified SEGI (*r*)	Contributor
PSME	Olympic Road	48.00	−124.00	267	WA	**0.22**	−0.03	0.06	Earle et al.
PSME	Newport	44.40	−124.03	20	OR	**0.38**	0.09	**0.19**	Black
PSME	Cape Perpetua	44.30	−124.07	244	OR	**0.29**	−0.01	0.12	Black
PSME	Fryday Ridge	40.75	−123.67	1560	CA	**0.19**	−0.10	**0.27**	Briffa & Schweingruber
PSME	Pt. Reyes	38.02	−122.80	120	CA	0.12	**0.28**	**0.22**	Brown
CHLA	Page Mountain	42.00	−123.57	1070	OR	**0.40**	−0.06	**0.42**	Carroll & Jules
JUOC	Sharp Mountain	41.72	−121.80	1417	CA	**0.29**	**0.38**	**0.40**	Holmes et al.
JUOC	Carson Pass East	38.70	−120.00	2591	CA	0.12	0.06	**0.39**	Meko et al.
JUOC	Kaiser Pass	37.28	−119.08	2731	CA	**0.20**	**0.24**	**0.42**	Holmes et al.
PIPO	Damon's Butte	41.50	−121.17	1448	CA	**0.19**	**0.33**	**0.48**	Graumlich
PIPO	Grizzly Peak	41.17	−122.03	1463	CA	**0.19**	−0.01	**0.30**	Graumlich
PIPO	St. Johns Mountain	39.43	−122.68	1555	CA	0.15	0.05	**0.34**	Holmes
QUDG	Dibble Creek	40.42	−122.63	218	CA	0.12	**0.47**	**0.30**	Stahle et al.
QUDG	Eel River	39.82	−123.07	610	CA	**0.17**	**0.39**	**0.44**	Stahle et al.
QUDG	Putah Creek	38.67	−122.45	180	CA	0.01	**0.60**	**0.33**	Stahle et al.
QUDG	Mt. Diablo	37.87	−121.95	245	CA	0.00	**0.66**	**0.33**	Stahle et al.
QUDG	North Fork Kaweah River	36.92	−118.90	701	CA	**0.16**	**0.56**	**0.37**	Stahle et al.
QUDG	Pinnacles National Monument	36.47	−121.18	350	CA	−0.05	**0.67**	**0.23**	Stahle et al.
PIJE	Blue Banks	39.67	−122.97	1598	CA	**0.20**	−0.08	**0.31**	Holmes et al.
PIJE	Buena Vista	36.72	−118.88	2280	CA	0.15	−0.02	**0.39**	Holmes et al.
PIJE	Kennedy Meadows	36.03	−118.18	2024	CA	0.01	**0.48**	**0.34**	Holmes et al.
PIJE	Piute Mountain	35.53	−118.43	1975	CA	0.15	**0.47**	**0.47**	Holmes et al.

Product-moment correlations (r) for the common period 1760 to 1980, with statistically significant correlations ($r \geq 0.16$, $P < 0.01$) highlighted in bold.
Species: PSME, Pseudotsuga menziesii (Douglas-fir); CHLA, Chamaecyparis lawsoniana (Port Orford cedar); JUOC, Juniperus occidentalis (Western juniper); PIPO, Pinus ponderosa (ponderosa pine); QUDG, Quercus douglasii (blue oak); PIJE, Pinus jeffreyi (Jeffrey pine).
Locations listed in descending order of latitude by species. States: WA, Washington; OR, Oregon; CA, California.
Northern SESE = JS, PC, RNP, HR; Southernmost SESE = LH; Unified SEGI = all SEGI locations.
All chronologies accessed via International Tree Ring Data Bank, except Black's two unpublished PSME chronologies.

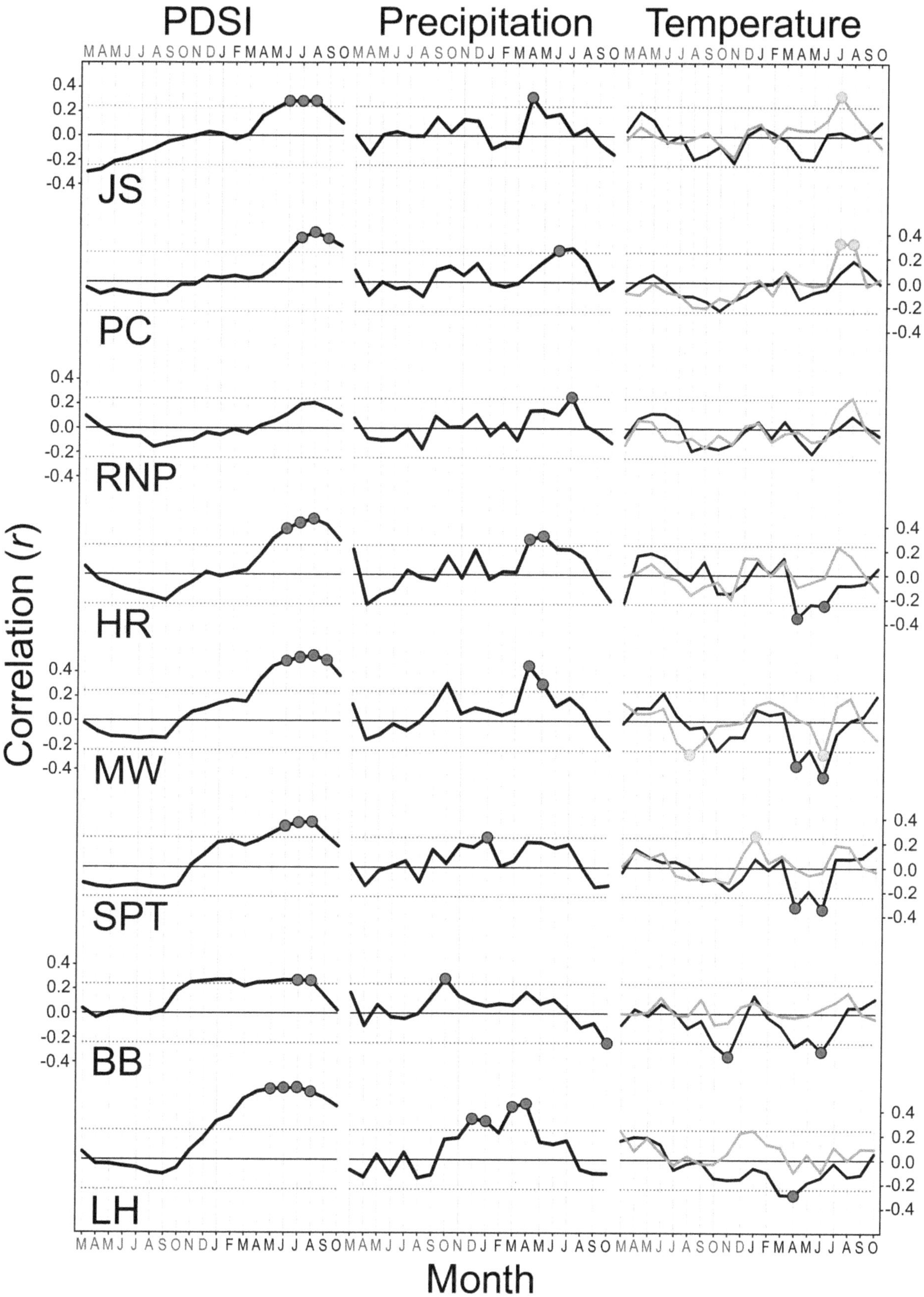

Figure 7. Summary of climate sensitivities for eight *Sequoia sempervirens* locations. Bootstrapped correlations and response functions of residual tree-ring chronologies against monthly Palmer Drought Severity Index (PDSI), precipitation, and maximum and minimum temperature conducted over a 114-year common period (1895–2008). Thicker black lines indicate correlation values. Thin dotted lines represent cutoff for statistical significance ($P<0.01$). Colored circles show months with significant correlations and response functions. Grey lines and blue circles

represent minimum temperature. Letters indicate 20 months from March of previous year to October of current year. Locations arranged by latitude from north (top) to south (bottom).

growth at all locations with only one location (RNP) lacking any significant response functions. Radial growth in the two northernmost locations (JS, PC) increased significantly with minimum July temperature (and August at PC) as did radial growth with minimum January temperature at SPT. The only significant negative responses to minimum temperature occurred at MW (July and prior August). Radial growth decreased significantly with maximum temperature in April and/or June at all locations from HR south to LH, as did radial growth with maximum prior November temperature at BB. SESE radial growth increased with decreasing summer cloudiness (i.e., airport fog), and correlations were significant ($P<0.01$) at three locations (JC, PC, MW) (Figure 8). The relationship between radial growth and summer cloudiness generally weakened towards the south, but inland and centrally located MW exhibited the strongest correlation ($P=0.0001$).

Significant dendroclimatic patterns were relatively few in SEGI (Figure 9). Growing season PDSI correlated positively with radial growth at all locations, but only two response functions were significant (May PDSI at WF, September PDSI at RMG). Precipitation also correlated positively with radial growth at all locations, but only three response functions were significant (prior October precipitation at WF and FC, January precipitation at WF). There were no significant correlations between radial growth and annual precipitation as snow, except for a positive correlation ($r=0.24$) in January at WF. June maximum and minimum temperatures correlated negatively with radial growth at all locations, and these response functions were significant at RMG, GF, and FC.

Radial growth of SESE and SEGI at all locations declined with increasing severity of summer drought, as captured by regional and state PDSI. Correlations between tree-ring chronologies and

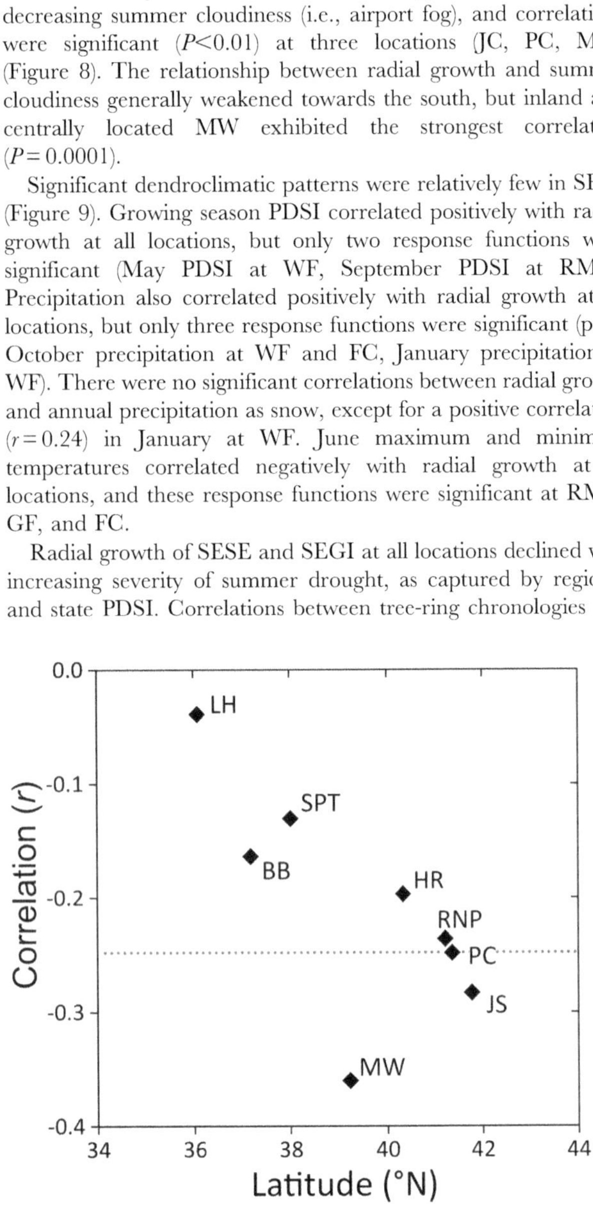

Figure 8. Correlations between residual tree-ring chronologies and reconstructed summer cloudiness for eight *Sequoia sempervirens* locations. Summer cloudiness (i.e., airport fog with cloud base ≤400 m elevation) reconstructed by Johnstone and Dawson [42] for June–September over a 108-year common period (1901–2008). Dotted line represents cutoff for statistical significance ($P<0.01$).

summer drought indices were statistically significant at ten locations, but by far the strongest correlation occurred at LH (Figure 10). The summer drought index for the central California coast was 46% coincident with the residual LH chronology (Figure 11).

Discussion

Crossdating

We successfully crossdated the majority of tree rings at all locations. Crossdating SESE was effective albeit more difficult than SEGI due to tight rings, missing rings, ring wedging, and spiral compression wood. These SESE growth anomalies were not pervasive, as some series provided continuous, visible rings spanning centuries while others had scores of missing rings (e.g., one tree at PC had 119 missing rings between 1659 and 1970). Our classification of nearly half a million rings by crossdating confidence will aid ongoing research. For example, areas that are *bounded with no annual resolution* are still useful for tree-level calculations of wood production and minimum age, while only *high crossdating confidence* will suffice for climate reconstructions and analyses of stable isotopes.

Unlike the situation in SEGI, crossdating success varied greatly among SESE trees. We confirmed previously reported difficulties crossdating suppressed SESE [13,14] and SESE with spiral compression wood [13,19]. Spiral compression wood is a relatively rare phenomenon [45] and may be more frequent in some portions of the SESE distribution. All nine of our study trees displaying spiral compression wood were from the three northernmost locations, though the phenomenon does occur occasionally throughout the range (Sillett, personal observation). Sampling trees of many ages facilitated millennium-scale crossdating. While younger trees often provided easy-to-decipher wider rings, older trees with tight outer rings allowed borers to sample discernible rings farther back in time. For example, a 35 m core from one SESE had 117 missing rings and *no annual resolution* from 1545–1971. This core had well-defined growth rings past this period extending to 1094 and thus provided much-needed replication for the earliest portion of the RNP chronology.

High within-tree variation in SESE crossdating success was partly attributable to more difficult lower trunk series, emphasizing the need for whole-trunk sampling. Despite avoiding abnormalities near the tree base, crossdating SESE was still more difficult in the lowest position sampled than higher on trunks, corroborating previous observations [13,14]. Series collected from relatively low on trunks tended to have diminished crossdating success, but they sometimes had wider rings associated with buttressing, often providing access to otherwise indecipherable periods of a tree's growth history. Growth irregularities characteristic of SESE were not limited to the lower trunk, however, and within-tree crossdating repeatedly relied on all samples collected from a given trunk. Regardless of height, certain series provided discernible rings where others did not. Sampling standing trees at multiple heights and azimuths along the trunk is therefore necessary to successfully crossdate SESE in many cases.

Chronology Characteristics and Synchrony

We established the first range-wide network of SESE chronologies. The only previous published SESE chronology extended to 1750 [10], while our longest chronology (RNP) extended to 328

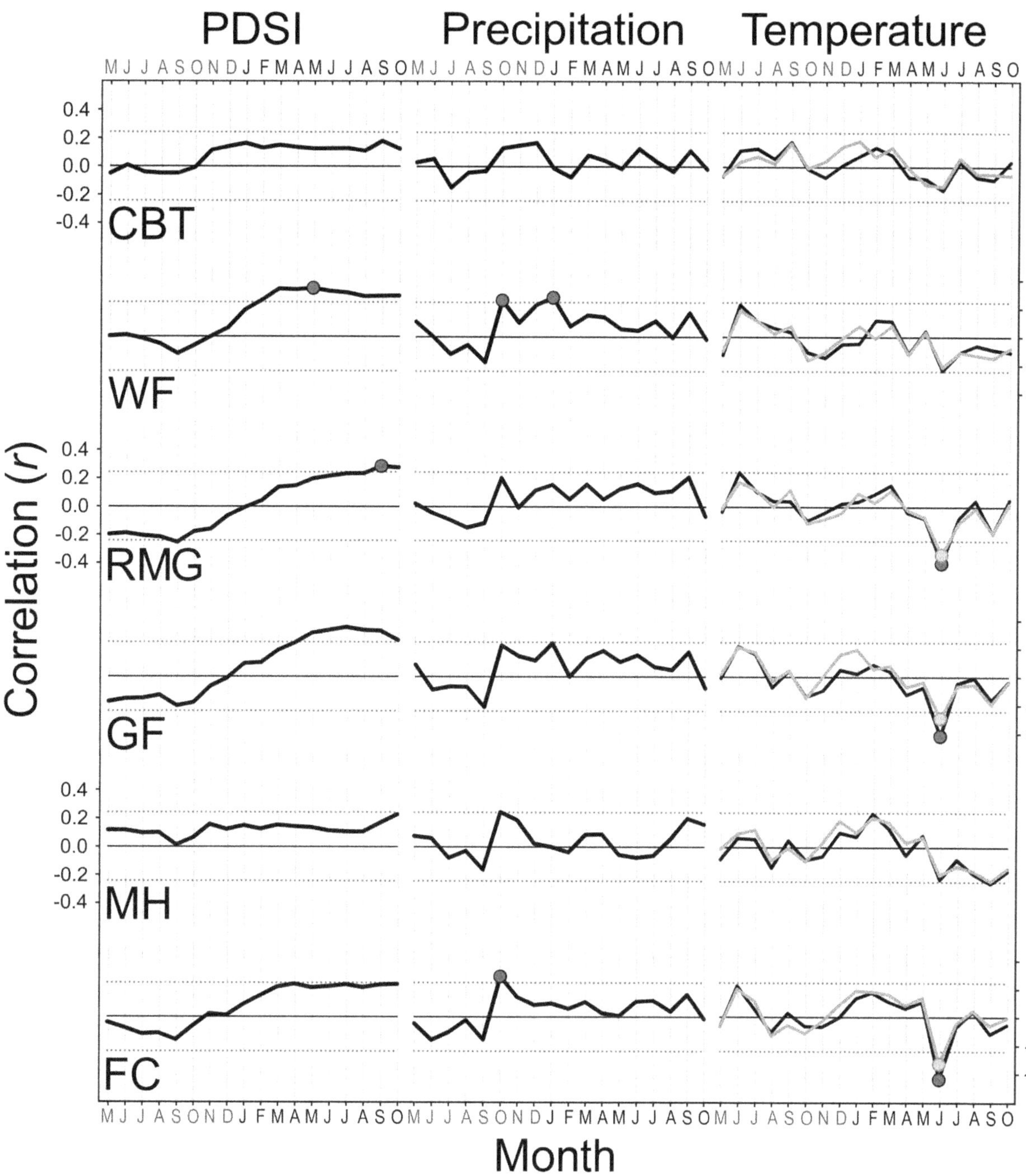

Figure 9. Summary of climate sensitivities for six *Sequoiadendron giganteum* locations. Bootstrapped correlations and response functions of residual tree-ring chronologies against monthly Palmer Drought Severity Index (PDSI), precipitation, and maximum and minimum temperature conducted over a 114-year common period (1895–2008). Thicker black lines indicate correlation values. Thin dotted lines represent cutoff for statistical significance ($P<0.01$). Colored circles show months with significant correlations and response functions. Grey lines and blue circles represent minimum temperature. Letters indicate 18 months from May of previous year to October of current year. Locations arranged by latitude from north (top) to south (bottom).

with two-tree replication to 1093. We crossdated independently, and upon review, our PC chronology correlated strongly ($r=0.54$; $P<0.001$) with the existing chronology derived from other trees in this reserve [10]. Two-tree replication exceeded 600 years in northern and central locations (JS, PC, RNP, HR, and MW) and 300–400 years in southern locations (SPT, BB, LH) (Table 2). Given our within-tree sampling strategy, which was often necessary to attain tree-level crossdating and associated research goals, tree-level replication was generally low by dendrochronological standards [46]. We present these SESE chronologies as proof of concept for crossdating and as a fruitful foundation upon which to build. In addition to early years with low replication, years with large differences between total and confidently crossdated series warrant further sampling (e.g., 1626–1632 at RNP, Figure 4).

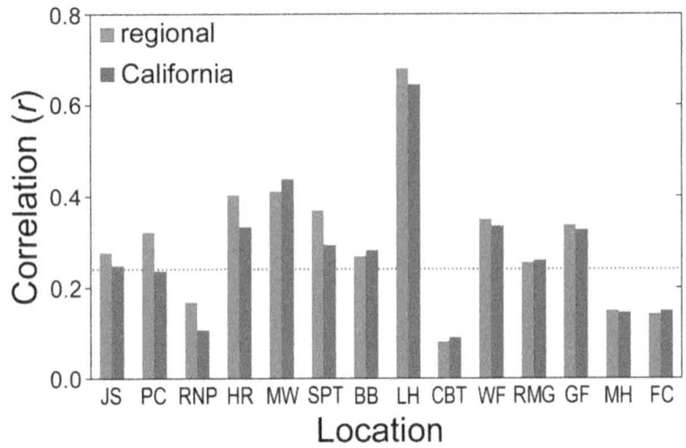

Figure 10. Summary of radial growth sensitivity to summer drought indices for two species at 14 locations. Correlations of residual *Sequoia sempervirens* and *Sequoiadendron giganteum* tree-ring chronologies against regional and state summer drought (average June–September PDSI) indices conducted over a 114-year common period (1895–2008). Dotted line represents cutoff for statistical significance ($P<0.01$). Locations arranged by latitude from north (left) to south (right) within species.

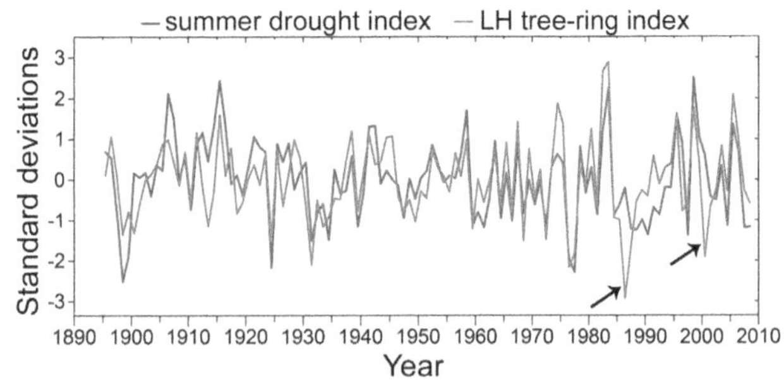

Figure 11. Visual comparison of southernmost *Sequoia sempervirens* (LH) tree-ring index and Central Coast California summer drought index. Summer drought is average June through September PDSI from the National Climatic Data Center for CA04 (Central Coast California). Both indices standardized by their standard deviates. Note discrepancies after known forest fires in 1985 and 1999 (arrows).

Our SEGI chronologies contributed to the breadth and robustness of tree-ring records for this species [4,47,48] and revealed commonalities with published marker years [6]. While much of the SEGI range is well-replicated, the northern extent is comparatively under-represented, and we improved it with the 1,226-year chronology at CBT. We augmented replication in the southern range with the 1,466-year chronology at FC, and we extended all SEGI chronologies well into the twenty-first century as many previous chronologies end in the early 1990s. These recent decades are important as tree-ring records provide a basis for long-term monitoring.

Range-wide chronologies permitted assessments of inter-annual climate variation across California, revealing latitudinal differentiation among SESE chronologies compared to a more unified SEGI signal. Northern SESE chronologies exhibited strong synchrony, and southern SESE chronologies often shared marker years and correlated more strongly with SEGI than northern SESE. Schulman alluded to such connections, noting that occasional small rings in SESE corresponded with SEGI marker years [49]. The distance separating distinctive northern and southern SESE tree-ring chronologies is far less than that observed in another western conifer, *Tsuga mertensiana* [50], which emphasizes the need to factor geographic location into models predicting SESE responses to climate change. While varying expressions of limiting climatic conditions were found among locations and species, SESE and SEGI chronologies often shared marker years reflecting large-scale events, such as historical (e.g., 1924) [51] and inferred (e.g., 1777, 1500) [6] droughts and El Niño years (e.g., large rings at 1982–1983) [52].

Regional Assessment

Comparisons with other western tree species helped place our SESE and SEGI chronologies into a regional context. The southernmost SESE chronology (LH) showed synchrony with several moisture-sensitive tree species in California, particularly QUDG, which exhibits a strong cool-season precipitation signal useful for climate reconstructions [53,54]. Indeed our LH chronology correlated slightly more than the closest QUDG chronology (Pinnacles National Monument) with annual California PDSI ($r=0.64$ vs. $r=0.62$) over the common period 1895–2003. Part of SESE's value in climatic reconstructions is the potential to extend chronologies farther back in time by virtue of decay-resistant heartwood.

Northern SESE rainforests have been considered the southern extent of coniferous forests in the Pacific Northwest [55] and are ecologically similar to coastal rainforests in Washington and Oregon [23]. Indeed, the northern SESE chronology was more similar to a PSME chronology (Olympic Road) from ~750 km away in Washington than to a PSME chronology from only ~375 km away in California. Northern SESE locations are an important addition to the dendrochronological record, as the southern Pacific Northwest is relatively underrepresented in tree-ring archives [35], and individual locations here influence gridded tree-ring networks more than in other regions (e.g., American Southwest) [56]. Although dendroclimatic signals can be species- and site-specific, correlations between SESE and other tree-ring chronologies reflect the spatial synchrony of extreme climate years across a broad geographic area [57–59].

The smallest ring in the replicated northern SESE chronology occurred in 1739 (Figure 1F). In the northernmost location (JS), this year's ring index was more than six standard deviations below the residual chronology mean for 698 years (1311–2008) (Figure 4). This ring also had the lowest index value and recorded cell damage in the CHLA chronology from southwestern Oregon [60] and was a strong marker year in PSME chronologies from coastal Oregon (Black, personal communication). Precipitation reconstruction based on six drought-sensitive conifer species in the Pacific Northwest showed 1739 as a severe, single-year drought event [57], while the more inland PDSI grid-point showed 1739 as a relatively dry, but not severe, soil moisture year [61]. Strength of the 1739 climate signal in tree-ring chronologies from northwestern California and southwestern Oregon highlights this region's distinctive climate.

The smallest ring in the unified SEGI chronology occurred in 1580 (Figure 1G) and has a well-documented regional extent. Of 22 SEGI trees with cores deep enough to reach this year, seven yielded at least one core missing 1580. The most extreme situation occurred at the highest elevation location (GF), where this year's ring index was more than six standard deviations below the residual chronology mean for 1538 years (474–2011) (Figure 5). This ring was missing in 11 of 12 cores collected from the largest SEGI tree sampled in GF. Douglass [3] made a special collection to confirm the existence of the 1580 ring, and it is the smallest ring in other SEGI chronologies as well [4]. Cores from LH were not deep enough, but those from two southern SESE locations (SPT, BB) also showed a strong low-growth marker for 1580. Multispecies dendroclimatic reconstructions for the region infer 1580 as a low water year [58,61] and the 16th century as a mega-drought [62]. One historical record from this time in California hints that low temperatures may also be involved. A written account of Chaplain Francis Fletcher from Sir Francis Drake's voyage (published in 1628) described extremely cold conditions on the coast just north of San Francisco Bay in July 1579 [63]. More radial growth occurred in 1579 than 1580 (standard ring index of 0.793 vs. 0.519, respectively) at all SEGI sites, so effects of a severe 1579/1580 winter may have lingered into the 1580 growing season, further restricting radial growth.

Dendroclimatic Relationships

As in other western North American forests, low moisture availability during dry summers appeared to constrain SESE radial growth [55,64,65]. The strength of this relationship was particularly noteworthy in the southernmost SESE location (Figure 11), where fires occurred in 1985 and 1999. Radial growth declined precipitously in the years immediately following these events, presumably as trees invested resources to rebuild fire-damaged crowns and restore leaf area. LH and other relatively dry SESE locations may thus be particularly useful for reconstruction of fire and drought histories. While SESE has generally been considered a complacent species [10] with low dendroclimatic potential, our results show that SESE ring widths do express meaningful climatic variation. Because our locations were all high-productivity forests with relatively ample access to soil water, sensitivities may be improved by sampling marginal locations and those closer to the inland and southern range limits [11,46]. However, SESE sampling must be balanced by the need to maintain crossdatability, which may be more difficult in less productive forests.

Observed climate sensitivities and synchrony among tree-ring chronologies along the latitudinal gradient were congruent with northern, central, and southern portions of SESE's geographic distribution [23]. SESE sub-regions showed different sensitivities to temperature: maximum spring temperatures correlated negatively with radial growth in central and southern locations, while minimum summer temperatures correlated positively with radial growth in rainforest locations. These relationships invoke a physiological interplay between maintenance costs, productivity, and optimum temperatures in tall forests [66]. For example, high spring temperatures may constrain radial growth by simultaneously increasing rates of maintenance respiration, elevating water stress, and decreasing gas exchange in relatively dry forests, whereas high summer temperatures may stimulate radial growth by promoting assimilation in cooler rainforests where water is not limiting. Such disparate sensitivities to temperature are particularly relevant for understanding the carbon sequestration capacity of SESE forests, because maximum and minimum temperatures have changed differentially over the last century with minimum temperature increasing at a faster rate, and overall temperatures in the region are projected to increase further during the 21st century [67–70]. Accurately forecasting SESE responses to climate change will likely depend on consideration of such temperature-growth dynamics.

While summer fog is an important source of moisture in SESE forests [21,22,42], it can also obscure sunlight. Limited long-term historical and site-specific data in addition to complex climatological dynamics associated with the marine fog layer [71] have constrained investigation of the relationship between fog and radial growth [but see 72,73]. Reconstructed fog frequency derived from airport visibility records [42] provided a century-long time series along which to examine SESE radial growth, but these measurements were not specific to our locations and did not distinguish low clouds from within-canopy fog capable of delivering moisture in the form of crown drip or foliar uptake. All of our SESE locations occurred below 400 m elevation, and all but one location (BB) occurred far enough below this elevation that even the highest leaves of the tallest tree in the forest are <400 m elevation, so when fog is a problem at airports, nearby SESE forests are often merely being shaded. The negative relationship between SESE radial growth and summer airport fog frequency suggests that cloudiness reduced SESE radial growth by decreasing light availability for photosynthesis at the height of the growing season. Recent isotopic analyses of SESE tree rings from three northern locations support this interpretation. Middle-wood cellulose was less enriched in ^{13}C, which indicates less stomatal regulation of water loss and lower rates of CO_2 assimilation, during years with higher summer fog (i.e., cloud base <200 m elevation) frequency, and this correlation strengthened and then stabilized with increasing cloud elevation, suggesting that the shading effects of cloudiness may influence radial growth more than canopy inundation by fog [20]. Latitudinal trends in correlations (Figure 8) imply that decreasing summer cloudiness (i.e., increasing light availability) has a significant positive effect on radial growth only in locations where water is least limiting. The fact that a central location (MW) exhibited the strongest negative correlation between radial growth and summer cloudiness makes sense, because this forest has perennial swamp-like conditions (personal observation) even though it receives only 60% of the average annual precipitation in northern rainforests (Figure 2). Clearly the relationship between fog and SESE growth warrants further study, especially in light of an inferred 33% reduction in summer airport fog frequency along the California coast since 1901 [42].

We confirmed findings of Hughes et al. [6,18] that inter-annual variation in SEGI radial growth does not tightly match that of temperature or precipitation, but instead the smallest rings record extreme events such as severe droughts. Compared to SESE, significant correlations between radial growth and drought severity were relatively few in SEGI. However, PDSI does not account adequately for snowmelt and may be less reflective of actual soil moisture in SEGI forests, especially during the first half of the growing season [74]. June was a key climate month, as three sites (RMG, GF, FC) showed significant negative responses to temperature. Hot June temperatures likely cause SEGI, which has higher leaf water-use efficiency than SESE [75], to close stomata midday to avoid stress-induced embolism, thus curtailing photosynthesis during the longest days of the year. Positive correlations between radial growth and prior-October precipitation at two locations (WF, FC) combined with the lack of association between prior-October precipitation and snowfall at any location (single degree-of-freedom χ^2, $P>0.5$) suggest that high rainfall in October extends the growing season, allowing SEGI to store more sugars overwinter for rapid trunk growth the following spring. Our analyses revealed only weak correlations between radial growth and snowfall with only one location (WF) showing significant responses to precipitation as snow and precipitation in January. We emphasize that these dendroclimatic relationships do not capture the potential lingering effects of snowpack, snowmelt, and the resultant elevated soil moisture, which demand research attention because snowmelt is occurring earlier in the year [76] and snowpack is projected to diminish in the Sierra Nevada even under low CO_2 emission scenarios [77].

Applications and Future Research

Crossdated SESE and SEGI tree-ring records have great potential to advance dendroclimatic, physiological, ecological, and archeological investigations of redwood forests. Our sampling technique of climbing and coring standing trees yielded sufficient replication to create baseline chronologies. Further replication will improve accuracy, and downed trees with ring-width records preserved in decay-resistant heartwood are now being utilized to extend sampling at multiple locations. For example, one very old remnant sample from HR has a ring count of 2,267 years with 742 years currently crossdated. Remnant wood can potentially be used to create multi-millennia tree-ring chronologies for climate reconstructions. In addition to the analyses presented here, our tree-ring data have been used to determine minimum tree ages [12] and crossdate branches to quantify branch growth dynamics [78]. Archeological applications of SESE chronologies are also emerging, including dating timber beams at the Presidio Officer's Club in San Francisco, CA (Worthington, personal communication).

Another application of our SESE tree-ring chronologies involves creating and improving fire histories to guide forest management. Fire scars were observed on some cores, often at high trunk positions indicating crown fires. For example, a tree at JS had eight fire scars at different years on cores collected 45 and 78 m above the ground (Figure 1H). These observations have exciting potential to explore legacy effects of crown fires, which damage crowns and create decaying wood habitats important to a wide assortment of arboreal biota, including epiphytes and lungless salamanders [79,80]. Extension of the northern SESE chronology back to the year 328 can enable accurate dating of fire scars previously observed in old samples [10,19]. Crossdated SESE chronologies can be used to quantify precisely fire return intervals inferred from un-crossdated ring counts [81].

Our cross-dated samples are available for study of complementary tree-ring parameters. Additional measurements of latewood width and maximum latewood density may provide proxies for summer temperature or precipitation [82,83]. Stable isotope composition of SESE wood cellulose can be assessed with inter-annual [20] and intra-annual [84] resolution, and crossdated samples provide dated material essential for such analyses. For this study, we detrended tree-ring chronologies to focus on inter-annual variation. We are also exploring lower-frequency variation in ring widths to quantify long-term trends in aboveground wood production and carbon sequestration, and we continue to monitor SESE and SEGI as part of ongoing research in a growing network of permanent plots.

Supporting Information

Figure S1 Average monthly precipitation and maximum and minimum temperature for northernmost and southernmost redwood locations. For each location, values are 114-year averages (1895–2008) at 800 m resolution using PRISM data [36].

Figure S2 Low-growth marker years in tree-ring chronologies from 14 locations for two species (1500–2008). Marker years were ten smallest ring widths per century at each location after removing years with three or fewer locations. LH chronology ended 1653.

Table S1 Tree-ring chronology and replication data for eight *Sequoia sempervirens* and six *Sequoiadendron giganteum* locations. Standard and residual chronologies using *high crossdating confidence* series >50 years detrended with 32-year spline.

Acknowledgments

We thank the following for help with field work or technical assistance: R. Van Pelt, A. Ambrose, W. Baxter, M. Antoine, J. Campbell-Spickler, G. Renzullo, C. Williams, E. Coonen, K. Scarla, and R. Naesborg. We thank R. Van Pelt for map creation; P. Brown, T. Swetnam, and B. Black for tree-ring data; and N. Pederson for advice on an earlier draft. We are grateful to California Department of Parks and Recreation, Redwood National Park, Sequoia and Kings Canyon National Parks, and Sequoia National Monument for granting permission to conduct research in these reserves and for logistical support.

Author Contributions

Conceived and designed the experiments: ALC SCS RDK. Performed the experiments: ALC SCS RDK. Analyzed the data: ALC SCS RDK. Wrote the paper: ALC SCS RDK.

References

1. Stephenson NL (2000) Estimated ages of some large giant sequoias: General Sherman keeps getting younger. Madroño 1: 61–67.
2. Van Pelt R (2001) Forest giants of the Pacific coast. Seattle: University of Washington Press. 200 p.
3. Douglass AE (1919) Climate cycles and tree-growth: a study of the annual rings of trees in relation to climate and solar activity, Vol I. Washington, D.C.: Carnegie Institute of Washington. Publ No 289.
4. Brown PM, Hughes MK, Baisan CH, Swetnam TW, Caprio AC (1992) Giant sequoia ring-width chronologies from the central Sierra Nevada, California. Tree-Ring Bulletin 52: 1–14.
5. Swetnam TW (1993) Fire history and climate change in giant sequoia groves. Science 262: 885–889.
6. Hughes MK, Brown PM (1992) Drought frequency in central California since 101 B.C. recorded in giant sequoias tree rings. Climate Dynamics 6: 161–167.
7. Douglass AE (1928) Climate cycles and tree-growth: a study of the annual rings of trees in relation to climate and solar activity, Vol II. Washington, D.C.: Carnegie Institute of Washington. Publ No 289.
8. Fritz E, Averill JL (1924) Discontinuous growth rings in California redwood. Journal of Forestry 22: 31–38.
9. Fritz E (1940) Problems in dating rings of California coast redwood. Tree-Ring Bulletin 6: 19–21.
10. Brown PM, Swetnam TW (1994) A cross-dated fire history from coast redwood near Redwood National Park, California. Can J For Res 24: 21–31.
11. Fritts HC (1976) Tree Rings and Climate. New York: Academic Press. 567 p.
12. Sillett SC, Van Pelt R, Koch GW, Ambrose AR, Carroll AL, et al. (2010) Increasing wood production through old age in tall trees. For Ecol Manage 259: 976–994.
13. Schulman E (1940) Climatic chronology in some coast redwoods. Tree-Ring Bulletin 5: 22–23.
14. Waring KM, O'Hara KL (2006) Estimating relative error in growth ring analyses of second-growth coast redwood (Sequoia sempervirens). Can J For Res 36: 2216–2222.
15. Brown PM, Kaye MW, Buckley D (1999) Fire history in Douglas-fir and coast redwood forest at Point Reyes National Seashore, California. Northwest Sci 73: 205–216.
16. Brown PM, Baxter WT (2003) Fire history in coast redwood forests of the Mendocino Coast, California. Northwest Sci 77: 147–158.
17. Antevs E (1925) The big trees as a climatic measure. Monthly Weather Review 53: 449–450.
18. Hughes MK, Richards BJ, Swetnam TW, Baisan CH (1990) Can a climate record be extracted from giant sequoia tree rings? Proceedings of Sixth Annual Pacific Climate (PACLIM) Workshop, 5–8 March 1989: 111–114.
19. Swetnam TW (1987) Fire history and dendroclimatic studies in coast redwood. Redwood National Park Report. P.O. No.8480–6–0875.
20. Johnstone JA, Roden JS, Dawson TE (2013) Oxygen and carbon stable isotopes in coast redwood tree rings respond to spring and summer climate signals. J Geophys Res Biogeosci 118: 1–13.
21. Dawson TE (1998) Fog in the California redwood forest: ecosystem inputs and use by plants. Oecologia 117: 476–485.

22. Burgess SSO, Dawson TE (2004) The contribution of fog to the water relations of Sequoia sempervirens (D.Don): foliar uptake and prevention of dehydration. Plant Cell Environ 27: 1023–1034.
23. Sawyer JO, Sillett SC, Popenoe JH, LaBanca A, Sholars T, et al. (2000) Characteristics of redwood forests. In Noss RF, editor. The redwood forest: History, ecology, and conservation of the coast redwoods. Covelo: Island Press. 39–79.
24. Harvey HT, Shellhammer HS, Stecker RE (1980) Giant sequoia ecology. Washington, D.C.: US Department of the Interior National Park Service. 182 p.
25. Willard D (2000) A guide to the Sequoia groves of California. Yosemite National Park: Yosemite Association. 124 p.
26. Rundel PW (1972) Habitat restriction in giant sequoia: the environmental control of grove boundaries. Am Midl Nat 87: 81–99.
27. Régent Instruments Canada Inc. (2009) WINDENDRO for tree-ring analysis. Québec, Canada.
28. Yamaguchi DK (1991) A simple method for cross-dating increment cores from living trees. Can J For Res 21: 414–416.
29. Holmes RL (1983) Computer-assisted quality control in tree-ring dating and measurement. Tree-Ring Bulletin 43: 69–75.
30. Herberich E, Sikorski J, Hothorn T (2010). A robust procedure for comparing multiple means under heteroscedasticity in unbalanced designs. PLOS ONE 5: 1–8.
31. R Core Team (2014) R: A language and environment for statistical computing. R Foundation for Statistical Computing, Vienna, Austria.
32. Cook ER (1985) A time series analysis approach to tree-ring standardization. Ph.D. Dissertation, The University of Arizona. Available: http://ltrr.arizona.edu/content/time-series-analysis-approach-tree-ring-standardization. Accessed 23 June 2014.
33. Grissino-Mayer HD (2001) Evaluating crossdating accuracy: a manual and tutorial for the computer program COFECHA. Tree-Ring Research 57: 205–221.
34. Osborn TJ, Briffa KR, Jones PD (1997) Adjusting variance for sample-size in tree-ring chronologies and other regional mean time series. Dendrochronologia 15: 89–99.
35. International Tree Ring Data Bank (2014) NOAA Paleoclimatology Program, Boulder, Colorado. Available: http://www.ncdc.noaa.gov/paleo/treering.html. Accessed 2014 Jan 31.
36. PRISM (2013) PRISM Climate Group, Oregon State University. Available: http://prism.oregonstate.edu. Accessed 2013 May.
37. Daly C, Gibson WP, Taylor GH, Johnson GL, Pasteris P (2002) A knowledge-based approach to the statistical mapping of climate. Climate Research 22: 99–113.
38. Daly C, Halbleib M, Smith JI, Gibson WP, Doggett MK, et al. (2008) Physiographically sensitive mapping of climatological temperature and precipitation across the conterminous United States. International Journal of Climatology 28: 2031–2064.
39. Wang T, Hamann A, Spittlehouse DL, Murdock TQ (2012) ClimateWNA–High-resolution spatial data for Western North America. Journal of Applied Meteorology and Climatology 51: 16–29.
40. Palmer WC (1965). Meteorologic drought. Research Paper No 45. U.S. Weather Bureau, Washington, D.C.
41. Abatzoglou JT (2013) Development of gridded surface meteorological data for ecological applications and modeling. International Journal of Climatology 33: 121–131.
42. Johnstone JA, Dawson TE (2010) Climatic context and ecological implications of summer fog decline in the coast redwood region. Proc Natl Acad Sci 107: 4533–4538.
43. Zang C, Biondi F (2013) Dendroclimatic calibration in R: The bootRes package for response and correlation analysis. Dendrochronologia 31: 68–74.
44. Wigley TML, Briffa KR, Jones PD (1984) On the average value of correlated time series, with applications in dendroclimatology and hydrometeorology. Journal of Climate and Applied Meteorology 23: 201–213.
45. Timell TE (1986) Compression wood in gymnosperms. New York: Springer-Verlag. 2150 p.
46. Speer JH (2010) Fundamentals of tree-ring research. Tuscon: The University of Arizona Press. 333 p.
47. Douglass AE (1945) Survey of sequoia studies. Tree-Ring Bulletin 11: 26–32.
48. Hughes MK, Touchan R, Brown PM (1996) A multimillenial network of giant sequoia chronologies for dendroclimatology. In Dean JS, Meko DM, Swetnam TW, editors. Tree rings, environment and humanity: Proceedings of the international conference, Tucson, Arizona, 17–21 May 1994. Tucson: Radiocarbon. 225–234.
49. Douglass AE (1933) Climatological researches. Washington, D.C.: Carnegie Institute of Washington. Publ No 32.
50. Gedalof Z, Smith DJ (2001) Dendroclimatic response of mountain hemlock (Tsuga mertensiana) in Pacific North America. Can J For Res 31: 322–332.
51. Dunning D (1925) The 1924 drought in California. The Timberman 26: 200–202.
52. Rienecker MM, Mooers CNK (1986) The 1982–1983 El Niño signal off northern California. J Geophys Res 91: 6597–6608.
53. Stahle DW, Griffin RD, Meko DM, Therrell MD, Edmondson JR (2013) The ancient blue oak woodlands of California: Longevity and hydroclimatic history. Earth Interactions 17: 1–23.
54. Cook ER, Seager R, Cane MA, Stahle DW (2007) North American drought: Reconstructions, causes, and consequences. Earth Sci Rev 81: 93–134.
55. Waring RH, Franklin JF (1979) Evergreen coniferous forests of the Pacific Northwest. Science 204: 1380–1386.
56. Meko DM, Cook ER, Stahle DW (1993) Spatial patterns of tree-growth anomalies in the United States and Southeastern Canada. J Clim 6: 1773–1786.
57. Graumlich LJ (1987) Precipitation variation in the Pacific Northwest (1675–1975) as reconstructed from tree rings. Ann Assoc Am Geogr 77: 19–29.
58. Meko DM, Therrell MD, Baisan CH, Hughes MK (2001) Sacramento River flow reconstructed to A.D. 869 from tree rings. J Am Water Resour Assoc 37: 1029–1039.
59. Knapp PA, Soule PT, Grissino-Mayer HD (2003) Occurrence of sustained droughts in the interior Pacific Northwest (A.D. 1733–1980) inferred from tree-ring data. J Clim 17: 140–150.
60. Carroll AL, Jules ES (2005) Climatic assessment of a 580-year Chamaecyparis lawsoniana (Port Orford cedar) tree-ring chronology in the Siskiyou Mountains, USA. Madroño 52: 114–122.
61. Cook ER, Woodhouse CA, Eakin MC, Meko DM, Stahle DW (2004) Long-term aridity changes in the Western United States. Science 306: 1015–1018.
62. Stahle DW, Cook ER, Cleaveland MK, Therrell MD, Meko D, et al. (2000) Tree-ring data document 16th century megadrought over North America. Eos (Washington DC) 81: 121–125.
63. Drake F (1628) The world encompassed by Sir Francis Drake, being his next voyage to that to Nombre de Dios. Collated with an unpublished manuscript of Francis Fletcher. London: The Hakluyt Society, 1854, 295 p.
64. Fritts HC (1974) Relationships of ring widths in arid-site conifers to variations in monthly temperature and precipitation. Ecol Monogr 44: 411–440.
65. Watson E, Luckman BH (2002) The dendroclimatic signal in Douglas-fir and ponderosa pine tree-ring chronologies from the southern Canadian Cordillera. Can J For Res 32: 1858–1874.
66. Larjavaara M (2013) The world's tallest trees grow in thermally similar climates. New Phytol 202: 344–349.
67. Karl TR, Jones PD, Knight RW, Kukla G, Plummer N, et al. (1993) A new perspective on recent global warming: Asymmetric trends of daily maximum and minimum temperature. Bulletin of the American Meteorological Society 74: 1007–1023.
68. Easterling DR, Horton B, Jones PD, Peterson TC, Karl TR, et al. (1997) Maximum and minimum trends for the globe. Science 277: 364–367.
69. Cayan DR, Maurer EP, Dettinger MD, Tyree M, Hayhoe K (2008) Climate change scenarios for the California region. Climate Change 87: S1–S42.
70. Mote PW, Salathé EP Jr. (2010) Future climate in the Pacific Northwest. Climate Change 102: 29–50.
71. Kelly KA (1985) The influence of winds and topography on the sea surface temperature patterns over the northern California slope. J Geophys Res 90: 11783–11798.
72. Biondi F, Cayan DR, Berger WH (1997) Dendroclimatology of Torrey Pine (Pinus torreyana Parry ex Carr.). Am Midl Nat 138: 237–251.
73. Williams PA, Still CJ, Fischer DT, Leavitt SW (2008) The influence of summertime fog and overcast clouds on the growth of a coastal Californian pine: a tree-ring study. Oecologia 156: 601–611.
74. Dai A, Trenberth KE, Qian T (2004) A global dataset of Palmer Drought Severity Index for 1870–2002: Relationship with soil moisture and effects of surface warming. Journal of Hydrometeorology 5: 1117–1130.
75. Ambrose AR, Sillett SC, Dawson TE (2009) Effects of tree height on branch hydraulics, leaf structure and gas exchange in California redwoods. Plant Cell Environ 32: 743–757.
76. Kapnick S, Hall A (2009) Observed changes in the Sierra Nevada snow pack: Potential causes and concerns. PIER Technical Report CEC-500-2009-016-F.
77. Mastrandrea MD, Luers AL (2012) Climate change in California: scenarios and approaches for adaptation. Climate Change 111: 5–16.
78. Kramer RD, Sillett SC, Carroll AL (2014) Structural development of redwood branches and its effect on wood growth. Tree Physiol 34: 314–330.
79. Sillett SC, Van Pelt R (2007) Trunk reiteration promotes epiphytes and water storage in an old-growth redwood forest canopy. Ecol Monogr 77: 335–359.
80. Spickler JC, Sillett SC, Marks SB, Welsh HH Jr. (2006) Evidence of a new niche for a North American salamander: Aneides vagrans residing in the canopy of old-growth redwood forest. Herpetol Conserv Biol 1: 16–27.
81. Stephens SL, Fry DL (2005) Fire history in coast redwood stands in the Northeastern Santa Cruz Mountains, California. Fire Ecology 1: 2–19.
82. Briffa KR, Jones PD, Schweingruber FH (1992) Tree-ring density reconstructions of summer temperature patterns across Western North America since 1600. J Clim 5: 735–754.
83. Meko DM, Baisan CH (2001) Pilot study of latewood-width of conifers as an indicator of variability of summer rainfall in the North American monsoon region. International Journal of Climatology 21: 697–708.
84. Roden JS, Johnstone JA, Dawson TE (2009) Intra-annual variation in the stable oxygen and carbon isotope rations of cellulose in tree rings of coast redwood (Sequoia sempervirens). The Holocene 19: 189–197.
85. Briffa KR (1995) Interpreting high-resolution proxy climate data – the example of dendroclimatology. In: Von Storch H, Navarra H, editors. Analysis of climate data variability, applications of statistical techniques. New York: Springer. 77–94.

12

Tree-Ring Based May-July Temperature Reconstruction Since AD 1630 on the Western Loess Plateau, China

Huiming Song[1,3], Yu Liu[1,2*], Qiang Li[1], Na Gao[4], Yongyong Ma[1,3], Yanhua Zhang[1,3]

1 The State Key Laboratory of Loess and Quaternary Geology, Institute of Earth Environment, Chinese Academy of Sciences, Xi'an, China, **2** Department of Environmental Science and Technology, School of Human Settlements and Civil Engineering, Xi'an Jiaotong University, Xi'an, China, **3** University of Chinese Academy of Sciences, Beijing, China, **4** Institute of Geology, China Earthquake Administration, Beijing, China

Abstract

Tree-ring samples from Chinese Pine (*Pinus tabulaeformis* Carr.) collected at Mt. Shimen on the western Loess Plateau, China, were used to reconstruct the mean May–July temperature during AD 1630–2011. The regression model explained 48% of the adjusted variance in the instrumentally observed mean May–July temperature. The reconstruction revealed significant temperature variations at interannual to decadal scales. Cool periods observed in the reconstruction coincided with reduced solar activities. The reconstructed temperature matched well with two other tree-ring based temperature reconstructions conducted on the northern slope of the Qinling Mountains (on the southern margin of the Loess Plateau of China) for both annual and decadal scales. In addition, this study agreed well with several series derived from different proxies. This reconstruction improves upon the sparse network of high-resolution paleoclimatic records for the western Loess Plateau, China.

Editor: Eryuan Liang, Chinese Academy of Sciences, China

Funding: This study was supported by grants from the National Basic Research Program of China (no. 2013CB955900), the Chinese Academy of Sciences (KZZD – EW – 04 – 01), the CAS/SAFEA International Partnership Program for Creative Research Teams (KZZD – EW – TZ – 03), the CAS Hundred Talents Program, the Key Project of the Institute of Earth Environment and the Project of State Key Laboratory of Loess and Quaternary Geology (SKLLQG). This research is an initiative of the Sino-Swedish Tree-Ring Research Center (SISTRR): Contribution (No. 025). The funders had no role in study design, data collection and analysis, decision to publish, or preparation of the manuscript.

Competing Interests: The authors have declared that no competing interests exist.

* E-mail: liuyu@loess.llqg.ac.cn

Introduction

Recently, the impacts of global warming on different regions especially on environmental fragile regions have received much attention [1–3]. The Loess Plateau of China, located at 100°54′–114°43′E and 33°43′–41°16′N, is one of the most climate-vegetation sensitive regions in China [4]. The study showed that extreme temperature events became more severe and frequent [5], which have significantly affected both the social economy and the people living in this area. Thus, there is a need to understand the mechanisms of climate change on the Loess Plateau. For this purpose, recent climate change must be studied in the context of the past one hundred to one thousand years. Instrumental weather records only provide limited data for approximately the last 60 years in China and are inadequate for examining the low-frequency variability that may underlie short-term climatic trends [6]. Tree-ring records are an important resource for understanding past climate change and can provide useful climate information for several centuries or even millennia beyond the instrumental record. Several dendroclimatological studies have been performed on the Loess Plateau in recent decades. Tree-ring based precipitation/drought reconstructions have been developed for the western Loess Plateau [7–10], and temperature reconstructions have been reported for the eastern and central Loess Plateau [11–13]. Nevertheless, temperature reconstructions remain limited for the western Loess Plateau.

The goals of this study were to reconstruct a seasonal temperature over the past 400 years using tree-ring widths from Mt. Shimen (MSM) and investigate the temperature variability at decadal to multi-decadal scale on the western Loess Plateau. Basing on the temperature reconstruction, the following questions need to be solved: (i) explore the relationship between seasonal temperature and drought events on the western Loess Plateau, (ii) find the affecting factor for the temperature variation.

Materials and Methods

Study area and climate data

Mt. Shimen (Fig. 1) is located on the northern slope of the western Qinling Mountains (QLM) and adjoins the Loess Plateau in the north. The climate, water resources and vegetation differ between the northern and southern slopes of the QLM due to its high altitude. The northern slopes are cold and dry, and the southern slopes are warm and humid. Therefore, the QLM is the transitional area between northern and southern China. The climate system for the study area is complicated due to its unique geography; the climate is arid to semiarid continental and is affected by the East Asia summer monsoon, the Indian monsoon and the Tibetan Plateau monsoon [14]. During the summer, the East Asia summer monsoon is the predominant factor and brings large quantities of heat and water vapor. During the winter, the weather throughout the entire Loess Plateau is controlled by the

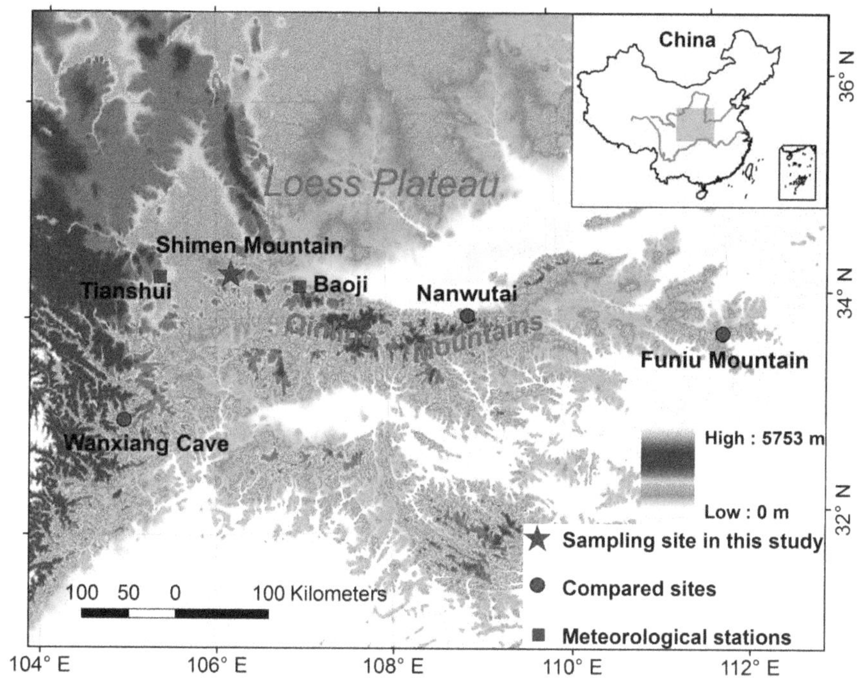

Figure 1. Map of the sampling site.

Mongolian High, which brings cold, dry air masses into the area [14].

The closest meteorological stations to the sampling site are at Tianshui (34°21′N, 105°27′E, 1141.7 m a.s.l., from AD 1951 to 2003) and at Baoji (34°12′N, 107°05′E, 612.4 m a.s.l., from AD 1952 to 2011). The Baoji station, which provides a relatively long record of observed data spanning 60 years, was chosen to investigate the relationships between tree-ring width and climate. The monthly mean temperature and total precipitation for both stations are presented in Figure 2. The Baoji station had more precipitation and higher temperature compared to the Tianshui station. However, both stations showed similar temperature and precipitation variation trends and showed the highest amount of rainfall from July to September. According to the records from the Baoji station, the mean annual precipitation from AD 1952 to 2011 was 675.18 mm (Fig. 2), of which 52.68% occurred from July

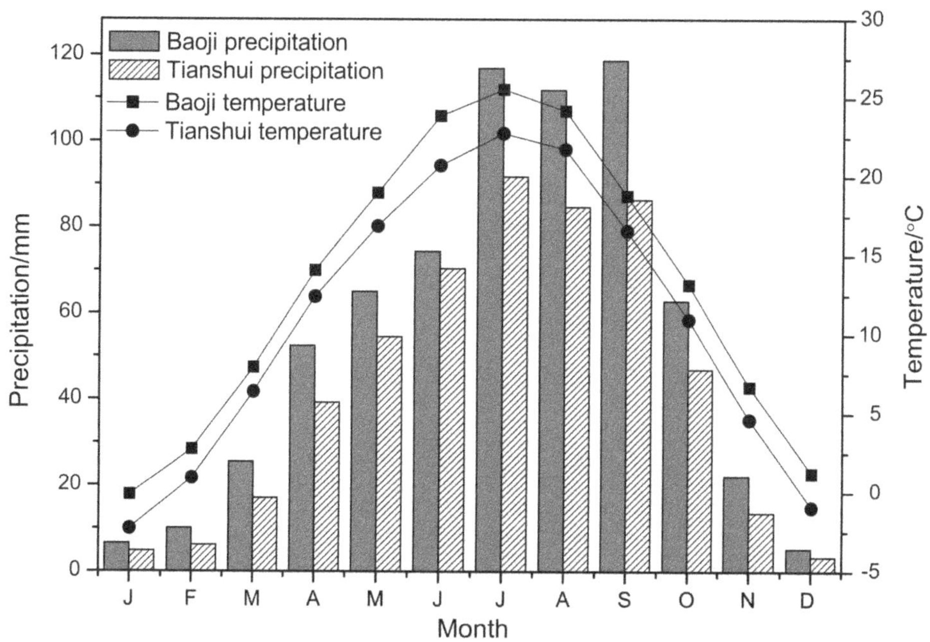

Figure 2. Average monthly temperature and precipitation for the Baoji (AD 1952–2011) and Tianshui (AD 1951–2003) meteorological stations.

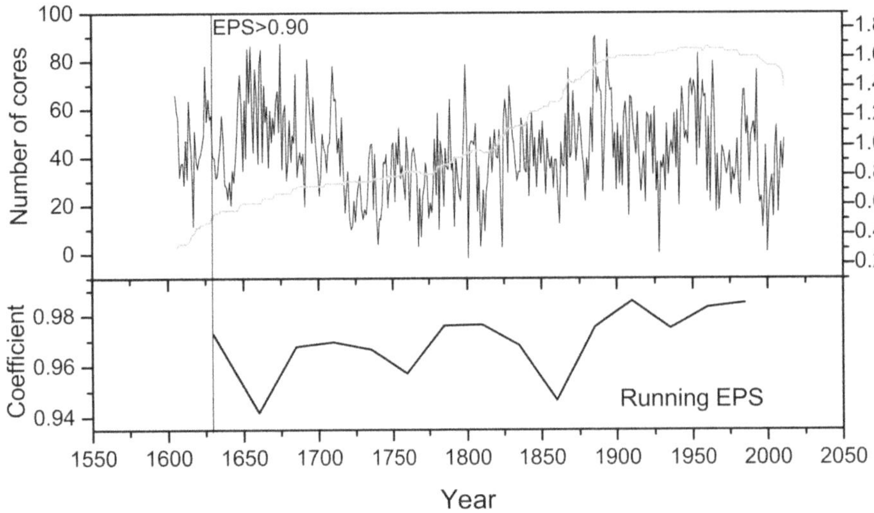

Figure 3. MSM tree-ring STD chronology, running EPS and sample size.

to September (Fig. 2). July (mean temperature of 25.59°C) and January (–0.15°C) were the warmest and the coldest months, respectively. The Palmer Drought Severity Index (PDSI) was obtained from the global PDSI data set developed by Dai et al. [15]. We used the PDSI data from the nearest grid point (33.75°N, 106.25°E).

Sampling site and chronologies

With the permission of local forest authority, we collected living *Pinus tabulaeformis* Carr. samples from the top of MSM (34°27′N, 106°09′E) at elevations ranging from 2050 to 2150 m (Fig. 1). The sampling site was located on steep rocky slopes having poorly developed soil, sparse ground cover and a low density of trees. Samples were fine-sanded and crossdated using standard dendrochronologial techniques [16]. The accuracy of the tree-ring dating and measurements were then checked and confirmed using the quality-control program COFECHA [17]. Cores with ambiguous results and those that were too short were excluded from further analysis; 64 cores from 34 trees were used to establish the chronology.

Measured ring-width series were standardized to a tree-ring chronology using the program ARSTAN. (http://www.ldeo.columbia.edu/res/fac/trl/public/publicSoftware.html) [18]. During this process, age-related trends were removed by fitting a

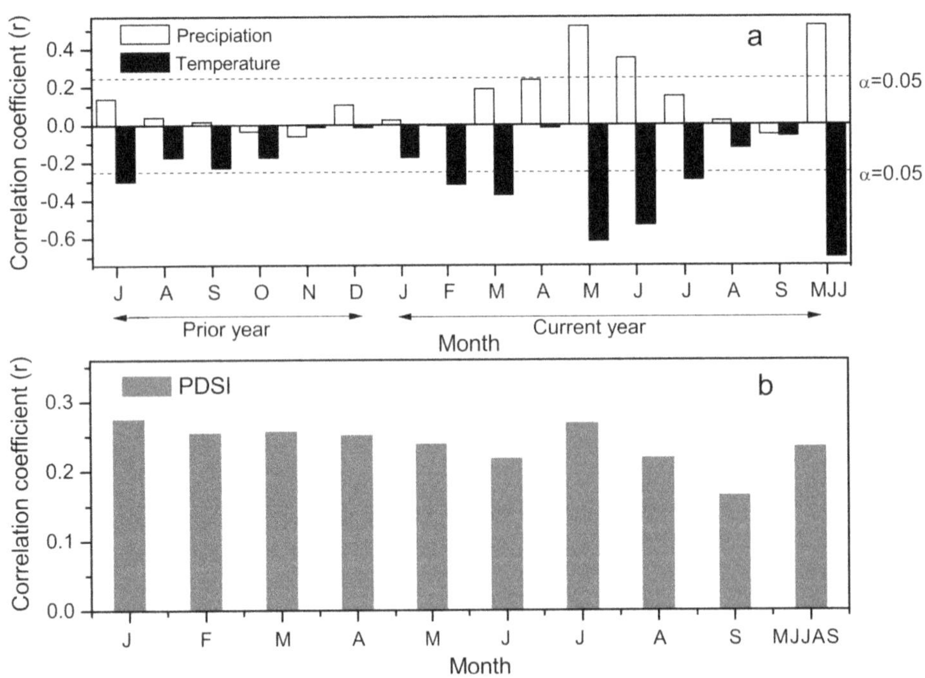

Figure 4. Correlations between the tree-ring STD chronology and monthly mean temperature (AD 1952–2011), monthly total precipitation (AD 1952–2011) and monthly mean PDSI (AD 1952–2005). Horizontal dashed lines represent the 95% confidence level.

Table 1. Statistics for a split calibration-verification procedure ($p<0.01$).

Calibration				Verification			
Period	r	R^2	R^2_{adj}	Period	r	RE	CE
1952–1981	−0.69	0.47	0.45	1982-2011	−0.71	0.46	0.29
1982–2011	−0.71	0.51	0.49	1952–1981	−0.69	0.48	0.15
1952–2011	−0.70	0.49	0.48	–	–	–	–

negative exponential curve or straight line function to the data. The individual index series were then combined into a single chronology by calculating a bi-weight robust means [18]. All subsequent analyses used the "standard chronology", which conserves more low frequency signals than other chronologies. The expressed population signal (EPS) was used to evaluate the reliability of the tree-ring chronology. Values exceeding 0.85 were considered acceptable [19]. The values of the running EPS from AD 1630 to 2011 were greater than 0.9 (Fig. 3), which affirmed the reliability of the chronology [18]. The series before AD 1630 could be regarded as reference (with 7 trees and 13 cores).

Method

To identify climate-growth relationships for the Chinese pine on MSM, a correlation analysis was performed for monthly mean temperature, monthly total precipitation and PDSI over the previous and current years of tree growth. The model stability and reliability were assessed using a split-sample method in which the model was divided into two subsets of equal length [20,21].The Pearson correlation coefficient (r), the positive RE (reduction of error) and CE (coefficient of efficiency) are the items to evaluate the results. Spatial correlations were used to assess the representativeness of our reconstruction by using the KNMI Climate Explorer (Royal Netherlands Meteorological Institute; http://climexp.knmi.nl).

Results and Discussion

Relationships between tree-ring widths and climate data

Correlation analysis showed that the ring-width STD chronology was negatively correlated with temperature for all months (Figure 4a). In contrast, the correlations between precipitation and tree growth were positive and slightly lower than those for temperature. This means that cool and wet conditions favor the growth Chinese pine in the study region. May-July precipitation and temperature are both significantly correlated with STD chronology, suggesting that May-July is a crucial period for the growth of Chinese pine on MSM. Negative correlations between tree-ring widths and temperature and positive correlations with precipitation were reported in other tree-ring studies on Loess Plateau [8,9,11,12], which suggests that the soil moisture regime during the onset of the growing season is important for tree's growth on the Loess Plateau. During May to July when the precipitation is not enough for tree's growth, high temperatures and strong radiation input intensify evaporation rates and then decrease moisture content in the topsoil further. The growth of fine roots gets inhibited and the uptake of nutrients might be hampered [22]. This moisture stress will lead to narrow or missing rings. Another study (Mt. Kongtong, near this study site) showed that the temperature played more important role in the moisture conditions (PDSI) than precipitation [23]. It has been observed

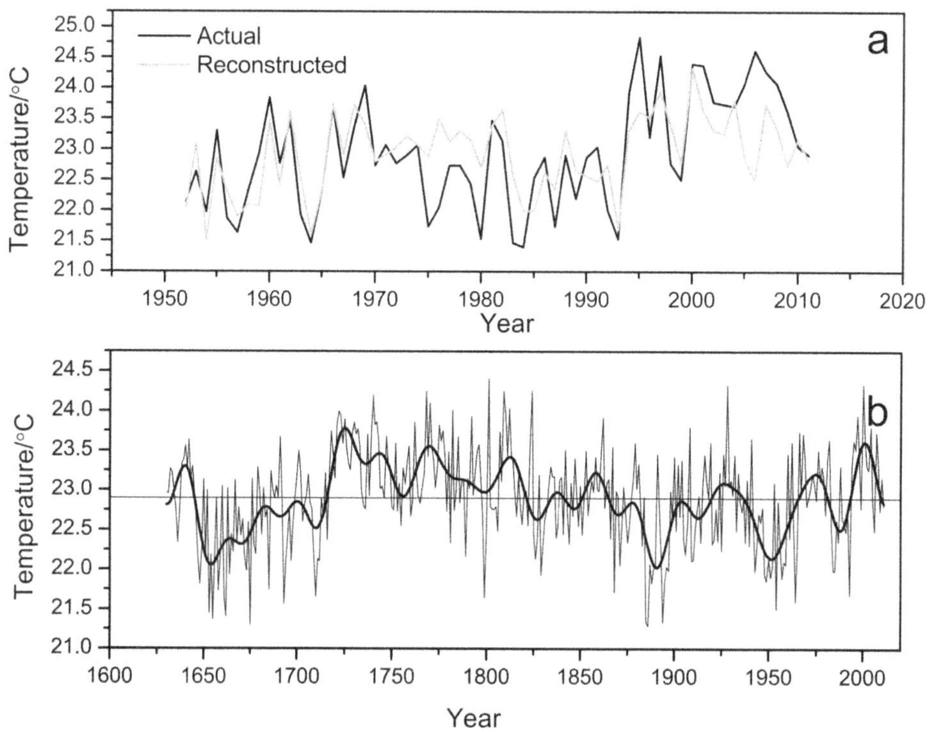

Figure 5. Comparison between observed and reconstructed mean temperatures from May to July. (a) The reconstructed (gray line) and observed (black line)temperature from May to July. (b)The reconstructed May–July mean temperatures after 20-year low pass filter for MSM since 1630.

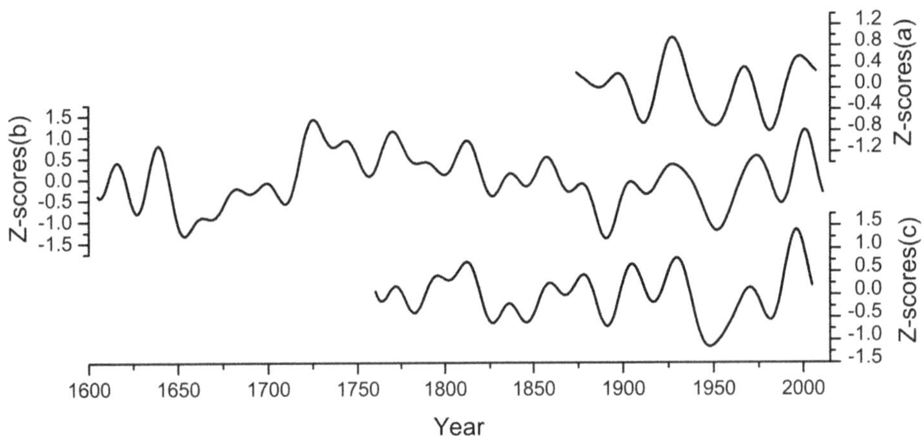

Figure 6. Tree-ring based temperature comparison among three sites from the northern slope of Qinling Mountains. (a)FNM [26], (b) MSM (this study) and (c) NWT [11].

that: (i) precipitation anomalies tend to dominate the change of PDSI in cold season when evaporation is minimal; (ii) the effect of temperature on PDSI becomes more important in warm seasons [24]. In addition, other studies showed that tree rings are significantly correlated with PDSI [8,9,25], whereas trees rings on MSM showed a weak correlation with PDSI (Fig. 4b), which is different from other study results. The strongest correlations were found between annual ring-width indices and May-July temperatures from AD 1952–2011 (r = –0.70). Basing on the analysis above, it makes sense that May-July mean temperature is the limiting factor for tree growth on MSM. The same response was observed in Nanwutai [11] and the Funiu Mountains [26], which are also located on the northern slope of the QLM.

May-July temperature reconstruction for Mt. Shimen

Based on the computational analysis presented above, the temperature from May–July was reconstructed from MSM tree rings using the linear regression model as follows:

$$T_{57} = -2.11 * STD + 24.89$$

$$(n = 60, r = -0.70, R^2 = 0.49, R^2_{adj} = 0.48, F = 55.631, DW = 1.17, p < 0.0001)$$

where T_{57} is the temperature from May to July (AD 1963–2011), and STD represents the standardized tree-ring chronology. The calibration and verification statistics are presented in Table 1. Correlation coefficients for the split-sample validation periods showed a strong relationship. The positive RE and CE results indicated that the reconstruction contributes unique paleoclimatic information. In addition, Figure 5a showed that the reconstruction closely tracked the observed temperature.

Temperatures on Mt. Shimen

Figure 5b shows the reconstructed May-July temperature variations since AD 1630 and its 20-year low-pass filter on MSM. This series shows interannual to multi-decadal variability (Fig. 5b). The long cooling period from AD 1640 to 1720 corresponds with well-known periods of solar minima (Maunder Minimum) during AD 1645–1715 [19]. Two obvious cool periods that occurred during the 1880s–1900s and the 1930s–1960s coincide with the 1900 minimum (AD 1880–1900) and a slight decrease in sun activity from AD 1940–1970, respectively [27]. The cool period during the reconstruction matched all solar minima except the Dalton minimum (AD 1800–1820), implying that solar forcing likely played an important role in past climate change for MSM. The longest warming trend occurred during AD 1720s–1810s and was followed by a mild climate for approximately 50 years. In addition, the filtered reconstruction indicates that the warming since AD 1970s is not significant and does not exceed the natural temperature variations that occurred over the past 400 years. AD 1928 stands out as an unusually warm year and corresponds with the drought event in AD 1928–1929, which has been noted as a dominant feature of the drought record throughout northern China [28,29]. Additionally, the warm year of AD 2000 is consistent with the severe drought that occurred in China [9].

Comparison with other tree-ring based temperature reconstructions

To evaluate the reliability of the reconstruction, we compared the reconstruction with two other tree-ring based temperature reconstructions from Nanwutai (NWT) [11] and the Funiu Mountains (FNM) [26]. MSM, NWT and FNM are all located on the northern slope of the QLM (Fig. 1). The Chinese pines from these three sites synchronously respond to May-July temperature variations, indicating that the growth patterns of trees throughout the region capture large-scale variations in the growing season due to the barrier function of the QLM. In addition to having the same growth patterns, the May-July temperature from MSM displays similar patterns of variation at high- and low-frequency to those in the NWT and FNM temperature records (Fig. 6). The temperature of MSM is highly correlated with those of NWT and FNM with r = 0.49 (n = 246, $p<0.0001$) and r = 0.32 (n = 132, $p<0.0001$) inter-annually and r = 0.49 and r = 0.60 after 20-year low-pass filtering, respectively. The most prominent and synchronous feature of the three reconstructions is the occurrence of cooler temperatures during the AD 1930s–1960s. The coherent temperature variation from the three sites indicates that the May–July temperature for the Northern slope of QLM has been consistent from east to the west during the last several centuries. In contrast, the tree-ring based temperature from Western Sichuan ([30] to the south of QLM) is

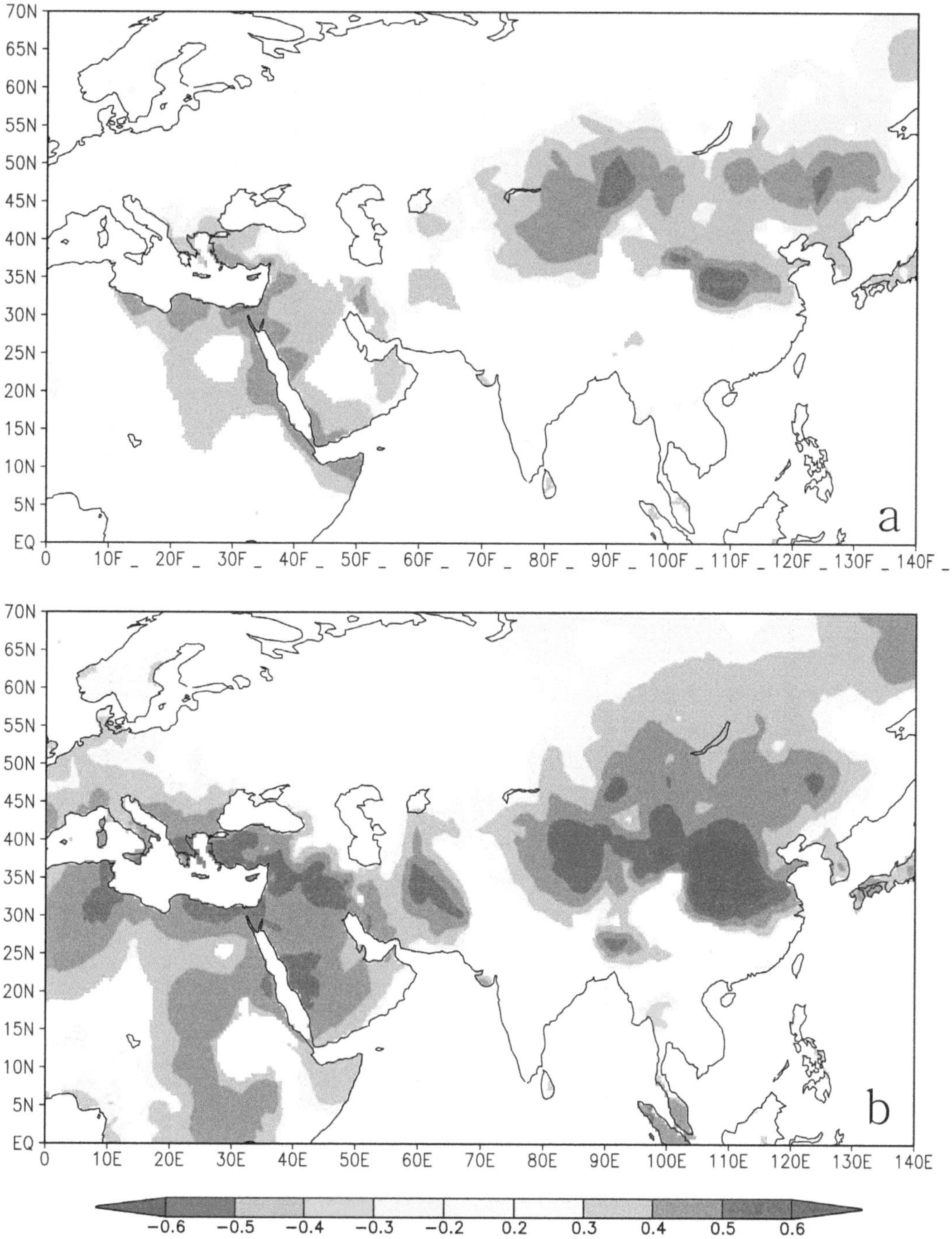

Figure 7. Patterns of field correlation in our study. (a) Correlations between the reconstructed temperatures with the CRU TS3 temperature during May–July (AD 1952–2009). (b) Correlations between the observed temperatures with the CRU TS3 temperature during May–July (AD 1952–2009).

significantly different because of the climate division function of QLM.

Spatial representativeness and comparisons with other records

Spatial correlations were computed between our temperature reconstruction, meteorological records and CRU TS3 temperature datasets over the calibration period (AD 1952–2009). The reconstruction and observational data showed a similar correlation pattern (Fig. 7), and they all correlated significantly with temperatures in north-central China; therefore, the MSM temperature reconstruction could represent the temperature variability for northern China to a certain extent. Additionally, the temperatures for the coastal regions of the Mediterranean and the Red Sea correlated significantly with the MSM temperature and the Baoji temperature during the period from May to July. The mechanism of this correlation between these geographically

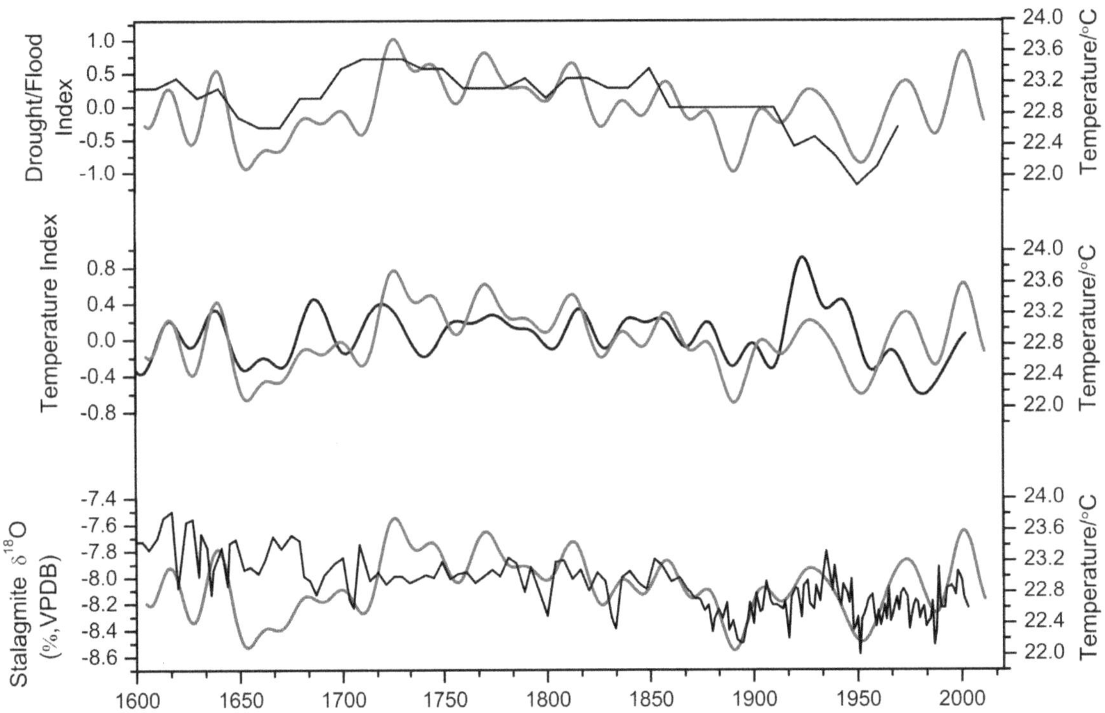

Figure 8. Comparison between the MSM temperatures (red) and other series (black). (a) Historical documents based on the Longxi Drought index [31]. (b) Regional temperature reconstruction for northern China [32] and (c) $\delta^{18}O$ time series of stalagmites from Wanxiang Cave, China [33].

separated areas is complicated and will be explored in future studies.

The series derived from the different proxies in the surrounding area provide the opportunity to validate our reconstruction and help us understand various aspects of the climate variation pattern for the western Loess Plateau. Figure 8 shows the comparison between the May–July temperature and series based on stalagmites, historical documents and multi-proxies. The variation trends between the MSM temperature and historical documents based on the Longxi Drought index [31] are highly similar(Fig. 8a), indicating that the drought and temperature patterns for the western Loess Plateau are synchronous with the long-term trend. The good agreement with regional temperature reconstruction for northern China [32] indicates that the MSM temperature reconstruction successfully captures the temperature variability for the large-scale region (Fig. 8b). The stalagmite record from Wanxiang Cave, China, characterizes the Asian Monsoon history over the past 1,810 years [33]. The May–July temperature for MSM maintained a high level of similarity to the stalagmite record since AD 1750, especially during the cool periods of AD 1880s–1900s and AD 1930s–1960s (Fig. 8c). Thus, the Asian Monsoon may influence the temperature variation of the MSM region to a certain extent at the low frequency variation. The study showed that Summer Asian Monsoon precipitation is driven by the thermal contrast between Asia and tropical Indo-Pacific and therefore should be linked to regional temperature change [33]. The 11-year running correlation analysis indicates that there is a positive correlation between the East Asian Summer Monsoon Index and summer surface air temperature in eastern China during 2/3 of the time span of 1880–2004, implying the impact of subtropical land-sea thermal contrast on the East Asian Summer Monsoon Index [34]. Although the monsoon series indicated by the stalagmite record was not weak during the period AD 1640–1720, the other three series show common decreased variations during the same period, which is consistent with the Maunder Minimum.

Conclusions

Here we described a reconstruction of the mean May–July temperatures for MSM on the western Loess Plateau based on tree-ring width data dating back to AD 1630. This reconstruction accounts for 49% of the actual temperature variance over the period from AD 1951 to 2011 (48% after adjustment for the loss of degrees of freedom). The reliability of our reconstruction was confirmed by comparing it with other records from nearby areas. The similarity among the tree-ring based temperature reconstructions obtained from the northern slope of the QLM reflects large-scale temperature variations over the past several centuries. The reconstruction compared well with series derived from other proxies, indicating that the drought and temperature variations were synchronous with the long-term trends on the western Loess Plateau. The reconstruction displayed cooler temperatures during periods of known solar minima except during the Dalton minimum. The East Asian summer monsoon may influence the temperature variation at MSM to some extent. However, additional research is needed to develop a tree-ring network that can reflect spatial climate variations on the Loess Plateau.

Acknowledgments

We thank Buli Cui, Li Xie and Baofa Shen for their great help.

Author Contributions

Conceived and designed the experiments: YL HMS QL. Performed the experiments: HMS QL YYM YHZ. Analyzed the data: HMS YL.

Contributed reagents/materials/analysis tools: HMS QL NG. Wrote the paper: HMS YL QL NG YYM YHZ.

References

1. Liang EY, Shao XM, Qin NS (2008) Tree-ring based summer temperature reconstruction for the source region of the Yangtze River on the Tibetan Plateau. Global and Planetary Change 61 (3–4): 313–320.
2. Rehman SFQ (2010) Temperature and rainfall variation over Dhahran, Saudi Arabia, (1970–2006). International Journal of Climatology 30:445–449.
3. Yang YH, Chen YN, Li WH, Yu SL, Wang MZ (2012) Climatic change of inland river basin in an arid area: a case study in northern Xinjiang, China. Theoretical and Applied Climatology 107:143–154.
4. Fan ZM, Li J, Yue TX (2012) Changes of Climate-Vegetation Ecosystem in Loess Plateau of China. Procedia Environmental Sciences 13:715–720.
5. Li Z, Zheng FL, Liu WZ, Flanagan D (2010) Spatial distribution and temporal trends of extreme temperature and precipitation events on the Loess Plateau of China during 1961–2007. Quaternary International 226: 92–100.
6. Cayan DR, Dettinger MD, Diaz HF, Graham NE (1998) Decadal variability of precipitation over western North America. Journal of Climate 11: 3148–3166.
7. Li JB, Chen FH, Cook ER, Gou XH, Zhang Y (2007) Drought reconstruction for north central China from tree rings: the value of the Palmer drought severity index. International Journal of Climatology 27: 903–909.
8. Fang KY, Gou XH, Chen FH, D'Arrigo R, Li JB (2010) Tree-ring based drought reconstruction for the Guiqing Mountain (China): linkages to the Indian and Pacific Oceans, International Journal of Climatology. International Journal of Climatology 30 (8): 1137–1145.
9. Song HM, Liu Y (2011) PDSI variations at Kongtong Mountain, China, inferred from a 283-year Pinus tabulaeformis ring-width chronology. Journal of Geophysical Research 116. doi: 10.1029/2011JD016220.
10. Liu Y, Lei Y, Sun B, Song HM, Sun JY (2012) Annual precipitation in Liancheng, China, since 1777AD derived from tree rings of Chinese pine (Pinus tabulaeformis Carr.). International Journal of Biometeorology. doi: 10.1007/s00484-012-0618-7.
11. Liu Y, Linderholm HW, Song HM, Cai QF, Tian QH, et al. (2009) Temperature variations recorded in Pinus tabulaeformis tree rings from the southern and northern slopes of the central Qinling Mountains, central China. Boreas 38 (2): 285–291.
12. Li Q, Liu Y, Song HM, Cai QF, Yang YK (2013) Long-term variation of temperature over North China and its links with large-scale atmospheric circulation. Quaternary International 283: 11–20.
13. Cai QF, Liu Y, Song HM, Sun JY (2008) Tree-ring-based reconstruction of the April to September mean temperature since 1826 AD for north-central Shaanxi Province, China. Science in China Series D: Earth Sciences 51 (8): 1099–1106.
14. Qian LQ (1991) Climate of Loess Plateau. China. Beijing: Meteorological Press. 368p.(in Chinese).
15. Dai AG, Trenberth AK, Qian T (2004) A global dataset of Palmar Drought Severity Index for 1870–2002: relationship with soil moisture and effects of surface warming. Journal of Hydrometeorology 5: 1117–1130.
16. Stokes MA, Smiley TL (1996) An introduction to tree-ring dating. Tucson: The University of Arizona Press. 73p.
17. Holmes RL (1983) Computer-assisted quality control in tree-ring dating and measurement. Tree-Ring Bulletin 43: 69–78.
18. Cook ER, Kairiukstis LA (1990) Methods of dendrochronology: applications in the environmental sciences. Dordrecht: Kluwer Academic Publishers. 394p.
19. Wigley TML, Briffa KR, Jones PD (1984) On the average value of correlated time series, with applications in dendroclimatology and hydrometeorology. Journal of Climate and Applied Meteorology 23: 201–213.
20. Fritts HC (1976) Tree Rings and Climates. London: Academic Press. 567p.
21. Snee RD (1977) Validation of regression models: methods and examples. Technometrics 19: 415–428.
22. Oberbuber W, Stumbock M, Kofler W (1998) Climate-tree-growth relationships of Scots pine stands (Pinus sylvestris L.) exposed to soil dryness. Trees 13: 19–27.
23. Song HM, Liu Y, Li Q, Linderholm H (2013) Tree-ring derived temperature records in the central Loess Plateau, China. Quaternary International 283: 30–35.
24. Mishra AK, Singh VP (2010) A review of drought concepts. Journal of Hydrology 391: 202–216.
25. Kang SY, Yang B, Qin C (2012) Recent tree-growth reduction in north central China as a combined result of a weakened monsoon and atmospheric oscillations. Climatic Change115: 519–536.
26. Tian QH, Liu Y, Cai QF, Bao G, Wang WP, et al. (2009) The maximum temperature of May-July inferred from tree-ring in Funiu Mountain since 1874 AD. Acta Geologica Sinica 64 (7): 879-887, (in Chinese, with English abstract).
27. Beer J, Mende W, Stellmacher R (2000) The role of the sun in climatic forcing. Quaternary Science Reviews 19: 403–415.
28. Liang EY, Shao XM, Kong ZC, Lin JX (2003) The extreme drought in the 1920s and its effect on tree growth deduced from tree ring analysis: a case study in North China. Annals of Forest Science 60 (2): 145–152.
29. Liu Y, Sun JY, Yang YK, Cai QF, Song HM, et al. (2007) Tree-ring-derived precipitation records from Inner Mongolia, China, since A.D. 1627. Tree-ring Research 63 (1), 3–14.
30. Deng Y, Gou XH, Gao LL, Yang T, Yang MX (2013) Early-summer temperature variations over the past 563 yr inferred from tree rings in the Shaluli Mountains, southeastern Tibet Plateau. Climate Research, in Press.
31. Tan LC, Cai YJ, Yi L, An ZS, Ai L (2008) Precipitation variations of Longxi, northeast margin of Tibetan Plateau since AD 960 and their relationship with solar activity. Climate of the Past 4: 19–28.
32. Yi L, Yu HJ, Ge JY, Lai ZP, Xu XY, et al.(2012) Reconstructions of annual summer precipitation and temperature in north-central China since 1470 AD based on drought/flood index and tree-ring records. Climatic Change 110: 469–498.
33. Zhang PZ, Cheng H, Edwards RL, Chen FH, Wang YJ, et al.(2008) A Test of Climate, Sun, and Culture Relationships from an 1810-Year Chinese Cave Record. Science 322: 940–942.
34. Xiang L, Zhu CW, Lv JM (2013) Decadal change of the East Asian summer monsoon and its related surface temperature in Asia-Pacific during 1880–2004. Chinese Science Bulletin 58 (35): 4497–4503.

Environmental Reconstruction of Tuyoq in the Fifth Century and Its Bearing on Buddhism in Turpan, Xinjiang, China

Ye-Na Tang[1], Xiao Li[2], Yi-Feng Yao[1]*, David Kay Ferguson[3,1], Cheng-Sen Li[1]*

1 State Key Laboratory of Systematic and Evolutionary Botany, Institute of Botany, Chinese Academy of Sciences, Beijing, China, **2** School of Chinese Classics, Renmin University of China, Beijing, China, **3** Department of Paleontology, University of Vienna, Vienna, Austria

Abstract

The Thousand Buddha Grottoes of Tuyoq, Turpan, Xinjiang, China were once a famous Buddhist temple along the ancient Silk Road which was first constructed in the Fifth Century (A.D.). Although archaeological researches about the Grottoes have been undertaken for over a century, the ancient environment has remained enigmatic. Based on seven clay samples from the Grottoes' adobes, pollen and leaf epidermis were analyzed to decipher the vegetation and climate of Fifth Century Turpan, and the environmental landscape was reconstructed in three dimensions. The results suggest that temperate steppe vegetation dominated the Tuyoq region under a warmer and wetter environment with more moderate seasonality than today, as the ancient mean annual temperature was 15.3°C, the mean annual precipitation was approximately 1000 mm and the temperature difference between coldest and warmest months was 24°C using Co-existence Approach. Taken in the context of wheat and grape cultivation as shown by pollen of *Vitis* and leaf epidermis of *Triticum*, we infer that the Tuyoq region was an oasis with booming Buddhism in the Fifth Century, which was probably encouraged by a 1°C warmer temperature with an abundant water supply compared to the coeval world that experienced the 1.4 k BP cooling event.

Editor: Michael D. Petraglia, University of Oxford, United Kingdom

Funding: This research was supported by China National Key Basic Research Program (2014CB954201). The funders had no role in study design, data collection and analysis, decision to publish, or preparation of the manuscript.

Competing Interests: The authors have declared that no competing interests exist.

* E-mail: lics@ibcas.ac.cn (CSL); yaoyf@ ibcas.ac.cn (YFY)

Introduction

The Xinjiang Uygur Autonomous Region in China is located in the heart of Central Eurasia (Fig. 1), where it has been governed by an arid, temperate continental climate since the uplift of the Tibetan Plateau [1,2]. In particular, the Turpan Basin is an extremely arid region of Xinjiang, with only 16.4 mm mean annual precipitation [3], but 3000 mm evaporation [4]. However, Turpan was once an important stop on the overland trade route 'Silk Road' linking China with Central Asia. The investigation of the historical vegetation and climate may provide a key to understand the people's lifestyle in the Turpan oasis in ancient time.

Turpan has been generally characterized by an arid climate with expanding temperate desert and steppe vegetation throughout the Holocene [5,6]. In recent years, there have been a series of archaeobotanical discoveries from the Turpan area, including Cannabis sativa [7], Capparis spinosa [8], Lithospermum officinale [9], Vitis vinifera from the Yanghai Tombs (2.5 kyrs BP) [10], *Sesamum indicum* from Boziklik Thousand Buddha Grottoes (700 yrs BP) [11], most of which were plants utilized by indigenous people for medicine, oil, spice etc. Moreover, six species of cereals including *Triticum aestivum* and Setaria italica from the Astana Cemeteries were recognized as food for local people in 3rd to 9th centuries [12]. Furthermore, Jiang et al. [13] and Yao et al. [14] reported the ancient Yanghai people of Turpan were living in an oasis with some swamps 2700 yrs BP ago based on an archaeological environmental study. However, there is still a shortage of research on the ancient climate and environmental reconstruction of Turpan.

The Tuyoq Grottoes of Turpan, first constructed in the Fifth Century and abandoned during the Fifteenth Century, were once a notable Buddhist pilgrimage site, even for the Tang royal family between the seventh to eighth centuries [15]. Since they were discovered by the Russian botanist E. A. Regel in the early 1870s, there have been continuous reports on the archaeological findings such as Buddhist frescoes, silk paintings, wood and pottery utensils and so on [16–18]. In 2010, approximately 2500 square meters of caves were excavated by a joint team of the Research Center for Frontier Archaeology of the Institute of Archaeology, Chinese Academy of Social Sciences and Academia Turfanica and Kizil Research Institute [19].

In the present study, we employ pollen and epidermis analyses to reconstruct the ancient vegetation, climate and environment of the archaeological site 'The Thousand Buddha Grottoes of Tuyoq' in Turpan in the Fifth Century, as well as to probe the possible historical signatures of the environment-human interaction there.

Materials and Methods

Ethics Statement

All necessary permits were obtained for the described field studies and were granted by the Academia Turfanica of Xinjiang Uygur Autonomous Region, China.

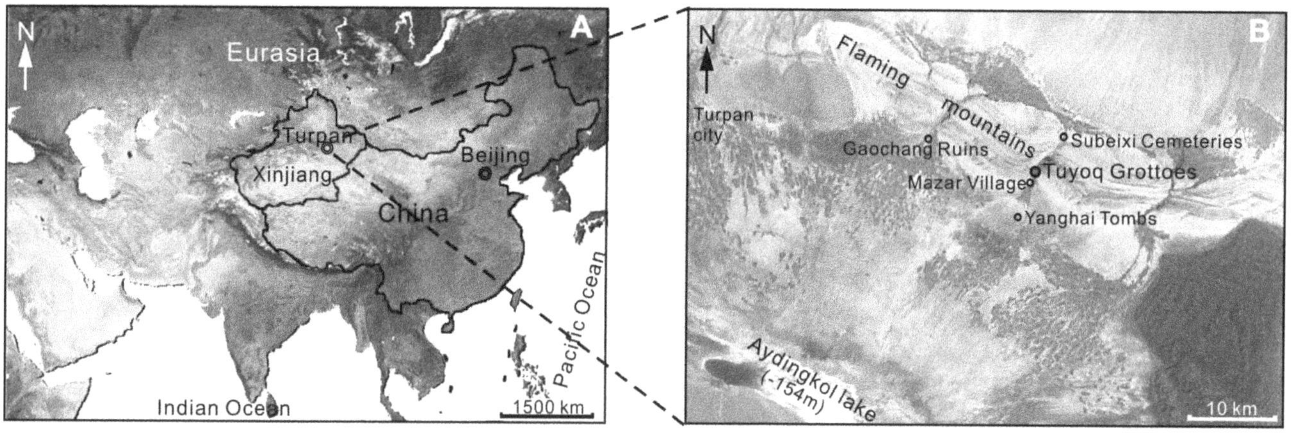

Figure 1. Map showing the location of Tuyoq Grottoes (revised from http://map.baidu.com/). (A) Location of Turpan in Eurasia, (B) Location of Tuyoq Grottoes in Turpan.

Research Site

Turpan (41°12′–43°40′N, 87°16′–91°55′E) is located in an intermontane basin enclosed by high mountains in Xinjiang, NW China. The basin covers 50,140 km^2 including a 7.8 km^2 oasis [20], with the highest Bogeda Peak (5445 m) in the north and the second lowest lake on the planet Aydingkol Lake (−154 m) in the central basin (Fig. 1). The basin is governed by an arid, temperate continental climate, and water resources in the basin come mainly from glacial melt in the Tian and Bogeda Mountains channeled by karez irrigation systems [4]. Modern plants in the basin include camel thorn (*Alhagi sparsifolia*), sea-buckthorn (*Hippophae rhamnoides*), tamarisk (*Tamarix chinensis*), artemisia (*Artemisia ordosica*), sand ilex (*Ammopiptanthus mongolicus*) and some cultivated grape (*Vitis vinifera*), cotton (*Gossypium hirsutum*), sweet melon (*Cucumis melo*), poplar (*Populus tomentosa*), mulberry (*Morus alba*) etc. [6].

The Thousand Buddha Grottoes of Tuyoq, comprising 94 caves, are situated in the eastern and western cliff faces of Tuyoq Valley cutting through the Flaming Mountains, Shanshan County near Turpan (Fig. 1). The Grottoes lie fifteen kilometers east of the Gaochang ruins, which are regarded as the earliest example of a combined central Asiatic and Chinese style of Buddhism [15]. The Tuyoq Valley is about eight kilometers in length, with Subeixi Cemeteries in the north and Yanghai Tombs in the south. About one kilometer from the Tuyoq Grottoes, an ancient, small Uighur village called 'Mazar Village' is located at the southern mouth of the valley, where yellow clay was the main source of material for local building.

Studied Materials and Methods

Seven soil samples from intact yellow-clay adobes in the northern portion of the eastern Tuyoq Grottoes (42°51.9′N, 89°41.7′E) were selected for this study. There were diverse plant stems and leaves mixed in the clay as architectural strengthening materials (Fig. 2). Samples nos.1 to 5 were collected from Grotto K18, the central Buddhist prayer room in the eastern grottoes. Samples no.6 and no.7 collected from Grotto K38, a cave with a Buddha niche in the center (Fig. 2). According to Chinese Bodhisattva frescoes classification and archaeological research, the grottoes must have been constructed in the Fifth Century [21,22].

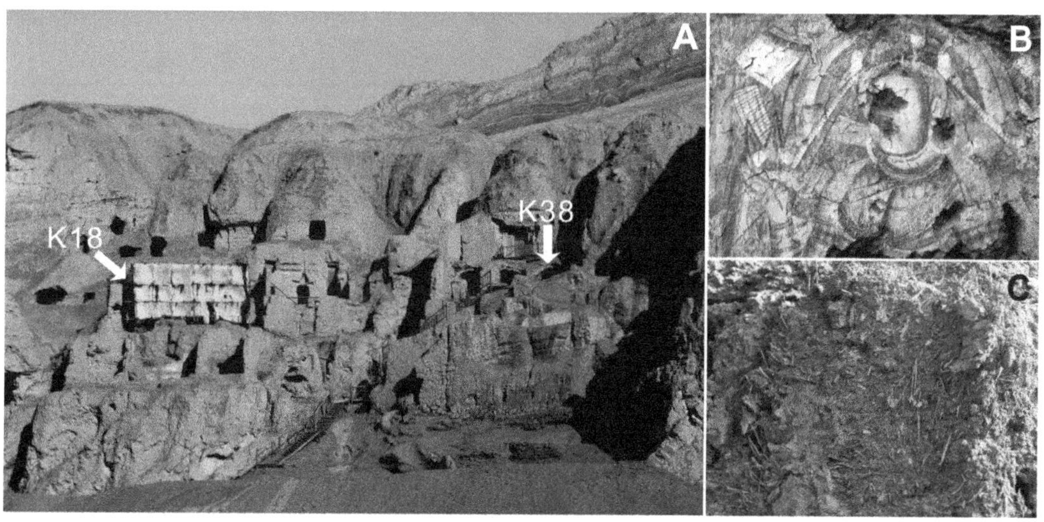

Figure 2. Tuyoq Grottoes sampling locality. (A) Sampling Grottoes K18 and K38, (B) Buddha fresco in the Grotto K18, (C) Picture showing plants fragments mixed in the clay adobe as strengthening materials.

Based on the technique of Heavy Liquid Separation (density: 2.0 g mL^{-1},) [23,24], 20 g soil samples were analyzed for pollen, which were preserved in glycerin, observed under a Leica DM 2500 light microscope and photographed at 200× or 400× magnifications. The single-grain technique of Ferguson *et al.* [25] was applied for pollen observation under a Hitachi S-4800 scanning electron microscope (SEM). 3500 pollen grains from the Tuyoq Grottoes were counted in all, 500 grains from each sample. Pollen identification was based on the monographs 'Pollen Flora of China', 'Pollen morphology of plants from dry and semi-arid areas in China' and other published literature [26–28], and compared to the pollen of modern plants from the Chinese National Herbarium (PE) if necessary. Simultaneously, pollen of nine common Poaceae species from the modern Tuyoq region were observed under SEM to assist ancient Poaceae pollen identification. The relative abundance (RA) of a pollen taxon is calculated by the equation: RA = N/Nt (N: pollen/spore number of a taxon, Nt: total pollen/spore number in the sample).

Leaf epidermis analysis was carried out by washing plant remains from the ancient clay adobe, then preparing the epidermis in a 1:1 mixture of 30% hydrogen peroxide (H_2O_2) and 99% glacial acetic acid (CH_3COOH) at 60°C in a water bath for one or two days [29]. Finally, the epidermides were cleaned with a fine brush and washed again in distilled water until the epidermis textures were clearly visible under a stereo microscope. The epidermides were mounted on microscopical slides with glycerin for observation and measurement, followed by the same procedure as pollen observation and photography. 35 blades of leaf epidermides were analyzed from the Tuyoq Grottoes, 5 blades from each sample. The epidermides were mainly identified using the monograph 'Micromorphological Atlas of Leaf Epidermis in Gramineae' and other published literature [30–33], and compared to the epidermides of modern plants from PE if necessary. Eight structural features were employed for leaf epidermis taxonomy, including the long cells, short cells, stomatal complexes and subsidiary cell shape, prickle-hairs, macro- and micro-hairs, and silica bodies. The terminology used to describe the leaf epidermis follows Ellis (1979) [34] and Chen (1993) [30].

The ancient climate of the Tuyoq Grottoes was reconstructed following the Coexistence Approach (CA) [35], whereby the following five climatic parameters were calculated: the mean annual temperature (MAT), the mean warmest monthly temperature (MWMT), the mean coldest monthly temperature (MCMT), the temperature difference between coldest and warmest months (DT) and the mean annual precipitation (MAP). On the assumption that the climatic tolerance of an ancient taxon was similar to its nearest living relatives (NLRs), the principle of CA is to estimate the ancient climatic parameters by using the modern climatic parameters of NLRs distribution [35]. In this work, each ancient pollen taxon in the assemblage is identified to a NLR. Modern distribution of NLRs in China follows Wu & Ding (1999) [36], and the climatic tolerances of NLRs were obtained from the modern meteorological data (1951–1980) recorded at various meteorological stations in China [3,37–41]. The 3D reconstruction of the Tuyoq landscape in the Fifth Century was completed using software of '3D Studio Max' and 'Adobe Photoshop'.

Results

Pollen Analyses

44 palynomorphs were extracted from the samples of Tuyoq Grottoes, consisting of 38 angiosperms (85.45%), 3 gymnosperms (13.83%), 1 pteridophyte (0.43%) and 2 unknowns (0.29%)

Table 1. The palynomorph relative abundance (RA) of Tuyoq Grottoes.

Palynomorph			RA
Pteridophytes	Polypodiaceae		0.43%
Gymnosperms	Ephedraceae	*Ephedra*	10.69%
	Pinaceae	*Pinus*	1.89%
		Picea	1.26%
Angiosperms	Poaceae	*Phragmites australis*	4.00%
		other Poaceae	26.11%
	Asteraceae	*Artemisia*	19.83%
		other Asteraceae	3.03%
	Chenopodiaceae		10.71%
	Fabaceae	*Alhagi*	3.83%
		Medicago	0.29%
		other Fabaceae	4.71%
	Moraceae		1.80%
	Ranunculaceae		1.54%
	Ulmaceae	*Ulmus*	1.29%
	Typhaceae	*Typha*	0.97%
	Alismataceae	*Alisma*	0.91%
	Apiaceae		0.71%
	Polygonaceae	*Polygonum*	0.66%
	Salicaceae	*Salix*	0.66%
	Caryophyllaceae		0.46%
	Lamiaceae		0.40%
	Potamogetonaceae		0.37%
	Rosaceae		0.34%
	Boraginaceae		0.31%
	Elaeagnaceae		0.31%
	Malvaceae		0.29%
	Rutaceae		0.29%
	Betulaceae	*Betula*	0.20%
		Corylus	0.14%
	Convolvulaceae		0.20%
	Liliaceae		0.20%
	Euphorbiaceae		0.17%
	Vitaceae	*Vitis*	0.17%
	Plantaginaceae		0.09%
	Fagaceae	*Castanea*	0.14%
		Quercus	0.09%
		other Fagaceae	0.06%
	Rubiaceae		0.09%
	Brassicaceae		0.03%
	Celastraceae		0.03%
	Juglandaceae		0.06%
Unknown-1			0.09%
Unknown-2			0.20%
Total	44		100%

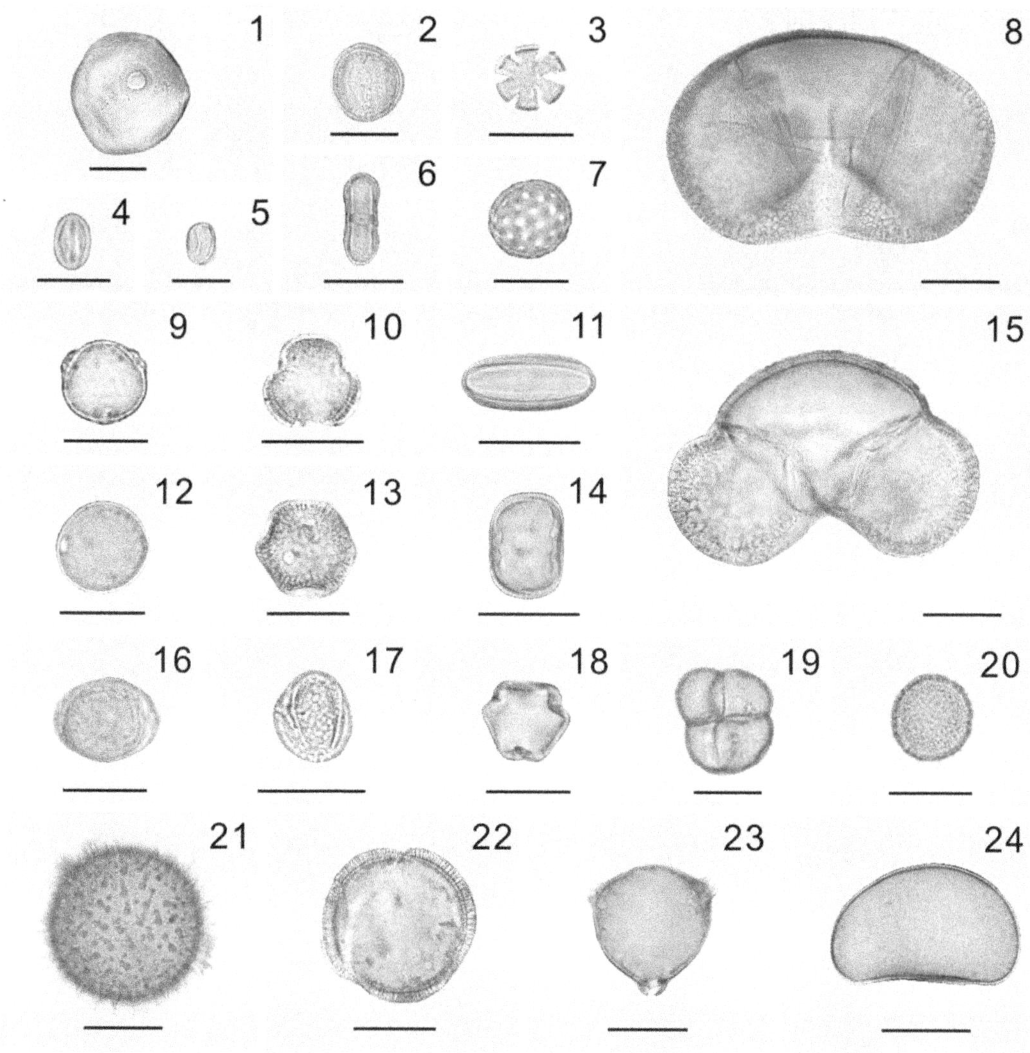

Figure 3. Main palynomorphs of Tuyoq Grottoes. 1. Poaceae 2. Ranunculaceae 3. Lamiaceae 4. *Castanea* 5. *Alhagi* 6. Apiaceae 7. Chenopodiaceae 8. *Picea* 9. *Betula* 10. *Artemisia* 11. *Ephedra* 12. Moraceae 13. Caryophyllaceae 14. Fabaceae 15. *Pinus* 16. *Ulmus* 17. *Salix* 18. *Vitis* 19. *Typha* 20. Potamogetonaceae 21. Malvaceae 22. Convolvulaceae 23. Elaeagnaceae 24. Polypodiaceae; Scale bar = 20 μm.

(Table 1, Figs. 3–4). The angiosperms were assigned to 31 families and the gymnosperms to 2 families.

In the spermatophytes, terrestrial plants comprised 93.02%, of which 70.11% were herbs (Poaceae 26.11% , *Artemisia* 19.83%, Chenopodiaceae 10.71%), 15.20% shrubs (*Ephedra* 10.69%, *Alhagi* 3.83%, Rosaceae 0.34%), 7.54% trees (*Pinus* 1.89%, Moraceae 1.80%, *Picea* 1.26%), 0.17% vines (*Vitis*), swamp plants (*Phragmites australis*) comprised 4.00%, aquatic plants only comprised 2.26% (*Typha* 0.97%, *Alisma* 0.91%, Potamogetonaceae 0.37%) (Fig. 5).

Epidermis Analyses

The leaf epidermis of *Triticum* sp. (wheat), *Agropyron mongolicum* and *Phragmites* sp. (reed) were identified from the samples of Tuyoq Grottoes (Table 2, Figs. 6–7).

1) *Triticum* sp. (Family: Poaceae, subfamily: Pooideae).

Description. Lower leaf epidermis composed of long cells, short cells, and stomatal complexes. Long cells narrowly oblong, 37–127 μm long (X_n = 78 μm, n = 20), 7–18 μm wide (X_n = 13 μm, n = 20), with wavy and thick walls, regularly or irregularly interspersed with short cells, non-papillate. Short cells present, paired. Stomatal complexes 19–39 μm long (X_n = 33 μm, n = 20), 9–25 μm wide (X_n = 20 μm, n = 20), subsidiary cells usually low dome-shaped or parallel-sided. Macro-hairs absent, micro-hairs present occasionally.

Identification. Unearthed epidermis of wheat (*Triticum* sp.) was identified by comparison with illustrations in Chen *et al.* (1993) [30], and Cai & Guo (1995) [32], and epidermis from modern wheat (*Triticum aestivum*) (specimen no.01941718). The thick cell-walls with constant width, remarkable and large stomatal complexes are characteristic features of *Triticum* epidermis (Fig. 6).

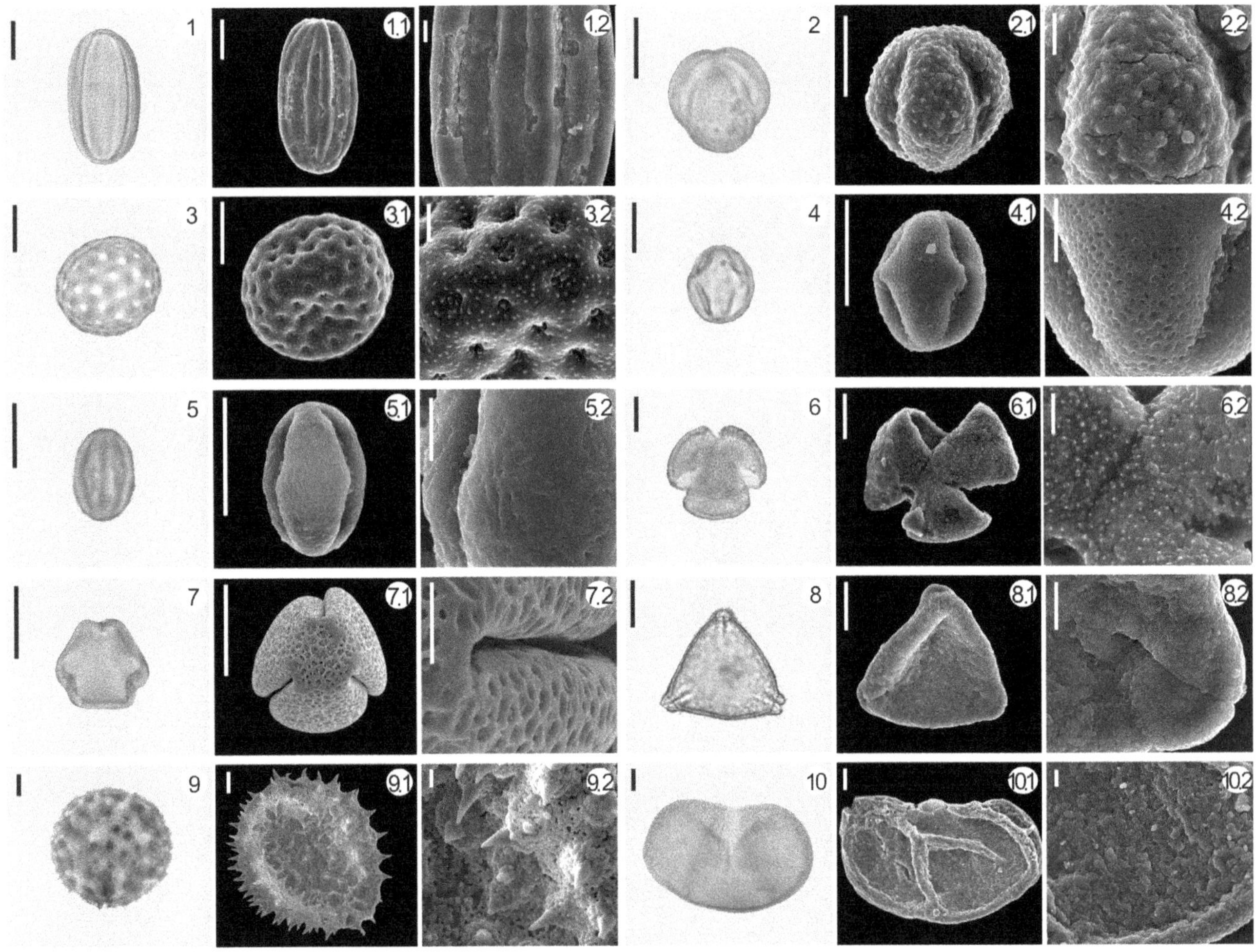

Figure 4. A selection of palynomorphs from Tuyoq Grottoes. 1. *Ephedra* 2. *Artemisia* 3. Chenopodiaceae 4. *Alhagi* 5. *Castanea* 6. Ranunculaceae 7. *Vitis* 8. Elaeagnaceae 9. Malvaceae 10. *Picea*; Scale bar in light microscope (LM) and scanning electron microscopic (SEM) overview 10 μm, in SEM close-up 2.5 μm.

2) *Agropyron mongolicum* (Family: Poaceae, subfamily: Pooideae).

Description. Lower leaf epidermis composed of long cells, short cells with pitted-cells, and stomatal complexes. Long cells narrowly oblong, 34–85 μm long (X_n = 62 μm, n = 10), 8–25 μm wide (X_n = 16 μm, n = 10), with wavy walls, irregularly interspersed with short cells, non-papillate. Short cells present, solitary or occasionally paired, pitted-cells distinctive with round shape having a diameter of 10–24 μm (X_n = 17 μm, n = 10). Stomatal complexes 18–32 μm long (X_n = 28 μm, n = 10), 15–27 μm wide (X_n = 23 μm, n = 10), subsidiary cells usually low dome-shaped or parallel-sided. Macro-hairs absent, micro-hairs present occasionally.

Identification. Unearthed epidermis of *Agropyron mongolicum* was identified by comparison with illustrations in Chen *et al.* (1993) [30] and Xie & Yang (1994) [31], and an epidermis from a modern plant (specimen no. 01821117). There are five species in the genus *Agropyron*, *Agropyron mongolicum* is characterized by pitted-cells, as well as paired short cells present occasionally (Fig. 6).

3) *Phragmites* sp. (Family: Poaceae, subfamily: Arundinoideae).

Description. Lower epidermis composed of long cells, short cells, stomatal complexes, and prickle-hairs. Long cells narrowly oblong with coarse wavy walls, non-papillate, 51–97 μm long (X_n = 61 μm, n = 20), 5–12 μm wide (X_n = 8 μm, n = 20), irregularly interspersed with short cells. Short cells single or paired, silica short cells usually saddle-shaped. Stomatal complexes 12–18 μm long (X_n = 15 μm, n = 20), 3–8 μm wide (X_n = 5 μm, n = 20), subsidiary cells usually parallel-sided, occasionally low dome-shaped. Prickle-hairs and micro-hairs present occasionally.

Identification. Unearthed epidermis of reed (*Phragmites* sp.) was identified by comparison with illustrations in Chen *et al.* (1993) [30], Cai & Guo (1995) [32] and Lin (2008) [42], and an epidermis from modern reed (*Phragmites australis*) (specimen no.0595279). Arundinoideae are distinct from other subfamilies of the Poaceae in having micro-hairs, no papillae, and saddle-shaped silica cells. *Phragmites* is characterized by the coarse wavy wall of the lower epidermis (Fig. 7).

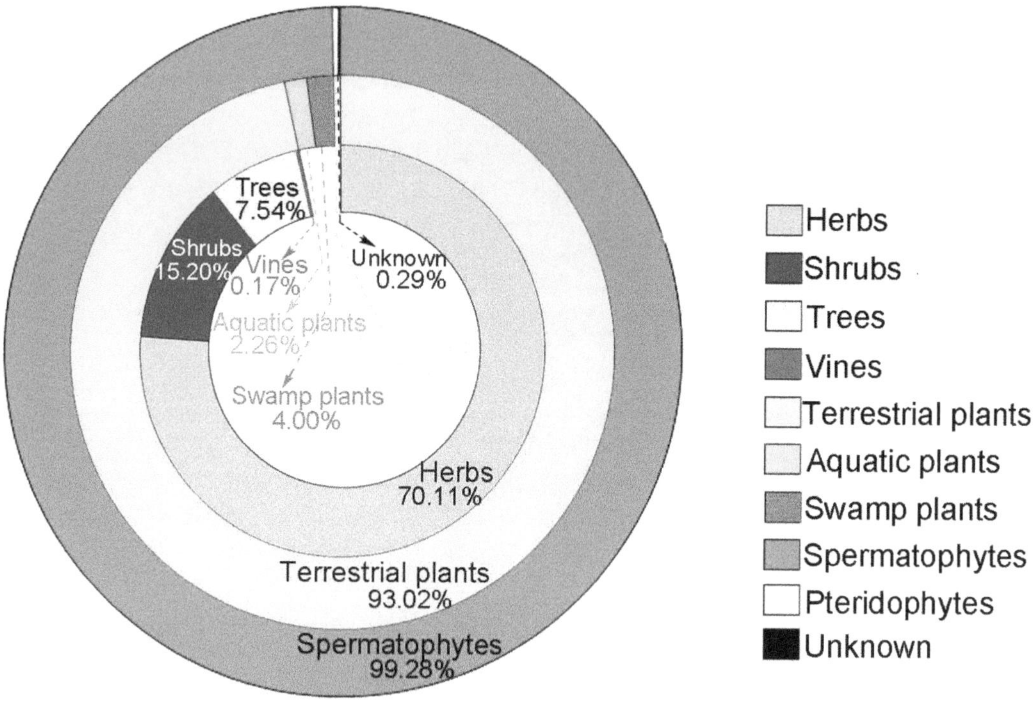

Figure 5. Relative abundance (%) of different plant categories of palynomorphs from Tuyoq Grottoes.

As a whole, the pollen assemblage of Tuyoq Grottoes is dominated by herbs (70.11%), whereas shrubs and trees are only represented by 15.20% and 7.54% respectively. Apart from widespread Poaceae and *Artemisia*, almost all palynotaxa with the greatest relative abundance are herbs and shrubs mainly found in arid or semiarid areas of the temperate zones: *Artemisia*, Chenopodiaceae, *Ephedra*, *Alhagi*. The leaf epidermis of *Agropyron mongolicum* also indicates arid steppe or sandy land. The few trees such as Moraceae, *Ulmus* and *Populus* in the palynoassemblage indicate the local presence of some deciduous broad-leaved plants. The coniferous pollen of *Pinus* and *Picea* was probably transported to the Turpan Basin by wind [43]. Based on the occurrence of aquatic plants (2.26%) such as *Typha* and Potamogetonaceae, swamp plants (4.00%) as *Phragmites australis* and the presence of Polypodiaceae (0.43%), the Tuyoq Valley is suggested to have represented a moist microenvironment with flowing water while the Turpan Basin was generally arid in the Fifth Century. At the same time, wheat and grape were cultivated in this region as pollen of *Vitis* and leaf epidermis of *Triticum* were found.

Ancient Vegetation

The A/C ratio (A: *Artemisia*, C: Chenopodiaceae) has been widely used in distinguishing different vegetation of arid areas [44–47]. Based on Luo *et al.*'s (2007) [48] study on surface soil pollen of Xinjiang, the A/C ratio (median value) of coniferous forest was 3.47, steppes 1.26, desert 0.57, thus an A/C ratio of 1.8 at the Tuyoq Grottoes is indicative of steppe. Furthermore, *Artemisia*, *Ephedra* and Chenopodiaceae are the most widespread pollen taxa throughout unforested areas of Xinjiang. When these taxa occurred with Asteraceae, Poaceae, Cyperaceae and Brassicaceae as the dominant pollen group, they suggest a meadow-steppe vegetation, whereas an alpine-steppe is indicated when they were found alongside Caryophyllaceae, Rosaceae, Ranunculaceae and Apiaceae [49]. Clearly, the dominant pollen group of Tuyoq Grottoes (Poaceae *Artemisia*, *Ephedra*, Chenopodiaceae, *Ephedra* and *Alhagi*) points to a temperate vegetation of meadow-steppe (Table 3).

Ancient Climate

Based on the 41 identified palynomorphs of spermatophytes, five ancient climatic parameters of the Tuyoq Grottoes were

Table 2. Plant taxa of leaf epidermis with quantitative distribution in the samples.

Plant taxa	Sample no. 1	no.2	no.3	no.4	no.5	no.6	no. 7	Total
Phragmites sp.	0		2	0	1	3	1	7
Triticum sp.	4	2	3	3	3	2	3	20
Agropyron mongolicum	0	2						2
Unknown	1	1		2	1		1	6
Total	5	5	5	5	5	5	5	35

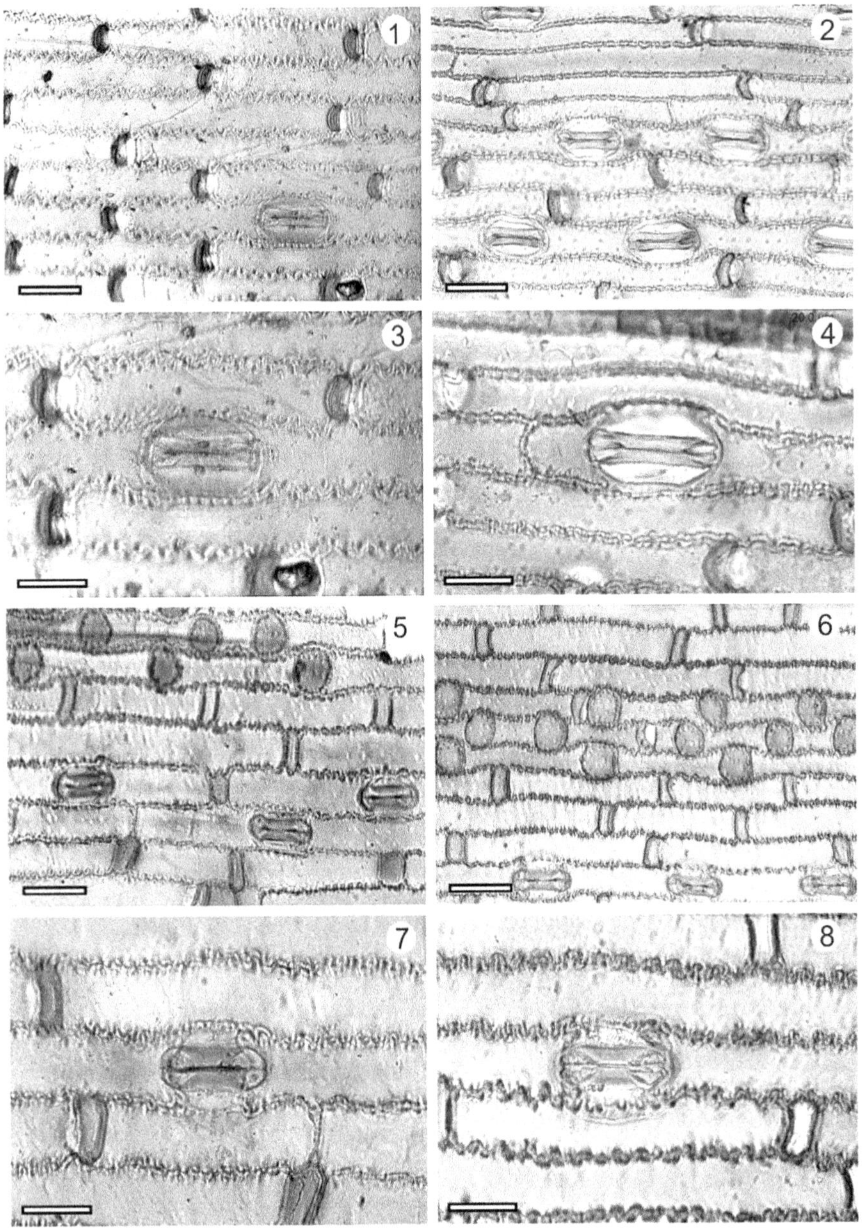

Figure 6. Comparison of epidermides from the Tuyoq Grottoes and modern plants. 1,3: Unearthed lower leaf epidermis of wheat (*Triticum* sp.) from Tuyoq Grottoes, 2,4: Lower leaf epidermis of modern wheat (*Triticum aestivum*), 5,7: Unearthed lower leaf epidermis of *Agropyron mongolicum* from Tuyoq Grottoes, 6,8: Lower leaf epidermis of modern *Agropyron mongolicum*; Scale bar in1,2,5,6 = 30 μm; in 3,4,7,8 = 20 μm.

reconstructed by CA as follows: MAT = 8.5–22.1°C (15.3°C median value), delimited by *Castanea* and *Alhagi*; MWMT = 23.2–27.5°C (25.4°C), delimited by *Alhagi* and *Betula*; MCMT = −9.4–5.9°C (−1.8°C), delimited by *Castanea* and *Ephedra*; DT = 13.8–34.2°C (24°C), delimited by *Alhagi* and *Castanea*; MAP = 618.9–1389.4 mm (1004 mm), delimited by *Castanea* and *Ephedra* (Fig. 8). Compared with the current meteorologic data (Table 4), the median value of the MAP was much higher 1500 years ago, the MAT, MCMT were respectively 1.4°C and 7.7°C higher 1500 years ago, whereas MWMT and DT were lower than today by 7.3°C and 18.2°C, which indicates that in the Tuyoq region the climate was warmer and wetter with weaker seasonality during the Fifth Century.

3D Reconstruction of Ancient Environment

The environment of Turpan in the Fifth Century was reconstructed and visualized in a 3D presentation based on the investigation of ancient vegetation and climate at the Tuyoq Grottoes (Fig. 9). In the Fifth Century, the landscape of the Tuyoq region was composed of valley, village, mountains and distant desert. Temperate vegetation of meadow-steppe predominated in this area with widespread herbs of *Artemisia*, Chenopodiaceae and numerous shrubs of *Ephedra* and *Alhagi*. The Tuyoq Grottoes were constructed in the cliffs on the sides of the valley. A river flowed through the valley, the village and on to the distant desert. *Phragmites australis*, *Typha* and other aquatic plants grew luxuriantly in the Tuyoq Valley. People lived in the basin oasis nourished by the river, wheat fields and vineyards around the village, along with

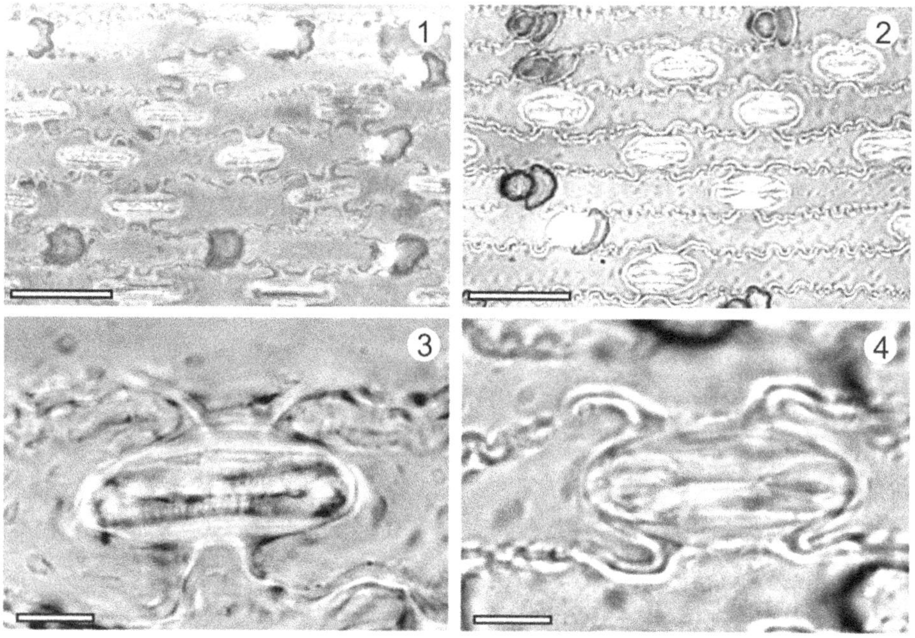

Figure 7. Comparison of epidermides from the Tuyoq Grottoes and modern plants. 1,3: Unearthed lower leaf epidermis of reed (*Phragmites* sp.) from Tuyoq Grottoes, 2,4: Lower leaf epidermis of modern reed (*Phragmites australis*); Scale bar in 1,2 = 20 µm; in 3,4 = 5 µm.

a few deciduous trees such as *Ulmus*, *Betula*, Moraceae, *Salix* and *Populus*. In the distance, the river is seen vanishing into the desert, while the conifers *Picea* and *Pinus* lived in the surrounding mountains.

Discussion

Ancient Pollen of Reed (*Phragmites australis*) Identified by SEM and its Archeological Significance

Poaceae pollen have never been used as a diagnostic trait for distinguishing between genera or species since they share the same characteristics (stenopalynous) under light microscopy [50]. In recent years, the scanning electron microscope (SEM) has been increasingly used in pollen identification of cereals and other grasses [51–55]. In the Tuyoq Grottoes pollen assemblage, the content of Poaceae pollen reached 26.11%. In order to improve the accuracy of pollen identification, pollen of nine common Poaceae species from the modern Tuyoq region were observed under a SEM, namely *Agropyron cristatum*, *Agropyron mongolicum*, *Festuca arundinacea* subsp. *orientalis*, *Festuca ovina*, *Sorghum bicolor*, *Stipa sareptana*, *Triticum aestivum*, *Zea mays*. In the sample from 1500 years ago, pollen which has the same characteristics as modern reed (*Phragmites australis*) was identified (Fig. 10). Although the identification of a wide range of grasses by their pollen remains a problem, our results are a useful reminder that there is a very real possibility of determining fossil or ancient grass palynotaxa under SEM observation when only a limited number of species is involved. This is promising and practicable for a higher resolution of crop pollen in archaeological research.

MAP and Water Supplied by Glacial Melt and Underground Water

We noticed that the reconstructed MAP of the Fifth Century was apparently more than sixty times higher than the current precipitation. Actually, in the arid Turpan Basin, most water is supplied by glacial melt from the Tian and Bogeda Mountains and underground water at present. We suppose that the actual rainfall in the Fifth Century could have either been higher than or similar to that of today. If it was similar to today, the reconstructed

Table 3. Pollen assemblage characteristics of different modern vegetations in Xinjiang with comparison to Tuyoq Grottoes.

Vegetation characteristics	Forest	Steppe		Desert	Tuyoq Grottoes
A/C ratio*	3.47	1.26		0.57	1.8
Vegetation subtype		meadow-steppe	alpine-steppe		
Dominant pollen group	*Picea*	A,C,E*	A,C,E*	A,C,E*	A,C,E*
	Larix	Asteraceae	Caryophyllaceae	*Nitraria*	Asteraceae
	Betula	Poaceae	Rosaceae		Poaceae
	Populus	Cyperaceae	Ranunculaceae	*Lycium*	*Alhagi*
		Brassicaceae	Apiaceae		

*A: *Artemisia*, C: Chenopodiaceae, E: *Ephedra* [48,49].

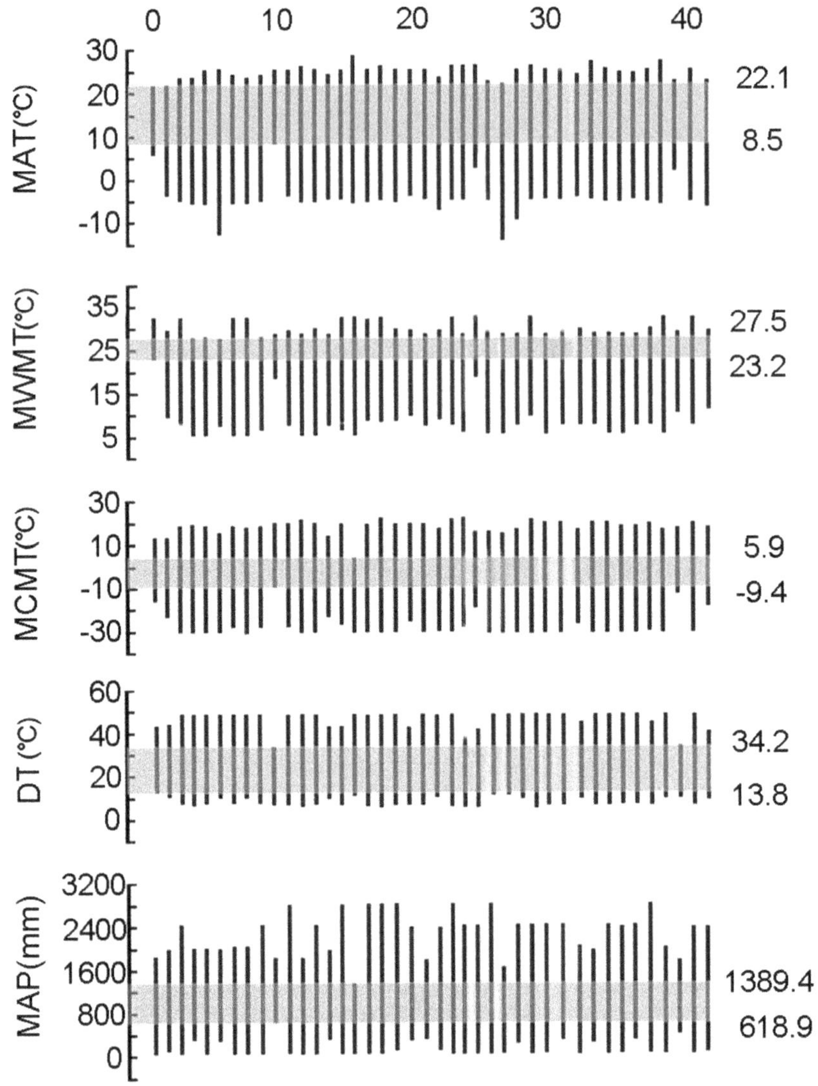

Figure 8. Coexistence intervals of climatic parameters of Tuyoq Grottoes palynoflora. 1. *Alhagi* 2. *Alisma* 3. Apiaceae 4. *Artemisia* 5. Asteraceae 6. *Betula* 7. Boraginaceae 8. Brassicaceae 9. Caryophyllaceae 10. *Castanea* 11. Celastraceae 12. Chenopodiaceae 13. Convolvulaceae 14. *Corylus* 15. Elaeagnaceae 16. *Ephedra* 17. Euphorbiaceae 18. Fabaceae 19. Fagaceae 20. Juglandaceae 21. Lamiaceae 22. Liliaceae 23. Malvaceae 24. *Medicago* 25. Moraceae 26. *Phragmites australis* 27. *Picea* 28. *Pinus* 29. Plantaginaceae 30. Poaceae 31. *Polygonum* 32. Potamogetonaceae 33. *Quercus* 34. Ranunculaceae 35. Rosaceae 36. Rubiaceae 37. Rutaceae 38. *Salix* 39. *Typha* 40. *Ulmus* 41. *Vitis.*

Table 4. The comparison between the median values of five climatic parameters in Fifth Century Tuyoq Grottoes and the current meteorological data.

Climatic parameters	5th Century	Modern*
MAT(°C)	15.3	13.9
MWMT(°C)	25.4	32.7
MCMT(°C)	−1.8	−9.5
DT(°C)	24.0	42.2
MAP(mm)	1004.2	16.4

*Land climatic data of Turpan Meteorological Station (N 42°56′,E 89°12′), about 47 km east of the Tuyoq Grottoes.

1000 mm MAP may not only have consisted of the actual rainfall, but also incorporated glacial melt or underground water, since plants can utilize any source of available water. Although the meteorological precipitation 1500 years ago is hard to evaluate, it is clear that the Tuyoq region had an abundant water supply as a river flowed in the Tuyoq Valley, which was crucial for the oasis development. At the same time, it indicates that a water source such as glacial melt and underground water have to be taken into account when the ancient MAP of an arid area such as Xinjiang is reconstructed using CA.

Oasis and Buddhism Nourished by Amicable Environment in Turpan

According to the epidermis analysis (Table 2), wheat (*Triticum* sp.) and reed (*Phragmites* sp.) were the most used materials mixed in the clay to strengthen the adobe, whereas *Agropyron mongolicum* and

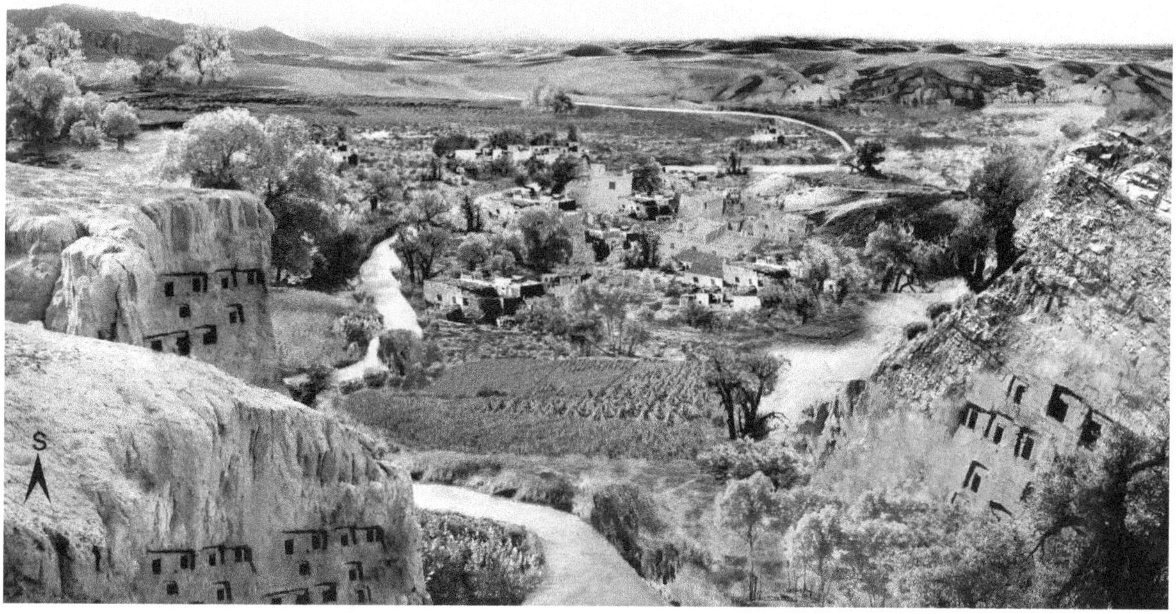

Figure 9. 3D reconstruction of environmental landscape of Tuyoq region in the Fifth Century. In this picture, the viewpoint is from north to south.

other plants were probably weeds for only a few of them were found. During the early period of the Subeixi Culture (around 700 BC), wheat (*Triticum aestivum*) was a supplemental crop of oases in Turpan [13]. Until the Fifth Century, the frequent occurrence of wheat (*Triticum* sp.) in the Tuyoq Grottoes suggests that it was cultivated on a certain scale,and it became one of the major crops during the Gaochang Period [56]. The development of wheat cultivation and the discovery of grape (*Vitis* sp.) pollen witnessed the blooming of oasis agriculture in this period.

In the Fifth Century, the Xinjiang area experienced a warming climatic tendency [57,58] whereas the global temperature was undergoing a cooling event around 1.4k BP [59,60]. The reconstructed MAT of Turpan 15.3°C was 1.1°C warmer than that of the northern hemisphere [60] and 7.4°C warmer than the winter half-year-mean temperature of eastern China [61] in the Fifth Century (Fig. 11), as well as 1.4°C warmer than the current temperature of Turpan 13.9°C [3]. Li (1985) believed that the MAT of the whole of Xinjiang in the Fifth Century was 1–2°C warmer than today [57]; our result is consistent with this

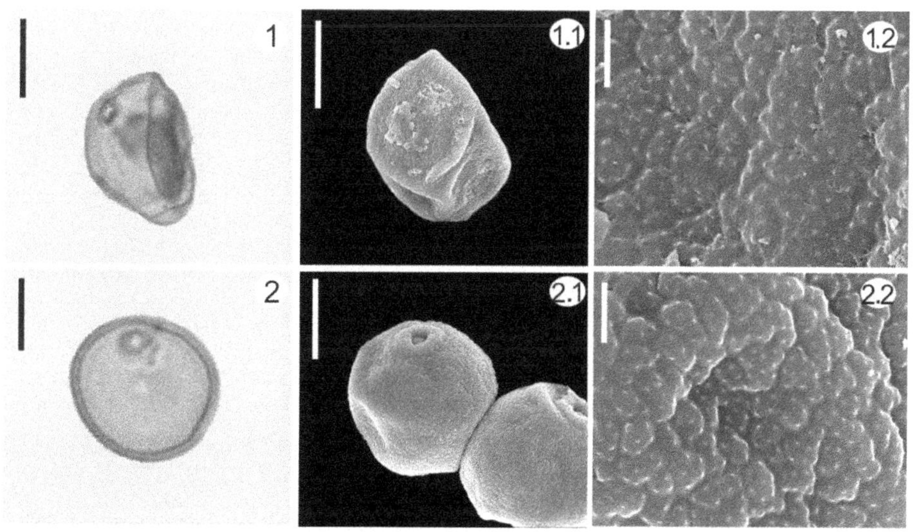

Figure 10. Comparison of ancient reed (*Phragmites australis*) pollen and modern reed pollen. 1. Unearthed pollen of reed (*Phragmites australis*) from Tuyoq Grottoes, 2. Pollen of modern reed (*Phragmites australis*); Scale bar in light microscope (LM) and scanning electron microscopic (SEM) overview = 10 μm, in SEM close-up = 1 μm.

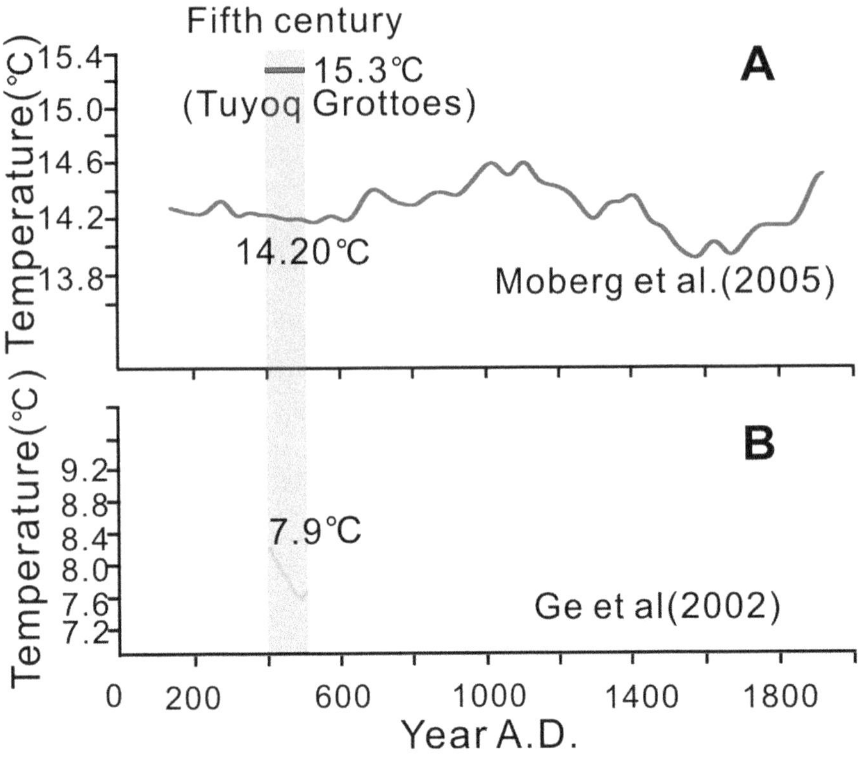

Figure 11. Temperature comparison of the Tuyoq Grottoes and the northern hemisphere (China) in the Fifth Century. (A) Temperature reconstruction of Northern Hemisphere from 133 to 1925 A.D., the mean temperature of fifth century (401–500 A.D.) was 14.20°C [60], (B) Winter half-year-mean temperature reconstruction of eastern China during the past 2000 years, the mean temperature of Fifth Century was 7.9°C [61].

conclusion. Since glacial melt was the main water source for the Turpan Basin [4], the alteration of water resources caused by climatic change was the crucial factor for oases development [62]. In the Fifth Century, the warmer climate at the Tuyoq Grottoes supplied abundant glacial meltwater to the valley, which enabled the oasis agriculture to flourish.

The Tuyoq Grottoes, included 94 caves along half a kilometer of cliff face in the Tuyoq Valley, were initiated during the Fifth Century. Construction continued until the Seventh Century [63]. If the oases were not well enough developed to furnish financial and material support, such large scale Buddhist grotto construction would have been impossible.

However, in the same period with a similar warm climate in the Taklamakan Desert [58,64], Niya and a series of ancient cities with Buddhist temples were abandoned (Fig. 12, Table 5). Possible reasons for the abandonment may include: 1) River diversion [65,66]. As glacial melt increased, increased flow of rivers in the desert basin was prone to cause river diversion but not in an enclosed mountain valley like Tuyoq. 2) River desiccation. When rivers flow across a desert, any increase in temperature would have strengthened evaporation, causing the lower part of the river to dry up. This is considered a possible reason for Niya abandonment [64,65,67]. 3) Desertification caused by an expansion of the Taklimakan Desert could explain the abandonment of Kaladun city [66]. 4) War or plague. Based on archaeological observations, we speculate that the abandonment of Niya was probably associated with a war threat or plague, since the unearthed Kharoshthi tablets repeatedly mentioned invasions by the Supis (probably Sump Kingdom of northern Tibet).

Table 5. Ancient cities of South Tarim Basin abandoned around the Fifth Century.

Ancient city	Geography	Time of abandonment	Probable reason	Reference
Niya	Lower part of Niya River	4th to 5th century A.D.	River drying up; War or plague (in present work)	[64,65,67]
Kaladun	Lower part of Keriya River	4th to 5th century A.D.	River diversion	[65,66]
Yixun	Middle part of Milan River	5th century A.D.	War	[65]
Ancient Pishan	Lower part of Pishan River	4th to 5th century A.D.	River drying up	[64]
Qianni	Middle part of Ruoqiang River	5th century A.D.	River drying up	[64]

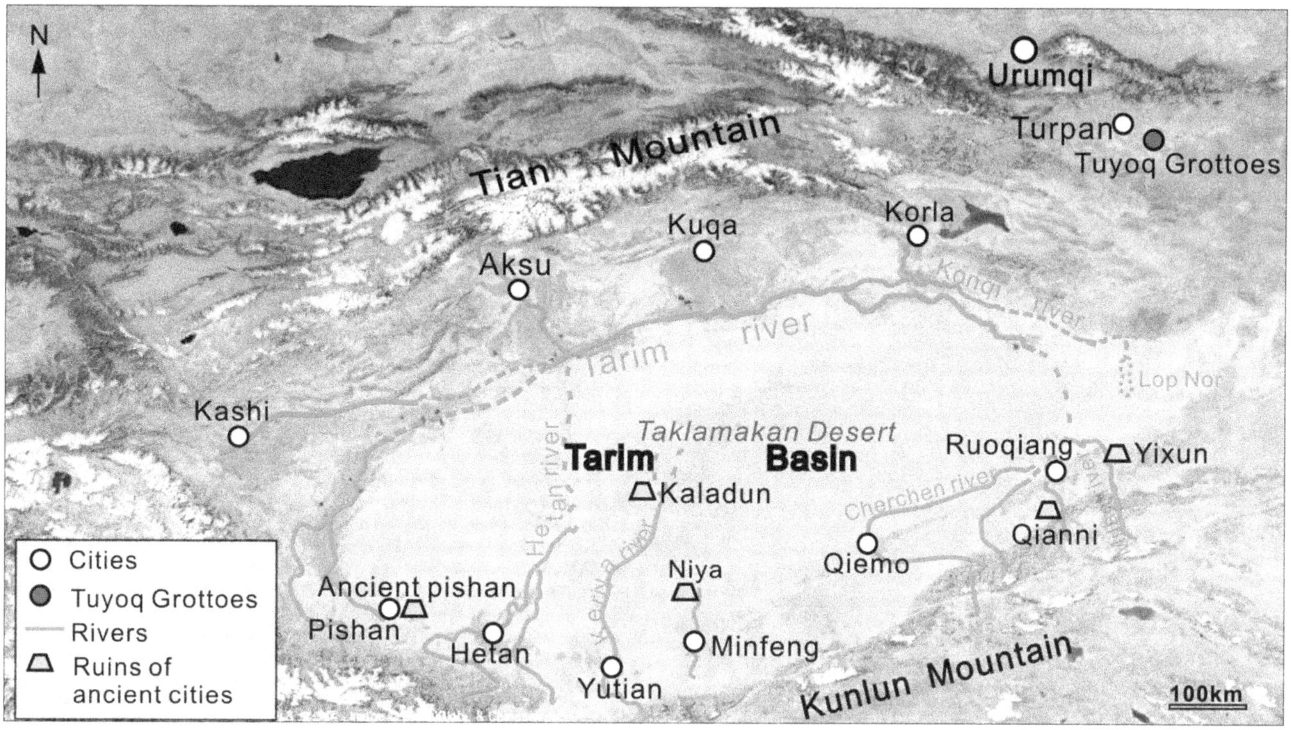

Figure 12. Distribution of ancient cities in southern Tarim Basin abandoned around the Fifth Century (revised from http://map.baidu.com/**).**

Acknowledgments

We thank Dr. Yun-Jun Bai, Institute of Botany, CAS, Beijing, China, for his kind help in identifying the leaf epidermides.

Author Contributions

Conceived and designed the experiments: YNT YFY CSL. Performed the experiments: YNT YFY. Analyzed the data: YNT YFY DKF CSL. Contributed reagents/materials/analysis tools: XL CSL. Wrote the paper: YNT.

References

1. Li JJ, Fang XM (1999) Uplift of the Tibetan Plateau and environmental changes. Chin Sci Bull 44: 2117–2124. doi:10.1007/BF03182692.
2. Zhang DF, Fengquan L, Jianmin B (2000) Eco-environmental effects of the Qinghai-Tibet Plateau uplift during the Quaternary in China. Environ Geol 39: 1352–1358. doi:10.1007/s002540000174.
3. IDBMC (Information Department of Beijing Meteorological Center) (1983c) Land Climate Data of China (1951–1980) (Part VI). Beijing: China Meteorological Press. 4–6, 126–128 p.
4. Li X, Yin JY (1993) Arid and dry characteristics of Turpan Basin. Arid Land Geogr 16: 63–69.
5. Pan W, Cheng J, Wang SH (2003) Quaternary climate change in Turpan basin and its significance to sandstone-type uranium mineralization. Uranium Geol 19: 332–338.
6. Cheng J, Zan LH, Zhang XW (2006) The Holocene vegetation and climate changes in the Turpan Basin of Xinjiang, China Beijing: China Ocean Press. pp. 283–294.
7. Jiang HE, Li X, Zhao YX, Ferguson DK, Hueber F, et al. (2006) A new insight into Cannabis sativa (Cannabaceae) utilization from 2500-year-old Yanghai Tombs, Xinjiang, China. J Ethnopharmacol 108: 414–422. doi:10.1016/j.jep.2006.05.034.
8. Jiang HE, Li X, Ferguson DK, Wang YF, Liu CJ, et al. (2007) The discovery of Capparis spinosa L. (Capparidaceae) in the Yanghai Tombs (2800 years b.p.), NW China, and its medicinal implications. J Ethnopharmacol 113: 409–420. doi:10.1016/j.jep.2007.06.020.
9. Jiang HE, Li X, Liu CJ, Wang YF, Li CS (2007) Fruits of Lithospermum officinale L. (Boraginaceae) used as an early plant decoration (2500 years BP) in Xinjiang, China. J Archaeol Sci 34: 167–170. doi:10.1016/j.jas.2006.04.003.
10. Jiang HE, Zhang YB, Li X, Yao YF, Ferguson DK, et al. (2009) Evidence for early viticulture in China: proof of a grapevine (Vitis vinifera L., Vitaceae) in the Yanghai Tombs, Xinjiang. J Archaeol Sci 36: 1458–1465. doi:10.1016/j.jas.2009.02.010.
11. Qiu Z, Zhang Y, Bedigian D, Li X, Wang C, et al. (2012) Sesame Utilization in China: New Archaeobotanical Evidence from Xinjiang. Econ Bot 66: 255–263. doi:10.1007/s12231-012-9204-5.
12. Chen T, Wu Y, Zhang YB, Wang B, Hu YW, et al. (2012) Archaeobotanical Study of Ancient Food and Cereal Remains at the Astana Cemeteries, Xinjiang, China. PLoS ONE 7: e45137. doi:10.1371/journal.pone.0045137.
13. Jiang HE, Li X, Li CS (2007) Cereal remains from Yanghai Tomb in Turpan, Xinjiang and their pelaeoenvrionmental significance. J Palaeogeogr 9: 551–558.
14. Yao YF, Li X, Jiang HE, Ferguson DK, Hueber F, et al. (2012) Pollen and Phytoliths from Fired Ancient Potsherds as Potential Indicators for Deciphering Past Vegetation and Climate in Turpan, Xinjiang, NW China. PLoS ONE 7: e39780. doi:10.1371/journal.pone.0039780.
15. Editorial office (2007) Thousand Buddha Grottoes of Tuyoq-the earliest confluence of Buddhist from Central Plains and the Western Regions. Chin Cult Herit: 80–82.
16. Klementz DA (1899) Nachrichten uber die von der Kaiserlichen Akademie der Wissenschaften zu St. Petersburg in Jahre 1898 ausgerustete Expedition nach Trufan, St. Petersburg.
17. Le Coq VA (1913) Chostscho.
18. Mokushiki K (1999) Xiyu kao gu tu pu. Beijing: Xuanyuan Press.
19. JAT (Joint Archaeological Team of Research Center for Frontier Archaeology of the Institute of Archaeology, Chinese Academy of Social Sciences and Academia Turfanica and Kizil Research Institute) (2011) The Excavation of Tuyoq Grottoes in Shanshan County, Xinjian. Archaeology 7: 27–32.
20. Abdirahman H, Bian ZF, Wahap H (2009) Discussion on the Water Resources and Their Rational Development and Utilization in Turpan Basin. J China Hydrol 29: 50–54.
21. The editorial board of Chinese Frescoes Corpus (1995) Corpus of Chinese Art Classification-Corpus of Xinjiang Frescoes (6) Tuyoq, Bezeklik. Liaoning Province Art Press.
22. JAT (Joint Archaeological Team of Research Center for Frontier Archaeology of the Institute of Archaeology, Chinese Academy of Social Sciences and Academia Turfanica and Kizil Research Institute) (2012) The Excavation of the North

Portion of the East Zone of Tuyoq Grottoes in Shanshan(Piqan) County,Xinjiang. Archaeology: 7–16.
23. Moore PD, Webb JA, Collinson ME (1991) Pollen Analysis, 2nd edn. Oxford: Blackwell scientific publications.
24. Li XQ, Du NQ (1999) The Acid-alkali-free analysis of Quaternary pollen. Acta Bot Sin 41: 782–784.
25. Ferguson DK, Zetter R, Paudayal KN (2007) The need for the SEM in palaeopalynology. Comptes Rendus Palevol 6: 423–430. doi:10.1016/j.crpv.2007.09.018.
26. IBCAS (Institute of Botany, the Chinese Academy of Sciences) (1976) Sporae Pterido-phytorum Sinicorum. Beijing: Science Press.
27. Xi YZ, Ning JC (1994) Pollen morphology of plants from dry and semidry area in China. Yushania 11: 119–191.
28. Wang FX, Chien NF, Zhang YL, Yang HQ (1995) Pollen Flora of China,2nd edn. Beijing: Science Press.
29. Ma QW, Zhang XS, Li FL (2005) Method of maceration and microscopical analysis on cuticle. Bull Bot Res 25: 307–310.
30. Chen SL, Jin YX, Wu ZJ (1993) Micromorphological atlas of leaf epidermis in Gramineae. Nanjing: Jiangsu Science Technology Press. 1–193 p.
31. Xie XM, Yang XL (1994) Taxonomic Values of the Leaf Anatomical Characters of Agropyron J.Gaertn.. J Inn Mong Norm Univ Nat Sci Ed 1: 53–58.
32. Cai LB, Guo YP (1995) study on constitute cells of leaf epidermis, systematic and phylogenetic path of the family Poaceae. Acta Bot Boreal- Occident Sin 15: 323–335.
33. Cai LB, Guo YP (1996) continue study on constitute cells of leaf epidermis, systematic and phylogenetic path of the family Poaceae. Acta Bot Boreal-Occident Sin 16: 65–72.
34. Ellis RP (1979) A procedure for standardizing comparative leaf anatomy in the Poaceae II: the epidermis as seen in surface view. Bothalia 12: 641–671.
35. Mosbrugger V, Utescher T (1997) The coexistence approach – a method for quantitative reconstructions of Tertiary terrestrial palaeoclimate data using plant fossils. Palaeogeogr Palaeoclimatol Palaeoecol 134: 61–86. doi:10.1016/S0031-0182(96)00154-X.
36. Wu ZY, Ding TY (1999) Seed Plants of China. Kunming: Yunnan Science and Technology Press.
37. IDBMC (Information Department of Beijing Meteorological Center) (1983a) Land Climate Data of China (1951–1980) (Part I). Beijing: China Meteorological Press. 4–6, 86–87 p.
38. IDBMC (Information Department of Beijing Meteorological Center) (1983b) Land Climate Data of China (1951–1980) (Part II). Beijing: China Meteorological Press. 4–5, 72–73 p.
39. IDBMC (Information Department of Beijing Meteorological Center) (1984a) Land Climate Data of China (1951–1980) (Part III). Beijing: China Meteorological Press. 4–6, 107–109 p.
40. IDBMC (Information Department of Beijing Meteorological Center) (1984b) Land Climate Data of China (1951–1980) (Part IV). Beijing: China Meteorological Press. 4–6, 99–101 p.
41. IDBMC (Information Department of Beijing Meteorological Center) (1984c) Land Climate Data of China (1951–1980) (Part V). Beijing: China Meteorological Press. 4–6, 108–110 p.
42. Lin XL (2008) Systematic Studies of the Cleistogenes(Poaceae) in China [Master Thesis]. Shandong Normal University.
43. Li WY (1998) Quaternary Vegetation and Environment of China. Beijing: Science Press. 1–253 p.
44. Yan S, Xu YQ (1989) Spore-pollen association in surface soils in the Altay Mountains, Xinjiang. Arid ZoneResearch 6: 26–33.
45. Yan S (1991) Quaternary pollen and spore assemblages and vegetation succession in Xinjiang. Arid Land Geogr 14: 1–9.
46. Weng CY, Sun XJ, Chen YS (1993) Numerical characteristics of pollen assemblages of surface samples from the West Kunlun mountains. Acta Bot Sin 35: 69–79.
47. Zhao Y, Liu HY, Li FR, Huang XZ, Sun JH, et al. (2012) Application and limitations of the Artemisia/Chenopodiaceae pollen ratio in arid and semi-arid China. The Holocene 22: 1385–1392. doi:10.1177/0959683612449762.
48. Luo CX, Zheng Z, Pan AD, An FZ, Beaudouin C, et al. (2007) Distribution of surface soil spore-pollen and its relationship with vegetation in X injiang, China. Arid Land Geogr 30: 536–543.
49. Luo CX, Zheng Z, Pan AD, An FZ, Beaudouin C, et al. (2008) Spatial Distribution of Modern Pollen in Xinjiang Region. Arid Land Geogr 28: 272–275.
50. Driessen MNBM, Willemse MTM, Van Luijn JAG (1989) Grass Pollen Grain Determination by Light- and UV-microscopy. Grana 28: 115–122. doi:10.1080/00173138909429962.
51. Andersen TS, Bertelsen F (1972) Scanning Electron Microscope Studies of Pollen of Cereals and other Grasses. Grana 12: 79–86. doi:10.1080/00173137209428830.
52. Page JS (1978) A Scanning Electron Microscope Survey of Grass Pollen. Kew Bull 32: 313–319. doi:10.2307/4117102.
53. Ma YX, Li Q, Cui DL (2004) Studies on the pollen morphology of nineteen species Poaceae from Northeast China. Bull Bot Res 24.
54. Perveen A (2006) A Contribution to the Pollen Morphology of Family Gramineae. World Appl Sci J 1: 60–65.
55. Mao LM, Yang XL (2011) Comparative poolen morphology of 7 species in Stipeae of Poaceae. Acta Micropalaeontologica Sin 28: 169–180.
56. Wang S (2000) Gaochang history: Transportation. Beijing: Cultural Relics Press. 85–102 p.
57. Li JF (1985) Climatic variation of Xinjiang in recent 3000 years. Proceeding of Quaternary Xinjiang Arid Area. Urumqi: Xinjiang people's publishing house. 1–7.
58. Zhong W, Shu Q, Xiong HG, Tashplati T (2001) Historic paleoenvironmental changes reflected by susceptibility in the sediment, southern margin of Tarim basin. Arid Land Geogr 24: 212–216.
59. Bond G, Showers W, Cheseby M, Lotti R, Almasi P, et al. (1997) A Pervasive Millennial-Scale Cycle in North Atlantic Holocene and Glacial Climates. Science 278: 1257–1266. doi:10.1126/science.278.5341.1257.
60. Moberg A, Sonechkin DM, Holmgren K, Datsenko NM, Karlén W (2005) Highly variable Northern Hemisphere temperatures reconstructed from low- and high-resolution proxy data. Nature 433: 613–617. doi:10.1038/nature03265.
61. Ge QS, Zheng JY, Fang XQ, Man ZM, Zhang XQ, et al. (2002) Temperature changes of winter-half-year in eastern China during the past 2000 years. Quat Sci 22: 166–173.
62. Zhang H, Wu JW, Zheng QH, Yu YJ (2003) A preliminary study of oasis evolution in the Tarim Basin, Xinjiang, China. J Arid Environ 55: 545–553. doi:10.1016/S0140-1963(02)00283-5.
63. Editorial office (2012) Witness Buddhist cultural confluence -The Tuyoq Grottoes Ruins of Shanshan County, Xinjiang. Hist Ref 22: 24–27.
64. Shu Q, Zhong W, Li C (2007) Distribution feature of ancient ruins in south edge of Tarim Basin and relationship with environmental changes and human activities. J Arid Land Resour Environ 21: 95–100.
65. Zhou XJ (1989) Probe into the historical period of Tarim Basin desertification. J Arid Reg Res 1: 9–17.
66. Fan ZL (1995) Study on the causes of formation and abandonment of ancient oasis in Taklimakan desert. Arid Zone Res Supp 12: 308–317.
67. Lin M (1996) Jingjue kingdom and ruins of Niya. Cult Relic 12: 54–55.

Ecological Structure of Recent and Last Glacial Mammalian Faunas in Northern Eurasia: The Case of Altai-Sayan Refugium

Věra Pavelková Řičánková*, Jan Robovský, Jan Riegert

Department of Zoology, Faculty of Science, University of South Bohemia, České Budějovice, Czech Republic

Abstract

Pleistocene mammalian communities display unique features which differ from present-day faunas. The paleocommunities were characterized by the extraordinarily large body size of herbivores and predators and by their unique structure consisting of species now inhabiting geographically and ecologically distinct natural zones. These features were probably the result of the unique environmental conditions of ice age ecosystems. To analyze the ecological structure of Last Glacial and Recent mammal communities we classified the species into biome and trophic-size categories, using Principal Component analysis. We found a marked similarity in ecological structure between Recent eastern Altai-Sayan mammalian assemblages and comparable Pleistocene faunas. The composition of Last Glacial and Recent eastern Altai-Sayan assemblages were characterized by the occurrence of large herbivore and predator species associated with steppe, desert and alpine biomes. These three modern biomes harbor most of the surviving Pleistocene mammals. None of the analyzed Palearctic Last Glacial faunas showed affinity to the temperate forest, taiga, or tundra biome. The Eastern part of the Altai-Sayan region could be considered a refugium of the Last Glacial-like mammalian assemblages. Glacial fauna seems to persist up to present in those areas where the forest belt does not separate alpine vegetation from the steppes and deserts.

Editor: Michael Hofreiter, University of York, United Kingdom

Funding: The study was supported by Czech Ministry of Education (MSM 6007665801) http://www.msmt.cz/index.php?lang = 2 and Czech Science Foundation (# P504/11/0454) http://www.gacr.cz/en/. The funders had no role in study design, data collection and analysis, decision to publish, or preparation of the manuscript.

Competing Interests: The authors have declared that no competing interests exist.

* E-mail: ricankova@seznam.cz

Introduction

The unique structure of Pleistocene mammalian communities has drawn the attention of scientists for many decades [1–7]. The extraordinary large body size of some Pleistocene mammals (e.g., mammoths, giant deer, or cave bear) and composition of the ice age communities have no analogies in the present-day faunas [8–9]. The Pleistocene communities consisted of species which now inhabit geographically and ecologically distinct natural zones (tundra, forest, steppe, savanna) [5,8]. Arctic reindeer and musk-ox thus lived in sympatry with e.g. steppe horse and bison and/or with forest elk and roe deer [4,10].

The reason for such a peculiar structure of Last Glacial (corresponding with Weichselian Glaciation) mammalian assemblages is probably associable with the unique environmental conditions of ice age ecosystems [8,10,11]. The non-taxonomic, ecological structure of a mammalian community (~ its guild composition according to e.g. Simberloff & Dayan [12]) is determined mainly by environmental factors such as climate, type of biome or vegetation heterogeneity [13–15]. Each type of ecosystem (e.g. woodland or arid ones) is characterized by a specific trophic-size structure of its mammalian community [5,13,16]. Mammalian communities from areas with similar regional climates tend to converge to similar community structures [15]. Historical factors also play an important role in forming community structure, but they operate on a different, probably much longer timescale than environmental factors [14–15].

The non-analogue Last Glacial communities evolved in a cold and dry continental climate, which supported highly heterogeneous vegetation and landscape structure, usually described as tundra-steppe or mammoth steppe [4,8,10,17].

The Pleistocene tundra-steppe ecosystem was quite heterogeneous locally but displayed a relatively high degree of homogeneity on the continental scale. This ecosystem covered wide areas of the northern part of the globe, thrived for approximately 100,000 years without major changes, and then suddenly went extinct about 12,000 years ago [4,10].

Surprisingly, environmental conditions similar to the Last Glacial period have been found in the Central Eurasian Altai-Sayan mountains [18]. This climatic analogue has recently been supported by biological data. Recent findings of the paleo-biome reconstruction [19–20] and pollen-analytical research [21–23] suggest that present-day Altai-Sayan landscapes could be considered the closest modern analogy to the Last Glacial environments. The area is currently inhabited by mollusc assemblages that were characteristic of full-glacial environments across large areas in Eurasia but went extinct in the regions that experienced considerable climatic change, namely in Europe [24]. Simulated paleovegetation maps based on paleoclimatic models and plant functional types have also suggested considerable stability in

Central Eurasia over the last 40,000 years [25,26]. Detailed analysis of the Altai Late Pleistocene assemblages of small mammals revealed that no significant changes occurred between the cold phase of the Pleistocene and the Holocene [27]. The environment of this region can thus be considered as conservative and stable.

In this study, we compared the ecological structure of Recent Altai-Sayan mammalian assemblages to the ecological structure of Last Glacial fauna of Altai-Sayan and several adjacent regions as well as to Recent mammalian communities from various natural zones of northern Eurasia. In order to examine the most important structural characteristics of Pleistocene assemblages, we assigned individual species according to their biome and the trophic-size categories. Given the analysis of Willis et al. [28] and Rodrigues et al.[7,29] showing that glacial vegetation consisted of steppe, tundra and forest and mammalian fauna was characterized by large herbivores and predators, we would expect that glacial communities are characterized (and differ from Recent communities) by the co-occurrence of steppe, tundra, and forest species [28], and by high a proportion of large herbivores and predators [29]. We hypothesize that, given the environmental stability of central Eurasia [18], Altai-Sayan Recent assemblages will be more similar to the glacial communities than to any other Recent community [22–23].

Materials and Methods

Regions and localities

To compare the ecological compositions of Recent and Last Glacial faunas, lists of mammalian species for 14 Recent and seven Last Glacial localities were collected (Fig. 1, see Table S1). The areas were selected in order to cover most of the Palearctic Realm above 35° N, to include well documented Last Glacial localities, and to be compatible with the WWF eco-regions (e.g., Altai-Sayan, Caucasus, Carpathian Mountains). Taking into account (i) the scarcity of paleontological localities, (ii) the general incompleteness of the fossil record, (iii) and relative homogeneity of the glacial fauna, larger regions were used as units for the analysis of Last Glacial faunas. This grouping of data, therefore, helps to average taphonomic biases [29]. The experimental mixture of modern communities tends to increase the taxonomic richness but does not significantly modify the overall ecological diversity [30].

For the analysis of Recent mammalian faunas we used smaller, ecologically homogeneous areas (see Table S1). The Recent fauna of Altai-Sayan region has been assigned to 12 areas, covering most of the region's heterogeneity (Fig. 2; Table S2).

Species

The Recent mammalian communities refer to the interval from now to approximately the 16th century AD in order to respect the IUCN definition of "recent extinction" [31] and to eliminate taxonomical uncertainties. The Last Glacial refers here to the last glacial period of the Late Pleistocene, corresponding with the Weichselian Glaciation. The Last Glacial communities were dated from approximately 125,000 to 12,000 BP, i.e. they included the time interval from MIS 5d to MIS 2 [32] which is well defined in the geological/fossil record [33,34]. The early-Holocene fauna was not included in our analyses.

Presence/absence of 379 mammalian species in Recent and Last Glacial regions were recorded.. The Recent species include extant (or extinct in the historical period) autochthonous elements but not allochthonous elements associated with human unintended or intentional activities (see Notes S1). The marine mammals (Cetacea, Pinnipedia) and island endemics (though geographically associated with the analyzed regions) were excluded from analyses. Bats (Chiroptera) have been removed from the list of examined species for trophic-size categories, following Rodríguez [7]. Domesticated mammals were not considered.

Some Recent species are not diagnosable in the fossil record owing to the lack of diagnostic morphological, cytogenetic, and/or molecular characters. We therefore fused some closely related mammalian species into single operational taxonomic units (for details see Notes S1). No permits were required for the described study, which complied with all relevant regulations.

Biome and trophic-size structure of the faunas

The ecological structure of a community is defined as the number of species in different ecological groups or categories, i.e. species with similar trophic habits, body size and ecological requirements [7]. The species were classified according to two schemes:

(i) The species' affinity to a particular biome (a group of terrestrial ecosystems with similar climates and vegetation structure): we recognized (1) tundra, (2) taiga, (3) deciduous forest, (4) steppe, (5) alpine grassland, and (6) desert species, following Duff & Lawson [35] and Wilson & Reeder [36].

(ii) Trophic and body-size categories combining information about trophic habits, locomotor abilities, microhabitat, and body size of the species examined cf. [7]): (1) aquatic predator (e.g. Eurasian otter *Lutra lutra*); (2) small terrestrial predator (e.g. red fox *Vulpes vulpes*); (3) large terrestrial predator (e.g. wolf *Canis lupus*); (6) aquatic predator of invertebrates (e.g. Russian desman *Desmana moschata*); (7) subterranean predator of invertebrates (e.g. Siberian mole *Talpa altaica*); (9) small terrestrial predator of invertebrates (e.g. Eurasian shrew *Sorex araneus*); (11) small terrestrial omnivore (e.g. Altai birch mouse *Sicista napaea*); (12) large terrestrial omnivore (e.g. brown bear *Ursus arctos*); (13) arboreal omnivore (e.g. forest dormouse *Dryomys nitedula*); (14) small terrestrial herbivore (e.g. common vole *Microtus arvalis*); (15) small sized foregut fermenter (e.g. goitered gazelle *Gazella subgutturosa*); (16) medium sized foregut fermenter (e.g. reindeer *Rangifer tarandus*); (17) large sized foregut fermenter (e.g. red deer *Cervus elaphus*); (18) small sized hindgut fermenter (e.g. mountain hare *Lepus timidus*); (20) large sized hindgut fermenter (e.g. Asiatic wild ass *Equus hemionus*); (21) subterranean herbivore (e.g. Siberian zokor *Myospalax myospalax*); (22) arboreal herbivore (e.g. Eurasian red squirrel *Sciurus vulgaris*); (23) aquatic herbivore (e.g. European water vole *Arvicola amphibius*).

Data analyses

To visualize the overall similarity of Recent and Last Glacial areas according to presence/absence of the mammalian species, we used all 21 regions/localities as "samples" and ecological categories as "species" for Principal Component Analysis (PCA). We also performed analyses complementary to PCA using non-metric multidimensional scaling (NMDS) with Bray-Curtis dissimilarity indices (CANOCO for Windows software [37]). We expressed the proportion of species associated with each biome and trophic-size category as percentages in the input data matrix (see Table S3). The percentages were log-transformed and standardized by species. The sensitivity of PCA was controlled by analyzing datasets after removing rare categories (biomes: tundra and desert, trophic-size: aquatic predator, aquatic predator of invertebrates, subterranean predator of invertebrates, arboreal omnivore, small-sized foregut fermenter, subterranean herbivore).

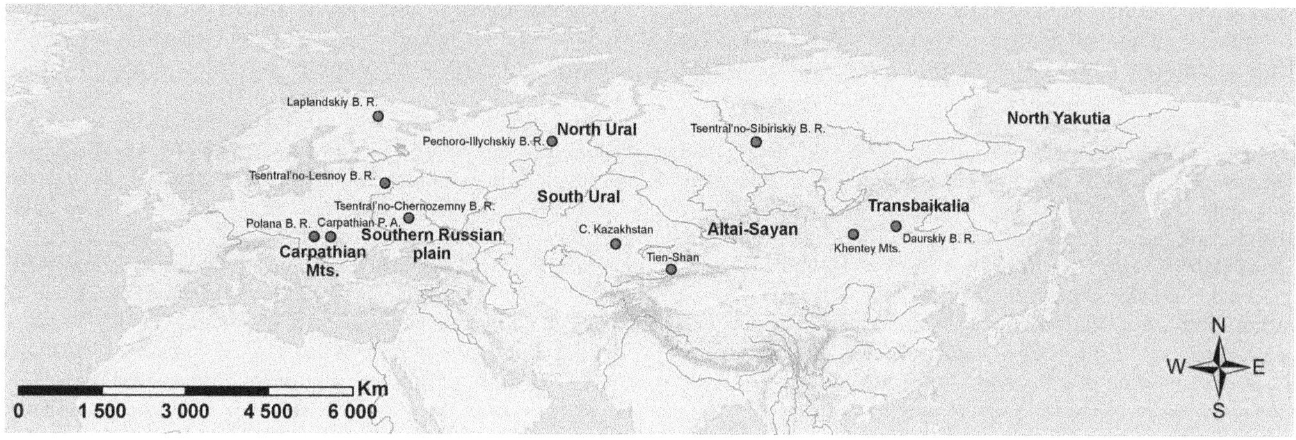

Figure 1. Location map of Recent and Pleistocene Palearctic localities used in the analyses. Abbreviations: B. R. - Biosphere Reserve; P. A. - Protected Area; Mts. - mountains

Results

(1) Faunas classified according to species-biome associations

Composition of Last Glacial faunas was characterized by the co-occurrence of steppe, desert, alpine, and tundra species (Fig. 3). The first axis of the ordination space was determined by the steppe/desert-to-taiga gradient, whereas presence/absence of the alpine faunas was strongly correlated with the second axis. The first two axes explain 89.9% of variance (see Table S4 for detailed results and comparison with NMDS analyses). The Last Glacial faunas were scattered in the area between the steppe/desert and

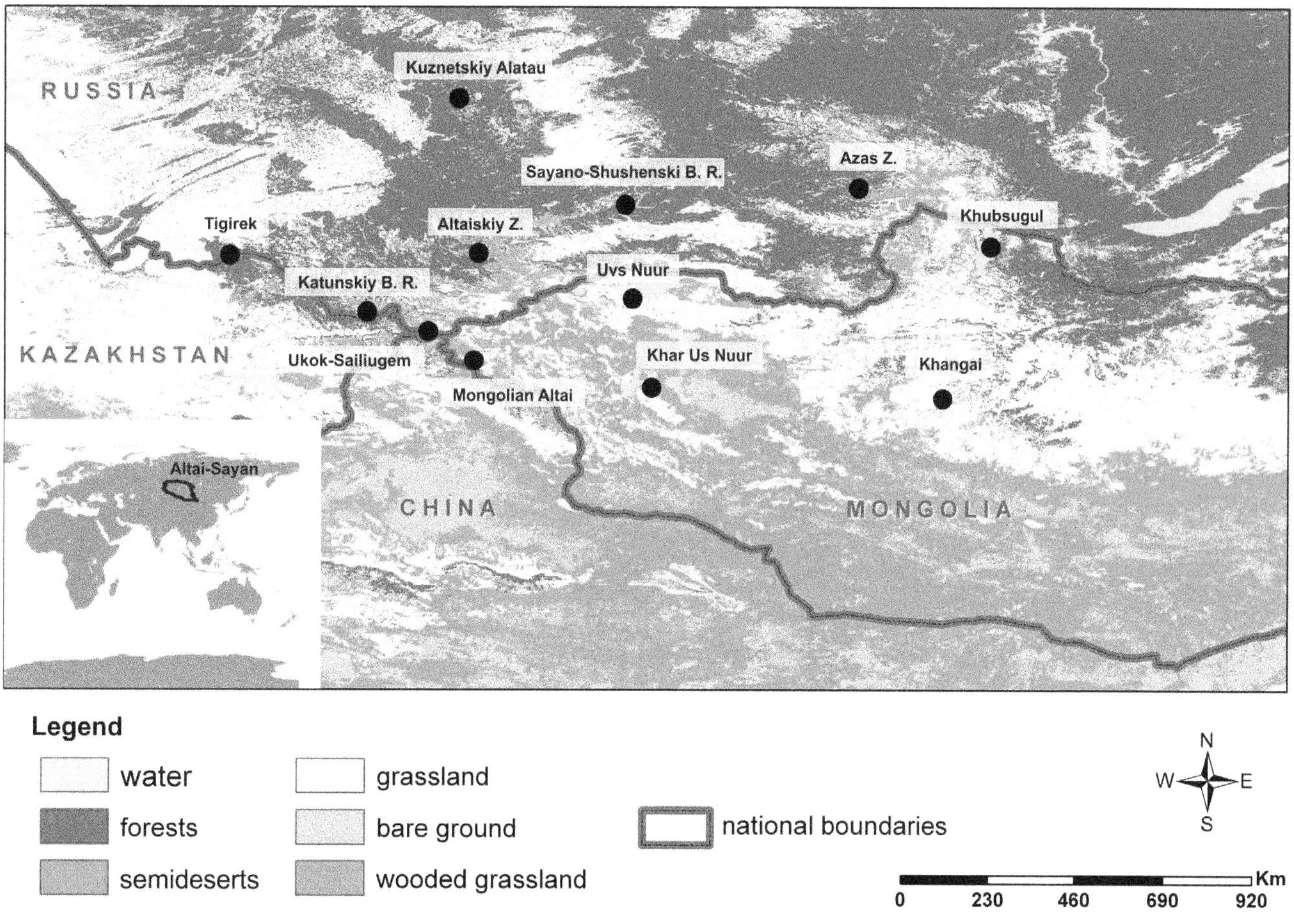

Figure 2. Location map of examined localities in the Altai-Sayan region. Abbreviations: B. R. - Biosphere Reserve; Z. – Zapovednik (National Park). Dashed line: state borders, solid line: Altai-Sayan region borders according to WWF ecoregion.

alpine faunas, with no affinities to the tundra and taiga ones. The Last Glacial faunas of North Yakutia and North Ural were characterized by the presence of alpine faunal elements, and did not resemble any of the Recent localities. The faunas of Recent eastern Altai-Sayan localities (with high proportion of grasslands) were characterized by the co-occurrence of steppe, desert, and alpine species (as well as their Last Glacial counterparts), and were more similar to some Last Glacial faunas than to faunas of any other Recent areas. The NMDS analysis and PCA without rare biomes showed similar results to PCA (see Fig. S1 and Fig. S2).

Comparison of the proportion of species assigned to biome categories between Altai-Sayan and North Yakutia regions (Fig. 4a) revealed that the main difference between the regions consisted of a high proportion of forest and taiga species among Holocene immigrant species in North Yakutia and desert species in Altai-Sayan.

(2) Faunas classified according to trophic-size structure

The gradient between Last Glacial and Recent faunas along the first axis (Fig. 5) was determined by the ratio of insectivores and aquatic predators to large ungulates and predators. The Last Glacial faunas were characterized by a high proportion of the large-sized terrestrial predators (e.g., lion, wolf), large-sized foregut fermenters (e.g., steppe bison, deer, camel), and large-sized hindgut fermenters (e.g., mammoth, horse). Recent communities were characterized by the presence of small terrestrial predators of invertebrates, aquatic predators, subterranean predators of invertebrates, and aquatic predators of invertebrates (Fig. 5a). The occurrence of arboreal and small terrestrial omnivores and subterranean herbivores was positively correlated, while the occurrence of small-sized foregut fermenters was correlated negatively to the above mentioned cluster along the second axis (Fig. 5a). The first two axes explain 63.4% of variance. (see Table S4 for detailed results and comparison with NMDS analyses).

All Last Glacial localities were situated on the right side of the diagram (Fig. 5b), together with the Recent faunas from Khar Us Nuur NP and Ukok-Sailiugem localities (both E Altai). The Khar Us Nuur NP is more similar to the Last Glacial Altai-Sayan fauna than to any Recent locality. The Recent fauna of Tigirek (SW Altai) differs from the other Altai-Sayan localities, as it is situated closer to the European temperate-forest localities (Fig. 5b) characterized by a larger proportion of small terrestrial and arboreal omnivores (Fig. 5a). The Last Glacial assemblage of the Carpathian Mountains was more similar to the Recent community of Central Kazakhstan than to any other assemblage. The NMDS analysis and PCA without rare trophic-sizes showed similar results to PCA (see Fig. S3 and Fig. S4).

The main difference between the proportion of species assigned to trophic-size categories in Altai-Sayan and North Yakutia regions (Fig. 4b) consisted of a high proportion of arboreal

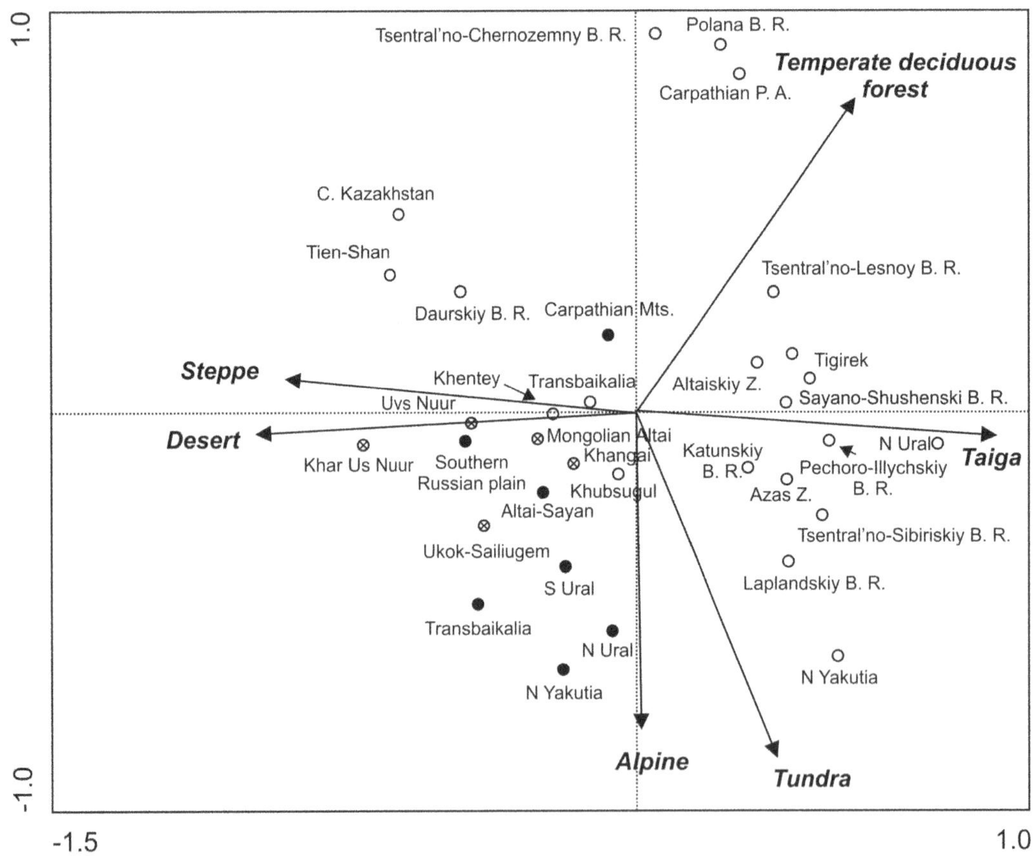

Figure 3. Projection scores of mammalian communities for studied localities (PCA). Species were classified according to their presence in a particular biome. The first two axes describe 89.8% of variance. Open circles – Recent assemblages; crossed circles – Recent eastern Altai assemblages; full circles – Last Glacial assemblages

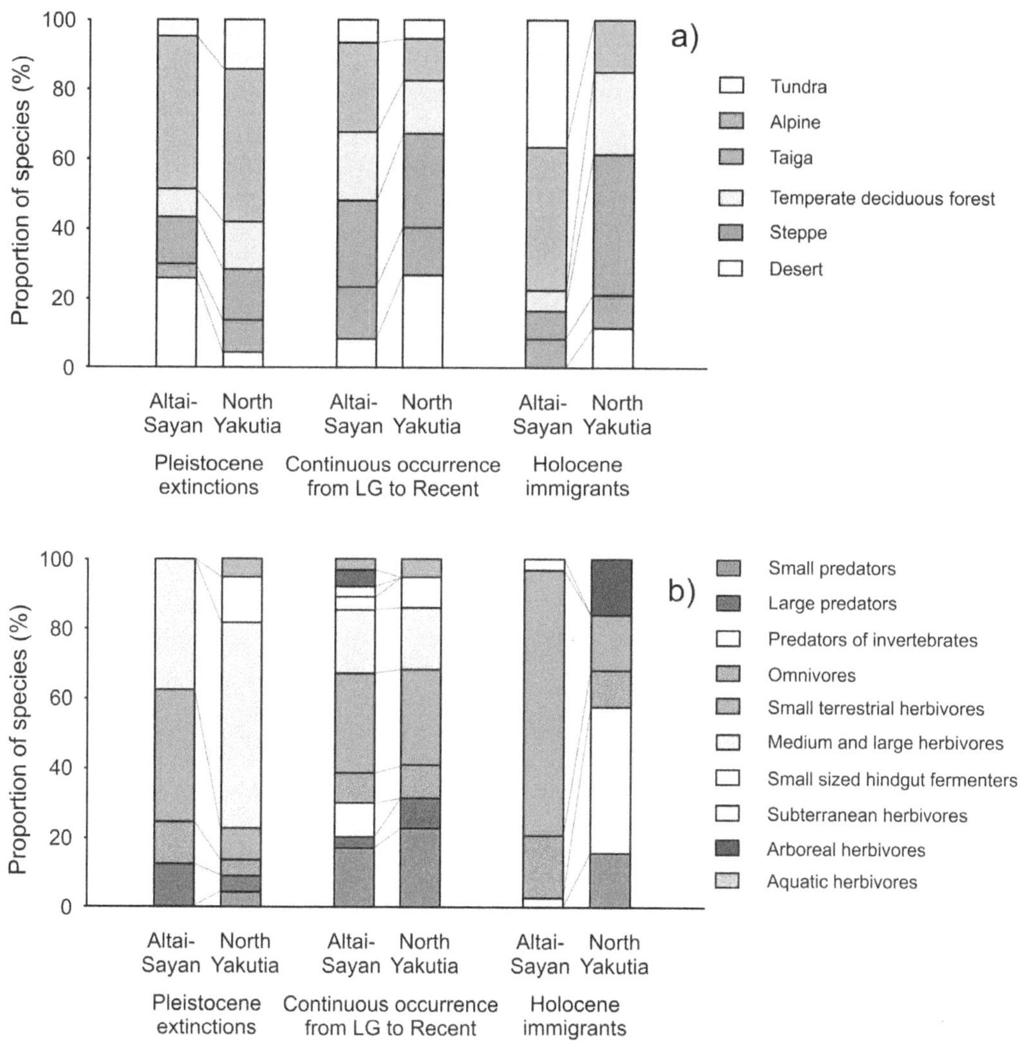

Figure 4. The differences in ecological structure between Altai-Sayan and North Yakutia faunas. Comparison of the proportion of species assigned to biome (a) and trophic-size (b) categories between the regions.

herbivores and predators of invertebrates among Holocene immigrant species in North Yakutia.

Discussion

We found a marked similarity in ecological structure between Recent Altai-Sayan mammalian assemblages and Last Glacial paleocommunities. The ecological structure of mammalian communities confirmed the possible persistence of the Last Glacial-like fauna in the present-day Altai-Sayan region, as the Recent assemblages of eastern Altai-Sayan (i.e. Ukok-Sailiugem, Khar Us Nuur, and Uvs Nuur) were more similar to various glacial localities than to the communities of any other Recent areas. Glacial communities have close modern analogues in the three eastern Altai-Sayan areas where e.g. reindeer and saiga antelope still live in sympatry [38].

Our results are congruent with other evidence supporting the persistence of Pleistocene biota in the Altai-Sayan region. Rodríguez [7] reported a similarity between the ecological structure of present-day mammalian communities from Central Eurasia (Uvs Nuur and Great Gobi) and Iberian Pleistocene communities. The Altai Mountains are an important refugium for full-glacial snail faunas, as recently documented by Horsák et al. [24] and Hoffmann et al. [39]. Kuneš et al. [22] and Pelánková & Chytrý [23] demonstrated a close similarity between glacial pollen samples from central Europe and modern surface-pollen spectra from the Altai-Sayan region. Fossil pollen spectra from the Altai and adjacent regions indicate little difference between modern biomes of this region and those reconstructed for the Last Glacial Maximum [20]. Similar conditions possibly occur in climatically stable areas of North American cold deserts. Fossil evidence from the Great Basin indicate that Pleistocene plant assemblages are comparable with the modern ones [40].

According to vegetation studies, three major biomes occurred widely in the Pleistocene: steppe, tundra, and taiga [28,41–42]. In contrast, the composition of the Last Glacial (and Recent eastern Altai-Sayan) mammalian faunas was characterized by the co-occurrence of steppe, desert and alpine species. These three modern biomes harbor most of the surviving Last Glacial mammals. The importance of the desert biome was probably more pronounced in the examined region of southern Russian plain and central Asia in comparison to the northern and western part of Eurasia.

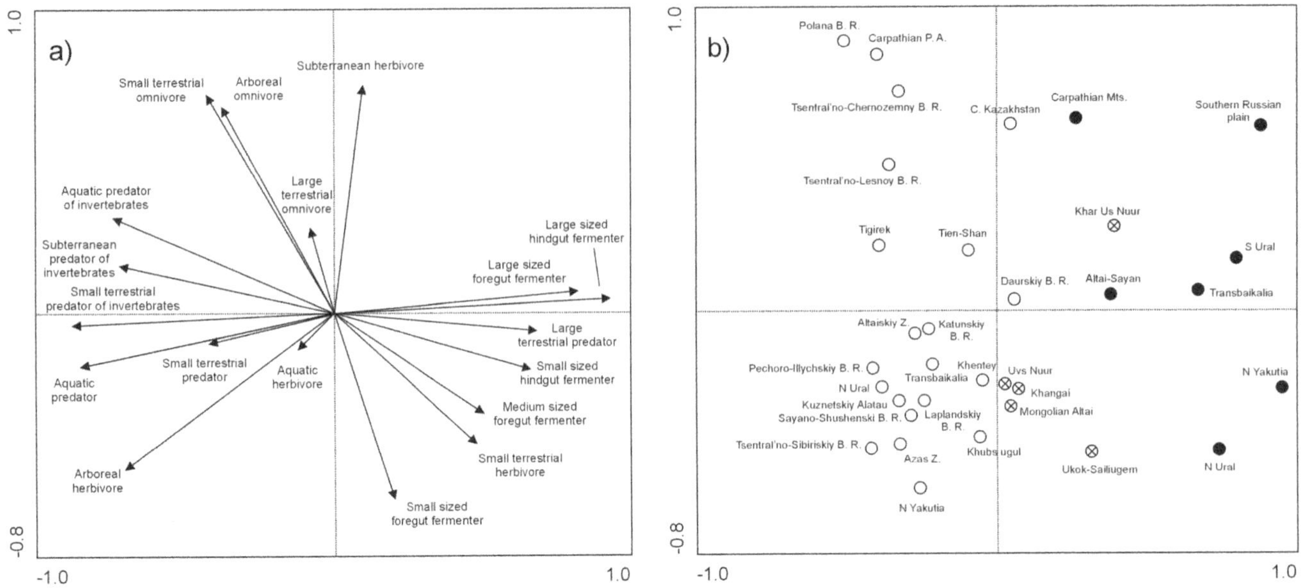

Figure 5. Projection scores of trophic-size community structure (PCA). The first two axes describe 63.4% of variance. Open circles – Recent assemblages; crossed circles – Recent eastern Altai assemblages; full circles – Last Glacial assemblages.

Glacial-like communities still persist in areas where the forest belt does not separate alpine vegetation from the steppes and (semi)deserts. High aridity in eastern Altai-Sayan restricts forests to isolated patches with higher soil moisture [43–44]. Holocene fragmentation of the alpine grasslands to isolated patches surrounded by forests could have lead to the extinction of some large mammals typical of the mammoth steppe. Sharp separation of originally intermixed faunal elements into distinct biomes seems to be the major pattern of the Last Glacial/Holocene transition [8,45].

Most previous attempts to find a modern analogue of the mammoth steppe have been focused on regions of the Arctic tundra, i.e., Yakutia, NE Russia, and Alaska [8,9,46,47]. However, present-day faunas of the Arctic regions have diverged greatly in their ecological structure from Last Glacial ones, showing more affinity to the taiga biome. The divergent ecological structure of modern Yakutian fauna is determined mainly by Holocene immigrant species, characterized by a high proportion of arboreal herbivores and predators of invertebrates associated with forest and taiga biomes (Fig. 4).

The modern arctic tundra is characterized by low productivity and relatively high homogeneity [4,25], in contrast to the alpine biome where tundra vegetation occurs in association with forests and steppes, e.g. [48–49]. During the Last Glacial, tundra vegetation was confined to places with higher precipitation or lower evapotranspiration, e.g. in the mountains or at non-glaciated higher latitudes, often in the close vicinity of steppe [20].

Taiga and temperate forest species occurred in all of the examined Last Glacial assemblages; however their percentage was very low, with the exception of the Last Glacial assemblage of the Carpathian Mountains. This assemblage holds a special position as the site resembles Recent rather than any other Last Glacial community. The Carpathian Mountains of eastern Europe possibly represented a glacial refugium of forest vegetation and forest-dwelling animals, as suggested by Sommer & Nadachowski [50], Jankovská and Pokorný [21], Markova et al. [51], Willis et al. [28], and Willis & van Andel [52].

The main difference in trophic-size structure between present-day and fossil assemblages is the higher richness of large mammals and proportionally lower richness of small mammals in the latter. The high proportion of large herbivores observed in the Last Glacial as well as in some present-day communities is generally typical of areas with low tree cover [15] and could be further supported by the year-round availability of high quality food in the glacial steppe [10,53]. Fossil assemblages are probably biased against small mammal species owing to fossil record incompleteness [54,7,29,55–56]. However, the low proportion of small mammals cannot be considered completely artifactual because the present-day assemblages from the eastern Altai-Sayan region show very similar composition to Last Glacial communities. The low species richness of insectivores and aquatic predators in Last Glacial-like Altai-Sayan assemblages (and probably in other Last Glacial assemblages as well) could be due to the dry and cold climate associated with permafrost which strongly limits insectivores' food sources [57, 48 58]. These results were confirmed using two different ecological classifications and using two independent statistical methods (NMDS, PCA). Moreover, our fossil datasets are based on large areas and a long time period to avoid taphonomic biases in species occurrence. Therefore, we suggest that potential bias in the Pleistocene data subset cannot significantly affect our results.

In general, Pleistocene assemblages were characterized by the occurrence of large herbivore and predator species associated with steppe, desert and alpine biomes. Mammalian paleocommunities classified according to biome type are relatively homogeneous, confirming the view of the mammoth steppe as a single Last Glacial biome [10]. In contrast to biome classification, the trophic-size structure of mammalian paleocommunities shows a degree of heterogeneity, comparable to the Recent localities. The trophic-size structure of communities is probably less influenced by historical factors [16]. Historical processes are considered to be the main factor promoting differences between communities from similar environments [15,29].

Our results open new research possibilities for many aspects of Quaternary paleoecology. The Altai-Sayan region offers a possibility to study factors shaping the structure of so called non-analogue communities and explore vegetation and faunal changes associated with Pleistocene/Holocene transition. Research of soil nutrient availability and cycling in glacial-like localities can provide an insight into the carrying capacity of ice age ecosystems supporting numerous large herbivore species. Modeling the impact of climate changes on the glacial-like landscape may elucidate the process of biome diversification in the Holocene. Our result can be confirmed by thorough paleontological research of the as yet unexplored eastern Altai-Sayan region as well as by phylogeographical analyses of typical glacial species (e.g. steppe lemming or pika) including Altai-Sayan populations.

Supporting Information

Figure S1 The projection scores of studied localities according to biome classification (NMDS analysis based on Bray-Curtis dissimilarity indices).

Figure S2 The projection scores of studied localities according to biome classification without rare categories (PCA analysis).

Figure S3 The projection scores of studied localities according to trophic-size classification (NMDS analysis based on Bray-Curtis dissimilarity indices).

Figure S4 The projection scores of studied localities according to trophic-size classification without rare categories (PCA analysis).

Table S1 Paleartic regions used in the analyses with associated references.

Table S2 Examined localities of Altai-Sayan region with associated references.

Table S3 Percentages of biomes and trophic-size categories for regions and localities (dataset).

Table S4 Component loadings of the PCA and NMDS analyses.

Notes S1 Taxonomic notes with associated references.

Acknowledgments

We thank David Storch, Martin Konvička, Daniel Frynta, Ivan Horáček, Vojtěch Novotný, Michal Horsák, Jan Zrzavý, Milan Chytrý, John Stewart and anonymous referee who helped greatly with useful comments. We thank Ingrid Steenbergen for language corrections and Martin Hais for maps.

Author Contributions

Conceived and designed the experiments: VPŘ JRi. Analyzed the data: JRi. Wrote the paper: VPŘ Jro JRi. Performed the literature search and data extraction: JRo VPŘ.

References

1. Romer AS (1933) Pleistocene vertebrates and their bearing on the problem of human antiquity in North America. In: Jenness D, editor. The American Aborigines: Their Origin and Antiquity. Toronto: University of Toronto Press. pp. 47–84.
2. Hibbard CW (1949) Pleistocene vertebrate paleontology in North America. Geol Soc Am Bull 60: 1417–1428.
3. Kurtén B (1968) Pleistocene mammals of Europe. London: Weidenfeld and Nicolson. 352 p.
4. Guthrie RD (1984) Mosaics, allelochemics and nutrients. An ecological theory of Late Pleistocene megafaunal extinctions. In: Martin PS, Martin RG, editors. Quaternary extinctions: A prehistoric revolution. Tuscon: University of Arizona Press. pp. 259–298.
5. Vereshchagin NK, Baryshnikov GF (1992) The ecological structure of the "Mammoth Fauna" in Eurasia. Ann Zool Fennici 28: 253–259.
6. Van Kolfschoten T (1995) On the application of fossil mammals to the reconstruction of the palaeoenviroment of northwestern Europe. Acta Zool Cracov 38: 73–84.
7. Rodríguez JS (2004) Stability in Pleistocene Mediterranean mammalian communities. Palaeogeogr Palaeoclimatol Palaeoecol 207: 1–22.
8. Guthrie RD (1990) Frozen fauna of the Mammoth steppe: The story of Blue Babe. Chicago: University of Chicago Press. 323 p.
9. Musil R (1999) The enviroment in the last Glacial on the territory of Moravia. Acta Musei Moraviae 84: 161–186.
10. Guthrie RD (2001) Origin and causes of the mammoth steppe: a story of cloud cover, woolly mammal tooth pits, buckles, and inside-out Beringia. Quaternary Review 20: 549–574.
11. Stewart JR (2008). The progressive effect of the individualistic response of species to Quaternary climate change: an analysis of British mammalian faunas. Quat Sci Rev 27: 2499-2508
12. Simberloff D, Dayan T (1991) The guild concept and the structure of ecological communities. Annu Rev Ecol Syst 22: 115–143.
13. Mendoza M, Goodwin B, Criado C (2004) Emergence of community structure in terrestrial mammal-dominated ecosystems. J Theor Biol 230: 203–214.
14. Rodríguez J (2006) Structural continuity and multiple alternative stable States in Middle Pleistocene European mammalian communities. Palaeogeogr Palaeoclimatol Palaeoecol 239: 355–373.
15. Louys J, Meloro C, Elton S, Ditchfield P, Bishop LC (2011) Mammal community structure correlates with arboreal heterogeneity in faunally and geographically diverse habitats: implications for community convergence. Global Ecol Biogeogr 20: 717–729.
16. Mendoza M, Janis CJ, Palmqvist P (2005) Ecological patterns in the trophic-size structure of large mammal communities: a 'taxon-free' characterization Evol Ecol Res 7: 505–530.
17. Blinnikov MS, Gaglioti BV, Walker DA, Wooller MJ, Zazula GD (2011) Pleistocene graminoid-dominated ecosystems in the Arctic. Quat Sci Rev 30: 2906–2929.
18. Frenzel B, Pécsi M, Velichko AA (1992) Atlas of paleoclimates and paleoenvironments of the Northern Hemisphere. JenaNew York: Geographical Research Institute, Hungarian Academy of Sciences Budapest and Gustav Fischer Verlag Stuttgart. 153 p.
19. Tarasov PE, Guiot J, Cheddadi R, Andreev AA, Bezusko LG, et al. (1999) Climate in northern Eurasia 6000 years ago reconstructed from pollen data. Earth Planet Sci Lett 171: 635–645.
20. Tarasov PE, Volkova VS, Webb III T, Guiot J, Andreev AA, et al. (2000) Last glacial maximum biomes reconstructed from pollen and plant macrofossil data from northern Eurasia. J Biogeogr 27: 609–620.
21. Jankovská V, Pokorný P (2008) Forest vegetation of the last full-glacial period in the Western Carpathians (Slovakia and Czech Republic). Preslia 80: 307–324.
22. Kuneš P, Pelánková B, Chytrý M, Jankovská V, Pokorný P, et al.(2008) Interpretation of the last-glacial vegetation of eastern-central Europe using modern analogues from southern Siberia. J Biogeogr 35: 2223–2236.
23. Pelánková B, Chytrý M (2009) Surface pollen-vegetation relationships in the forest-steppe, taiga and tundra landscapes of the Russian Altai Mountains. Rev Palaeobot Palyno 157: 253–265.
24. Horsák M, Chytrý M, Pokryszko BM, Danihelka J, Ermakov N, et al. (2010) Habitats of relict terrestrial snails in southern Siberia: lessons for the reconstruction of palaeoenvironments of full-glacial Europe. J Biogeogr 37: 1450–1462.
25. Allen JRM, Hickler T, Singarayer JS, Sykes MT, Valdes PJ, et al. (2010) Last glacial vegetation of northern Eurasia. Quat Sci Rev 29: 19–20.
26. Huntley B, Allen JRM, Collingham YC, Hickler T, Lister AM, jet al. (2013) Millennial climatic fluctuations are key to the structure of Last Glacial ecosystems. PloS ONE 8: e61963. doi: 10.1371/journal.pone.0061963
27. Agadjanian AK, Serdyuk NV (2005) The history of mammalian communities and paleogeography of the Altai Mountains in the Paleolithic. Paleontological Journal 39: 645–821.
28. Willis KJ, Rudner E, Sumegi P (2000) The full-glacial forests of central and southeastern Europe. Quaternary Research 53: 203–213.
29. Rodríguez J, Hortal J, Nieto M (2006) An evaluation of the influence of environment and biogeography on community structure: the case of Holarctic mammals. J Biogeogr 33: 291–303.
30. Fur SL, Fara E, Vignaud P (2011) Effect of simulated faunal impoverishment and mixture on the ecological structure of modern mammal faunas: Implications for the reconstruction of Mio-Pliocene African palaeoenvironments. Palaeogeogr Palaeoclimatol Palaeoecol 305: 295–309.

31. IUCN Red List Categories and Criteria version 3.1. Available: http://www.iucnredlist.org/technical-documents/categories-and-criteria/2001-categories-criteria. Accessed January 2012.
32. Shackleton NJ (1969) The last interglacial in the marine and terrestrial records. Proc Biol Sci 174:135–154.
33. Williams DF, Thunell RC, Tappa E, Rio R, Raffi I (1988) Chronology of the Pleistocene isotope record 0-1.88 m.y. B.P. Palaeogeogr Palaeoclimatol Palaeoecol 64: 221–240.
34. Horáček I, Ložek V (1988) Palaeozoology and the Mid–European Quaternary past: scope of the approach and selected results. Rozpravy ČSAV, řada matematických a přírodních věd 98: 1–102.
35. Duff A., Lawson A (2004) Mammals of the World: A checklist. London: A & C Black. 312 p.
36. Wilson DE, Reeder D-AM (2005) Mammal species of the World. A taxonomic and geographic reference. Baltimore: Johns Hopkins University Press. 2142 p.
37. Ter Braak CJF, Šmilauer P (2012) Canoco reference manual and user's guide: software for ordination, version 5.0. (Ithaca, NY, Microcomputer Power).
38. Yudin BS, Galkina LI, Potapkina AF (1979) Mammal s of the Altai-Sayan mountainous part. Novosibirsk: Nauka. 296 p. (In Russian).
39. Hoffman MH, Meng S, Kosachev PA, Terechina TA, Silanteva MM (2011) Land snail faunas along an enviromental gradient in the Altai mountains (Russia). J Mollusan Stud 77: 76–86.
40. Wilson JS, Pitts JP (2010) Illuminating the lack of consensus among descriptions of earth history data in the North American deserts: a resource for biologists. Prog in Phys Geogr 34: 419–441.
41. Jankovská V, Chromý P, Nižnianská M (2002) "Šafárka" - first palaeobotanical data on vegetation and landscape character of Upper Pleistocene in West Carpathians (North East Slovakia). Acta Palaeobot 42: 29–52.
42. Simakova AN (2006) The vegetation of the Russian Plains during the second part of the Late Pleistocene (33–18 ka). Quat Int 149: 110–114.
43. Hoffmann MH, Telyatnikov MYu., Ermakov N (2001) Phytogeographical analysis of plant communities along an altitudinal transect through the Kuraiskaya basin (Altai, Russia). Phytocoenologia 31: 401–426.
44. Dulamsuren C, Hauck M, Muhlenberg M (2005) Ground vegetation in the Mongolian taiga forest-steppe ecotone does not offer evidence for the human origin of grasslands. Applied Vegetation Science 8: 149–154.
45. Stewart JR, Lister AM, Barnes I, Dalén L (2010) Refugia revisited: individualistic responses of species in space and time. Proc. R. Soc. B 277: 661-671.
46. Zimov SA, Chuprynin VI, Oreshko AP, Chapin III FS, Reynolds JF, cet al. (1995) Steppe-tundra transition: a herbivore-driven biome shift at the end of the Pleistocene. Am Nat 146: 765–794.
47. Boeskorov GG (2006) Arctic Siberia: refuge of the Mammoth fauna in the Holocene. Quat Int 142–143: 119–123.
48. Smith RL (1986) Elements of Ecology. Second Edition. Harper & Row Publishers 704 p.
49. Wells AG, Rachlow JL, Garton EO, Rice CG (2012) Mapping vegetation communities across home ranges of mountain goats in the North Cascades for conservation and management. Applied Vegetation Science 15: 560–570.
50. Sommer RS, Nadachowski A (2006) Glacial refugia of mammals in Europe: evidence from fossil record. Mamm Rev 36: 251–265.
51. Markova AK, Simakova AN, Puzachenko AYu (2009). Ecosystems of Eastern Europe at the time of maximum cooling of the Valdai glaciation (24–18 kyr BP) inferred from data on plant communities and mammal assemblages. Quat Int 201: 53–59.
52. Willis KJ, van Andel TH (2004). Trees or no trees? The enviroments of central and eastern Europe during the Last Glaciation. Quat Sci Rev 53: 203–213.
53. Faith JT (2011) Late Pleistocene climate change, nutrient cycling, and the megafaunal extinctions in North America. Quat Sci Rev 30: 13–14.
54. Alberdi MT, Caloi L, Dubrovo I, Palombo MR, Tsoukala E (1998) Large mammal faunal complexes and palaeoenviromental changes in the late Middle and Late Pleistocene: a preliminary comparison between the Eastern European and the Mediterranean areas. Geologija 25: 8–19.
55. Brugal J-P, Croitor R (2007) Evolutoin, ecology and biochronology of herbivore associations in Europe during the last 3 million years. Quaternaire 18: 129–151.
56. Meloro C, Clauss M (2012). Predator-prey biomass fluctuations in the Plio-Pleistocene. Palaios 27: 90–96.
57. Smidt S, Oswood MW (2002). Landscape patterns and stream reaches in the Alaskan taiga forest: potential roles of permafrost in differentiating macroinvertebrate communities. Hydrobiologia 468: 1–3.
58. McCain CM (2007) Could temperature and water availability drive elevational species richness patterns? A global case study for bats. Global Ecol Biogeogr 16: 1–13.

15

180,000 Years of Climate Change in Europe: Avifaunal Responses and Vegetation Implications

Sandra Ravnsbæk Holm*, Jens-Christian Svenning

Ecoinformatics and Biodiversity, Department of Bioscience, Aarhus University, Aarhus C, Denmark

Abstract

Providing an underutilized source of information for paleoenvironmental reconstructions, birds are rarely used to infer paleoenvironments despite their well-known ecology and extensive Quaternary fossil record. Here, we use the avian fossil record to investigate how Western Palearctic bird assemblages and species ranges have changed across the latter part of the Pleistocene, with focus on the links to climate and the implications for vegetation structure. As a key issue we address the full-glacial presence of trees in Europe north of the Mediterranean region, a widely debated issue with evidence for and against emerging from several research fields and data sources. We compiled and analyzed a database of bird fossil occurrences from archaeological sites throughout the Western Palearctic and spanning the Saalian-Eemian-Weichselian stages, i.e. 190,000–10,000 years BP. In general, cold and dry-adapted species dominated these late Middle Pleistocene and Late Pleistocene fossil assemblages, with clear shifts of northern species southwards during glacials, as well as northwards and westwards shifts of open-vegetation species from the south and east, respectively and downwards shifts of alpine species. A direct link to climate was clear in Northwestern Europe. However, in general, bird assemblages more strongly reflected vegetation changes, underscoring their usefulness for inferring the vegetation structure of past landscapes. Forest-adapted birds were found in continuous high proportions throughout the study period, providing support for the presence of trees north of the Alps, even during full-glacial stages. Furthermore, the results suggest forest-dominated but partially open Eemian landscapes in the Western Palearctic, including the Northwestern European subregion.

Editor: Michael Hofreiter, University of York, United Kingdom

Funding: JCS was supported by the European Research Council (ERC-2012-StG-310886-HISTFUNC). The authors additionally consider this article a contribution by Centre for Biocultural History, funded by Aarhus University and Aarhus University Research Foundation under the AU IDEAS program. The funders had no role in study design, data collection and analysis, decision to publish, or preparation of the manuscript.

Competing Interests: The authors have declared that no competing interests exist.

* E-mail: s.ravnsbaekholm@gmail.com

Introduction

A main response of species to climatic changes has been to move by niche tracking i.e. following the shifting climate to remain in favorable living conditions. Numerous examples of this response are reported in the literature; both from the present and past, on short time scales and on long geological time scales, and covering a wide range of organism groups [1–9]. Studies of past communities and range shifts provide important insights for understanding and predicting current and future biotic responses to the ongoing global warming and for guiding conservation management in the face of climate change (cf. [10]).

Past climatic cycles of the Quaternary have had great impact on species ranges. The Late Pleistocene in the Western Palearctic has attracted much attention in relation to the effects of climate change on biotic dynamics. This period covers the Last (Eemian) Interglacial and the Last (Weichselian) Glacial terminating with the beginning of the present interglacial (Holocene). Stable oxygen isotope variations in ice cores and deep-sea sediments have provided a detailed record of past changes in global climate [11–13], but regional vegetation responses to these changes remains a matter of debate. Generally the vegetation oscillated on a north-south and east-west gradient between two vegetation extremes; coniferous and deciduous forests under warm, oceanic conditions, and open subarctic steppe-tundra under cold, continental conditions [14]. Paleontological studies indicate that many organisms shifted southwards during glacial periods, as a response to cold temperatures and aridification [15]. Similarly, studies on present range shifts show that the opposite is happening now in response to current global warming [4,16]. Palynological evidence show that during the warm Eemian there was a drop in open vegetation and an increase of forest, which terminated with an increasing dominance of first cold-tolerant tree species and finally herbaceous species, marking the breakup of European forests in response to the beginning of colder conditions [17–20]. This view of a forest-covered interglacial Europe is widely accepted, albeit the exact vegetation structure and the degree of openness is debated (e.g. [21]). The extent of forest during the cold stages of the Weichselian has attracted more attention from paleoecologists. The focal point of discussion is the contradictory evidence from pollen studies versus macrofossil studies and climate reconstruction models. Traditionally, most forest species were believed to have survived the full-glacial periods in mountainous belts of favorable conditions in the Iberian, Italian and Balkan peninsulas, the so-called glacial refugia hypothesis [22–25]. This was supported by palynological studies indicating an absence of forest north of the Alps [18,26], even during some of the milder periods [20], and phylogeographic studies reporting genetic patterns indicating isolation of temperate species in the southern refugia followed by post-glacial recolonization of the north [27,28]. This traditional view has been challenged in recent years by tree charcoal remains in Central and Eastern Europe, indicating the presence of boreal forest tree species and even thermophilous deciduous tree species during full-glacial conditions [29,30]. The presence of more northerly glacial forests is supported by other studies on plant [31] and vertebrate fossils [32], as well as genetic evidence from boreal trees [33] and temperate animals [34], as well as distribution hind-casting and vegetation simulation studies, suggesting a possibility for the presence of tree species at mid-to-high latitudes in Europe even during the Last Glacial Maximum (LGM) [35,36]. Notably, forest tree species could have survived at low densities in discontinuous so-called cryptic refugia [25,31,37].

This study adds to these current discussions on the European Late Pleistocene by investigating the paleoecological implications of the Western Palearctic avian fossil record, with birds being a little-studied group in this context. Some earlier studies using avian fossils as proxies for past environments exist, but mainly on local scales [38–43], albeit with some regional-scale studies [44–47]. By assuming that stratigraphic bird assemblages reflect the relative weight of different biotypes over time, most of these studies have provided snapshots of local environments in a given time and a continental-scale paleoecological study remains to be done. The use of birds as proxies has several advantages: 1) They are mostly easily identified to species level due to morphological skeletal characteristics [48,49], though passerines are often perceived as a group where species-level identification can be difficult. 2) Most birds respond mainly to auditory and visual stimulants in a relatively similar manner as humans, making habitat parameters easily defined [48], an advantage that is strengthened by the fact that most bird species have specific – and in the Western Palearctic, well-known – habitat and often vegetation-defined requirements [46]. 3) The great mobility of most bird species furthermore make them more likely to fill their potential range relatively well as they can more easily track habitat shifts caused by climate change, than many less mobile organism groups with slower migration rates [50]. The main focus in this study is on the extent to which climate-driven range shifts have occurred in the period 190,000–10,000 years BP, either directly via abiotic effects or indirectly via vegetation effects. This period covers the Eemian and Weichselian as well as the latest part of the penultimate glacial, the Saalian (190,000–130,000 years BP) as well as the earliest Holocene (11,700–10,000 years BP). Furthermore, the implications for the much-discussed issues of the extent of forest cover in Europe during full-glacial periods and – vice versa – the degree of vegetation openness during the interglacial are also considered. The following specific questions are investigated: Firstly, is there direct evidence for range shifts in the Western Palearctic avifauna in the past Saalian-Eemian-Weichselian time span, as seen in other organism groups? Secondly, to what extent can these shifts be linked to the known past changes in climate and vegetation? More specifically, species associated with cold, dry and open vegetation environments are expected to dominate assemblages from glacial stages, while species of warm, humid and wooded environments are expected to dominate during the interglacial stage. Thirdly, how does the evidence provided by the avian fossil record relate to the ongoing debate on the degree of woodland presence outside traditional southern glacial refugia during the glacial stages? Fourth and lastly, bird compositional dynamics and questions 1–3are also investigated for the Northwestern European subregion within the Western Palearctic. There are clear subregional differences in the strength of the glacial-interglacial climate changes, and the Northwestern subregion has experienced relatively large climatic fluctuations during the Late Pleistocene [24,36,51]. How has these changes affected the bird communities in this area compared to the entire Western Palearctic?

Methods

Database

A database on Western Palearctic bird fossils for the last 190,000–10,000 years BP was compiled based on the information contained in the monograph "Pleistocene Birds of the Palearctic: A Catalogue" and its revised supplement by Tyrberg [52,53]. This collection makes up the most extensive synthesis of avian fossils in Pleistocene Europe and contains information on fossiliferous sites collected from a large quantity of original work and secondary references.

The database was comprised of information on the Western Palearctic, defined as Europe south to the Mediterranean and east along the Caucasus, bordered to the west by the Atlantic Ocean and to the east by the 40 degree longitude. Coordinates for each site was obtained through extensive internet and journal searches, and compared with the maps provided by Tyrberg [52]. They were recorded as decimal degrees and mapped using Quantum GIS version 1.8.0 with a European base map obtained from Natural Earth [54] (Fig. 1). The sites and sometimes even individual stratigraphic layers of sites included in Tyrberg's catalogues have been dated using a variety of different methods ranging from absolute radiometric dating (such as 14-C and U/Th) to relative dating using stratigraphic and archaeological methods. Consequently, care was taken to infer calendar dates for the sites used in the database and 14-C calendar dates were calibrated using CalPal-2007^{online} [55]. For the purpose of subsequent analyses, the sites were then categorized according to the Marine Isotope Stages (MIS's) defined in Table 1. These definitions were used to hinder the appearance of individual site entries in more than one category in the database, as some sites would otherwise have dating estimates that overlapped more than one MIS category. The Late-Glacial border category between MIS 1 and 2 (denoted "1/2") was created in order to further accommodate this transition period, notably because many sites clustered within this short period which contain much warmer episodes than the rest of MIS 2. Furthermore, due to small sample sizes, the moderately cold MIS 3 and 4 as well as the relatively warm MIS 5a to 5e (the Last Interglacial) were pooled together as MIS 3-4 and MIS 5, respectively. Individual stratigraphic layers in the catalogues, with dating estimates that overlapped multiple stages of opposite climatic regimes (e.g. spanning MIS 5a and MIS 4) were excluded. We also excluded layers older than 190,000 and younger than 10,000 years BP due to scarcity of the former and the sporadic coverage of the latter in the source material.

The fossil bird species for each site were indexed in the database as follows: Fossils that had been taxonomically identified to a single species were scored a 2 and fossils that had not been confidently identified (e.g. denoted "?", "cf." or "aff.") were scored a 1. Fossils identified as belonging to one of two or more closely related species were scored with a 1 for each species (e.g. *Corvus corone/frugileus*). Fossils that had only been identified to family level or higher were excluded and fossils identified to subspecies were scored under their respective species. 16 non-extant species were excluded given uncertainty regarding their climatic and habitat requirements. As the exception, the recently exterminated *Pinguinus impennis* (Great Auk) was not excluded as its requirements are relatively well-known. Furthermore, only sites and stratigraphic layers containing a minimum of 10 species were included in the database. This was done in order to minimize errors by focusing on sites where birds had received clear attention and to reduce

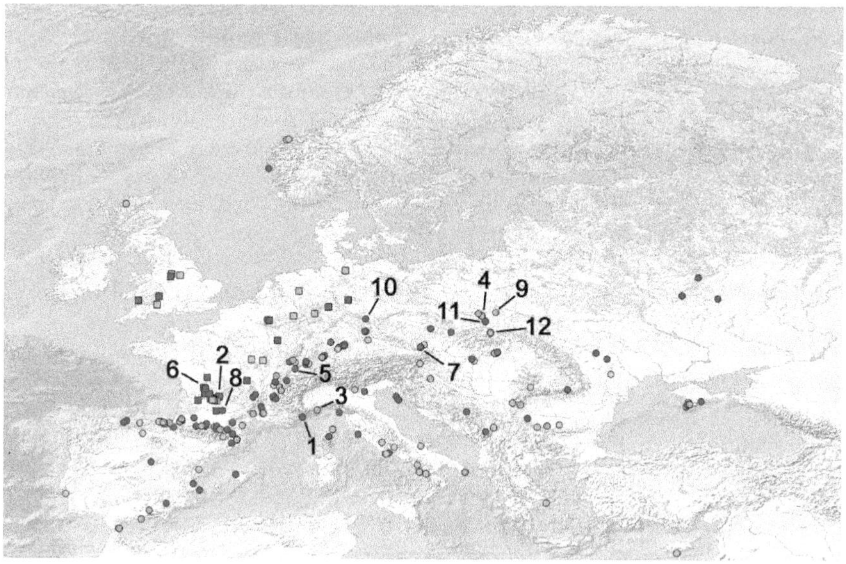

Figure 1. Position of sites included in the database. All sites are included in the statistical analyses of the bioclimatic properties of birds of the Western Palearctic, while only square sites are included in the analyses of the Northwestern subregion. The color of each site corresponds to the temperature trend of the stage it has been dated to, inferred from Table 1; Dark red = MIS 5, Light red = MIS 1/2, Light blue = MIS 3–4, Dark blue = MIS 2 and 6, Grey = Sites with multiple stratigraphic layers assigned to different stages. Numbers designate sites specifically mentioned in the text: 1) Grotte du Lazaret, France, 2) La Fage, France, 3) Arene Candide, Italy, 4) Krucza Skala, Poland, 5) Grotte de Cotencher, Switzerland, 6) Fontéchevade, France, 7) Schusterlucke Cave, Austria, 8) Abri de Fontales/Saint-Antonin Noble-Val, France, 9) Raj Cave, Poland, 10) Zwergloch, Germany, 11) Mamutowa, Poland, 12) Oblazowa Cave, Poland.

stochastic noise due to small sample size. The individual species were categorized according to their present distributional association with three climate and habitat attributes, namely temperature, humidity and vegetation, as defined by Finlayson [14]; Temperature: According to Finlayson each species' breeding distribution was compared to a bioclimatic map of the World and given a rank on a temperature gradient based on which bioclimatic area it occupies. The gradient went from 1% indicating the coldest conditions to 100% indicating the warmest conditions. From this rank, each species was classified in a group from A to E: A = 1–20%, B = 21–40%, C = 41–60%, D = 61–80%, E = 81–100%. Humidity: As with temperature, this gradient went from 1% indicating the most xeric conditions to 100% indicating the most humid conditions: A = 1–20%, B = 21–40%, C = 41–60%, D = 61–80% and E = 81–100%. Vegetation: We focused on foraging habitat rather than nesting habitat, since the source material, did not indicate maturity of the fossil individuals and therefore there was no way of knowing if a fossil came from a breeding site or not. The classifications were: F = Forest, with a high density of trees; O = Open; M = Mixed, including savannah, scrubland and tree-open-habitat-mosaics; R = Rocky; W = Wetland, all kinds except marine; Ma = Marine; A = Aerial. Of these habitat categories provided by Finlayson, only forest, mixed and open were used here to infer temporal changes in vegetation. Twelve species in the database were not included in Finlayson's work and habitat information for these were obtained from the Birdlife International webpage [56] and classified according to the habitat categories provided by Finlayson [14]. These species were excluded from the climate-related part of the analyses.

The final database consisted of 361 species from 61 families distributed among 227 fossil sites on a total of 474 stratigraphic layers (Dataset S1 in Supporting Information).

Analysis

To investigate if the fossil record show evidence of species range shifts during the 190,000–10,000 years BP period or not, we

Table 1. Marine Isotope Stage definitions and sample sizes in the study.

MIS	From (yr BP)	To (yr BP)	Sample size Western Palearctic	Sample size northwestern Europe	Temperature characteristics
1/2	13,000	10,000	95	8	Transition, cold to warm
2	28,000	10,000	181	17	Fully glacial
3–4	75,000	28,000	97	9	Cool
5	130,000	75,000	12	5	Warm, Fully interglacial
6	190,000	130,000	10	-	Fully glacial

MIS definitions was made loosely under influence by Taylor & Aitken [89], though constricted by sample size issues. The short transitional period MIS 1/2 was created due to a high concentration of sites dated to this time and is overlapped by a few sites in MIS 2 that have large date range estimates, causing both MIS categories to have the same end date. Sample size states the number of individual sites or stratigraphic layers for each period included in the analyses of the Western Palearctic and Northwestern European subregion, respectively.

compared the distribution of fossil occurrences to species' present-day distribution as estimated by Birdlife International and NatureServe [57]. Many species have distributions covering large latitudinal spans suggesting high bioclimatic tolerances [14]. Given that fossils represent presence-only data so that observed absences cannot with certainty be inferred as true absences, range shifts for such widely distributed species cannot easily be concluded with certainty. We therefore focused on species with present ranges restricted to more localized areas when assessing range shifts.

To explore climate and vegetation associations of the fossil bird species in relation to different stages, the species index number for each site in the database was summarized and from this, the proportions of birds within each temperature, humidity and vegetation category were calculated. A few sites consisted of more than 10 stratigraphic layers dated to the same MIS and proportions for these were averaged in order to avoid pseudoreplication. Grotte du Lazaret (FR147, FR148), La Fage (FR171), sites dated to MIS 2 of Arene Candide (IT3) and Krucza Skala (PO37) with 20, 11, 17, 16 and 17 layers respectively, were all averaged (Fig. 1: site 1–4). Using R Studio (version 0.97.332) Kruskal-Wallis one-way analyses of variances between means of proportions of the different MIS's were performed for each climate and vegetation category followed by posthoc Wilcoxon rank sum tests.

The above questions were re-assessed just for Northwestern Europe by repeating the statistical analyses described above using entries in the database from sites of Belgium, non-alpine Germany, non-alpine and non-Pyrenean France, the Netherlands and England only (Fig. 1). MIS 6 were excluded in this part of the analysis due to low sample size.

To assess the robustness of the results to uncertainties in the identification of fossils to species level, two supplementary analyses were made. One in which each species' habitat and climate attribute were replaced by the average of the genus (AvGenus) and one in which Passeriformes were excluded from the analysis (NoPas), as this group is probably the most challenging for species-level identifications. All of the analyses described above were repeated on these alternative datasets.

Results

Range Shifts

A number of latitudinal shifts in range are apparent. Southward glacial shifts are found for the forest species *Perisoreus infaustus* (Siberian Jay), *Loxia pytyopsittacus* (Parrot Crossbill), *Pinicola enucleator* (Pine Grosbeak), *Surnia ulula* (Northern Hawk Owl) and *Strix nebulosa* (Great Grey Owl), as well as for the open vegetation species *Lagopus lagopus* (Willow Ptarmigan), *Falco rusticolus* (Gyr Falcon), *Plectrophenax nivalis* (Snow Bunting) (Fig. 2) and *Bubo scandiaca* (Snowy Owl) (Fig. 3). All fossils of these species are from cold stages or stadials, except from one occurrence of *Lagopus lagopus* at Grotte de Cotencher (CH10) (Fig. 1: site 5) dated to the Odderade interstadial. Northward shifts under both cold and warm stages, are seen in *Gyps fulvus* (Griffon Vulture), *Melanocorypha calandra* (Calandra Lark), *Buteo rufinus* (Long-legged Buzzard), *Pyrrhocorax pyrrhocorax* (Red-billed Chough) and *Tetrax tetrax* (Little Bustard) (Fig. 3). Some species also exhibited longitudinal range shifts, with a tendency for westward expansions relative to today, e.g. *Falco vespertinus* (Red-footed Falcon), *Sturnus roseus* (Rosy Starling) (Fig. 3), *Circus macrourus* (Pallid Harrier) and *Anthropoides virgo* (Demoiselle Crane) (Fig. 4), which are all open-vegetation species. Fossils of these species have been dated to belong to cold stages, the Late-Glacial and to the Eemian. Furthermore, downwards altitudinal range shifts relative to today are seen for *Pyrrhocorax graculus* (Yellow-billed Chough), *Lagopus muta* (Rock Ptarmigan) and *Montrifringilla nivalis* (White-winged Snowfinch) (Fig. 4). Most of these fossils are from cold stages or stadials, except for a single occurrence of *Lagopus muta* at the Eemian-dated site Fontéchevade cave (FR84) (Fig. 1: site 6). Lastly, *Aegypius monachus* (Cinereous Vulture), *Tetrao urogallus* (Western Capercaillie) and *Tetrao tetrix* (Black Grouse) that today have patchy or disjunct distributions have fossil occurrences indicative of widespread occurrences in the past (Fig. 4).

Climate And Vegetation-Related Patterns For The Western Palearctic

Avian communities were dominated by cold-adapted species throughout the 190,000–10,000 years BP period (Table S1 in Supporting Information), with species of temperature categories A and B constituting more than half of the fossils. Similarly, there was also a general dominance of xeric-adapted species consistent with the cold, dry glacial climate. Species of open vegetation fluctuated in proportion between cold (average of MIS 2, 3–4 and 6) and warm (MIS 5) stages with 41.8% and 32.6%, respectively. For forest species the proportions were 18.0% for average cold stages and 28.5% for the warm stage (Table S1). Statistically, the strongest differences between stages were between open and forest vegetation categories, while there were no significant differences among climate categories (Table 2 and Fig. 5). Open-vegetation species constituted higher proportions of assemblages during the two full-glacial stages, MIS 2 and 6, compared with the warm and partly interglacial MIS 5 and the Late-Glacial MIS 1/2, while forest species conversely peaked during MIS 5. The temporal distributions of proportions were similar in the two supplementary analyses (Table S2, Fig. S1, Table S3 and Fig. S3), indicating that these findings should not be sensitive to species-level identification uncertainties.

Climate And Vegetation-Related Patterns For The Northwestern Subregion

In contrast to the patterns for the Western Palearctic, considering only Northwestern Europe, bird communities exhibited significant differences between stages in terms of temperature associations (Table 2, Fig. 6). Overall, warm-adapted species predominated during MIS 5 and 3–4, while cold-adapted species predominated during MIS 2 and 1/2. Of the vegetation categories, the only significant difference was that MIS 3–4 had a smaller proportion of forest-associated bird species than other periods. The results of the supplementary analyses were overall consistent with these results (Table S2, Fig. S2, Table S3 and Fig. S4).

Discussion

Numerous bird species have experienced range shifts in the Western Palearctic over the 180,000 years study period from the late Middle Pleistocene to the Late Pleistocene/Holocene transition. These shifts can primarily be linked to vegetation changes in response to climate change, with a dominance of species of open vegetation during cold periods and forest-adapted species during warm periods. However, there was a strong link to climate per se in the more climatically exposed subregion; Northwestern Europe.

Avian Responses To Climate Change

Studies on avifaunal responses to recent climate change have provided insight into the effects of change on bird species distribution and community composition. Generally, the strongest effects are found with regard to phenology and population

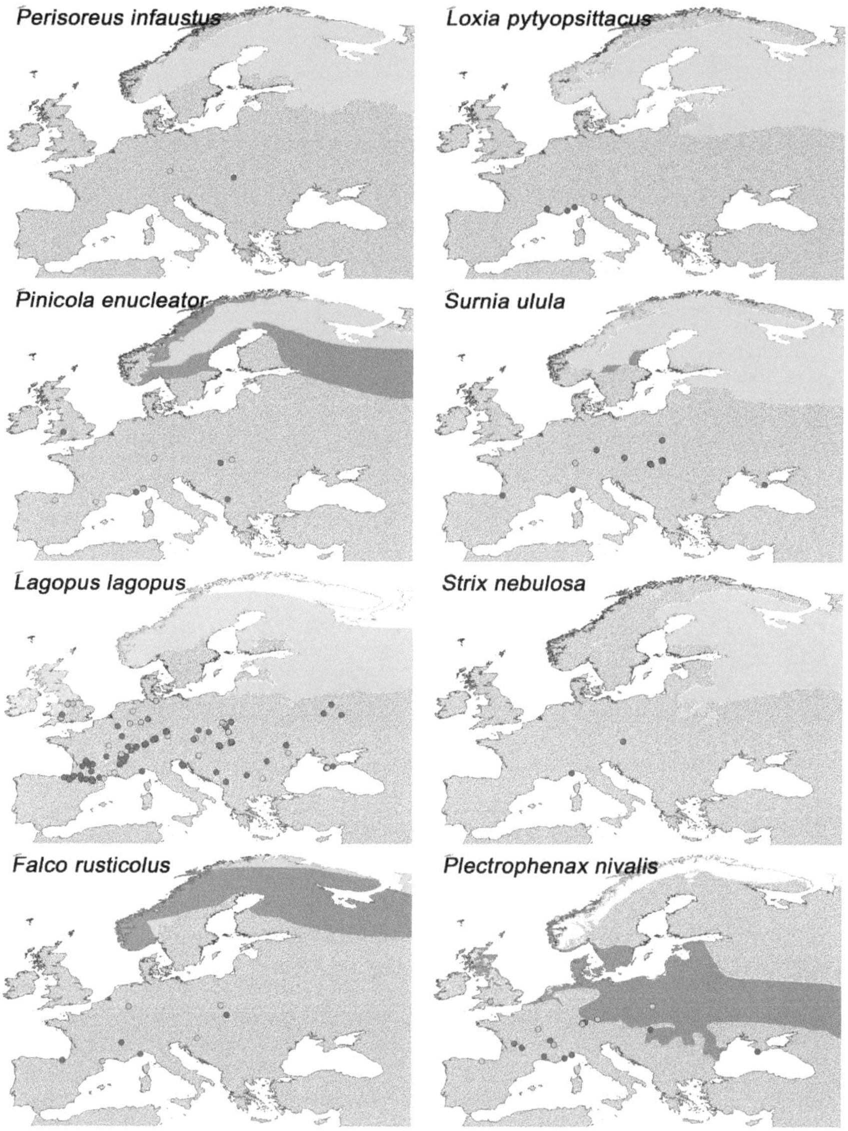

Figure 2. Present distribution and fossil locations of selected species in the analysis I. Present distribution is indicated by native resident (Green), native breeding (Yellow) and native non-breeding (Blue) ranges. For explanation on color codes for fossils sites see Fig. 1. These species all have fossil occurrences further south of their present distribution indicating latitudinal range change in the past.

dynamics, though shifts in range are also apparent [58]. A local study in California, reported range changes for several bird species over the course of a century in response to temperature and precipitation changes [59], and a study in England found climate change effects even within a short time span of 20 years [60]. Given the emerging present trends, it is not surprising to find that range shifts have occurred in response to the dramatic Quaternary glacial-interglacial changes in climate and vegetation, as reported here. Such shifts were especially evident for species with a presently northern distribution (Fig. 2). During periods of the Last Glacial, when Northern Europe was inhospitable due to severe climate and extensive glaciation, these species were forced south to Central and Southern Europe. The opposite scenario in which birds of the south move northwards during warm stages is not nearly as clear, at least with respect to temperature responses per se (Fig. 3). Instead many of the observed northwards extensions are from cold periods as well. Apart from *Pyrrhocorax pyrrhocorax*, these latter species are all dry-adapted and associated with open-vegetation, matching the general environment that existed during the cold stages and it is probably aridity and openness rather than temperature that allowed these species to spread northwards. Similar glacial expansions of tundra and steppe species have also been observed for plants, insects and mammals [61–65]. The rich megafauna of the glacial steppe-tundra would have provided ample food for the scavengers such as *Gyps fulvus* (Fig. 3) and *Aegypius monachus* (Fig. 4), providing an explanation for the larger ranges observed for these species in the past, similar to what have been proposed for the large range changes observed for Late Pleistocene North American scavenging birds [66–68]. Evidence of longitudinal range shifts as a response to glaciations has previously been shown for beetles [7,62] and mammals [69] and is generally seen along a longitudinal moisture gradient, which in Europe is created by a strong west-east precipitation gradient. It is thus likely that the longitudinal range shifts observed here for birds are also due to the increased aridity and openness discussed above. This climate would have also favored alpine species and the

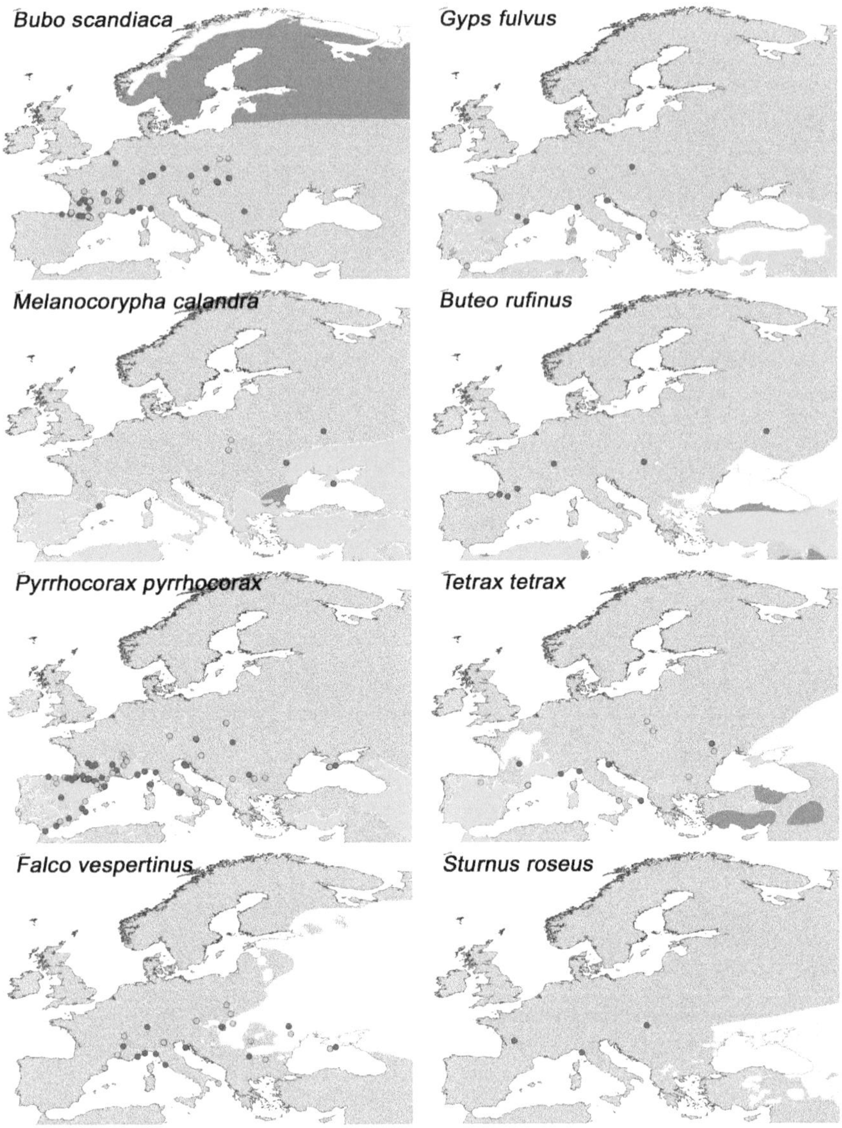

Figure 3. Present distribution and fossil locations of selected species in the analysis II. For explanation on distribution area, see Fig. 2. For explanation on color codes for fossils sites see Fig. 1. These species have fossil occurrences north (south in the case of *Bubo scandiaca*) of their present distribution indicating latitudinal range change in the past, except for *Sturnus roseus* that exhibits longitudinal range change.

downward expansion of *Montifringilla nivalis*, *Lagopus muta*, *Pyrrhocorax graculus* and *Pyrrhocorax pyrrhocorax* into the European lowlands is consistent with this (Fig. 3 and Fig. 4). The present disjunct distributions of these species thereby exemplify southern topographic refugia for cold-adapted species during interglacials, a phenomenon also observed in plants and mammals [22,25,70]. Finally we note that a lot of species did not exhibit marked past range shifts, indicating that they were not strongly affected by the climate changes and were able to survive in place.

The Late Pleistocene avifaunal range shifts nicely illustrate individualistic responses to environmental changes, as broadly reported for other organism groups [9,28,71]. These species-specific responses probably reflect that the individual species are limited by different abiotic factors and biotic interactions as well as having varying dispersal abilities. An important outcome of this is the creation of non-analogue assemblages, which is also seen in the Late Pleistocene avifaunal record. In Schusterlucke cave (AU16) (Fig. 1: site 7), dated ca. 115,000 years BP, for example, *Surnia ulula* and other northern species like *Lagopus lagopus*, *Bubo scandiaca* and *Strix nebulosa* are found together with South European mountain species like *Prunella collaris* (Alpine Accentor), *Pyrrhocorax graculus* and *Pyrrhocorax pyrrhocorax* and temperate species like *Perdix perdix* (Grey Partridge), *Corturnix coturnix* (Common Quail), *Crex crex* (Corncrake), *Cuculus canorus* (Common Cuckoo) and *Picus viridis* (Eurasian Green Woodpecker). Similarly, 100,000 years later during the Late-Glacial in France at Abri de Fontalés cave (FR16) (Fig. 1: site 8) *Lagopus lagopus* and *Bubo scandiaca* are also found together with *Pyrrhocorax graculus*, *Prunella collaris*, *Perdix perdix* and *Hirundo rupestris* (Eurasian Crag-martin). The examples from Tyrberg's catalogue are numerous and indicate that such non-analogue assemblages were not rare occurrences. This phenomenon has also been observed for other taxa [72,73] and has important implications for both paleoecology and future ecological forecasting [74].

An important caveat to this interpretation is that some localities may sample from time periods spanning varying climates or

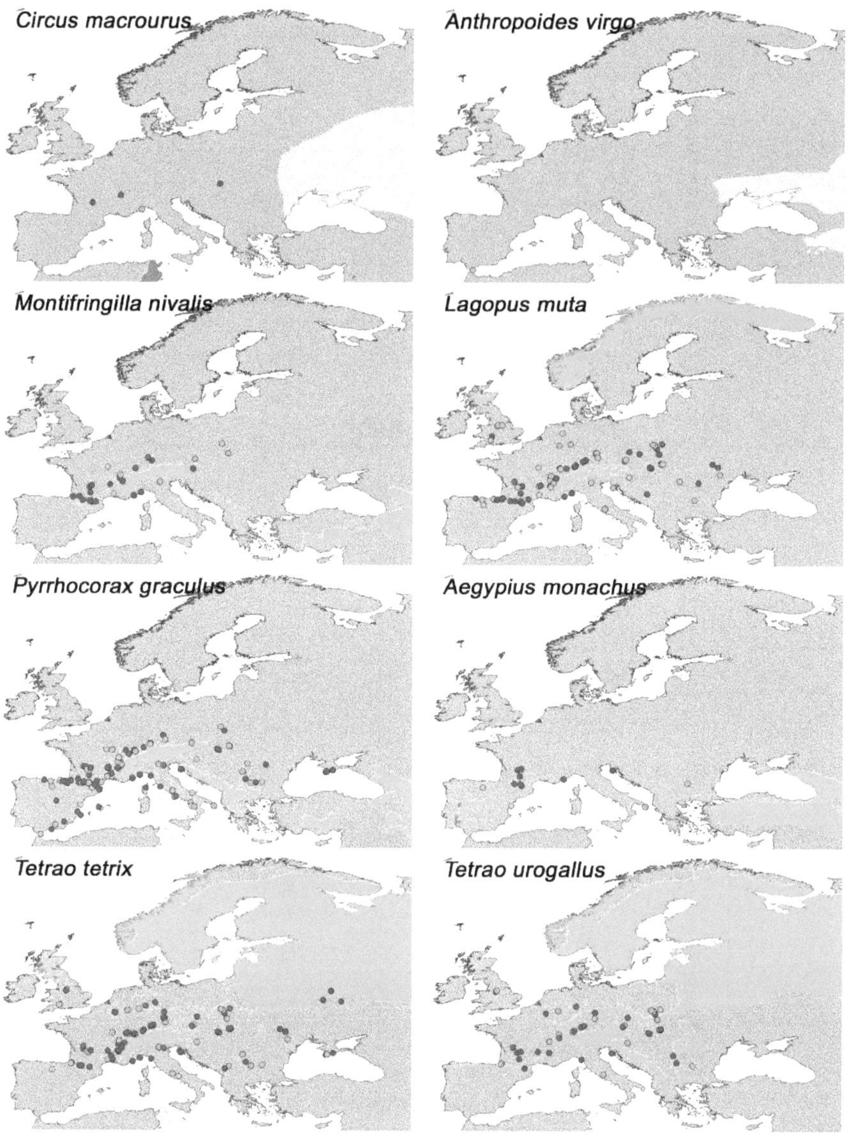

Figure 4. Present distribution and fossil locations of selected species in the analysis III. For explanation on distribution area, see Fig. 2. For explanation on color codes for fossils sites see Fig. 1. These species exhibit westwards longitudinal (*Circus macrourus and Anthropoides virgo*) and downwards altitudinal (*Montifringilla nivalis, Lagopus muta and Pyrrhocorax graculus*) range changes as well as evidence of past larger ranges relative to today (*Aegypious monachus, Tetrao tetrix and Tetrao urogallus*).

contain mixed stratigraphic layers. For example, the apparent occurrences of cold-adapted species like *Lagopus muta*, *Buteo lagopus* and *Plectrophenax nivalis* at the Eemian site Fontéchevade Cave (FR84) (Fig. 1: site 6) together with species like *Sturnus roseus*, *Hirundo rupestris* and *Coturnix coturnix* and others associated with a high to medium temperature could reflect mixing of the stratigraphic layers [75].

Bird species preferences for temperature and humidity did not differ significantly between stages, when considering the entire Western Palearctic (Table 2), perhaps reflecting that this broad region spans much climatic heterogeneity and encompasses both climatically unstable and relatively stable areas, buffering against overall changes. In Northwestern Europe, however, the proportion of birds in the different temperature preference categories differed between stages (Table 2 and Fig. 6), indicating that this smaller subregion did not have the sufficient conditions for all climatic types to persist through all stages. A similar heterogeneous pattern in community composition between stages has been found for Middle Pleistocene mammals, with marked shifts in Northern Europe due to the strongly disruptive effects of glacial climate here and continuity in Southern Europe [76]. In contrast to the patterns for climate preferences, there were clear shifts in general vegetation preferences between stages both for the Western Palearctic overall and for Northwestern Europe (Table 2, Fig. 5 and Fig. 6). In Northwestern Europe, the proportion of forest species during MIS 3-4 is surprisingly small compared to the colder, more arid, MIS 2 (Fig. 6). This pattern is probably artefactual, reflecting that most of the MIS 2 data belong to the end of the period when there were several warm episodes. Another interesting pattern that emerged is the high proportion of warm-adapted (Temp. D) birds in combination with the low proportion of forest-adapted birds during MIS 3–4 in Northwestern Europe

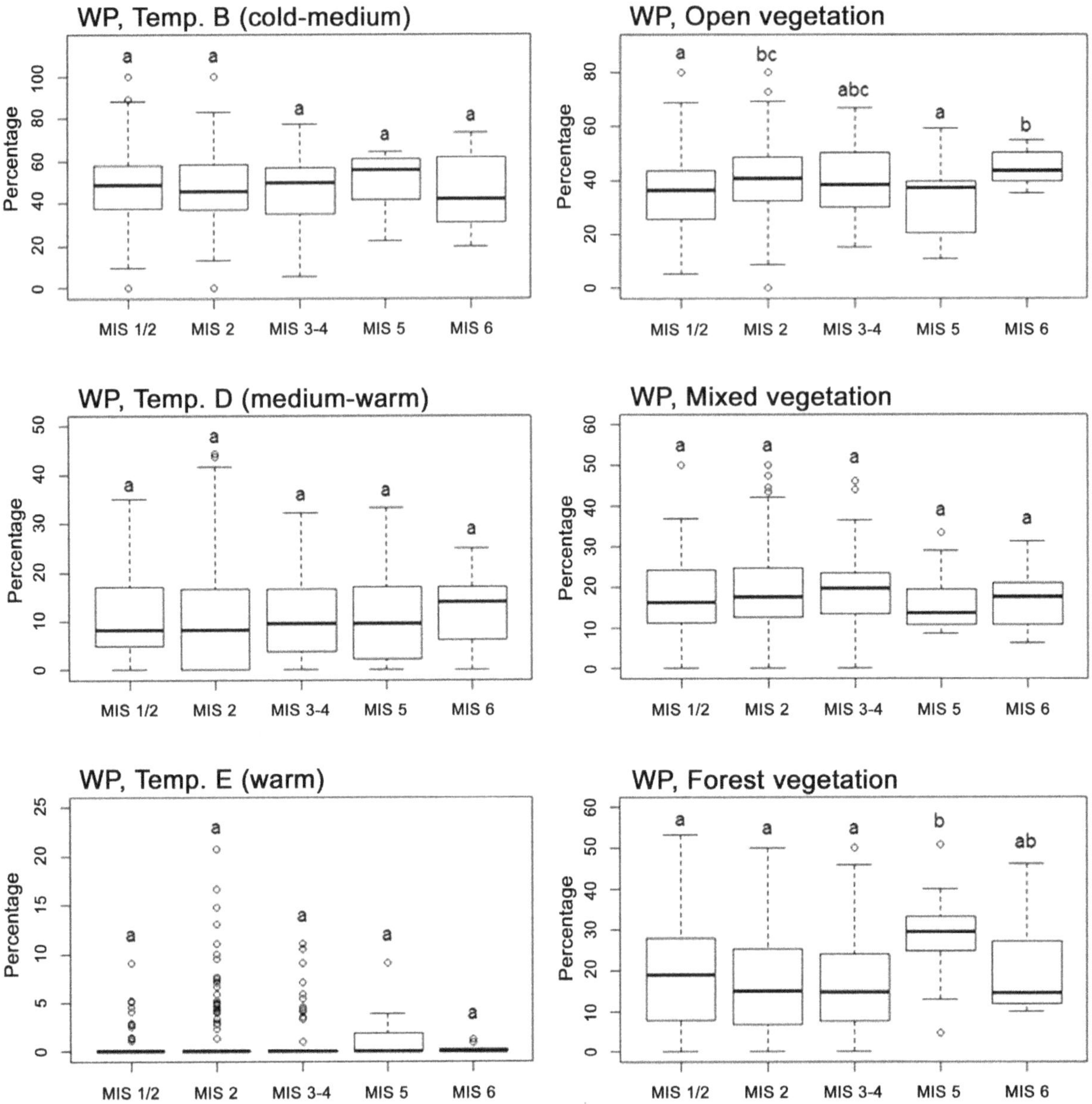

Figure 5. Boxplots of vegetation and temperature properties for birds of the Western Palearctic (WP). Boxplots shows the percentage of fossil bird species preferring cold-medium (Temp. B), medium-warm (Temp. D) and warm (Temp. E) conditions (Left) and open, mixed and forest vegetation (Right) in the Western Palearctic, for each MIS. Boxes show the median, 25th and 75th percentile and whiskers extending 1.5 interquartile range (IQR). Dot symbols identify outliers. Letters indicate significant relationships according to Wilcoxon signed-rank tests ($p \leq 0.05$). Open and Forest vegetation are the only significant variables according to variance analyses (Table 2) and show significant higher proportions of open vegetation species during the glacial periods MIS 2 and 6 compared with the partly interglacial MIS 5 and the transitional period MIS 1/2. Furthermore, MIS 5 had significantly higher proportions of forest species compared with later periods of colder temperatures.

(Fig. 6). This suggests that although temperatures were relatively high (a majority of sites within this category dates to MIS 3, the warmer of the two stages), tree species were not able to colonize Northwestern Europe at this time, possibly due to migration lag, as has been suggested by Coope based on fossil beetles [77]. Lags in treeline shifts and tree species ranges in response to climate change are widely reported in the literature [10]. The continuous presence of open, mixed and forest-adapted species throughout the study period are in agreement with simulation studies that found a similar continuous presence of both non-arboreal and different arboreal vegetation types, throughout the middle to late Weichselian in Europe [35,36] and the direct fossil evidence for the presence of both boreal and temperate tree species in Central and Eastern Europe [30].

Vegetation Implications Of The Avian Fossil Record

The current debate on the degree of forest cover during the full-glacial and interglacial stages is treated in detail in the following by considering the observed stage-by-stage pattern in forest- and open-habitat-associated bird species, as well as the more specific habitat requirements of the various species found.

The Saalian glacial stage (MIS 6) surpassed the Weichselian both in ice sheet extension and duration [78]. Permafrost prevailed north of the Alps and the vegetation here was mainly a steppe-

Table 2. Results of Kruskal–Wallis one-way analysis of variance.

	Western Palearctic			Northwestern Europe		
Variable	χ^2	**Df**	**p**	χ^2	**Df**	**p**
Temperature A	3.1243	4	0.5372	0.8493	3	0.9317
Temperature B	1.5118	4	0.8246	11.3243	3	**0.02315**
Temperature C	4.2025	4	0.3793	4.659	3	0.3241
Temperature D	2.8698	4	0.5799	14.4951	3	**0.005872**
Temperature E	3.0851	4	0.5437	16.437	3	**0.002486**
Humidity A	8.6433	4	0.0707	3.1268	3	0.5368
Humidity B	5.2386	4	0.2637	1.8566	3	0.7621
Humidity C	3.0629	4	0.5474	3.5202	3	0.4748
Humidity D	5.6540	4	0.2265	1.7719	3	0.7776
Humidity E	6.7712	4	0.1485	2.1658	3	0.7053
Vegetation Open	14.159	4	**0.0068**	8.2334	3	0.08339
Vegetation Mixed	2.9086	4	0.5732	4.3537	3	0.3602
Vegetation Forest	10.7363	4	**0.0297**	11.015	3	**0.0264**

χ^2 = value of H statistic for the adopted significance level; Df = degree of freedom; p = significance level ($p \leq 0.05$). Bold values indicate significance. The table is divided between the Western Palearctic and the Northwestern European subregion and shows the results of variance analyses between means of MIS's for several temperature, humidity and vegetation categories and is based on the proportions of fossil bird species assigned to each category. Temperature categories are based on a gradient going from the coldest (Temperature A) to the warmest (Temperature E) conditions and humidity follows a similar gradient going from the driest (Humidity A) to the wettest (Humidity E) conditions.

tundra mixture, probably reflecting the low temperatures and low CO_2 levels [78]. Palynological evidence suggest that tree species were mainly found south of the Alps in localized refugia were the moisture levels were adequate [20]. The avian fossil representation for this stage is sparse and localized to Central and Western Europe, but suggests, though not significantly so, a higher degree of open landscape compared with any later period (Fig. 5). A spatial presentation of the forest bird species shows that they are primarily located south of the Alps (Fig. 7), in agreement with the palynological evidence. Among the most commonly found species in these areas are *Turdus viscivorus* (Mistle Thrush), *Turdus merula* (Eurasian Blackbird), *Garrulus glandarius* (Eurasian Jay), *Scolopax rusticola* (Eurasian Woodcock), *Columba palumbus* (Common Wood-pigeon) and *Otus scops* (Common Scops-owl), suggesting temperate mixed woodlands. However, finds of species like *Turdus iliacus* (Redwing) and *Aegolius funerous* (Boreal Owl) points toward boreal elements in Southern Europe as well. The Eastern European landscape cannot be deduced from the present results due to data scarcity, but a paleoecological study on avian fossils from the Bisnik Cave in Poland report *Tetrao tetrix*, *Corvus monedula* (Eurasian Jackdaw) and *Lagopus lagopus* [38], pointing towards the presence of open woodlands.

European interglacial vegetation structure has been much discussed (e.g. [21]). Pollen studies suggest that coniferous and deciduous forests dominated the vegetation during the warmer periods, with dwarf shrubs and herbaceous plants prevalent during colder intervals [17–19,79,80]. One third of the sites from MIS 5 date to the Eemian and are truly interglacial while the rest stem from subsequent stadials and interstadials within MIS 5. The results show that during the entire MIS 5 the Western Palearctic had significant higher proportion of forest-adapted bird species (Table S1, 28.5%) compared with later stages of colder character, somewhat in agreement with traditional views of a forest-covered interglacial landscape. The high proportion of open-adapted bird species (Table S1, 32.6%) for this stage, however, also suggests a substantial degree of openness, even so within the interglacial per se (Fig. 7). Species of temperate grasslands like *Crex crex*, *Perdix perdix* and *Coturnix coturnix* are frequent in the fossil record for the entire MIS 5, including the Eemian, along with the more cold-tolerant *Lagopus lagopus*, which was, however, only found from sites dated to the stadials and interstadials after MIS 5e. Northwestern Europe exhibited equally strong patterns regarding openness, despite one out of three sites being from the Eemian. *Crex crex*, *Perdix perdix* and *Coturnix coturnix* are also found in this region along with *Corvus corone* (Carrion Crow) *Carduelis carduelis* (European Goldfinch) and *Alauda arvensis* (Eurasian Skylark), all suggesting at least partially open landscapes. Throughout the Western Palearctic, these open-vegetation species could have survived in open areas maintained by grazing megafauna as well as marginal edaphic conditions, fluvial activity, windthrows and forest fires [21].

The Early and Middle Weichselian (MIS 3-4) trended towards colder temperatures and during the coldest periods, polar desert and steppe-tundra prevailed with tree species primarily found in Southern European refugia [20]. During warmer intervals shrubs and tree species expanded. Forest bird species are found south of a tilted line from the Pyrenees in the west across Central Europe just north of the Alps and in Eastern Europe as high as the 50°N latitude at Raj Cave in Poland (PO21) (Fig. 1: site 9 and Fig. 7). This spatial distribution corresponds well with the many findings of tree macrofossils in Eastern Europe [29,30], reporting both boreal coniferous and temperate deciduous tree species from the region during the late MIS 3 when conditions were mildest. In Northwestern Europe, pollen studies indicates that the landscape was dominated by polar desert to the north and steppe-tundra to the south during MIS 4 while the warmer parts of MIS 3 had open woodland of *Picea* (Spruce), *Pinus* (Pine) and *Betula* (Birch) to the south and treeless shrub tundra to the north [18,20,81], something which is supported by mammalian fossil assemblages [72]. There have, however, been pollen findings of *Betula* sp., *Juniperus* sp. and *Pinus* sp. dated to early MIS 3 as far north as Denmark [82]. The avifaunal results presented here suggest a relatively open landscape in Northwestern Europe at this time with only a few occurrences of woodland species, notably *Turdus philomelos* (Song Trush) *Turdus viscivorus* and *Turdus merula*, all from central France (Fig. 6 and Fig. 7).

During the Last Glacial Maximum (MIS 2: LGM, 18,000–22,000 years BP), the Fennoscandian ice sheet reached as far south as the 50°N latitude in Poland and Germany [83], close to some of the sites dated to this period. Pollen evidence indicates a prevalence of tundra north of the Alps and steppe-like vegetation with few trees in lowland Southern Europe, with most trees mostly restricted to mountain belts of suitable conditions. However, findings of macrofossils of coniferous and deciduous tree species in Hungary (dated 32,500–16,500 years BP) [29], as well as conifer species in southern Poland and deciduous tree species in Austria (dated 30,000–25,000 years BP) [30], indicates a presence of trees further north. This is further supported by findings of *Pinus* sp., *Picea* sp., *Alnus* sp. (Alder), shrubby *Betula* sp. and *Salix* sp. (Willow) further north and east in Belarus and Ukraine [31] and findings of a continuous presence of forest species of birds and mammals in the aforementioned Bisnik Cave in Poland [38]. Additionally, a species distribution modeling study have found suitable LGM conditions for several boreal tree species close to the Fennoscandian ice-sheet in Eastern Europe [35] and a genetic study on *Picea*

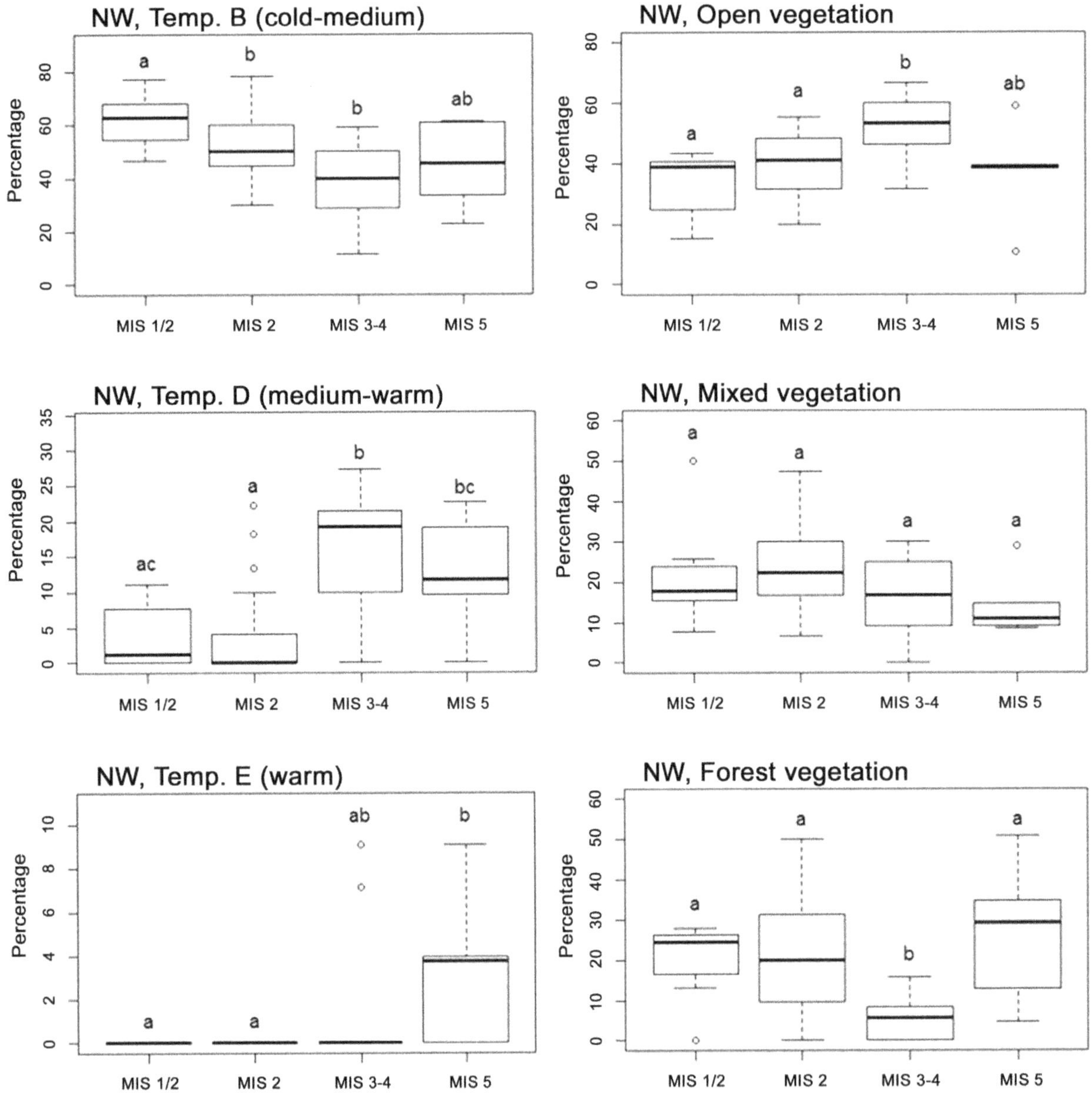

Figure 6. Boxplots of vegetation and temperature properties for birds of the Northwestern European subregion (NW). Boxplots shows the percentage of fossil bird species preferring cold-medium (Temp. B), medium-warm (Temp. D) and warm (Temp. E) conditions (Left) and open, mixed and forest vegetation (Right), in Northwestern Europe, for each MIS. Boxes show the median, 25th and 75th percentile and whiskers extending 1.5 interquartile range (IQR). Dot symbols identify outliers. Letters indicate significant relationships according to Wilcoxon signed-rank tests ($p \leq 0.05$). In this analysis the three temperature variables presented here, as well as the forest vegetation variable were significant according to variance analyses (Table 2). Species of warmer temperatures are present in larger proportions at MIS 5 and to some degree MIS 3-4, while cold adapted species are more prominent in MIS 1/2. The vegetation trends follow that of the Western Palearctic.

abies (Norway Spruce) suggested that this species survived the full-glacial in microenvironmentally favorable pockets in western Norway [84]. LGM occurrences in Northwestern Europe are poorly represented in the avian fossil record, making it impossible to infer landscape characteristics in this area during this period. For the whole of the Western Palearctic, the proportion of forest birds for this particular cold period was on average 11% (Table S1), but as high as >20% for some sites in Spain, Italy, Germany and Poland (Fig. 7). The location of forest birds in Eastern and Central Europe suggests that some forest tree species were able to survive the full-glacial climate in places north of the traditionally recognized southern refugia. The three sites closest to the Fennoscandian ice sheet (Zwergenloch (BRD156) in Germany and Mamutowa (PO13) and Oblazowa Cave (PO35) in Poland) are interesting in relation to the debate on glacial forests (Fig. 1: site 10–12), as they all had bird species from various forest types. Among these were *Tetrao urogallus*, *Aegolius funereus* and *Asio otus* (Long-eared Owl) which prefers mature coniferous forest with some degree of open ground or clearings. Species of mixed as well as broad-leaved deciduous forest and woodland, like *Dendrocopos medius* (Middle Spotted Woodpecker), *Turdus merula* and *Coccothraustes coccothraustes* (Hawfinch) are also reported. It has been suggested that boreal trees like *Pinus*, *Picea*, *Betula*, *Juniperus* (Juniper), *Salix* and *Larix* (Larch) predominated in patchy

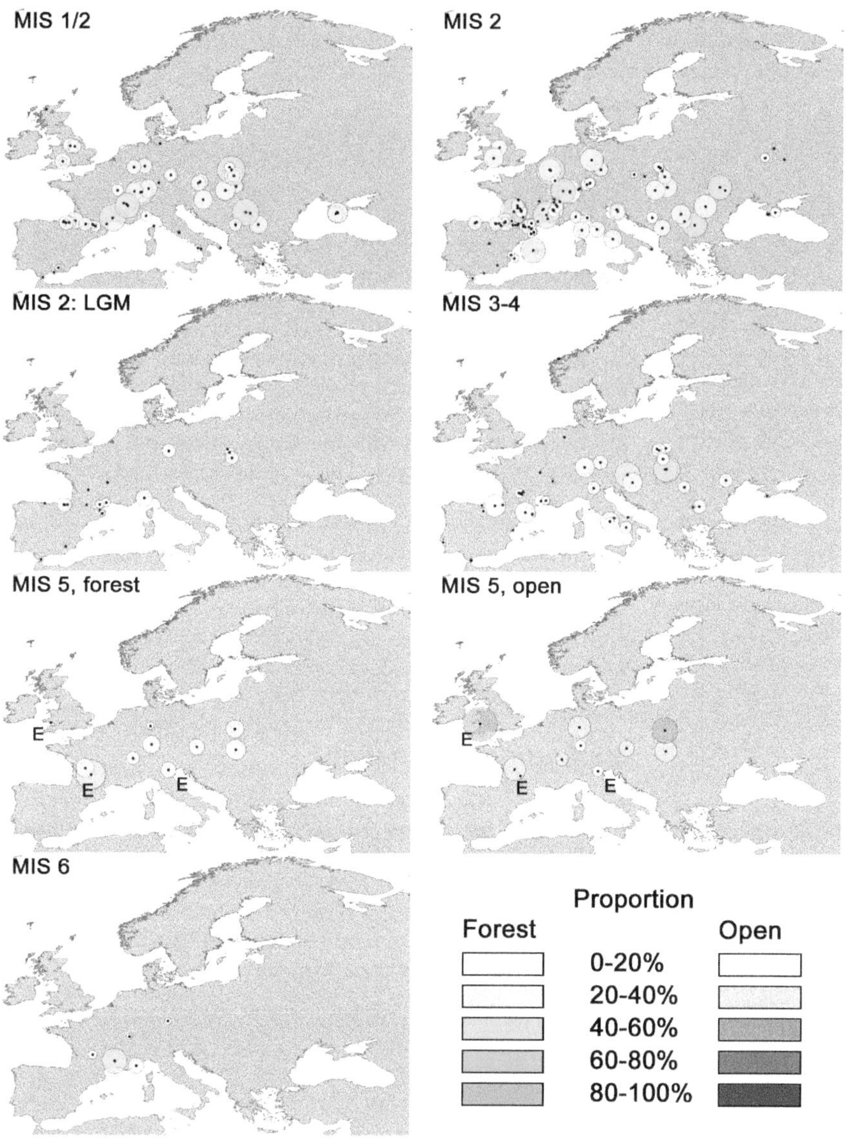

Figure 7. Spatial distribution of site proportions of forest bird species for each MIS in the study. The size and color scales of individual sites indicate the proportion of forest species for each site. Centers are marked with a black dot. E denotes sites of Eemian origin. The maps show that the sites with the highest proportion of forest birds are found within the warm MIS 5, but that MIS 1/2, 2 and 3–4 all have considerable proportions as well despite their overall cold temperature trends. Fossils from the LGM (22,000–18,000 years BP) within MIS 2 show that forest species were present in Poland and Germany as well as the Mediterranean. In addition, the proportion of open vegetation species is shown for MIS 5 indicating that there was a high degree of openness in the landscape, also during the Eemian.

woodland stands in Central and Eastern Europe, while more thermophilous species like *Quercus*, *Corylus* (Hazel), *Ulmus* (Elm) and *Tilia* (Linden) may have survived in localized pockets of favorable mild and humid conditions [29], for example in north/south oriented river valleys [26,31], and the mixture of fossilized birds of different forest types presented here is in accordance with this hypothesis.

In the late Weichselian (post LGM MIS 2, but before the Late-Glacial) forest birds are found throughout Europe including areas further northwest in continental Europe and southern England (Fig. 7). The high proportion of forest birds in Northwestern Europe is in contrast to the impression made by pollen studies that found largely non-arboreal vegetation species in this region [26]. Importantly, the forest-bird proportion in this subregion is largely made up by *Turdus* spp., which, despite forests being considered a habitat of major importance, are able to breed in areas outside this type of vegetation, for example in scrublands [56]. Two other species are also well represented namely *Tetrao urogallus* and *Garrulus glandarius*, which are found in southern England and Belgium. Their presence indicates that well-developed forests were present in this subregion at times during the period. Simulation studies have shown that boreal tree species could have been present in Northwestern Europe at the LGM [35,36], while the climate would have permitted both boreal and especially temperate types at 14,000 and 10,000 years BP [36], the latter also well-documented by paleoecological studies [85,86].

The Late-Glacial (MIS 1/2) is a short period relative to the other stages, that spans the relatively warm Bølling-Allerød interstadials as well as the cold Younger Dryas stadial and is thus a period of rapid changes in climate, the effect of which depended heavily on geographic location, even within Northwestern Europe [51]. Consequently, vegetation responses differed geographically [86], but were sometimes very rapid [85]. At the start of the Late-Glacial, the continuous permafrost in Northwestern Europe became more sporadic or disappeared [82] and plant fossil data show a development of dense vegetation, initially by a rise in *Artemisia*, followed by woodlands of varying density of *Betula* spp., *Salix* spp. and *Juniperus communis* [85,86] and appearance of *Populus tremula* (Aspen) in northwest Germany [87]. At the transition to the Younger Dryas, forests became increasingly open, transgressing to heath and tundra landscapes in Northwestern Europe [86]. For the Late-Glacial overall, the forest bird proportion in Northwest Europe increased compared with MIS 2 with the presence of species such as *Dendrocopos major* (Great Spotted Woodpecker), *Scolopax rusticola* (Eurasian Woodcock) and *Coccothraustes coccothraustes*. *Turdus* spp. are still represented, but make up smaller parts of assemblages compared with MIS 2. At the same time, there was still a high proportion of open-adapted species in Northwestern Europe, in particular *Lagopus lagopus*, *Lagopus muta*, *Corvus monedula*, and *Perdix perdix*, suggesting that assemblages within this MIS category stems from both warm and cold periods. However, the temporal classification of the data hinders any close inspection of the species' presences in relation to specific interstadial or stadial periods.

Methodological Considerations

Applying the uniformitarian principle to deduce past environments by utilizing modern analogues require important assumptions. Firstly, the climate and vegetation attributes of each species in the study were inferred from their present ranges, it was thus assumed that these ranges in broad terms reflect the bioclimatic spaces that the species tolerate. Secondly, one needs to assume that the species had the same climatic tolerances and vegetation requirements as today to use their presence in space and time to infer environmental conditions during past periods. Although some niche evolution may have occurred in some species we have no reason to suspect large, general changes over this time period. Furthermore, fossils in general are subject to collection and preservation biases of both temporal and spatial character [88]. Reflecting this, the dataset was predominated by fossils younger than 75,000 years (Table 1) and by fossils from mountain cave sites in Western Europe, especially France and northern Spain (Fig. 1). The latter biases the overall spatial distribution of the results towards Western Europe, although Eastern Europe did have some coverage as well. In addition, the dataset were predominated by large-bodied species (Corvidae appeared most frequently with 14.6% of the fossils belonging to this family, followed by Anatidae (10.6%), Tetraonidae (8.0%), Turdidae (6.5%), Phasanidae (6.4%) and Strigidae (6.4%)) supporting the general view that fossil records are taxonomically biased towards species with robust bones. From the paleoecological perspective of the present study, this should not affect the results much as it should not lead to spatiotemporal biases. Furthermore, the species used here represents a fairly even share of open and woodland vegetation types. For investigating the range shifts, a mixture of sedentary and migratory species with relatively restricted modern distributions was considered. While the past distribution of sedentary species can be reliably inferred from fossils, the shifts of migrating species should be inferred with caution. This is especially evident with species like *Plectrophenax nivalis* (Fig. 2) that have fossil specimens in close proximity to present winter ranges, but also for sedentary species with irruptive behavior that presently are observed as vagrants in areas far away from native ranges (e.g. *Pinicola enucleator* and *Surnia ulula* [56] (Fig. 2)). Fossils of these species could therefore be remnants of vagrant individuals, although it seems unlikely that such vagrants could dominate the observed patterns.

Conclusion

The avian fossil record from the latest Middle Pleistocene to the Late Pleistocene/Holocene transition documents sometimes strong range shifts in response to past climate change, linked to the climatic shifts per se, but even more so to vegetation changes. Present-day northern species of different vegetation types moved southwards, while open-vegetation species of the south and east moved north and westwards, respectively, during cold stages, the latter presumably due to increased aridity and openness. These conditions also favored alpine species causing them to shifts downwards into the European lowlands. The direct link to climate was primarily clear in Northwestern Europe, probably reflecting the particularly disruptive glacial climatic effects in this part of the Western Palearctic region. The close association between many bird species and vegetation types makes them useful for inferring woodland characteristics of past landscapes. Interestingly, the presence of forest birds in areas outside Southern Europe during the coldest stages, especially in East and Central Europe, are in agreement with the increasingly supported presence of trees in northern glacial refugia. In addition, the avifaunal record also suggests forest-dominated, but partially open landscapes during the Last Interglacial.

Supporting Information

Figure S1 Boxplots of vegetation and temperature properties for birds in the AvGenus supplementary analysis of the Western Palearctic (WP). Boxplots shows the percentage of fossil bird species preferring cold-medium (Temp. B), medium-warm (Temp. D) and warm (Temp. E) conditions (Left) and open, mixed and forest vegetation (Right) in the Western Palearctic, for each MIS. Boxes show the median, 25th and 75th percentile and whiskers extending 1.5 interquartile range (IQR). Dot symbols identify outliers. Letters indicate significant relationships according to Wilcoxon signed-rank tests ($p \leq 0.05$). Proportions are overall similar to the main analysis, especially regarding the Forest variable.

Figure S2 Boxplots of vegetation and temperature properties for birds in the AvGenus supplementary analysis of the Northwestern European subregion (NW). Boxplots shows the percentage of fossil bird species preferring cold-medium (Temp. B), medium-warm (Temp. D) and warm (Temp. E) conditions (Left) and open, mixed and forest vegetation (Right) in the Western Palearctic, for each MIS. Boxes show the median, 25th and 75th percentile and whiskers extending 1.5 interquartile range (IQR). Dot symbols identify outliers. Letters indicate significant relationships according to Wilcoxon signed-rank tests ($p \leq 0.05$). The proportions of the vegetation variables are similar to the main analysis while the proportions for the temperature variables are less consistent.

Figure S3 Boxplots of vegetation and temperature properties for birds in the NoPas supplementary analysis of the Western Palearctic (WP). Boxplots shows the percentage of fossil bird species preferring cold-medium (Temp. B), medium-warm (Temp. D) and warm (Temp. E) conditions (Left) and open, mixed and forest vegetation (Right) in the Western Palearctic, for each MIS. Boxes show the median, 25th and 75th percentile and whiskers extending 1.5 interquartile range (IQR). Dot symbols identify outliers. Letters indicate significant relationships according to Wilcoxon signed-rank tests ($p \leq 0.05$). The proportions are similar to the main analysis. Forest is the only significant variable, with MIS 2 having significantly lower proportions than MIS 1/2, 5 and 6.

Figure S4 Boxplots of vegetation and temperature properties for birds in the NoPas supplementary analysis of the Northwestern European subregion (NW). Boxplots shows the percentage of fossil bird species preferring cold-medium (Temp. B), medium-warm (Temp. D) and warm (Temp. E) conditions (Left) and open, mixed and forest vegetation (Right) in the Western Palearctic, for each MIS. Boxes show the median, 25th and 75th percentile and whiskers extending 1.5 interquartile range (IQR). Dot symbols identify outliers. Letters indicate significant relationships according to Wilcoxon signed-rank tests ($p \leq 0.05$). The proportions of both temperature and vegetation variables are consistent with the main analysis.

Table S1 Average proportion of birds for each climate and vegetation category. Proportions for cold stages consist of averaged data from MIS 2, 3–4 and 6 (though MIS 6 is not included in proportions for Northwestern Europe), while proportions for warm stage consist of data from MIS 5 only. Proportions for LGM consists of data from MIS 2 that are dated between 22–18 kya BP. The remainder of the habitat categories (Marine, Wetlands, Rocky and Aerial) included in the database are not treated in this study, hence Open, Mixed and Forest does not amount to 100%.

Table S2 Results of Kruskal–Wallis one-way analysis of variance for the AvGenus supplementary analyses. AvGenus = Average genus. In these analyses the habitat and climate attributes of each bird species were replaced with the genus average providing a conservative estimate of the potential impact of uncertainty in species-level fossil identification, most of which pertain species within the same genus.

Table S3 Results of Kruskal–Wallis one-way analysis of variance for the NoPas supplementary analyses. NoPas = No Passerines. These analyses excluded Passerines, which are often perceived as a group with problematic species-specific fossil identification.

Dataset S1 Dataset of fossilized birds in Europe used in this study. Based on Tyrberg (1998, 2008).

Acknowledgments

We are grateful for the provision of data in the published literature and online. We thank freeware enthusiasts around the world, for providing the necessary tools like R and Quantum GIS, without which this project could not have been completed in its present form.

Author Contributions

Conceived and designed the experiments: SRH J-CS. Performed the experiments: SRH. Analyzed the data: SRH J-CS. Contributed reagents/materials/analysis tools: SRH J-CS. Wrote the paper: SRH J-CS.

References

1. Huntley B, Webb T (1989) Migration: species' response to climatic variations caused by changes in the Earth's orbit. J Biogeogr 16: 5–19.
2. Jackson ST, Overpeck JT, Webb-III T, Keattch SE, Anderson KH (1997) Mapped plant-macrofossil and pollen records of Late Quaternary vegetation change in Eastern North America. Quat Sci Rev 16: 1–70.
3. Hughes L (2000) Biological consequences of global warming: is the signal already apparent? Trends Ecol Evol 15: 56–61.
4. Walther G-R, Burga CA, Edwards PJ (2001) "Fingerprints" of climate change - adapted behavior and shifting species ranges. New York: Kluwer Academic/Plenum Publishers.
5. McCarty JP (2001) Ecological consequences of recent climate change. Conserv Biol 15: 320–331.
6. Walther G-R, Post E, Convey P, Menzel A, Parmesan C, et al. (2002) Ecological responses to recent climate change. Nature 416: 389–395.
7. Abellán P, Benetti CJ, Angus RB, Ribera I (2010) A review of quaternary range shifts in European aquatic coleoptera. Glob Ecol Biogeogr 20: 87–100.
8. Davis MB, Shaw RG (2001) Range shifts and adaptive responses to Quaternary climate change. Science 292: 673–679.
9. Graham RW, Grimm EC (1990) Effects of global climate change on the patterns of terrestrial biological communities. Trends Ecol Evol 5: 289–292.
10. Svenning J-C, Sandel B (2013) Disequilibrium vegetation dynamics under future climate change. Am J Bot 100: 1–21.
11. Bowen DQ (1978) Quaternary geology: a stratigraphic framework for multidisciplinary work. Oxford: Pergamon Press.
12. Johnson SJ, Vinther BM (2007) Greenland stable isotopes. In: Elias AS, editor. Encyclopedia of Quaternary Science. Elsevier Ltd., Vol. 26. pp. 1250–1258.
13. Bassinot FC (2007) Oxygen isotope stratigraphy of the oceans. In: Elias AS, editor. Encyclopedia of Quaternary Science. Elsevier Ltd., Vol. 18. pp. 1740–1748.
14. Finlayson C (2011) Avian Survivors: the history and biogeography of Palearctic birds. London: Bloomsbury Publishing Plc.
15. Prentice IC, Jolly D (2000) Mid-Holocene and glacial-maximum vegetation geography of the northern continents and Africa. J Biogeogr 27: 507–519.
16. Parmesan C, Yohe G (2003) A globally coherent fingerprint of climate change impacts across natural systems. Nature 421: 37–42.
17. Zagwijn WH (1996) An analysis of Eemian climate in western and central Europe. Quartenary Sci Rev 15: 451–469.
18. Behre K-E (1989) Biostratisgraphy of the last glacial period in Europe. Quartenary Sci Rev 8: 25–44.
19. Shackleton NJ, Sánchez-Goñi MF, Pailler D, Lancelot Y (2003) Marine isotope substage 5e and the Eemian Interglacial. Glob Planet Change 36: 151–155.
20. Van Andel TH, Tzedakis PC (1996) Palaeolithic landscapes of Europe and environs, 150,000–25,000 years ago: an overview. Quartenary Sci Rev 15: 481–500.
21. Svenning J-C (2002) A review of natural vegetation openness in north-western Europe. Biol Conserv 104: 133–148.
22. Birks HJB, Willis KJ (2008) Alpines, trees, and refugia in Europe. Plant Ecol Divers 1: 147–160.
23. Taberlet P, Cheddadi R (2002) Quaternary refugia and persistence of biodiversity. Science (80-) 297: 2009–2010.
24. Bennett KD, Tzedakis PC, Willis KJ (1991) Quaternary refugia of north European trees. J Biogeogr 18: 103–115.
25. Stewart JR, Lister AM, Barnes I, Dalén L (2010) Refugia revisited: individualistic responses of species in space and time. Procedings R Soc B Biol Sci 277: 661–671.
26. Huntley B (1988) Glacial and Holocene vegetation history -20 ky to present: Europe. Vegetation History. Dordrecht, The Netherlands: Kluwer Academic Publishers.
27. Hewitt G (2000) The genetic legacy of the Quaternary ice ages. Nature 405: 907–913.
28. Taberlet P, Fumagalli L, Wust-Saucy A-G, Cosson J-F (1998) Comparative phylogeography and postglacial colonization. Mol Ecol 7: 453–464.
29. Willis KJ, Rudner E, Sümegi P (2000) The full-glacial forests of central and southeastern Europe. Quat Res 53: 203–213.
30. Willis KJ, van Andel TH (2004) Trees or no trees? The environments of central and eastern Europe during the last glaciation. Quat Sci Rev 23: 2369–2387.
31. Binney HA, Willis KJ, Edwards ME, Bhagwat SA, Anderson PM, et al. (2009) The distribution of late-quaternary woody taxa in northern Eurasia: evidence from a new macrofossil database. Quat Sci Rev 28: 2445–2464.
32. Stewart JR, van Kolfschoten T, Markova A, Musil R (2003) The mammalian faunas of Europe during oxygen isotope stage Three. Neanderthals and Modern Humans in the European Landscape during the Last Glaciation. Cambridge: McDonald Institute for Archeological Research, University of Cambridge. pp. 103–130.
33. Petit RJ, Aguinagalde I, de Beaulieu J-L, Bittkau C, Brewer S, et al. (2003) Glacial refugia: hotspots but not melting pots of genetic diversity. Science (80-) 300: 1563–1565.
34. Schmitt T (2007) Molecular biogeography of Europe: pleistocene cycles and postglacial trends. Front Zool 4.

35. Svenning J-C, Normand S, Kageyama M (2008) Glacial refugia of temperate trees in Europe: insights from species distribution modelling. J Ecol 96: 1117–1127.
36. Allen JRM, Hickler T, Singarayer JS, Sykes MT, Valdes PJ, et al. (2010) Last glacial vegetation of northern Eurasia. Quat Sci Rev 29: 2604–2618.
37. Stewart JR, Lister AM (2001) Cryptic northern refugia and the origins of the modern biota. Trends Ecol Evol 16: 608–613.
38. Tomek T, Bocheński ZM, Socha P (2012) Continuous 300,000-year fossil record: changes in the ornithofauna of Biśnik Cave, Poland. Palaeontol Electron 15: 1–20.
39. Tomek T, Bocheñski Z (2005) Weichselian and Holocene bird remains from Komarowa Cave, Central Poland. Acta Zool cracoviensia 48A: 43–65.
40. Bedetti C, Pavia M (2007) Reinterpretation of the late pleistocene Ingaro deposit based on the fossil bird associations (Apulia, south-eastern Italy). Riv Ital di Paleontol e Stratigr 113: 487–507.
41. Bocheñski Z, Tomek T (2004) Bird remains from a rock-shelter in Krucza Skala (Central Poland). Acta Zool cracoviensia 47: 27–47.
42. Bocheñski Z (2002) Bird remains from Oblazowa – zoogeographical and evolutionary remarks. Acta Zool cracoviensia 45: 239–252.
43. Finlayson G, Finlayson C, Giles Pacheco F, Rodriguez Vidal J, Carrión JS, et al. (2008) Caves as archives of ecological and climatic changes in the Pleistocene—The case of Gorham's cave, Gibraltar. Quat Int 181: 55–63.
44. Sánchez-Marco A (2004) Avian zoogeographical patterns during the Quartenary in the Mediterranean region and paleoclimatic interpretation. Ardeola 51: 91–132.
45. Sánchez-Marco A (1999) Implications of the avian fauna for paleoecology in the Early Pleistocene of the Iberian Peninsula. J Hum Evol 37: 375–388.
46. Finlayson C, Carrión J, Brown K, Finlayson G, Sánchez-Marco A, et al. (2011) The *Homo* habitat niche: using the avian fossil record to depict ecological characteristics of palaeolithic Eurasian hominins. Quat Sci Rev 30: 1525–1532.
47. Tyrberg T (1999) Seabirds and Late Pleistocene marine environments in the northeast Atlantic and the Mediterranean. Smithson Contrib to Paleobiol 89: 139–157.
48. Baird RF (1989) Fossil bird assemblages from Australian caves: precise indicators of late Quaternary environments? Palaeogeogr Palaeoclimatol Palaeoecol 69: 241–244.
49. Yalden DW, Albarella U (2009) The bird in the hand. The History of British Birds. Oxford, New York: Oxford University Press.
50. Parmesan C (2001) Detection of range shifts: general methodological issues and case studies of butterflies. "Fingerprints" of Climate Change: Adapted Behaviour and Shifting Species Ranges. New York: Kluwer Academic/Plenum Publishers. pp. 57–76.
51. Coope GR, Lemdahl G (1995) Rapid communication regional differences in the Late Glacial climate of northern Europe based on coleopteran analysis. J Quat Sci 10: 391–395.
52. Tyrberg T (1998) Pleistocene birds of the Palearctic: a catalogue. Cambridge: Nuttall Ornithological Club.
53. Tyrberg T (2008) Pleistocene birds of the Palearctic. Available: http://web.telia.com/~u11502098/pleistocene.html. Accessed 2013 March 6.
54. Natural Earth (n.d.) Free vector and raster map data @ naturalearthdata.com. Available: http://www.naturalearthdata.com/. Accessed 2013 Feb 25.
55. Danzeglocke U (n.d.) CalPal the cologne radiocarbon calibration & palaeoclimate research package. Available: http://www.calpal-online.de/. Accessed 2013 Jan 25.
56. BirdLife International (2013) IUCN red list for birds. Available: http://www.birdlife.org. Accessed 2013 July 1.
57. BirdLife International and NatureServe (2012) Bird species distribution maps of the world. BirdLife International, Cambridge, UK and NatureServe, Arlington, USA.
58. Crick HQP (2004) The impact of climate change on birds. Ibis (Lond 1859) 146: 48–56.
59. Tingley MW, Monahan WB, Beissinger SR, Moritz C (2009) Birds track their grinnellian niche through a century of climate change. Proc Natl Acad Sci U S A 106: 19637–19643.
60. Thomas CD, Lennon JJ (1999) Birds extend their ranges northwards. Nature 399: 213.
61. Bell FG (1969) the occurrence of southern, steppe and halophyte elements in Weichselian (last-glacial) floras from southern Britain. New Phytol 68: 913–922.
62. Coope GR (1973) Tibetan species of Dung Beetle from Late Pleistocene deposits in England. Nature 245: 335–336.
63. Zimina RP, Gerasimov IP (1973) The periglacial expansion of marmots (*Marmota*) in middle Europe during Late Pleistocene. J Mammal 54: 327–340.
64. Dalén L, Fuglei E, Hersteinsson P, Kapel CMO, Roth JD, et al. (2005) Population history and genetic structure of a circumpolar species: the arctic fox. Biol J Linn Soc 84: 79–89.
65. Flagstad Ø, Røed KH (2003) Refugial origins of reindeer (*Rangifer tarandus* L.) inferred from mitochondrial DNA sequences. Evolution (N Y) 57: 658–670.
66. Steadman DW, Miller NG (1987) California Condor associated with Spruce-Jack Pine woodland in the Late Pleistocene of New York. Quat Reseach 28: 415–426.
67. Chamberlain CP, Waldbauer JR, Fox-Dobbs K, Newsome SD, Koch PL, et al. (2005) Pleistocene to recent dietary shifts in California condors. Proc Natl Acad Sci U S A 102: 16707–16711.
68. Fox-Dobbs K, Stidham TA, Bowen GJ, Emslie SD, Koch PL (2006) Dietary controls on extinction versus survival among avian megafauna in the Late Pleistocene. Geology 34: 685–688.
69. Lyons SK (2003) A Quantitative assessment of the range shifts of Pleistocene mammals. J Mammal 84: 385–402.
70. Bennett K, Provan J (2008) What do we mean by "refugia"? Quat Sci Rev 27: 2449–2455.
71. Parmesan C (2006) Ecological and evolutionary responses to recent climate change. Annu Rev Ecol Evol Syst 37: 637–669.
72. Markova AK, Puzachenko AY, van Kolfschoten T (2010) The north Eurasian mammal assemblages during the end of MIS 3 (Brianskian–Late Karginian–Denekamp Interstadial). Quat Int 212: 149–158.
73. Jackson ST, Williams JW (2004) Modern analogs in Quaternary paleoecology: here today, gone yesterday, gone tomorrow? Annu Rev Earth Planet Sci 32: 495–537.
74. Williams JW, Jackson ST (2007) Novel climates, no-analog communities, and ecological surprises. Front Ecol Environ 5: 475–482.
75. Chase PG, Debénath A, Dibble HL, McPherron SP, Schwarcz HP, et al. (2007) New dates for the Fontéchevade (Charente, France) Homo remains. J Hum Evol 52: 217–221.
76. Rodríguez J (2006) Structural continuity and multiple alternative stable states in Middle Pleistocene European mammalian communities. Palaeogeogr Palaeoclimatol Palaeoecol 239: 355–373.
77. Coope GR (2002) Changes in the thermal climate in Northwestern Europe during marine oxygen isotope stage 3, estimated from fossil insect assemblages. Quat Res 57: 401–408.
78. Roucoux K, Margari V, Lawson IT, Tzedakis PC (2010) Vegetation responses to climate changes during the penultimate glacial period (marine isotope stage 6) in southern Europe. Geophys Res Abstr 12: 15216.
79. Huntley B, Alfano MJ., Allen JRM, Pollard D, Tzedakis PC, et al. (2003) European vegetation during marine oxygen isotope stage-3. Quat Res 59: 195–212.
80. Watts WA (1988) Late-tertiary and Pleistocene vegetation history – 20 My to 20 ky: Europe. Vegetation History. Dordrecht, The Netherlands: Kluwer Academic Publishers.
81. Zagwijn WH (1989) Vegetation and climate during warmer intervals in the Late Pleistocene of western and central Europe. Quat Int 3–4: 57–67.
82. Kolstrup E (1992) Danish pollen records radiocarbon-dated to between 50,000 and 57,000 yr BP. J Quat Sci 7: 163–172.
83. Ehlers J, Astakhov V, Gibbard PL, Mangerud J, Svendsen JI (2007) Late Pleistocene glaciations in Europe. In: Elias AS, editor. Encyclopedia of Quaternary Science. Elsevier Ltd. pp. 1085–1095.
84. Parducci L, Jørgensen T, Tollefsrud MM, Elverland E, Alm T, et al. (2012) Glacial survival of boreal trees in northern Scandinavia. Science 335: 1083–1086.
85. Hoek WZ (1997) Late-glacial and early Holocene climatic events and chronology of vegetation development in the Netherlands. Veg Hist Archaeobot 6: 197–213.
86. Walker MJC, Bohncke SJP, Coope GR, O'Connell M, Usinger H, et al. (1994) The Devensian/Weichselian Late-glacial in northwest Europe (Ireland, Britain, north Belgium, The Netherlands, northwest Germany). J Quat Sci 9: 109–118.
87. Bittmann F (2006) Reconstruction of the Allerød vegetation of the Neuwied Basin, western Germany, and its surroundings at 12,900 cal B.P. Veg Hist Archaeobot 16: 139–156.
88. Varela S, Lobo JM, Hortal J (2011) Using species distribution models in paleobiogeography: a matter of data, predictors and concepts. Palaeogeogr Palaeoclimatol Palaeoecol 310: 451–463.
89. Taylor RE, Aitken MJ (1997) Chronometric dating in archeology. New York: Plenum Press.

Coral Luminescence Identifies the Pacific Decadal Oscillation as a Primary Driver of River Runoff Variability Impacting the Southern Great Barrier Reef

Alberto Rodriguez-Ramirez[1]*, Craig A. Grove[2], Jens Zinke[3], John M. Pandolfi[4], Jian-xin Zhao[1]

1 Radiogenic Isotope Facility, School of Earth Sciences, The University of Queensland, Brisbane, Queensland, Australia, **2** NIOZ Royal Netherlands Institute for Sea Research, Department of Marine Geology, Den Burg, Texel, The Netherlands, **3** School of Earth and Environment, The University of Western Australia and the UWA Oceans Institute, Australia and the Australian Institute of Marine Science, Perth, Western Australia, Australia, **4** Australian Research Council Centre of Excellence for Coral Reef Studies, Centre for Marine Science, School of Biological Sciences, The University of Queensland, Brisbane, Queensland, Australia

Abstract

The Pacific Decadal Oscillation (PDO) is a large-scale climatic phenomenon modulating ocean-atmosphere variability on decadal time scales. While precipitation and river flow variability in the Great Barrier Reef (GBR) catchments are sensitive to PDO phases, the extent to which the PDO influences coral reefs is poorly understood. Here, six *Porites* coral cores were used to produce a composite record of coral luminescence variability (runoff proxy) and identify drivers of terrestrial influence on the Keppel reefs, southern GBR. We found that coral skeletal luminescence effectively captured seasonal, inter-annual and decadal variability of river discharge and rainfall from the Fitzroy River catchment. Most importantly, although the influence of El Niño-Southern Oscillation (ENSO) events was evident in the luminescence records, the variability in the coral luminescence composite record was significantly explained by the PDO. Negative luminescence anomalies (reduced runoff) were associated with El Niño years during positive PDO phases while positive luminescence anomalies (increased runoff) coincided with strong/moderate La Niña years during negative PDO phases. This study provides clear evidence that not only ENSO but also the PDO have significantly affected runoff regimes at the Keppel reefs for at least a century, and suggests that upcoming hydrological disturbances and ecological responses in the southern GBR region will be mediated by the future evolution of these sources of climate variability.

Editor: Hans G. Dam, University of Connecticut, United States of America

Funding: A Graduate School International Travel Award (GSITA) from The University of Queensland and the PADI foundation to ARR supported the luminescence analysis of the cores at the Royal Netherlands Institute for Sea Research (NIOZ). This work was partially funded by the Marine and Tropical Science Research Facility (MTSRF) Project 1.1.4 granted to JXZ, JMP and others; the ARC discovery project DP0773081 granted to JXZ, and others; the NERP Tropical Ecosystems Hub Project 1.3 to JXZ, JMP., and others. Sample collection was conducted under the GBRMPA permit number G10/33402.1. JZ was supported by an IOMRC UWA/AIMS/CSIRO postdoctoral fellowship. The funders had no role in study design, data collection and analysis, decision to publish, or preparation of the manuscript.

Competing Interests: The authors have declared that no competing interests exist.

* E-mail: alberto.rodriguez@uq.edu.au

Introduction

Understanding past climate variability and the historical occurrence of extreme weather events, such as tropical cyclones and floods, is critical when predicting the ecological consequences of future climate change as well as preparing for their impacts on human coastal settlements. Although the effects of the El Niño-Southern Oscillation (ENSO) and the Pacific Decadal Oscillation (PDO) on Australian hydrological regimes are relatively well understood [1–6], the 2010–2011 La Niña event, one of the strongest on record [7], severely impacted human communities and coastal ecosystems along the Queensland coast of Australia. At the start of 2011, heavy rainfall caused one of the most significant floods in Australia's recorded history, followed by severe Tropical Cyclone Yasi, which was the strongest cyclone to make landfall in Queensland since 1918 [7]. More recently, by the end of January 2013, ex-Tropical Cyclone Oswald strongly affected human populations along the east coast of Australia due to the extreme rainfall that broke historical records (precipitation and flood) at several localities [8]. Comprehensive historical analyses of such extreme events are scarce because of the lack of long-term instrumental and proxy climate records [9–12]. Therefore, predictions of their frequency and intensity remain uncertain.

Natural archives, such as annually-banded coral skeletons, can be used to derive proxy climate data on seasonal to centennial time-scales, extending far beyond instrumental records [13,14]. For instance, luminescent lines in coral skeletons, which are caused by the incorporation of terrestrial humic acids carried to the reef during flood events [15], are a reliable proxy for reconstructing freshwater inputs to coastal ecosystems and regional precipitation variations [16,17]. While the use of coral luminescence has increased our understanding of how climatic cycles influence rainfall, flood regimes and hurricane activity [18–23], recent advances in luminescence controls and application techniques [24] have revealed previously unidentified relationships with climate phenomena, such as the PDO, contributing to local and regional analyses of past, present, and future climate variability [25].

For the east coast of Australia, ENSO is the dominant driver of inter-annual rainfall variability [26], yet this ENSO–rainfall teleconnection is, in turn, modulated by the Interdecadal Pacific

Oscillation (IPO, similar to PDO) [27,28]. Consequently, historical analysis analyses of rainfall, river discharge and flood risk modelling have identified that during negative/cool PDO phases, the impact of La Niña events on rainfall/floods is greater than during positive/warm PDO phases [6,29–32]. For the GBR, it is known that ENSO events unevenly affect the system but the influence of other large-scale sources of climate variability has not been fully assessed [33]. While some studies on the GBR have verified the relationship between ENSO and coral luminescence [19] and river flow and rainfall reconstructions based on luminescence [17,34], the modulating effect of the PDO on such records has received little attention. Thus far, varying correlations between river flow reconstructions and ENSO indices using warm (1925–1946) and cold (1947–1976) phases of the PDO confirm the non-stationary ENSO-river flow teleconnection for the GBR [20]. Therefore the nature and extent of the relationship between the PDO and coral luminescence records or luminescence-based rainfall/runoff reconstructions remain poorly constrained for the GBR. Determining primary drivers of inter-annual and decadal luminescence will not only allow better rainfall/runoff reconstructions but also improve the predictability and management of hydrological-related disturbances impacting human populations and reef ecosystems along the GBR catchment area.

Here, we present the first decadal-scale (90-year) composite record (1921–2011) of luminescence spectral ratios from multiple coral colonies as an indicator of the Fitzroy River discharge to the Keppel reefs, southern GBR. We also examine potential environmental (runoff and rainfall) and climatic (ENSO and PDO) drivers of luminescence variability on monthly to multi-decadal time scales and discuss key implications for the hydroclimatology of the southern GBR. Our results support growing indications that the future evolution of ENSO and the PDO will determine the frequency and intensity of extreme climatic events affecting Australia's east coast (i.e. floods) and provide new insights into the significant role that the PDO cycles play in coral reef dynamics of the southern GBR.

Methods

Ethics Statement

Sample collection was conducted under the Great Barrier Reef Marine Park Authority (GBRMPA) permit number G10/33402.1.

Study Site and Sampling

Coral cores were collected with a pneumatic drill along the growth axis from six massive *Porites* sp. colonies at water depths between 3–7 m from four locations in the Keppel islands, inshore Great Barrier Reef (23°05′-04′ S and 150°54′-53′ E; Figure 1, Table S1). These continental islands are surrounded by fringing reefs [35] with relatively high coral cover (52%), and are dominated by extensive stands of branching *Acropora* spp [36]. The reefs are influenced by terrestrial run-off from the Fitzroy catchment, the largest seaward-draining catchment discharging to the GBR lagoon with an area of ~144 000 km^2 [37,38]. All sampling sites were located within 50 km of the mouth of the Fitzroy River (Figure 1, Table S1), which is a major source of terrestrial material to the GBR lagoon [39–41] delivering more than 3400 ktonnes/yr of total suspended solids, only second to the Burdekin River [42]. The mean annual river discharge measured at the closest gauging station to the river mouth (Rockhampton) is 4.8×10^6 ML, reaching up to 22×10^6 ML during large flood events [43,44]. The climate of the region is characterized by a

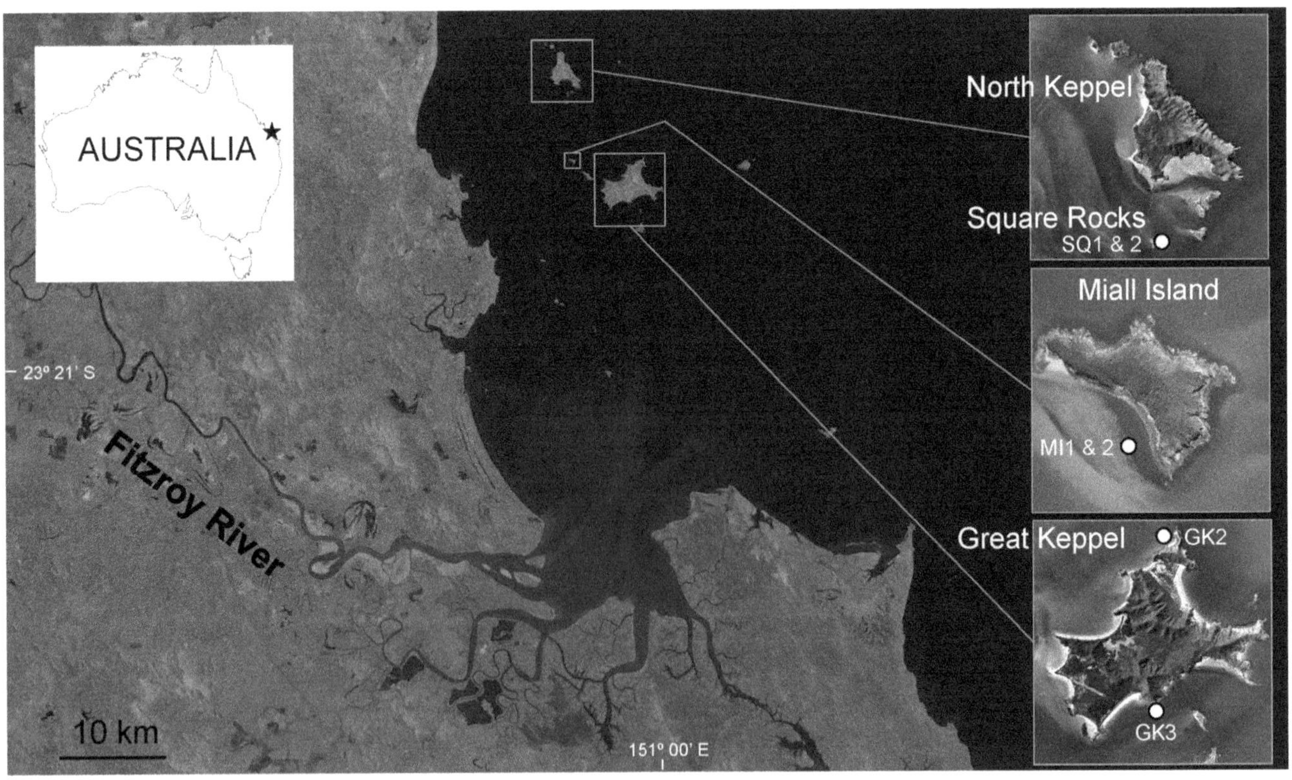

Figure 1. Map displaying the location of the Keppel Islands, sampling sites (white dots) and the Fitzroy River. Additional information about sampling sites is provided in Table S1. Satellite image obtained from http://glovis.usgs.gov/and inset aerial photos courtesy of P. Willams.

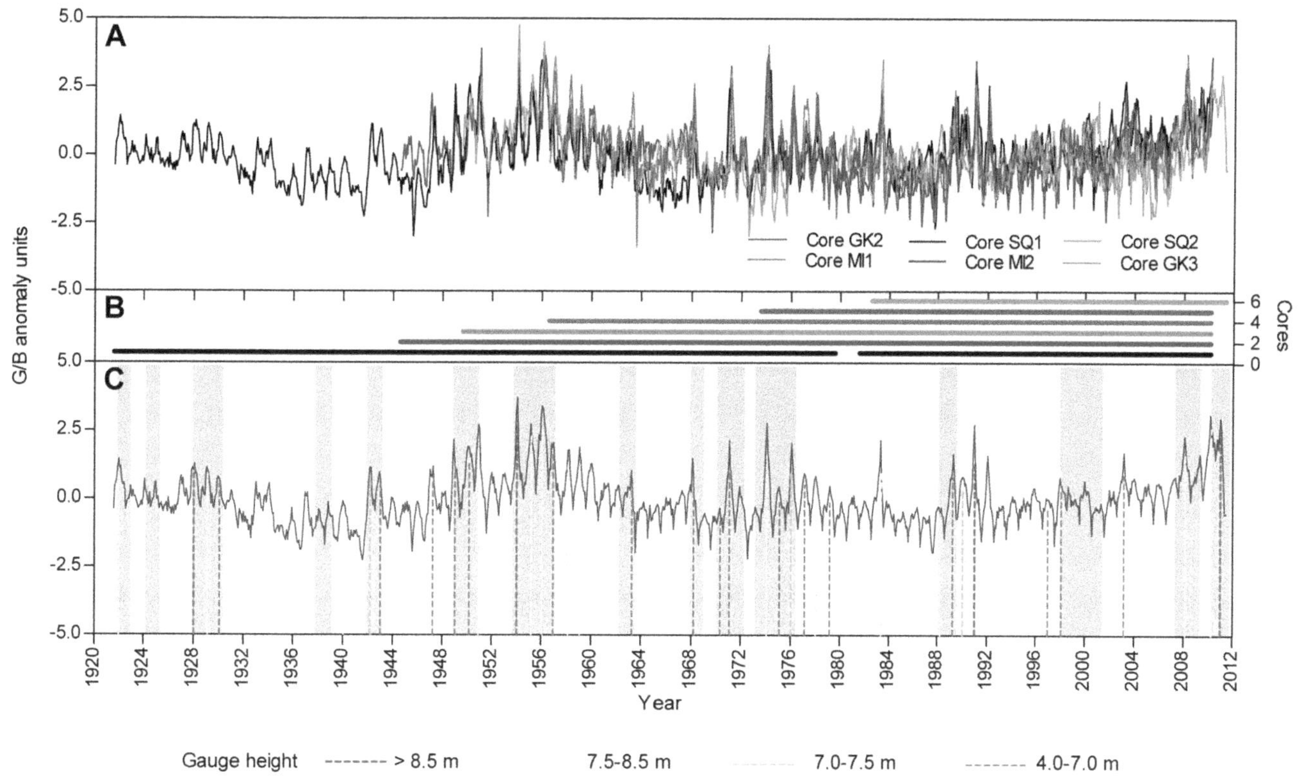

Figure 2. Time series of G/B anomalies for the period 1921–2011. (A) Monthly G/B anomalies for all six cores. (B) Number of cores used to construct the composite record. (C) Long term G/B composite record, with major flood events registered by instrumental records. Colour dashed lines under the profile denote the highest individual flood events registered at Rockhampton, the nearest gauging station to the river mouth (data from Water Division Brisbane, Bureau of Meteorology, Station 039264). Colours refer to gauge height in m. Shaded areas correspond to La Niña periods.

winter dry season (April to September) and a summer wet season (October to March) [20,45]. As a result, river discharge mainly occurs in the summer-wet season [41,46]. Although the Fitzroy River discharges to the south of the Keppel islands, predominant south-east winds and currents promote a north-flowing movement of flood plumes generated during high flow events [47]. Further, the Coriolis Force contributesto diverting flood plumes northwards on the GBR [48].

Luminescence Analysis

Coral cores were cut longitudinally using a circular saw into 7 mm thick slabs. Before luminescence analysis, coral slabs were X-rayed and then treated with NaOCl for 24 h to remove organic contaminants that can quench the luminescence signal [49]. The slabs were ultrasonically rinsed several times with ultrapure water (18.2.MΩ.cm) and subsequently oven-dried for 24 h (50°C).

The Spectral Luminescence Scanning (SLS) technique was applied to the six coral cores in order to quantify skeletal luminescence [24] and to reconstruct the historical river influence in the Keppel reefs. This method scans coral slabs under a UV light source using a line-scan camera (Figure S1), which records luminescence emission intensities into three spectral ranges, blue (B), green (G) and red (R). RGB intensities were acquired using the Line Scan Software Version 1.6 (Avaatech). The spectral G/B ratio was used as a proxy for river runoff as this relationship normalises the humic acid (G) signal derived from hinterland soils to the skeletal aragonite (B) signal [24,50].

Table 1. Correlation coefficients (R) of monthly and annual G/B anomalies between the composite record and each core from 1982 to 2011.

	Composite record (monthly)		Composite record (annual)	
GK2	**0.75**	$p<0.001$	**0.83**	$p<0.001$
SQ1	**0.73**	$p<0.001$	**0.87**	$p<0.001$
SQ2	**0.60**	$p<0.001$	**0.56**	$p=0.002$
MI1	**0.81**	$p<0.001$	**0.91**	$p<0.001$
MI2	**0.78**	$p<0.001$	**0.82**	$p<0.001$
GK3	**0.58**	$p<0.001$	**0.57**	$p=0.002$

Bold values significant at $p<0.05$.

Age Model and Coral Composite G/B Record

Age models were constructed by counting density bands using digital X-rays. In addition to X-rays, crossdating, using conspicuous luminescent lines, validated the age models for all cores [19,24]. Inter-annual chronologies were based on the seasonal cycle of G/B ratios. We assigned the G/B minima values to the driest month (August), according to historical rainfall and river discharge records (see below - Environmental and climatic data). As SLS provides data at sub-weekly resolution [24], G/B time series from each core were linearly interpolated to 12 points per year to obtain monthly chronologies using AnalySeries 2.0 [51].

A composite G/B record spanning 90 years (1921–2011) was created by standardizing the six coral cores by the mean (monthly

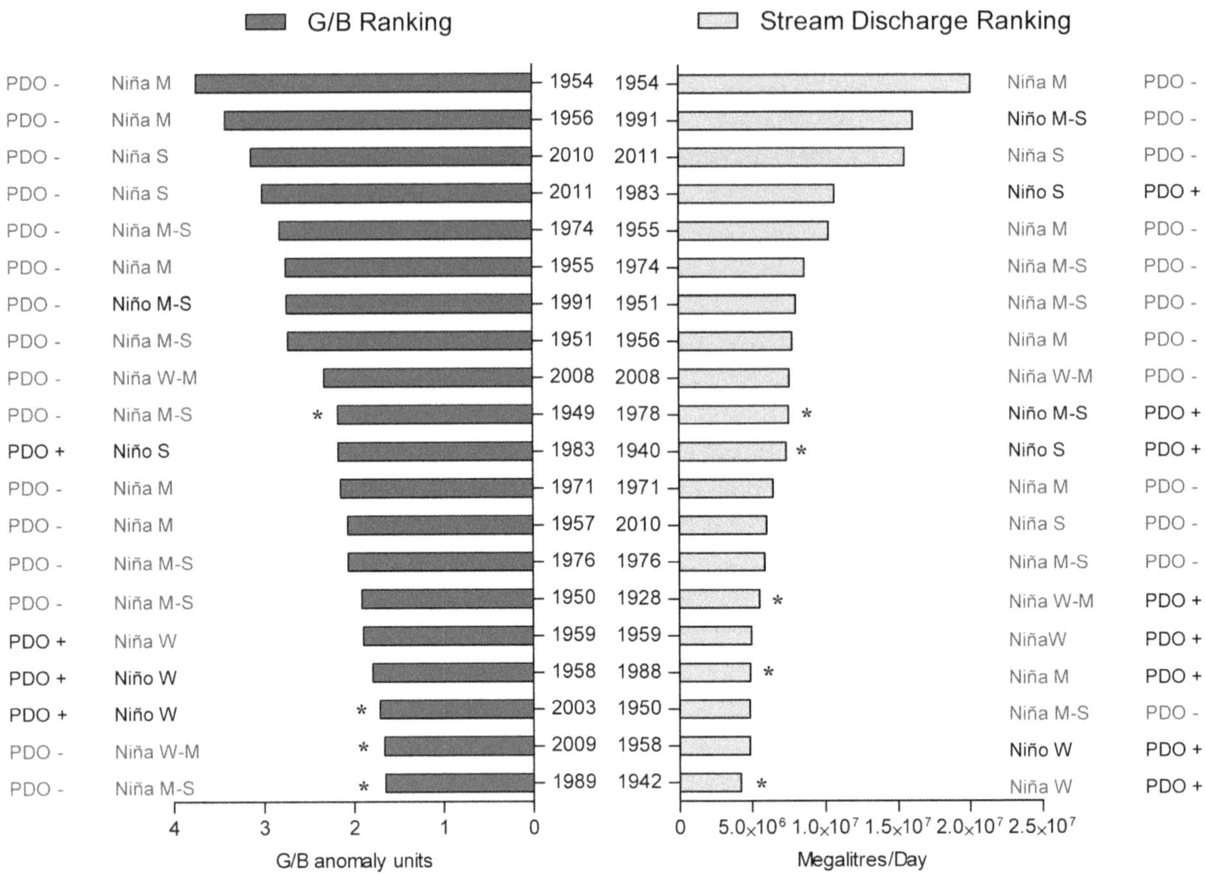

Figure 3. Comparison of the 20 highest G/B anomalies and monthly stream discharge rates for the period 1921–2011. Year, PDO phase (positive/negative) and ENSO state (Niño/Niña) is indicated for each record. Asterisks denote if a record is present only in one ranking. Stream discharge data from Queensland Department of Environment and Resource Management gauging stations on the Fitzroy River at The Gap (Station number 130005A) and Riverslea (130003A) (http://watermonitoring.derm.qld.gov.au/host.htm). W = weak, M = moderate, S = strong.

values) and SD of the time period common to all the cores (1982–2010) and then averaging the standardized records. This procedure reduces the intrinsic variability of individual records and enables the identification of a common regional climatic signal [9,19,25,52]. The agreement between cores was verified by correlating (Pearson linear correlation) G/B monthly and annual anomalies over five common periods for cores with overlapping records (Table S2). Annual anomalies were calculated by averaging all monthly G/B anomalies from August through to July to account for river flow and rainfall extremes during the summer (October-March) in north-eastern Australia [20]. Additionally, the quality of crossdating was assessed by applying a similar approach used for tree-ring chronologies by the program COFECHA [53]. Each individual standardized G/B record was correlated with the average of all other standardized G/B records (the composite record minus the record being tested). A positive and significant correlation indicates that the tested record is crossdated precisely [53].

Table 2. Correlation coefficients (R) of monthly and annual G/B anomalies with environmental and climatic records.

	G/B (monthly)		G/B (annual)		PRECDS
Stream water level (m)	**0.54**	p<0.001	**0.66**	p<0.001	1922–2011*
Stream discharge (ML/day)	**0.47**	p<0.001	**0.64**	p<0.001	1922–2011**
Rainfall (mm)	**0.39**	p<0.001	**0.46**	p<0.001	1921–2011**
SOI	**0.26**	p<0.001	**0.35**	p<0.001	1921–2011**
Niño 3.4	**−0.24**	p<0.001	**−0.29**	p = 0.008	1921–2011**
PDO	**−0.38**	p<0.001	**−0.55**	p<0.001	1921–2011**
CPIPO			**−0.41**	p<0.001	1921–2004***

Bold values significant at p<0.05. Abbreviations: SOI, Southern Oscillation Index; PDO, Pacific Decadal Oscillation; CPIPO, combined paleo IPO-PDO; IPO, Interdecadal Pacific Oscillation. PRECDS, period of record for environmental and climatic data sets. * = incomplete monthly record for the period indicated. ** = complete monthly record for the period indicated. *** = only complete annual values for the period indicated. Further information on data sets is provided in Methods (section Environmental and climatic data).

Luminescence Drivers

To verify the influence of the Fitzroy River on luminescence variability, the highest flood peaks were plotted against the long-term G/B composite record. To validate the composite G/B record, monthly and annual averages of stream water level, stream discharge, and rainfall (see below - Environmental and climatic data) were correlated with monthly and annual G/B values over the period of 1921 to 2011. In addition, agreement between each individual coral record and environmental dataset was assessed by

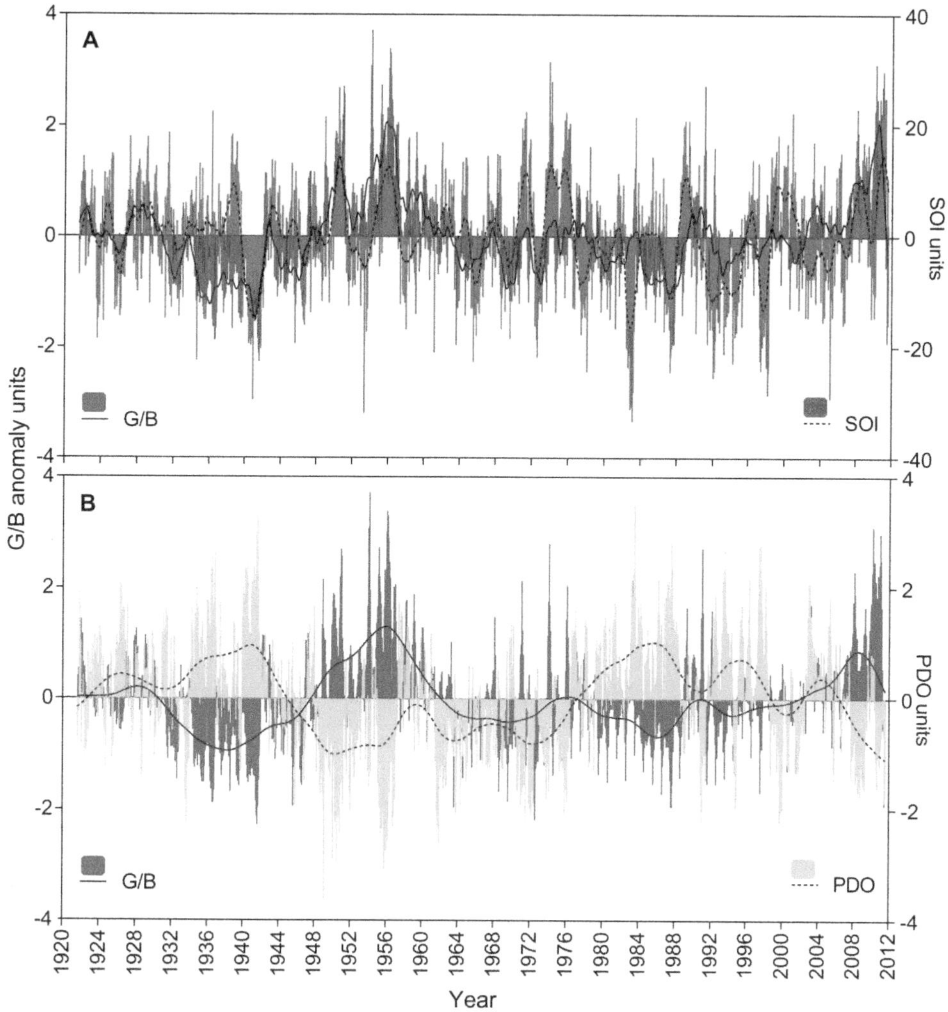

Figure 4. Relationship between monthly values of G/B anomalies and climatic oscillations for the period 1921–2011. (A) Southern Oscillation Index (SOI) and (B) Pacific Decadal Oscillation (PDO). Solid and dashed lines are 12-month and 85-month moving averages respectively. Data at annual resolution is provided in supplementary information (Figure S3).

correlating annual averages of stream water level, stream discharge, and rainfall data with annual G/B anomalies.

To explore potential climatic drivers of luminescence variability, the Southern Oscillation Index (SOI), Niño 3.4, PDO, and CPIPO (combined paleo Interdecadal Pacific Oscillation and Pacific Decadal Oscillation) indices were compared with the long-term G/B composite record using monthly and annual averages over the period 1921 to 2011 (see below Environmental and climatic data). To further examine the temporal variability in luminescence, spectral analysis (REDFIT) [54] was applied to annual G/B anomalies, and correlations between G/B and stream discharge, rainfall, SOI and the PDO index during the Australian climatic seasons were calculated. Monthly values of the G/B anomalies and environmental and climatic variables were averaged accordingly to summer (December to February), autumn (March to May), winter (June to August) and spring (September–November).

Finally, to identify potential drivers explaining most of the variability in the composite record, a distance-based linear model (DISTLM), using a resemblance matrix of G/B (based on Euclidean distance) and the forward procedure, was applied [55,56]. Forward selection adds one variable at a time to the model, choosing the variable at each step which produces the greatest improvement in the value of the selection criterion. We used adjusted R^2 as selection criterion instead of R^2 as we aimed to include only predictor variables that significantly explained the variation in the model. Predictor variables comprised river discharge, rainfall, the SOI and PDO indices. DISTLM outcomes provide a marginal test, fitting each variable individually (ignoring all other variables), and a sequential test, fitting each variable one at a time, conditional on the variables that were already included in the model [55,57]. Analyses were done using PERMANOVA+ for PRIMER v6.

Environmental and Climatic Data

Historical flood peaks (m) were obtained from Rockhampton, the nearest gauging station to the Fitzroy river mouth. This manual recording station registers only when the Fitzroy River exceeds a minimum height threshold (http://www.bom.gov.au/hydro/flood/qld/brochures/fitzroy). River flow data (Megaliters/day) and stream water level (m) were obtained from the Queensland Department of Environment and Resource Management gauging station on the Fitzroy River at The Gap (Station number 130005A) and Riverslea (130003A) (http://watermonitoring.derm.qld.gov.

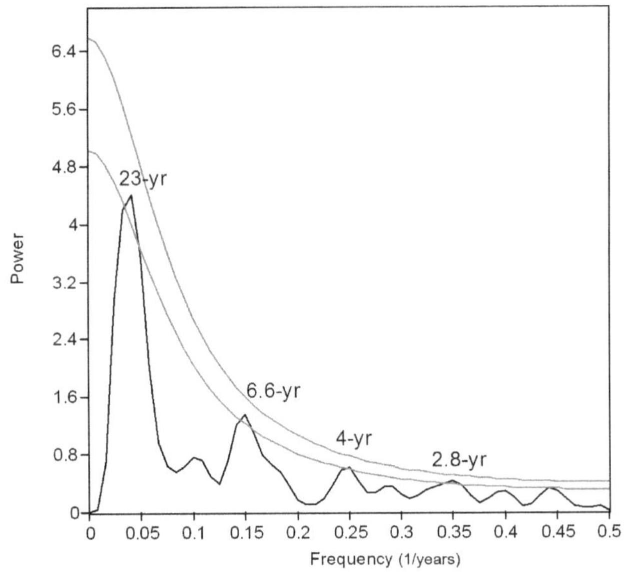

Figure 5. Spectral analysis (REDFIT) of annual G/B anomalies. Significant spectral peaks are indicated. Green solid lines show false-alarm levels of 95% and 99%.

au/host.htm), and rainfall (mm) from the Australian Bureau of Meteorology at Pacific Heights (Station number 033077) (http://www.bom.gov.au/climate/data). These river and rainfall data sets provided the longest records for comparative purposes. The SOI and PDO data were obtained from the Australian Bureau of Meteorology (http://www.bom.gov.au/climate/data), and from the Joint Institute for the Study of the Atmosphere and Ocean (http://jisao.washington.edu/pdo/PDO.latest), respectively. The combined paleo Interdecadal Pacific Oscillation (IPO) and PDO index (CPIPO) was obtained from Henley et al. [58]. The Niño 3.4 data was obtained from the Climate and Global Dynamics Division (CGD) of the National Center for Atmospheric Research (http://www.cgd.ucar.edu/cas/catalog/climind/TNI_N34/index.html#Sec5). We will now refer to the PDO and IPO collectively as the PDO-IPO, since they are considered the same broad-scale climatic phenomenon [28,58–60], unless the distinction is necessary.

Results

The six cores analysed here showed excellent reproducibility in terms of luminescence (G/B ratios). The six G/B time series showed similar variations over the time period common to all the cores 1982–2010 (Figure 2A). A strong agreement in G/B profiles was also evident, even for the longest coral records (core GK2 and SQ1; Figure 2A). Significant correlations of both monthly and annual G/B anomalies for cores with overlapping records showed the high consistency between cores (Table S3; Table S5–S9). Only one non-significant correlation of annual G/B anomalies was observed between cores SQ2 and MI2 (1973–2010) (Table S6). The correlation was significant, however, when considering monthly data (Table S6). Strong significant correlations were observed between (1) the G/B composite record and each G/B record for the time period common to all the cores (Table 1) and (2) the rest of the time periods with overlapping records (Table S6–S7), confirming that the cores were crossdated correctly and therefore, the common environmental signal (runoff) was optimized.

Table 4. Results of the marginal test performed by DISTLM-forward analysis.

Variable	SS(trace)	Pseudo-F	p	PROP
Stream discharge	26.072	67.357	0.001	**0.44**
PDO	17.771	36.834	0.001	**0.30**
SOI	7.2578	12.03	0.003	**0.12**
Rainfall	12.241	22.417	0.001	**0.20**

Significant proportions of explained variation (PROP) are given in bold.

The long-term composite record displayed a seasonal cycle in monthly G/B values that was strongly influenced by the Fitzroy River discharge (Figure 2C). The highest G/B luminescence peaks corresponded to significant flood events registered at Rockhampton gauge. Moreover, the ranking of the 20 highest G/B anomaly peaks were very similar (though not identical in order) to the ranking of the 20 highest rates of stream discharge recorded during the wet season from 1921 to 2011 (Figure 3). This relationship was confirmed by the significant correlations observed between the G/B composite record (monthly and annual G/B anomalies) and all instrumental records available (water level, stream discharge, and rainfall; Table 2). The strongest relationships observed were between the annual values of G/B and the stream flow indicators (stream water level, $R = 0.66$, $p < 0.001$; stream discharge, $R = 0.64$, $p < 0.001$; Table 2; Figure S2). Individual G/B chronologies (GK2, SQ1, SQ2, MI1, MI2 and GK3) also showed significant correlations with the environmental data over the six different periods covered by cores (Table S4).

The long-term composite record exhibited strong interactions with sources of inter-annual and decadal climate variability (Figures 4 and S3). While monthly G/B and SOI anomalies tended to co-vary (Figure 4A), positive and negative monthly G/B anomalies corresponded remarkably well to negative and positive

Table 3. Correlation coefficients (R) of seasonal G/B anomalies with environmental and climatic records.

	G/B Summer		G/B Autumn		G/B Winter		G/B Spring	
Stream discharge (ML/day)	**0.69**	p<0.001	**0.58**	p<0.001	**0.44**	p<0.001	**0.29**	p = 0.007
Rainfall (mm)	**0.56**	p<0.001	**0.30**	p = 0.004	0.13	p = 0.23	0.20	p = 0.06
SOI	**0.31**	p = 0.003	0.20	p = 0.06	**0.39**	p<0.001	**0.38**	p<0.001
PDO	**−0.55**	p<0.001	**−0.43**	p<0.001	**−0.39**	p<0.001	**−0.55**	p<0.001

Bold values significant at p<0.05. Correlation for period 1921–2011. Abbreviations: SOI, Southern Oscillation Index; PDO, Pacific Decadal Oscillation. Information on data sets is provided in Methods (section Environmental and climatic data).

Table 5. Results of the sequential test performed by DISTLM-forward analysis.

Variable	Adjusted R^2	SS(trace)	Pseudo-F	p	PROP	CUMUL
Stream discharge	0.4299	26.072	67.357	0.001	**0.44**	0.44
PDO	0.5283	6.1315	19.145	0.001	**0.10**	0.54
SOI	0.5235	4.78E-02	0.14762	0.689	7.99E-04	0.54
Rainfall	0.5180	8.62E-03	2.63E-02	0.882	1.44E-04	0.54

Significant proportions of explained variation (PROP) are given in bold. CUMUL = cumulative proportion of explained variation.

PDO phases, respectively (Figure 4B). Indeed, 80% of the 20 highest G/B anomalies occurred during La Niña events (positive values of SOI) and negative PDO phases (Figure 3). Similarly, within the 20 highest records of stream discharge, 75% matched La Niña events while 60% corresponded to negative phases of the PDO (Figure 3). Correlations between monthly and annual G/B with ENSO (SOI and Niño 3.4) and PDO-IPO indices (PDO and CPIPO) showed significant relationships (Table 2). The strongest of these relationships were observed between G/B annual anomalies and the PDO index ($R = -0.55$, $p<0.001$; Table 2) and with the SOI index ($R = 0.35$, $p<0.001$; Table 2). Spectral analysis indicated cycles of 23 and 2.8–6.6 years dominated luminescence variability in the reconstructed record (Figure 5).

Correlations between G/B anomalies and environmental and climatic data (stream discharge, rainfall, SOI and PDO), using seasonal averages, revealed that the strength of relationships varied between seasons yet remained significant for most variables (Table 3). During summer (wet season), G/B anomalies showed the most robust correlations with all variables (Table 3).

DISTLM analysis indicated that individually, all predictor variables (see marginal test, Table 4) explained a significant amount of the variation in the composite record, in agreement with the results obtained by correlations. Yet, the variability of the composite record was mainly explained by stream discharge and the PDO, which together accounted for 54% of the total variation (see sequential test, Table 5). While the PDO added 10% to the explained variation when stream discharge was fitted (44%), the contribution of rainfall and ENSO (SOI index) was negligible and non-significant (Table 5).

Discussion

Here we provide a 90-year record of coral luminescence (G/B) based on multiple cores from the Keppel reefs, which included the 2011 flood event, one of the most significant floods recorded in the history of the southern GBR (Figure 3). Strong correlations between individual G/B records, as well as between the composite record, and each G/B series demonstrate the presence of a regional signal. Consequently, our luminescence composite record was consistent with historical flood peaks and showed significant correlations with rainfall, stream water level and discharge data. Further, DISTLM analysis revealed that the most important driver of luminescence was stream discharge as this predictor explained most of the variability in the record (44%). These linkages between luminescence and environmental variables confirmed that the Fitzroy River catchment largely influences the Keppel reefs, and therefore corals from these reefs are suitable for reconstructing regional river discharge and/or precipitation variability on monthly to decadal time scales. Similar to earlier studies from inshore reefs [16,17,34], we found that coral luminescence captured the variability of hydrological regimes in the southern GBR.

Spectral analysis confirmed that our composite record has cycles consistent with the typical periodicities for the PDO-IPO (15–25 years) [61], and ENSO (3–6 years) [62], yet luminescence data also indicated that river runoff in the Keppel reefs is differentially influenced by these climatic forces. Clearly, G/B variability followed the asymmetric teleconnection of ENSO with Australian climate [5,6,29], whereby positive G/B anomalies were commonly amplified during strong/moderate La Niña phases of ENSO, and negative G/B anomalies were occasionally enhanced during extreme El Niño events (Figure 4A). Our results, along with previous evidence on the non-stationary relationship between luminescence (runoff/rainfall) and ENSO [19,20], confirm the widespread influence of this climatic oscillation on inshore reefs of the GBR. However, the luminescence record (G/B) consistently showed stronger correlations with the PDO than the SOI irrespective of whether monthly, seasonal or annual averages were examined. Further, DISTLM analysis indicated that the PDO explained a significant proportion (10%) of variation in luminescence, and is the only climatic variable contributing significantly to the variation explained by stream discharge. These results do not detract from the ENSO-luminescence/runoff relationship but highlight the significant role of the PDO in modulating river runoff in the Keppel reefs. A recent study documented that ENSO is not the primary source of inter-annual climate variability on the GBR and recognized that other sources such as the PDO need to be further investigated [33].

The most significant finding of our study is the relationship between coral luminescence and the PDO, which was consistently verified by both simple correlations over a range of time scales and the multivariate model. Lough [19] previously used warm and cool PDO phases as reference periods to verify links between reconstructed rainfall and river flow series based on coral luminescence and ENSO. Our study, however, is the first to directly establish the link between coral luminescence and the PDO for the southern GBR region. Further, we identified that the luminescence composite record closely mirrored the PDO index over the last century (Figures 4B and S3B). Prevalent negative G/B anomalies (reduced runoff) coincided with positive PDO phases, while periods of predominant positive G/B anomalies (increased runoff) coincided with negative PDO phases. In addition, the magnitude of G/B anomalies within these periods varied consistently with the magnitude of the PDO (Figures 4B and S3B). Therefore, our study reveals the extent of the PDO influence on the southern GBR, whilst also supporting earlier studies showing how the hydroclimatology of eastern Australia is profoundly influenced by the PDO-IPO [3–6,63,64].

Since G/B is an indicator of humic acid runoff [24,25], its variability can also support historical analysis of human-induced erosion impacting the GBR. Decadal frequency of extreme luminescence anomalies showed that major runoff events increased after 1950 and were particularly confined to the 50's (Figure S4), suggesting a higher influence of continental runoff (increased sediment loads) on the Keppel reefs. While instrumental data and river flow reconstructions along the GBR region show a similar variability, with increased flow conditions in the 1950s and 1970s [20], pointing to a large-scale climatic driver, the intensification of anthropogenic activities in the Fitzroy catchment by the mid-20th century [65] may also explain extreme luminescence anomalies. Historic records of human activities in the Fitzroy River catchment indicate that major shifts were related to intensive clearing of the native forests (*Acacia harpophylla*) from the 1960s to 1980s [66], increase in beef cattle numbers from 1955 [67], and expansion in coal mining since the 1970s [68]. Indeed, total suspended solids load from the Fitzroy River to the GBR is estimated to have increased 3.1×(1100–3400 ktonnes/yr) since European settlement [42]. While the length of our record (1921–2011) precludes the interpretation of luminescence changes related to European colonisation (after about 1870), the occurrence of the highest G/B peaks during the period of increased catchment modification suggests that the strong climatic signal on the luminescence record may have been enhanced by land-use changes. Increases in baseline values of runoff proxies from corals (Ba/Ca, Y/Ca) have been linked to human settlement periods and subsequent modification of river catchments by the mid to late 19th century in the central GBR [69–71]. Luminescence records and trace element analysis from coral cores predating European settlement are certainly required to unequivocally decouple the potential human component from the climatic signal.

Implications of Coral Luminescence-PDO Relationship

The clear link between coral luminescence and the PDO-IPO documented here may assist studies in the historical variability of the PDO-IPO, which are critical to improving and understanding the predictability of climate change impacts in the Pacific region [58,59,72,73]. Thus far, only eight paleo PDO-IPO time series of centennial scale exist [58, and references therein], and few studies have developed a reconstruction of the PDO using corals [58,74]. Fingerprints of the PDO-IPO in corals from the South Pacific Convergence Zone [74–77] and the North Pacific region [78] have previously been derived from temperature/salinity proxies ($\delta^{18}O$, Sr/Ca, U/Ca). For the GBR, Calvo et al. [79] found agreement between Sr/Ca and $\delta^{18}O$ records and the IPO index, Pelejero et al. [80] reported covariation of $\delta^{11}B$ record (proxy for paleo-pH) and the PDO-IPO, and Lough [20] found that during the 1947–1976 PDO negative phase, the instrumental and reconstructed river flow and rainfall records were significantly correlated with ENSO indices. Thus, our study reveals a novel signature of the PDO in a non-temperature based coral proxy for the western South Pacific. While modern massive corals under the influence of rivers may contain invaluable century-length records of the PDO-IPO, fossil corals from the late Quaternary preserving luminescence lines [81] may also be useful for PDO-IPO reconstructions at millennial timescales.

The significant relationship between coral luminescence and the PDO-IPO provided insights into past disturbance regimes at the Keppel reefs. Recent studies documented extensive coral mortalities of 85% in 1991, around 40% in 2008 and 2010, and up to 100% in 2011 after severe floods from the Fitzroy River [17,82 85]. Because these floods imprinted marked G/B peaks on the cores (Figure 2), they provide a basis to infer flood-induced disturbances in earlier decades prior to coral reef monitoring and instrumental records. Similarly, luminescence and runoff proxies have been used in retrospective analysis of historical disturbances and coral reef community changes on the GBR [86] Thus, based on the magnitude of G/B anomalies, we suggest that Keppel reefs likely faced coral mortality of varying extents in the wet seasons of 1951, 1954–1956, 1974 and 1983. In those years along with 1991, 2008, 2010 and 2011 (years of documented flood-induced mortality), the highest monthly discharge records of the Fitzroy River were also registered (Figure 3). Importantly, most of these events coincided with moderate-strong La Niña events during the negative phase of the PDO (Figure 3). Hence, we propose that La Niña-PDO/IPO cycles have played a significant role in the reef dynamics of the southern GBR.

Conclusions

The luminescence record presented here displayed the temporal variability of river runoff on Keppel reefs and how such variability is influenced by the strength of La Niña and significantly modulated by the negative phase of PDO-IPO. Because recent evidence indicates that the effects of climatic phenomena (i.e. ENSO events) may vary spatially in the GBR [33], the strong relationship between PDO and coral luminescence documented here suggests that the southern GBR is particularly sensitive to the PDO influence. This implies that corals and other palaeoclimate archives from the Keppel Islands have great potential for studying the historical link between the PDO and climate variability for the southern GBR.

This study demonstrates that the PDO is a primary climatic driver of river runoff affecting the southern region of the GBR and supports growing indications that the PDO-IPO directly and indirectly influences biological communities, marine ecosystems and climate-related extreme events such as cyclones and floods/drought [31,58,87–97]. Given that emerging evidence points to a transition towards a negative PDO-IPO phase, which could increase rainfall and risk to flooding during La Niña events in Australia [32,59,74], we emphasize the critical role of the PDO-IPO and La Niña events in mediating disturbance and ecological processes for the GBR region in the context of rapid climate change.

Supporting Information

Figure S1 (A) An example (core MI1) of the digital image obtained by the Spectral Luminescence Scanning (SLS) technique. The orange lines indicate transects used to extract the down-core luminescence data. (B) Monthly G/B time-series obtained from core MI1.

Figure S2 Relationship between annual G/B anomalies and stream discharge rates anomalies for the period 1921–2011. Stream discharge data from Queensland Department of Environment and Resource Management gauging stations on the Fitzroy River at The Gap (Station number 130005A) and Riverslea (130003A) (http://watermonitoring.derm.qld.gov.au/host.htm).

Figure S3 Relationship between annual G/B anomalies and climatic oscillations for the period 1921–2011. (A) Southern Oscillation Index (SOI) and (B) Pacific Decadal Oscillation (PDO). Solid and dashed lines are 6-year moving averages.

Figure S4 Number of extreme G/B anomalies (>1.5 units) per decade for the period 1921–2011.

Table S1 Location and details of corals used in luminescence reconstructions from Keppel islands on the Great Barrier Reef.

Table S2 Common periods and the cores associated.

Table S3 Summary of correlation results for monthly and annual G/B anomalies among cores at different overlapping periods.

Table S4 Correlation coefficients (R) between annual G/B anomalies and environmental records, SOI and PDO for each individual coral core.

Table S5 Correlation coefficients (R) of monthly (upper) and annual (lower) G/B anomalies between cores sharing records from 1982 to 2010.

Table S6 Correlation coefficients (R) of monthly (upper) and annual (lower) G/B anomalies between cores sharing records from 1973 to 2010. Last column includes correlation coefficients between the composite record and each core for the same period.

Table S7 Correlation coefficients (R) of monthly (upper) and annual (lower) G/B anomalies between cores sharing records from 1956 to 2010. Last column includes correlation coefficients between the composite record and each core for the same period.

Table S8 Correlation coefficients (R value) of monthly (upper) and annual (lower) G/B anomalies cores sharing records from 1949 to 2010. Last column includes correlation coefficients between the composite record and each core for the same period.

Table S9 Correlation coefficients (R value) of monthly (upper) and annual (lower) G/B anomalies between cores sharing records from 1944 to 2010. Last column includes correlation coefficients between the composite record and each core for the same period.

Acknowledgments

ARR thanks Tara Clark, Sander Scheffers and Peter Williams for fieldwork assistance. We specially thank Catalina Reyes-Nivia for help during field, laboratory work, and valuable comments on this paper. We thank the NIOZ for technical support with luminescence scanning. Cores were sectioned at the RSES, Australian National University, with the support of Michael Gagan and Heather Scott-Gagan, and Oceans Institute and School of Earth and Environment, The University of Western Australia, with the support of Malcolm McCulloch and Juan Pablo D'Olivo. We also thank to Ben Henley for providing CPIPO data.

Author Contributions

Conceived and designed the experiments: ARR CAG JMP JXZ. Performed the experiments: ARR CAG. Analyzed the data: ARR CAG JZ. Contributed reagents/materials/analysis tools: ARR CAG JZ JMP JXZ. Wrote the paper: ARR CAG JZ JMP JXZ.

References

1. Chiew FHS, Piechota TC, Dracup JA, McMahon TA (1998) El Nino/Southern Oscillation and Australian rainfall, streamflow and drought: Links and potential for forecasting. Journal of Hydrology 204: 138–149.
2. Holbrook NJ, Davidson J, Feng M, Hobday AJ, Lough JM, et al. (2009) El Niño-Southern Oscillation. In: E.S Poloczanska, Hobday AJ, Richardson AJ, editors. A Marine Climate Change Impacts and Adaptation Report Card for Australia 2009: NCCARF Publication 05/09. pp. 1–25.
3. Arblaster JA, Meehl GM, Moore AM (2002) Interdecadal modulation of Australian rainfall. Clim Dynam 18: 519–531.
4. Micevski T, Franks SW, Kuczera G (2006) Multidecadal variability in coastal eastern Australian flood data. Journal of Hydrology 327: 219–225.
5. Power S, Casey T, Folland C, Colman A, Mehta V (1999) Inter-decadal modulation of the impact of ENSO on Australia. Clim Dynam 15: 319–324.
6. Verdon DC, Wyatt AM, Kiem AS, Franks SW (2004) Multidecadal variability of rainfall and streamflow: Eastern Australia. Water Resour Res 40: W10201.
7. BOM (2012) Annual Climate Summary 2011. Australian Goverment Bureau of Meteorology. 24 p.
8. BOM (2013) Queensland in January 2013: Extreme heat in the west and extreme rainfall and flooding on the coast. Australian Goverment Bureau of Meteorology Monthly Climate Summary for Queensland. Available: http://www.bom.gov.au/climate/current/month/qld/archive/201301.summary.shtml#summary: Accessed 25 Feb 2013.
9. Jones PD, Briffa KR, Osborn TJ, Lough JM, van Ommen TD, et al. (2009) High-resolution palaeoclimatology of the last millennium: a review of current status and future prospects. The Holocene 19: 3–49.
10. McGregor S, Timmermann A, Timm O (2010) A unified proxy for ENSO and PDO variability since 1650. Clim Past 6: 1–17.
11. Nott J, Hayne M (2001) High frequency of 'super-cyclones' along the Great Barrier Reef over the past 5,000 years. Nature 413: 508–512.
12. Rayner NA, Parker DE, Horton EB, Folland CK, Alexander LV, et al. (2003) Global analyses of sea surface temperature, sea ice, and night marine air temperature since the late nineteenth century. J Geophys Res 108: 4407.
13. Grottoli AG, Eakin CM (2007) A review of modern coral $\delta^{18}O$ and $\Delta^{14}C$ proxy records. Earth Sci Rev 81: 67–91.
14. Lough JM (2010) Climate records from corals. Wiley Interdisciplinary Reviews: Climate Change 1: 318–331.
15. Isdale P (1984) Fluorescent bands in massive corals record centuries of coastal rainfall. Nature 310: 578–579.
16. Lough JM (2011) Measured coral luminescence as a freshwater proxy: comparison with visual indices and a potential age artefact. Coral Reefs 30: 169–182.
17. Lough JM (2011) Great Barrier Reef coral luminescence reveals rainfall variability over northeastern Australia since the 17th century. Paleoceanography 26: 1–14.
18. Ayliffe LK, Bird MI, Gagan MK, Isdale PJ, Scott-Gagan H, et al. (2004) Geochemistry of coral from Papua New Guinea as a proxy for ENSO ocean-atmosphere interactions in the Pacific Warm Pool. Cont Shelf Res 24: 2343–2356.
19. Hendy EJ, Gagan MK, Lough JM (2003) Chronological control of coral records using luminescent lines and evidence for non-stationary ENSO teleconnections in northeast Australia. The Holocene 13: 187–199.
20. Lough JM (2007) Tropical river flow and rainfall reconstructions from coral luminescence: Great Barrier Reef, Australia. Paleoceanography 22: 1–16.
21. Maina J, de Moel H, Vermaat JE, Henrich Bruggemann J, Guillaume MMM, et al. (2012) Linking coral river runoff proxies with climate variability, hydrology and land-use in Madagascar catchments. Mar Pollut Bull 64: 2047–2059.
22. Nyberg J (2002) Luminescence intensity in coral skeletons from Mona Island in the Caribbean Sea and its link to precipitation and wind speed. Philos Transact A Math Phys Eng Sci 360: 749–766.
23. Nyberg J, Malmgren BA, Winter A, Jury MR, Kilbourne KH, et al. (2007) Low Atlantic hurricane activity in the 1970s and 1980s compared to the past 270 years. Nature 447: 698–701.
24. Grove C, Nagtegaal R, Zinke J, Scheufen T, Koster B, et al. (2010) River runoff reconstructions from novel spectral luminescence scanning of massive coral skeletons. Coral Reefs 29: 579–591.
25. Grove CA, Zinke J, Peeters F, Park W, Scheufen T, et al. (2013) Madagascar corals reveal a multidecadal signature of rainfall and river runoff since 1708. Clim Past 9: 641–656.
26. Risbey JS, Pook MJ, McIntosh PC, Wheeler MC, Hendon HH (2009) On the Remote Drivers of Rainfall Variability in Australia. Monthly Weather Review 137: 3233–3253.

27. Klingaman NP, Woolnough SJ, Syktus J (2013) On the drivers of inter-annual and decadal rainfall variability in Queensland, Australia. Int J Climatol 33: 2413–2430.
28. Power S, Haylock M, Colman R, Wang X (2006) The predictability of interdecadal changes in ENSO activity and ENSO teleconnections. J Clim 19: 4755–4771.
29. Cai W, van Rensch P, Cowan T, Sullivan A (2010) Asymmetry in ENSO teleconnection with regional rainfall, its multidecadal variability, and impact. J Clim 23: 4944–4955.
30. Franks SW, Kuczera G (2002) Flood frequency analysis: Evidence and implications of secular climate variability, New South Wales. Water Resour Res 38: 1062.
31. Kiem AS, Franks SW, Kuczera G (2003) Multi-decadal variability of flood risk. Geophys Res Lett 30: 1035.
32. King AD, Alexander LV, Donat MG (2013) Asymmetry in the response of eastern Australia extreme rainfall to low-frequency Pacific variability. Geophys Res Lett 40: 2271–2277.
33. Redondo-Rodriguez A, Weeks SJ, Berkelmans R, Hoegh-Guldberg O, Lough JM (2012) Climate variability of the Great Barrier Reef in relation to the tropical Pacific and El Niño-Southern Oscillation. Mar Freshw Res 63: 34–47.
34. Isdale PJ, Stewart BJ, Tickle KS, Lough JM (1998) Palaeohydrological variation in a tropical river catchment: a reconstruction using fluorescent bands in corals of the Great Barrier Reef, Australia. The Holocene 8: 1–8.
35. Hopley D, Smithers SG, Parnell KE (2007) The geomorphology of the Great Barrier Reef. New York: Cambridge University Press. 532 p.
36. Jones AM, Berkelmans R, Houston W (2011) Species richness and community structure on a high latitude reef: Implications for conservation and management. Diversity 3: 329–355.
37. Douglas GB, Ford PW, Palmer M, Noble RM, Packett R (2006) Fitzroy River, Queensland, Australia. II. Identification of Sources of Estuary Bottom Sediments. Environ Chem 3: 377–385.
38. Neil DT, Orpin AR, Ridd PV, Yu B (2002) Sediment yield and impacts from river catchments to the Great Barrier Reef lagoon: a review. Mar Freshw Res 53: 733–752.
39. Bostock HC, Brooke BP, Ryan DA, Hancock G, Pietsch T, et al. (2007) Holocene and modern sediment storage in the subtropical macrotidal Fitzroy River estuary, Southeast Queensland, Australia. Sediment Geol 201: 321–340.
40. Devlin MJ, McKinna LW, Álvarez-Romero JG, Petus C, Abott B, et al. (2012) Mapping the pollutants in surface riverine flood plume waters in the Great Barrier Reef, Australia. Mar Pollut Bull 65: 224–235.
41. Furnas M (2003) Catchments and corals, terrestrial runoff to the Great Barrier Reef. TownsvilleQld: Australian Institute of Marine Science CRC Reef Research Center. 334 p.
42. Kroon FJ, Kuhnert PM, Henderson BL, Wilkinson SN, Kinsey-Henderson A, et al. (2012) River loads of suspended solids, nitrogen, phosphorus and herbicides delivered to the Great Barrier Reef lagoon. Mar Pollut Bull 65: 167–181.
43. Douglas GB, Ford PW, Palmer MR, Noble RM, Packett RJ, et al. (2008) Fitzroy River Basin, Queensland, Australia. IV. Identification of flood sediment sources in the Fitzroy River. Environ Chem 5: 243–257.
44. Smith J, Douglas GB, Radke LC, Palmer M, Brooke BP (2008) Fitzroy River Basin, Queensland, Australia. III. Identification of sediment sources in the coastal zone. Environ Chem 5: 231–242.
45. Lough JM (1991) Rainfall variations in Queensland, Australia: 1891–1986. Int J Climatol 11: 745–768.
46. Furnas M, Mitchell A (2000) Runoff of terrestrial sediment and nutrients into the Great Barrier Reef World Heritage Area. In: Wolanski E, editor. Oceanographic processes of coral reefs: physical and biological links in the Great Barrier Reef. United States of America: CRC Press. pp. 37–51.
47. van Woesik R (1991) Immediate impact of the January 1991 floods on the coral assemblages of the Keppel Islands, Great Barrier Reef Marine Park Authority; Great Barrier Reef Marine Park Authority, editor. Townsville: Great Barrier Reef Marine Park Authority. 24 p.
48. King B, McAllister F, Wolanski E, Done T, Spagnol S (2000) River Plume Dynamics in the Central Great Barrier Reef. In: Wolanski E, editor. Oceanographic Processes of Coral Reefs: Physica land Biological Links in the Great Barrier Reef: CRC Press. pp. 145–159.
49. Nagtegaal R, Grove CA, Kasper S, Zinke J, Boer W, et al. (2012) Spectral luminescence and geochemistry of coral aragonite: Effects of whole-core treatment. Chem Geol 318: 6–15.
50. Grove CA, Zinke J, Scheufen T, Maina J, Epping E, et al. (2012) Spatial linkages between coral proxies of terrestrial runoff across a large embayment in Madagascar. Biogeosciences 9: 3063–3081.
51. Paillard D, Labeyrie L, Yiou P (1996) Macintosh Program performs time-series analysis. Eos 77: 379.
52. Pfeiffer M, Dullo W-C, Zinke J, Garbe-Schönberg D (2009) Three monthly coral Sr/Ca records from the Chagos Archipelago covering the period of 1950–1995 A.D.: reproducibility and implications for quantitative reconstructions of sea surface temperature variations. Int J Earth Sci 98: 53–66.
53. Grissino-Mayer HD (2001) Evaluating crossdating accuracy:a manual and tutorial for the computer program COFECHA. Tree-Ring Research 57: 205–221.
54. Schulz M, Mudelsee M (2002) REDFIT: estimating red-noise spectra directly from unevenly spaced paleoclimatic time series. Comput Geosci 28: 421–426.
55. Anderson MJ, Gorley RN, Clarke KR (2008) PERMANOVA+ for PRIMER: Guide to Software and Statistical Methods. UK: PRIMER-E Ltd. 214 p.
56. Legendre P, Anderson MJ (1999) Distance-based redundancy analysis: testing multispecies responses in multifactorial ecological experiments. Ecol Monogr 69: 1–24.
57. Anderson MJ (2003) DISTLM Forward: A FORTRAN computer program to calculate a distance-based multivariate analysis for a Linear Model using forward selection. New Zealand: Department of Statistics, University of Auckland. 10 p.
58. Henley BJ, Thyer MA, Kuczera G, Franks SW (2011) Climate-informed stochastic hydrological modeling: Incorporating decadal-scale variability using paleo data. Water Resour Res 47: W11509.
59. Cai W, van Rensch P (2012) The 2011 southeast Queensland extreme summer rainfall: A confirmation of a negative Pacific Decadal Oscillation phase? Geophys Res Lett 39: L08702.
60. Folland CK, Renwick JA, Salinger MJ, Mullan AB (2002) Relative influences of the Interdecadal Pacific Oscillation and ENSO on the South Pacific Convergence Zone. Geophys Res Lett 29: 1643.
61. Minobe S (1999) Resonance in bidecadal and pentadecadal climate oscillations over the North Pacific: Role in climatic regime shifts. Geophys Res Lett 26: 855–858.
62. Meehl GA, Gent PR, Arblaster JM, Otto-Bliesner BL, Brady EC, et al. (2001) Factors that affect the amplitude of El Nino in global coupled climate models. Clim Dynam 17: 515–526.
63. Heinrich I, Weidner K, Helle G, Vos H, Lindesay J, et al. (2009) Interdecadal modulation of the relationship between ENSO, IPO and precipitation: insights from tree rings in Australia. Clim Dynam 33: 63–73.
64. McGowan HA, Marx SK, Denholm J, Soderholm J, Kamber BS (2009) Reconstructing annual inflows to the headwater catchments of the Murray River, Australia, using the Pacific Decadal Oscillation. Geophys Res Lett 36: L06707.
65. Lloyd PL (1984) Agricultural and pastoral land use in the Brigalow Belt of Queensland. In: Bailey A, editor. The Brigalow Belt of Australia. Brisbane: Royal Society of Queensland. pp. 81–96.
66. Fensham RJ, Fairfax RJ (2003) A land management history for central Queensland, Australia as determined from land-holder questionnaire and aerial photography. J Environ Manage 68: 409–420.
67. Gilbert M, Brodie J (2001) Population and major land use in the Great Barrier Reef catchment area spatial and temporal trends. Townsville: Great Barrier Reef Marine Park Authority. 72 p.
68. Johnston N, Peck G, Ford P, Dougall C, Carroll C (2008) Fitzroy basin water quality improvement report. Rockhampton: The Fitzroy Basin Association. 124 p.
69. Lewis SE, Brodie JE, McCulloch MT, Mallela J, Jupiter SD, et al. (2012) An assessment of an environmental gradient using coral geochemical records, Whitsunday Islands, Great Barrier Reef, Australia. Mar Pollut Bull 65: 306–319.
70. Lewis SE, Shields GA, Kamber BS, Lough JM (2007) A multi-trace element coral record of land-use changes in the Burdekin River catchment, NE Australia. Palaeogeogr Palaeoclimatol Palaeoecol 246: 471–487.
71. McCulloch M, Fallon S, Wyndham T, Hendy E, Lough J, et al. (2003) Coral record of increased sediment flux to the inner Great Barrier Reef since European settlement. Nature 421: 727–730.
72. Meehl GA, Hu A, Tebaldi C (2010) Decadal Prediction in the Pacific Region. J Clim 23: 2959–2973.
73. Shen C, Wang W-C, Gong W, Hao Z (2006) A Pacific Decadal Oscillation record since 1470 AD reconstructed from proxy data of summer rainfall over eastern China. Geophys Res Lett 33: L03702.
74. Linsley BK, Zhang P, Kaplan A, Howe SS, Wellington GM (2008) Interdecadal-decadal climate variability from multicoral oxygen isotope records in the South Pacific Convergence Zone region since 1650 A.D. Paleoceanography 23: PA2219.
75. Evans MN, Cane MA, Schrag DP, Kaplan A, Linsley BK, et al. (2001) Support for tropically-driven pacific decadal variability based on paleoproxy evidence. Geophys Res Lett 28: 3689–3692.
76. Gedalof Z, Mantua NJ, Peterson DL (2002) A multi-century perspective of variability in the Pacific Decadal Oscillation: new insights from tree rings and coral. Geophys Res Lett 29: 2204.
77. Linsley BK, Wellington GM, Schrag DP (2000) Decadal Sea Surface Temperature Variability in the Subtropical South Pacific from 1726 to 1997 A.D. Science 290: 1145–1148.
78. Felis T, Suzuki A, Kuhnert H, Rimbu N, Kawahata H (2010) Pacific Decadal Oscillation documented in a coral record of North Pacific winter temperature since 1873. Geophys Res Lett 37: L14605.
79. Calvo E, Marshall JF, Pelejero C, McCulloch MT, Gagan MK, et al. (2007) Interdecadal climate variability in the Coral Sea since 1708 A.D. Palaeogeogr Palaeoclimatol Palaeoecol 248: 190–201.
80. Pelejero C, Calvo E, McCulloch MT, Marshall JF, Gagan MK, et al. (2005) Preindustrial to modern interdecadal variability in coral reef pH. Science 309: 2204–2207.
81. Klein R, Loya Y, Gvirtzman G, Isdale PJ, Susic M (1990) Seasonal rainfall in the Sinai Desert during the late Quaternary inferred from fluorescent bands in fossil corals. Nature 345: 145–147.
82. Tan JCH, Pratchett MS, Bay LK, Baird AH (2012) Massive coral mortality following a large flood event. Proceedings of the 112th International Coral Reef Symposium: 11E.

83. van Woesik R, De Vantier LM, Glazebrook JS (1995) Effects of Cyclone 'Joy' on nearshore coral communities of the Great Barrier Reef. Mar Ecol Progr Ser 128: 261–270.
84. Thompson A, Costello P, Davidson J, Logan M, Schaffelke B, et al. (2011) Reef Rescue Marine Monitoring Program. Report of AIMS Activities 2011– Inshore Coral Reef Monitoring Townsville: Australian Institute of Marine Science (AIMS). 128 p.
85. Berkelmans R, Jones AM, Schaffelke B (2012) Salinity thresholds of Acropora spp. on the Great Barrier Reef. Coral Reefs 31: 1103–1110.
86. Jupiter S, Roff G, Marion G, Henderson M, Schrameyer V, et al. (2008) Linkages between coral assemblages and coral proxies of terrestrial exposure along a cross-shelf gradient on the southern Great Barrier Reef. Coral Reefs 27: 887–903.
87. Chiba S, Batten S, Sasaoka K, Sasai Y, Sugisaki H (2012) Influence of the Pacific Decadal Oscillation on phytoplankton phenology and community structure in the western North Pacific. Geophys Res Lett 39: L15603.
88. Francis RC, Hare SR, Hollowed AB, Wooster WS (1998) Effects of interdecadal climate variability on the oceanic ecosystems of the NE Pacific. Fisheries Oceanography 7: 1–21.
89. Hare SR, Mantua NJ, Francis RC (1999) Inverse Production Regimes: Alaska and West Coast Pacific Salmon. Fisheries 24: 6–14.
90. Kiem AS, Franks SW (2004) Multi-decadal variability of drought risk, eastern Australia. Hydrol Process 18: 2039–2050.
91. Lee HS, Yamashita T, Mishima T (2012) Multi-decadal variations of ENSO, the Pacific Decadal Oscillation and tropical cyclones in the western North Pacific. Prog Oceanogr 85: 67–80.
92. Mantua NJ, Hare SR, Zhang Y, Wallace JM, Francis RC (1997) A Pacific Interdecadal Climate Oscillation with impacts on salmon production. Bull Am Meteorol Soc 78: 1069–1079.
93. Martinez E, Antoine D, D'Ortenzio F, Gentili B (2009) Climate-Driven Basin-Scale Decadal Oscillations of Oceanic Phytoplankton. Science 326: 1253–1256.
94. Maue RN (2011) Recent historically low global tropical cyclone activity. Geophys Res Lett 38: L14803.
95. McGowan JA, Cayan DR, Dorman LM (1998) Climate-Ocean Variability and Ecosystem Response in the Northeast Pacific. Science 281: 210–217.
96. Menge BA, Gouhier TC, Freidenburg T, Lubchenco J (2011) Linking long-term, large-scale climatic and environmental variability to patterns of marine invertebrate recruitment: Toward explaining "unexplained" variation. J Exp Mar Biol Ecol 400: 236–249.
97. Peterson WT, Schwing FB (2003) A new climate regime in northeast pacific ecosystems. Geophys Res Lett 30: 1896.

17

Cenozoic Planktonic Marine Diatom Diversity and Correlation to Climate Change

David Lazarus[1]*, John Barron[2], Johan Renaudie[1], Patrick Diver[3], Andreas Türke[1,4]

1 Museum für Naturkunde, Berlin, Germany, **2** United States Geological Survey, Menlo Park, California, United States of America, **3** Divdat Consulting, Wesley, Arkansas, United States of America, **4** Department of Geosciences, University of Bremen, Bremen, Germany

Abstract

Marine planktonic diatoms export carbon to the deep ocean, playing a key role in the global carbon cycle. Although commonly thought to have diversified over the Cenozoic as global oceans cooled, only two conflicting quantitative reconstructions exist, both from the Neptune deep-sea microfossil occurrences database. Total diversity shows Cenozoic increase but is sample size biased; conventional subsampling shows little net change. We calculate diversity from a separately compiled new diatom species range catalog, and recalculate Neptune subsampled-in-bin diversity using new methods to correct for increasing Cenozoic geographic endemism and decreasing Cenozoic evenness. We find coherent, substantial Cenozoic diversification in both datasets. Many living cold water species, including species important for export productivity, originate only in the latest Miocene or younger. We make a first quantitative comparison of diatom diversity to the global Cenozoic benthic $\partial^{18}O$ (climate) and carbon cycle records ($\partial^{13}C$, and 20-0 Ma pCO_2). Warmer climates are strongly correlated with lower diatom diversity (raw: rho = .92, $p < .001$; detrended, r = .6, p = .01). Diatoms were 20% less diverse in the early late Miocene, when temperatures and pCO_2 were only moderately higher than today. Diversity is strongly correlated to both $\partial^{13}C$ and pCO_2 over the last 15 my (for both: $r > .9$, detrended $r > .6$, all $p < .001$), but only weakly over the earlier Cenozoic, suggesting increasingly strong linkage of diatom and climate evolution in the Neogene. Our results suggest that many living marine planktonic diatom species may be at risk of extinction in future warm oceans, with an unknown but potentially substantial negative impact on the ocean biologic pump and oceanic carbon sequestration. We cannot however extrapolate our my-scale correlations with generic climate proxies to anthropogenic time-scales of warming without additional species-specific information on proximate ecologic controls.

Editor: Moncho Gomez-Gesteira, University of Vigo, Spain

Funding: German Federal Research Agency grant DFG LA1191/8; database porting partial support by Center for Ecological and Evolutionary Synthesis (CEES), Oslo. Both grants to senior author (DBL). The funders had no role in study design, data collection and analysis, decision to publish, or preparation of the manuscript.

Competing Interests: PD is owner of Divdat Consulting. There are no patents, products in development or marketed products to declare.

* E-mail: david.lazarus@mfn-berlin.de

Introduction

Marine planktonic diatoms (hereafter 'diatoms') are major components of the phytoplankton and are most common in regions of high productivity (upwelling zones) and in high latitudes [1–3]. Diatoms are important for the carbon cycle, generating ca 20% of global primary productivity [4], and, are key components of the ocean carbon pump via rapid sinking of large cells and aggregates [5]. Diatoms are thought to have diversified over the Cenozoic, and their evolutionary history is of great interest. Paleoceanographic studies frequently examine the role changing abundances of diatoms have had on the evolution of ocean environments and the carbon cycle. While studies may sometimes make use of quantitative estimates of past diatom export productivity from measurements of sedimentary opal abundance, over longer time periods and global scales, sedimentary opal abundance estimates are not available, and recourse is often made to diatom diversity (diversity here means species richness) as a proxy for diatom ecologic significance and export productivity (e.g. [6,7]). The Cenozoic history of diatom diversity is thus of interest, not only to understand processes of evolution in plankton, but also for its use as a proxy for Cenozoic diatom ecologic influence and export productivity. How diatoms respond to future global warming (2–4° warmer by 2100 [8]) is also of considerable interest [3]. Most studies have used the living flora and focus on changing biogeography or ecosystem function [2,9–12]. These studies have generally concluded that global warming, by reducing global latitudinal wind stress, will lead to more highly stratified, oligotrophic oceans, reduced abundances of diatoms, and possibly reduce the effectiveness of the ocean carbon pump. Studies of extinction risk in marine biota by contrast have concentrated on benthos or nekton: there are essentially no studies of marine plankton extinction risk.

Diatom fossils in Cenozoic pelagic sediments [1,13,14] provide in principle an unusually good record of diversity [15]. However, although there have long been qualitative statements about diatom diversification over the Cenozoic in the literature, the first quantitative species-level assessment was only made by Spencer-Cervato [16] (Fig. 1), as part of the first systematic analyses of the Neptune database [17]. Spencer-Cervato calculated range-through diversity, which compensates for uneven data quality in individual time-bins, but which is sensitive to data outliers [15].

She evaluated the data for taxonomic problems (synonyms, etc) and eliminated the bulk of outliers by removing data adjacent to hiatuses in the database's age model library. Although she considered the effects of differing data amounts on diversity and calculated a simple ratio of diversity vs number of sections, a full analysis using standard diversity/subsampling theory was not attempted. All subsequent citations of her diatom diversity estimate use her simple range-through curve, not adjusted to number of sections. Rabosky and Sorhannus [18] (Fig. 1) later made use of subsampling algorithms, derived from ecology and newly popularized by their use in studying Phanerozoic fossil diversity, to provide a better estimate of diatom diversification after adjustment for differences in sample sizes over time. Uneven sampling intensity can bias observed relative diversity in comparisons between samples [19,20]. Although aspects of Rabosky and Sorhannus' study are not documented (age scale, taxonomy, outlier detection and removal) the subsampling procedures employed are clearly defined: simple rarefaction, and three variations of sampling by lists (a 'list' = a 'sample', in the micropaleontologic terminology of this current study) and all gave very similar results. Subsampling does not give absolute but only relative values, ranging from zero to (at maximum) the arbitrary number of individuals subsampled, so a reference level is needed to compare curves. In Fig. 1 we have adjusted the scale of Rabosky-Sorhannus' simple rarefaction curve so that the Paleogene levels, on average, match those of the Spencer-Cervato curve. This makes the two main differences between the two analyses apparent. The Rabosky-Sorhannus curve shows a substantial decline in diversity during the late Oligocene-early Miocene, a period of gradual global warming, while the Spencer-Cervato curve shows nearly no change. The Spencer-Cervato curve displays a strong increase in diversity in the Neogene, a period of strong global cooling, relative to the Paleogene; the Rabosky-Sorhannus curve does not. These differences are critical to understanding how diatom diversity has responded to past global temperature changes, how changing marine export productivity is linked to changes in the global carbon cycle and changes in silicate

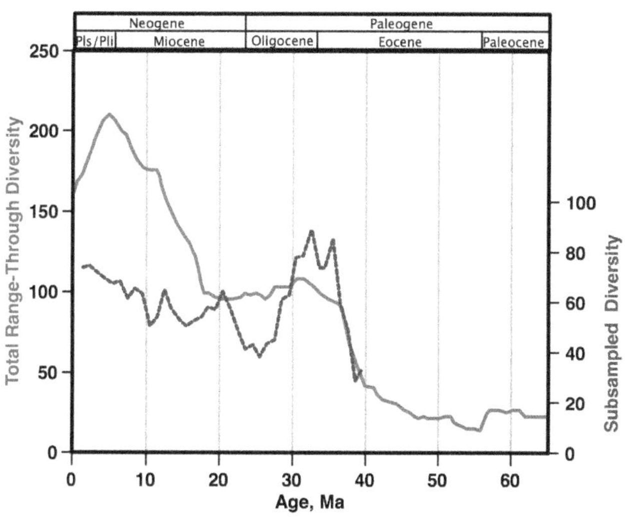

Figure 1. Published Cenozoic diatom diversity estimates, both from the Neptune database. Solid blue line - Spencer-Cervato (1) based on range-through simple (not subsampled) diversity; dashed red line - Rabosky and Sorhannus (4) based on rarefaction subsampling. Age scale: [42].

weathering on land, and how these systems may respond to future global warming.

While diversity estimates compiled from raw data can have major data size biases, subsampling can also easily produce misleading results when underlying assumptions are violated (e.g., homogeneity of geographic structure and constancy of evenness of occurrence of species in samples), producing in such cases artificially 'flattened' diversity vs time curves. The Rabosky and Sorhannus results have already been questioned for this reason [21]. A new, robust estimate of Cenozoic diatom diversity history is thus needed, which takes into account sample size biases in occurrence data, the effects of data outliers, and potential biases due to changing geography and evenness. Ideally, alternate approaches to diversity estimation, such as catalogs [15] should also be considered. Deriving such a diversity history is the primary goal and result of this paper. In addition, we explore how our new diversity history correlates to environmental parameters such as paleoceanographic change, climate, carbon cycle and marine export productivity as derived from sedimentary opal. In these comparisons we do not attempt full analyses of possible causal mechanisms, as these are generally complex and require consideration of many factors, often with detailed time-series analysis and/or modeling, which is beyond the scope of our current study. We hope however by examining these correlations to point out possible important relationships between the evolutionary development of Cenozoic diatoms and environments, and thereby to stimulate further research.

Methods

Our analytic strategy to estimate diatom diversity history is two-fold. First, we re-analyze the Neptune data. In contrast to prior studies, we explicitly test for sample size bias, and correct subsampled diversity estimates for both changing geographic endemism and changing mean evenness. Second, we use the complementary nature of catalog-derived diversity estimates [15] to confirm the robustness of our results using a new separate catalog ('BDC') of diatom species ranges by J. Barron. We also particularly consider the history of living taxa as most relevant to future responses to global warming, as extinct taxa may have had different biologic responses.

To test the sensitivity of diatom diversity to climate state, the resultant diversity estimates are compared to the global compilations of Cenozoic marine benthic foraminiferal isotope data for $\partial^{18}O$ and $\partial^{13}C$ of Zachos et al. [22], and to the Cenozoic record of biogenic opal in marine sediments [23–26]. Benthic $\partial^{18}O$ is an often used proxy for Cenozoic climate in studies of climate and evolution, e.g. [27]. This proxy reflects change in continental ice together with a strong high latitude/deep water ocean temperature signal [22,28] and reflects many, for diatoms important, changes in the physical ocean environment that are correlated to changing polar/deep-sea temperatures over the Cenozoic (e.g. frontal systems, water column stratification, productivity). Benthic $\partial^{13}C$ reflects many different factors, but Cenozoic changes are usually interpreted as reflecting change in either the fraction of carbon sequestered in organic form into sediments, or changes in global organic isotope fractionation ratios (e.g., global increases in low C12-enriched plants: C4 grasses, diatoms) [29,30]. We also compare diatom diversity history directly to the paleo-atmospheric pCO2 estimate of van de Wal et al. [31] for the interval 0–20 Ma. While all paleo-atmospheric pCO2 estimates have uncertainties (for the dataset used, ca 20–40 ppm) [31], this estimate provides continuous, high resolution values with a consistent methodology, and thus suffers less from the variability in other compilations due to systematic differences between estimation methods, e.g. [32].

Data sources and analytic tools

The Neptune database was originally developed in the early 1990s [17], initially analyzed in the late 1990s [16] and ported to the internet by the Chronos project (USA) in the early 2000s [33]. The Chronos version has for some years now been in a somewhat unstable post-funding maintenance state, and one recent user [34] reported finding errors in the calcareous nannofossil content. Errors are important if they create outliers that extend ranges of species in time, causing incorrect diversity estimates. The NSB version of Neptune (for 'Neptune Sandbox Berlin') used in this study is a new implementation of the database hosted at the Museum für Naturkunde in Berlin. It was forked off the Chronos version of the Neptune database in early/mid 2010. Analyses of NSB content [35,36] have specifically checked, and not found data errors like those reported by Lloyd et al. [34], suggesting that these, if present in Chronos Neptune, postdate the NSB fork. As a further check on possible data errors we apply a Pacman analysis [35] to examine the effect of outliers in our diatom data (see below). NSB is otherwise similar in content to the Chronos Neptune database. It contains data for individual species according to the original name published, plus synonym lists that allow data to be linked that were published under different names. The Neptune database is restricted to deep-sea drilling sources, and its age models and taxonomy are a mixture from numerous authors [16,17]. For our study we used the built-in synonym information to extract all occurrence data for 662 valid species and their synonyms, for a total of 63,675 occurrence records.

The species range catalog (Barron Diatom Catalog, 'BDC', Table S1a) is a first and last occurrence database similar to others used in paleontology and micropaleontology for evolutionary analyses [27,37–39]. This database is new and the first such compilation available for diatoms. It was compiled from a variety of literature sources (Table S1b, total = 62), including the deep-sea drilling reports used to compile Neptune (ca half of the sources used), additional primary studies, many land sections (majority of the remainder), plus a few papers that themselves are syntheses of other literature. It uses a uniform taxonomy, and ranges are evaluated for various other errors e.g. [15]. It records 529 species' age ranges and their biogeography (tropical, North Pacific, Southern Ocean). Species first and last occurrences were judged when possible by examination of the actual occurrence data and careful evaluation of the accuracy of the age information available for the occurrences, with the single best source being chosen to provide the age value for first or last occurrence. This is admittedly more subjective but is much better at identifying and filtering out questionable data than a purely automatic computerized procedure, and is the data compilation method frequently used in other micropaleontologic studies, e.g. [27].

A small number (ca 10%) of the taxa in the BDC have ranges determined in part from the Chronos Neptune database. The two data sources are thus not 100% independent of each other. However virtually all taxa present in the BDC are present as well in the primary literature used in the compilation, so use of NSB at most provides a minor adjustment to the range of a small number of species rather than significantly affecting diversity itself.

Data errors and choice of time bin size

All marine microfossil diversity data has potential sources of error [15]. These include uneven completeness of diversity recording by different authors, non-uniform data coverage by time, geographic region and author, differing amounts of fixed-list biostratigraphic vs diversity survey data, age model errors, reworking, and others. Diversity data is drawn for example from a much broader number of deep-sea sections, with on average poorer quality age models, than the much smaller number of sites with high-quality age models typically used for high-resolution paleoceanography, and compilation of stable isotope curves. The complexity of the sources of error precludes formal analysis, but based on our own experience, we feel that binning such diverse global diversity data much below .5 my or 1 my is unlikely to bring much improved real resolution. Age model mismatches are mostly <.5 my but can sometimes be >0.5 my or even 1 my, particularly between low and high latitude sites [40,41], while smaller bins both increase random effects of other aspects of data quality due to smaller data pools, and increase the number of bins with too little data in subsampling procedures. Larger bins are not desirable, decreasing our ability to compare more rapid (often significantly less than 1 my) changes in climate state to diversity. In smaller e.g. regional, more homogeneous microfossil diversity data sets Renaudie and Lazarus [36] explored the effect of bin size on diversity studies and find no significant effect between 0.5 and 1 my bins. Their study also points out that the incompleteness and/or biostratigraphic bias of data available in databases such as NSB also limits effective resolution of time-series change in primary biodiversity signals to at best ca .5 my. We therefore analyzed our data using .5 and 1 my bins. Comparisons of selected identical analyses run at these two different resolutions (not presented) showed no significant differences.

Age scale and chronology

The large majority of the data in our study, including the NSB data and most published literature data was originally calibrated to the Berggren et al. 1995 timescale [42], including all NSB data and stable isotope environmental data. Only the BDC ranges were in the more recent Gradstein et al. 2004 timescale [43]. For this study some early analyses exploring the nature of sampling bias, or comparison to published literature, were carried out using the older Berggren scale. For later analyses, including all comparisons of diversity to environmental parameters, the Gradstein et al. scale was used. Differences between scales are mostly <1 my and have no significant effect on the results of the study.

Subsampling

As biotas usually consist mostly of rare species, a sample's diversity is usually much less than the total diversity of the sampled biota; thus sampled diversity increases with sample size. If sample size varies in a data series, sampled diversities, and also range-through diversity from catalogs compiled from such samples, can be biased [19]. Standardized subsampled Neptune data is needed to correct for the strong variation of sample sizes in Neptune with time. Subsampling though yields biased results if either taxa ranked abundance distributions are not constant between samples; or if clustering of taxa by region change. Both of these phenomena occur in Neptune Cenozoic diatom data. A variety of subsampling algorithms have been proposed, including sampling a fixed number of individuals (simple rarefaction) or sampling a fixed or variable number of samples (lists, in subsampling terminology). We use both sampled-in-bin classical rarefaction, and a new method, SQS [44]. Subsampling by list was not used in our study for several reasons. When the average number of species in a list is short compared to the total diversity in a bin, which is true of NSB data [15] the effectiveness of list methods is reduced (and collapses as list size approaches 1 taxon to simple rarefaction); modeling subsampling by lists is more complex and requires information on taxon clustering in lists (e.g. samples), which would make modeling the effect of evenness on subsampling (below) much more difficult; the newest SQS subsampling algorithm [44] is based on (size adjusted) simple rarefaction, making it easier to compare results by using (our own alternate size adjusted-see below) simple rarefaction in our analysis; lastly, in the prior study by Rabosky and Sorhannus [18], simple rarefaction and list type analyses of Neptune data gave very similar results.

Geographic correction

The SQS algorithm was specifically developed to compensate for the inability of classical subsampling methods to adequately capture change in diversity associated with the development of geographic differences in distributions (e.g. endemism) [44]. Rather than taking a fixed number of occurrences, SQS varies the sample size to compensate for diversity underestimation due to increased geographic structure [44]. As the SQS method is still very new, we use as well an alternate method to correct for changing endemism, making use of the well resolved geographic affinities information for each species in the BDC database. The ratio of the largely mutually endemic polar to tropical species in the BDC data for each time interval is used to correct a simple classical rarefaction estimate of diversity for each time bin, as the geographic diversity ratio in time bins in the BDC should not be sample size biased; the resultant curve is our 'PTR' sampling method.

Evenness correction

To correct for the effect of changing evenness (degree of similarity in relative abundances of different species) over time, a simple correction factor was computed based on 1) a simple ranked abundance data shape metric 'D(80)' (Fig. 2a) that quantified the changes in evenness patterns in each time bin (Fig. 2b), and 2) a scaling of this metric according to the results of a simulation of the effects that different evenness patterns in the data have on rarefaction. The D(80) metric, here defined as the fraction of total diversity reached in ranked frequency occurrence data at 80% of the total area, is only one of any number that could be used, but is linear to cumulative frequency, is not weighted towards a few most common taxa, and is similar to the cumulative frequency adjustments used in SQS geographic correction. We quantified a major shift over time, from very even occurrence frequencies in the Paleogene, to highly uneven distributions in the Neogene (Fig. 2b).

The pooled data for species ranked occurrences used were from two major time intervals (44-31 Ma: Paleogene, and 14-1 Ma: Neogene). Each data set was scaled to the same total area (= number of occurrences or total sample size) for the ranked abundance curves, interpolated down to the same number (100) of virtual species, and these two virtual data sets - having the same diversity and total sample size, differing only in the relative frequencies of taxa - were subsampled using simple rarefaction at a range of sample sizes. Each sample size was repeated 50 times and averaged. The mean diversity found by subsampling for a given subsample size in each of the two virtual populations was used to calculate the degree to which the diversity of the low evenness Neogene population was being underestimated compared to the Paleogene population. The results for a wide range of sample sizes are shown in Figs. 3a and b. For sample sizes within the range used in rarefaction subsampling of NSB, both in our study and in that of Rabosky and Sorhannus [18] (100 and 96 individuals, respectively) the correction factor needed to make the diversity obtained by rarefaction the same is ca 1.55. Given the mean D(80) values for these two pooled data sets (0.551 and 0.282 for respectively, Paleogene and Neogene), and no correction factor for the

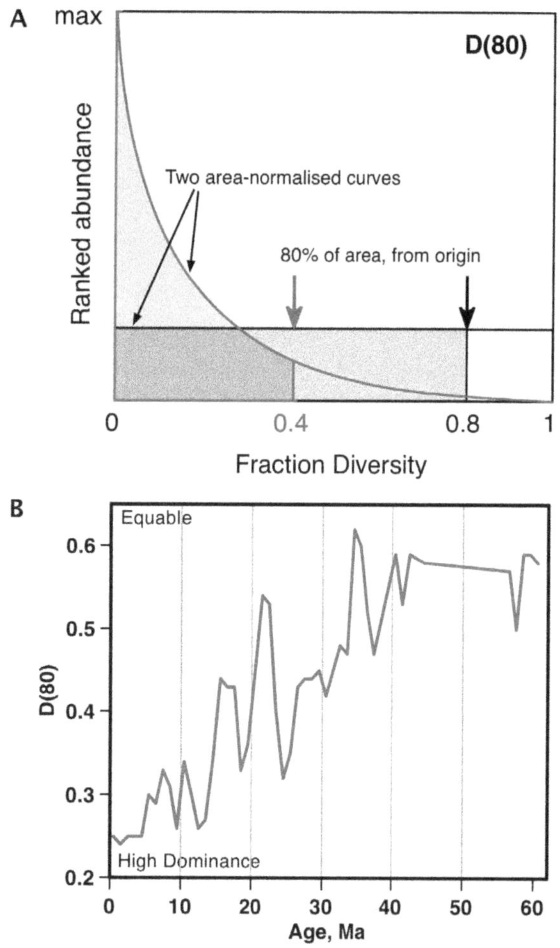

Figure 2. Equability (increasing dominance) in relative frequencies of diatom species in NSB occurrence data over the Cenozoic. a) Dimensionless shape metric D(80), defined as fraction of total diversity reached in ranked frequency occurrence data at 80% of total area. For data with all frequencies equal (flat black line) D(80) = 0.8; D(80) decreases with increasingly unequal frequencies. b) D(80) vs time in Cenozoic NSB diatom data. Age scale: [43].

Paleogene pooled data (i.e. = 1) a linear equation giving the required correction factor (y) as a function of D(80) can easily be computed: y = 2.127–2.04*D(80). This correction function result was applied to both of our subsampled results using the bin specific D(80) evenness metric values (Fig. 3b) to correct for evenness effects. Note that we did not use the simpler method of evenness correction proposed by Alroy [44], of just leaving the most common species out: we found this very ineffective with the diatom data of our study. This is not surprising since differences in evenness are seen over a large fraction of the species in the diatom data, not just the very most abundant one.

Outliers and other errors in occurrence data

Outliers can have major effects on range-through calculations of diversity [15]. However, so long as they are distributed at random and are uncommon, they have little effect on subsampled diversity estimates. Outliers are an unavoidable aspect of large data compilations, and also exist in the NSB database due to intrinsic aspects of the data such as reworking, taxonomic mis-identifications and age model errors [15]. While trimming fixed percentages of the range ends of age-composited occurrence data ('pacman

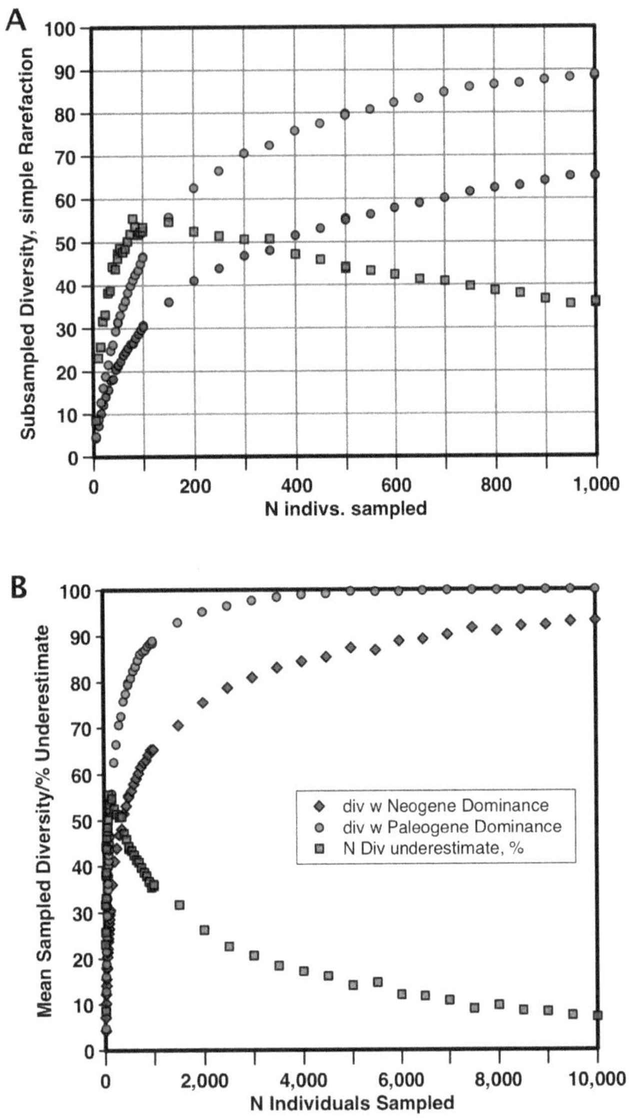

Figure 3. Results of simple rarefaction subsampling two virtual computer populations of equal diversity (100 taxa) and total sample size but with averaged Neogene or Paleogene relative abundance distributions as seen in NSB diatom occurrence data. a) for subsample sizes between 0–1,000; b) subsample sizes 0–10,000. Blue dots - Paleogene subsampled diversity; red - Neogene subsampled diversity; green squares - percentage underestimation of Neogene diversity relative to Paleogene diversity. Each point is an average of 50 subsampling trials. Age scale: [43].

trimming') can remove most outliers [35] this procedure can also create edge effects, particularly for the youngest time bin(s), and thus complicate comparisons of diversity and environmental history in the late Neogene. For this reason we did not apply a Pacman trim in our main diversity reconstruction. We checked however for the possible effect of outliers on our sampled-in-bin diversity estimates by running successively stronger Pacman trims on the data and applying simple rarefaction to the output, examining the curves for changes in shape (Fig. 4). Even with very strong trim levels (25–45% of the total occurrence data, vs ca 10% in the study of Lazarus et al [35]) the basic rarefaction curve was largely unaltered, although the dynamics were stronger, particularly the relative increase in diversity over the Neogene. This may be an artifact of strong trim levels depressing diversity in bins where many taxa are jointly near the ends of their true ranges, e.g. a gradual turnover between relatively homogeneous late Paleogene vs early Neogene floras. This analysis shows that our 'untrimmed' subsampled NSB diversity curve is not strongly affected by outliers, and if anything is a conservative estimate of the degree of dynamics of diversification over the Neogene.

Detrended analyses

In comparing time-series data, one problem occurs when both variables have a trend, resulting in a correlation being seen even if there is no causal relationship between them. A second problem arises when one variable cannot respond fully to changes in the forcing variable due to internal limits (e.g. stochastic effects), which will affect the observed correlation between variables. There is considerable a-priori evidence from diatom biology and biogeography to expect diatom diversity to respond to climate, and thus to expect the long-term as well as short term correlations to be at least in part causal, not chance. There is also no evidence in our diversity-climate comparisons for significant mis-correlation due to limitations in the freedom of response in the diatom diversity data: even short-term changes in climate state are partially mirrored by diatom diversity change, and maximum bin-to-bin step size changes in diatom diversity, both increasing and decreasing, match in amplitude those of the presumed climate forcing function. Such observations however cannot fully exclude the existence of correlation artifacts. One established method of testing for the robustness of a correlation is to use linear regression to detrend, or to first-difference the time series and compare only residuals. This has the major disadvantages of potentially removing one of the main (long-term) components of the signal, and by using only residuals, unless errors in individual values (noise) are low, a significant loss of ability to detect correlations, and thus correctly evaluate a hypothesis.

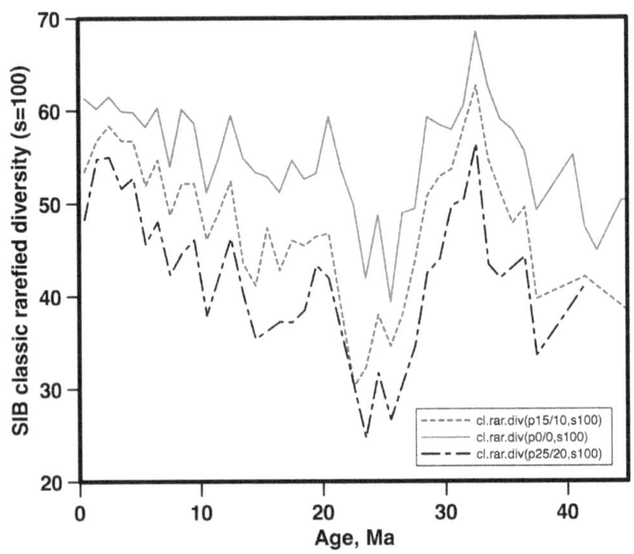

Figure 4. Subsampled diatom diversity (sample size = 100) from the NSB database, using different levels of Pacman trimming of data from the ends of species ranges (7) to remove outliers due to possible data errors in NSB. Blue solid line - no data trimming; dashed red - 15% of the youngest and 10% of the oldest occurrences for each species removed; alternate dashed black - 25% of the youngest and 20% of the oldest. Age scale: [43].

Given the existence of a variety of errors limiting individual bin estimates for diversity (described above under Data Sources) we do not feel that first-differencing the data would yield a time-series with a usable signal. Hannisdal [45,46] for example has very recently proposed new methods for analysis of correlation between time series, but their robustness to different types of complex sources of data error (the methods use first differences residuals) is not yet, in our opinion, sufficiently documented to justify employing in our study. We believe that a somewhat less severe detrending by simple linear regression should retain a significant signal, and at least allow us to test for co-incidental correlation of unrelated primary trends. Detrended versions of all significant results were thus also analyzed for statistical significance, and the results of these tests are considered in our discussion and conclusions.

Results

Simple diversity vs sample size

Although the primary justification given by Rabosky and Sorhannus [18] for a new analysis of the Neptune data was the expected correlation of diversity to sample size, they surprisingly did not explicitly examine this effect. Here we provide an analysis of the correlation between simple diversity and sample size for both sources of data used in our study - the NSB version of the Neptune database, and the BDC listing. NSB occurrence data is strongly correlated to the number of samples in the time bin (Fig. 5a), although the relationship is not linear. The correlation is particularly pronounced in detrended data (Fig. 5b). Data compilations such as the BDC are done at the level of publications, and underlying correlations to sample size could in theory be masked. To see if this is true for the micropaleontologic literature, we compared simple NSB diversity per bin to the number of Legs (ca 2-month long individual DSDP/ODP expeditions) with data in that bin. Due to long standing policies on leg staffing and sample access, the large majority of deep-sea drilling legs have only a single primary diatom paper reporting occurrence data. There is still a strong correlation between number of legs/bin and diversity/bin (Fig. 5c), suggesting that using numbers of papers as a proxy for number of samples does not mask a sample size-diversity correlation.

By contrast, similar analysis shows that there is only a weak overall correlation between the BDC diversity and the number of papers used in its compilation (Figs. 6a and 6b). Most importantly, there is no discernible correlation between numbers of papers used and diversity in the Neogene, either in the raw or the detrended data, and raw BDC data may thus provide (vs. raw Neptune) a less biased estimate of diversity, particularly in the Neogene.

In both data sources it is also clear that the quantity of data declines with time, and in particular, very little data is available to estimate diatom diversity below ca. 40 Ma (late middle Eocene), and particularly between ca 42–55 Ma.

Diversity history of diatoms

Our analyses yielded three different estimates of Cenozoic diatom diversity, using different methods and different data sources: the SQS and PTR estimates from the NSB database, and the BDC estimate from the range catalog (Fig. 7). For comparison purposes all diversity results are z-score normalized, as SQS/PTR yield relative change only. Cenozoic BDC range-through diversity (total: curve 'BDC'; and by region, Fig. 8) increases strongly toward the present, with a transient late Eocene peak; and a larger, rapid Neogene-Recent rise (ca 15-0 Ma), mostly from developing endemic polar floras (Fig. 8; [13,14]). The NSB estimates, despite occasional implausibly high rates of change over short time intervals, are very similar to the BDC estimate. All

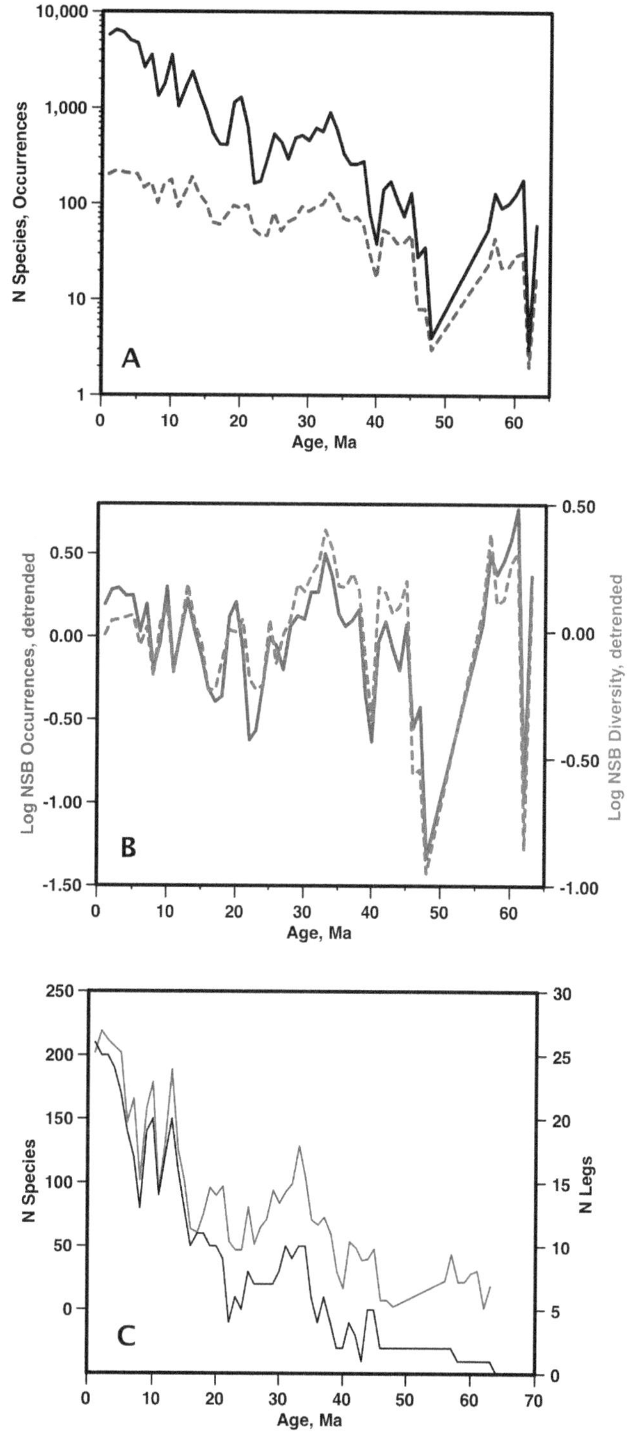

Figure 5. Neptune (NSB) diatom diversity vs data density. a) Diversity (red) vs number occurrences (black). Note log scale and different slopes. b) detrended NSB diversity (blue) vs number occurrences (red) (detrending by residuals of linear regressions vs. time). c) NSB total in-bin diversity (red) vs number of distinct drilling program source Legs (black). Due to DSDP/ODP/IODP staffing/publication policies, Legs are a very good proxy for number of papers. Age scale: [42].

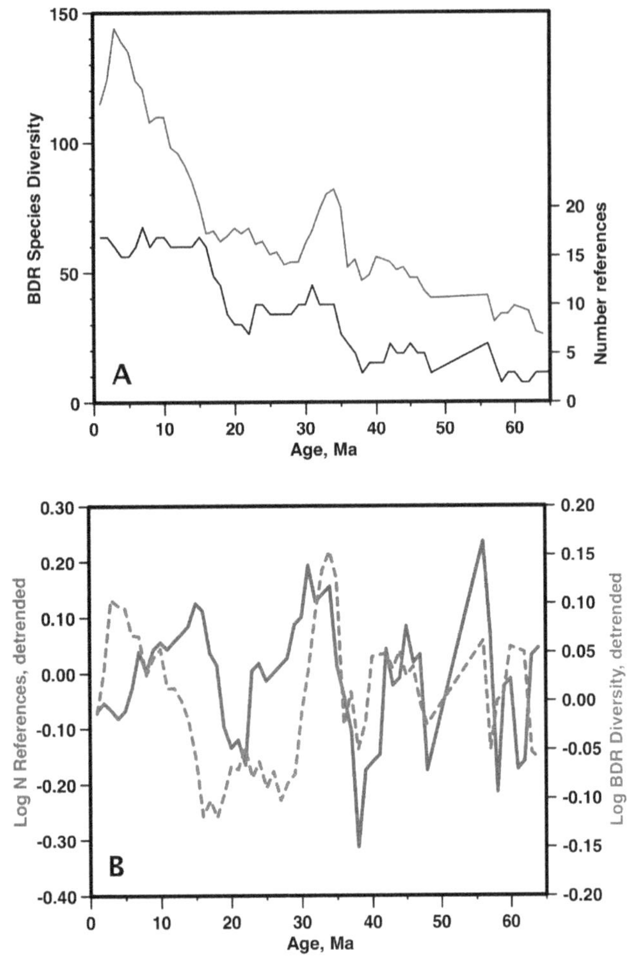

Figure 6. BDC diversity vs data density. a) BDC diversity (red) vs number of sources (black - papers or other as given in table ST1b). b) detrended BDC diversity (blue) vs number of sources (red). Note lack of correlation in 0–30 Ma interval. Age scale: [42].

three estimates are remarkably similar for the Neogene, showing rapid increase 15-0 Ma. Given this similarity, we use the average of the three results as our new Cenozoic diatom diversity estimate ('LBR'). The LBR average reflects the combined common signal from three different methods/datasets, while the average removes much of the discrepancies (noise) found only in a single estimate. Below ca 42 Ma the LBR estimate is not very robust, as for the most part, too few data were available in the NSB database to allow estimates of diversity using the SQS or PTR methods. The LBR curve in this interval is thus primarily derived only from the BDC data.

Origin of modern flora

The first occurrences of all living species with origins in the Neogene from the BDC are shown in a range plot in Fig. 9, colored by biogeography. A shift towards higher origination rates for living species, and an increase in the contribution of polar taxa can be seen in the latest Miocene (ca 7-5 Ma) immediately preceding the early Pliocene warm interval (ca 4.5-3.5 Ma). Nearly 50% of living species originated <5 Ma; ca. 80% (and 50% of genera) in the Neogene (Figs. 9, 10). This suggests that Neogene, particularly late Miocene-Recent, diversity-environment relationships are most important in estimating future responses to climate change. Also, the recent flora is dominated by species originating in very cold latest Cenozoic conditions, and many of these are from the coldest (polar) regions (Figs. 8, 9).

Diatom diversity and Cenozoic climate change

Diatom diversity and Cenozoic benthic oxygen isotopes. The diversity curve is strongly visually correlated to climate (Fig. 11a), with both the primary trend and many secondary fluctuations being similar between the two curves. A direct comparison (Fig. 11b) reveals the strong, but changing relationship between climate and diatom diversity. Paleocene-Eocene diversity is largely insensitive to climate, Oligocene-Miocene diversity is highly sensitive, while Pliocene-Recent diversity appears largely insensitive to climate, with the early Pliocene being a transition between very low and high sensitivity regimes. The correlation is statistically highly significant for the entire curve ($r = 0.82$, $rho = -.88$, $p<.001$), and even more so for the Miocene interval of rapid change (15-5 Ma, $r = .98$, $p<.001$).

Cenozoic carbon cycle (benthic carbon isotopes, pCO_2) and diatom diversity. Comparison of the LBR diversity curve to the Cenozoic record of $\partial^{13}C$ (Fig. 12) shows that over most of the Cenozoic, diatom diversity and $\partial^{13}C$ were not correlated. Paleocene and Eocene variability in carbon isotopes was significant but not matched by equivalent variation in diatom diversity; more variable diversity in the Oligocene and early Miocene did not match shifts in carbon isotopes. Mid Miocene-Recent diversity (from ca 15-0 Ma) however is strongly correlated to the $\partial^{13}C$ record ($r = .92$, $p<.001$). Diatom diversity is also strongly correlated to the pCO_2 record over the last 15 my ($r = .92$, $p<.001$; [31]; Fig. 13).

Robustness of correlation between diversity and geochemical parameters. Detrended analyses and analyses using alternate datasets (individual diversity estimates vs averaged; full Cenozoic vs only the 40-0 or 15-5/0 Ma time intervals) results are summarized along with undetrended analyses in Table 1. These show that the correlation between diatom diversity and climate is highly significant even in detrended data, or using variations of calculated diversity. Correlations are less for the full Cenozoic data but all significant ($p<.05$), and stronger for both the 40-0 Ma interval where the diversity estimates are most robust, and for the Miocene-early Pliocene interval during which most living species originated (p values all $<.001$). Various other tests (not listed), such as not excluding time bins with less than three diversity estimates in calculating average bin diversities adds data values primarily to the Paleogene, and has no significant effect on the observed degree of correlation, nor are there (with one exception - PTR in the Miocene-early Pliocene) significant differences in the p values for individual or combined diversity estimates.

Table S3 contains the main results of our analyses for use in further studies.

Discussion

Methodology

We argue that we have been able to extract a coherent diversity signal from incomplete, biased occurrence data, as demonstrated by the similarity between diversity estimates made using different methods, and using very different types of source data (raw occurrence data or range estimates from a catalog). An essential component of our reconstruction method from occurrence data is the use not only of SQS subsampling to address changing geographic structure in the occurrence data, but also the application of an evenness correction factor based on modeling.

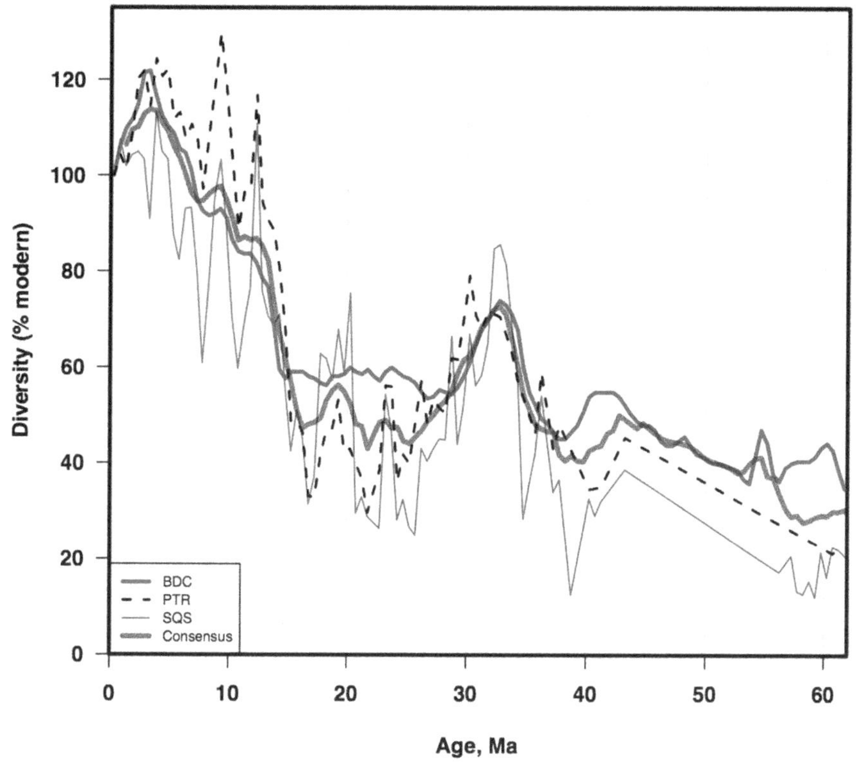

Figure 7. Cenozoic diatom diversity. BDC - total range-through diversity from Barron diatom catalog; PTR - evenness corrected subsampled NSB diversity with geographic correction computed from polar-tropical diversity ratio in BDC; SQS - evenness corrected diversity using SQS subsampling. Percent scale (right) from BDC values. Age scale: [43].

This latter correction is not normally used in paleodiversity studies, but in our data is shown to be very significant and necessary in order to obtain a coherent diversity signal. In a recent study of similar deep-sea microfossil data, Renaudie and Lazarus [36] implicated changing evenness in occurrence data as the primary reason why standard subsampling methods, including SQS, failed to recover an unbiased diversity history. Our modeling method is straightforward and may offer an effective solution for this problem, improving the accuracy of diversity reconstructions using occurrence data.

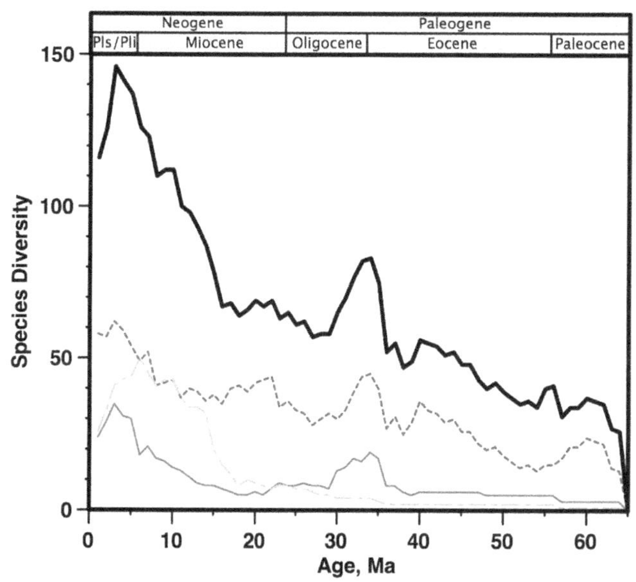

Figure 8. Cenozoic global and regional diatom diversity (range-through) from the BDC catalog. Bold solid black - total; red dash - tropical; green alternate dash - North Pacific; solid blue - Southern Ocean. Age scale: [42].

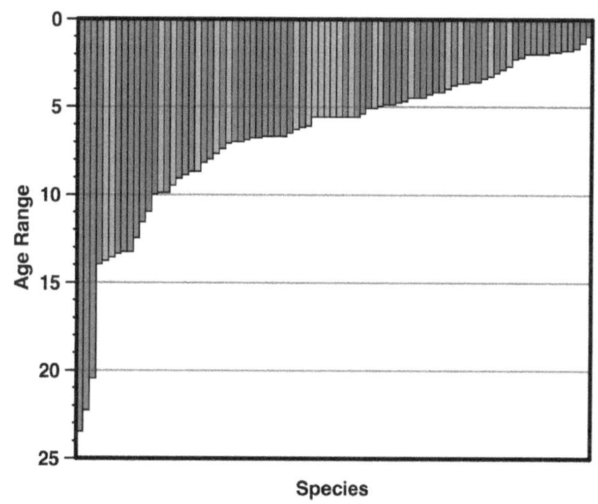

Figure 9. Origin of modern diatom flora, from stratigraphic ranges in the BDC. red - tropical; light blue - Southern Ocean; purple - North Pacific. Age scale: [42].

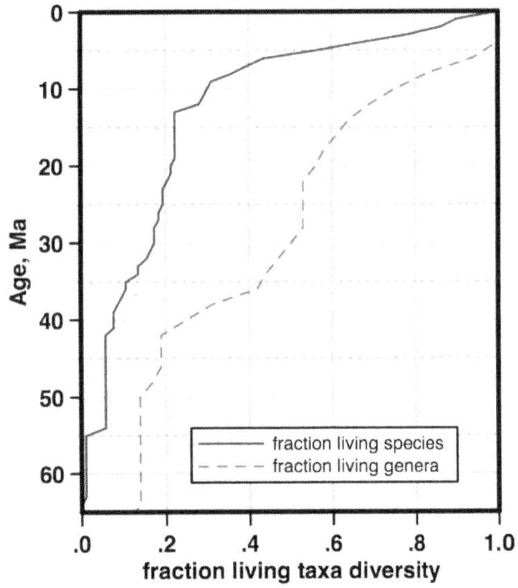

Figure 10. Fraction of living diversity in the BDC catalog present in older time intervals. Solid red line - fraction living species; dashed blue line - fraction living genera. Age scale: [42].

Systematic biases in data

Despite obtaining a coherent diversity signal, our results may still be inaccurate if there are systematic biases in the data which would produce coherent biases in our diversity estimates. We specifically consider changes in preservation over time which might bias our results. The height of the late Eocene diversity peak may in part be a preservation artifact: older Paleogene diatoms are often diagenetically altered [13,14] to widespread cherts [47], while decreasing silicification in Oligocene and younger species [14], and more corrosive waters due to declining oceanic silica concentrations, particularly since the Eocene-Oligocene boundary [48] may have reduced preserved diversity in these younger sediments. We conclude that our estimate of substantial increase in Cenozoic diversity is conservative, and may even underestimate the true total relative diversity rise.

Diversity and productivity in living systems

Our results have implications for several fields. Many of these implications depend not on diatom diversity itself but diatom export productivity. We distinguish two distinct types of relationship between diversity and export productivity: naturally occurring behavior that has developed over evolutionary time; and perturbed relationships due to rapid extirpation or extinction, in which evolution has not had time to operate.

The relationship between diversity and productivity in living natural systems is controversial, and scale dependent. However there are several reasons to believe, on the global scale and over longer (but not geologic) time-scales, that biotic diversity and productivity are in general, positively related, even for groups where local diversity-productivity relationships show low diversity with very high productivity [49]. This appears to be true also for diatoms. Although on short time scales (weeks to seasonal), or in local environments diatom export productivity is associated with blooms dominated by very few species, over broader scales diversity and productivity appear to be correlated. In the modern Atlantic ocean living diatom diversity is strongly linked to diatom abundance/export of carbon (diversity:log biomass correlation $r = .864$, from data in [50], Fig. S1); productivity is significantly correlated to diversity also in freshwater plankton [51]. These results are in accord with global models of marine plankton diversity [52] in which high diversity (despite a slight geographic offset due to the inclusion of other phytoplankton functional groups) is associated with regions of high diatom export as indicated by sedimentary opal deposition [15].

Perturbed systems - To our knowledge, no experimental studies of short-term (annual to decades-centuries) pelagic ecosystem

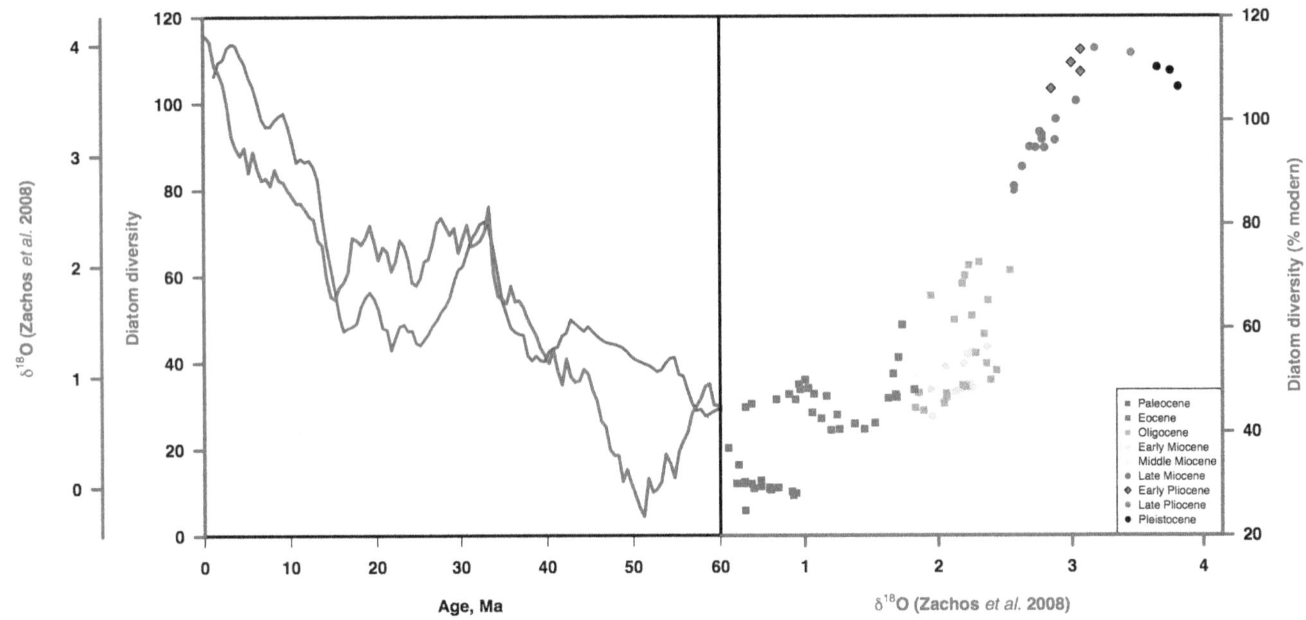

Figure 11. Diatom diversity and global climate proxy mean benthic $\partial^{18}O$. a) Three point moving average of three diatom diversity time series estimates from figure 7 (blue line) vs time series of $\partial^{18}O$ [22] (red line). b) Diatom diversity vs $\partial^{18}O$ over the Cenozoic. Squares - Paleogene; circles - Neogene. Early Pliocene - diamond. Age scale: [43].

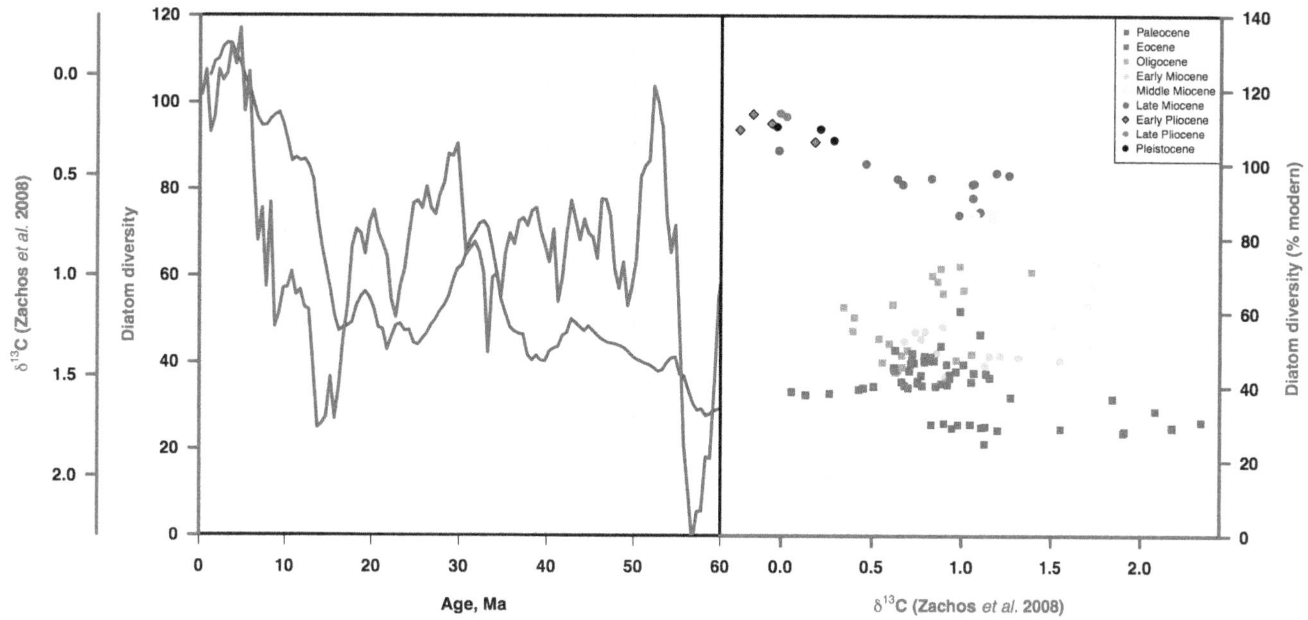

Figure 12. Diatom diversity vs carbon cycle proxy mean global benthic $\partial^{13}C$. a) Three point moving average of three diatom diversity time series estimates from figure 7 (blue line) vs time series of $\partial^{13}C$ [22] (red line). Diatom diversity vs $\partial^{13}C$ over the Cenozoic. Symbols as in figure 11. Age scale: [43].

responses to species loss exist. However, hundreds of experimental studies with a variety of other systems, including not only terrestrial but also aquatic environments, have shown that ecosystem services and diversity are significantly correlated, with diversity loss reducing productivity and nutrient cycling [53,54].

Comparison of new diversity result to prior estimates

Having an independently derived estimate (LBR) for Cenozoic diatom diversity history, which we argue is reliable, we can compare this to prior estimates of diversity: the CSC and RS curves.

The RS diversity estimate is based, as the LBR estimate, on subsampling of the same occurrence database. Between ca 40 and 20 Ma the two estimates are quite similar, with an initial rapid rise from 40 Ma to a peak at the Eocene-Oligocene boundary, a decline throughout most of the Oligocene, and a more modest increase in the early Miocene. The curves are very different over remainder of the Neogene. The LBR curve shows a dramatic,

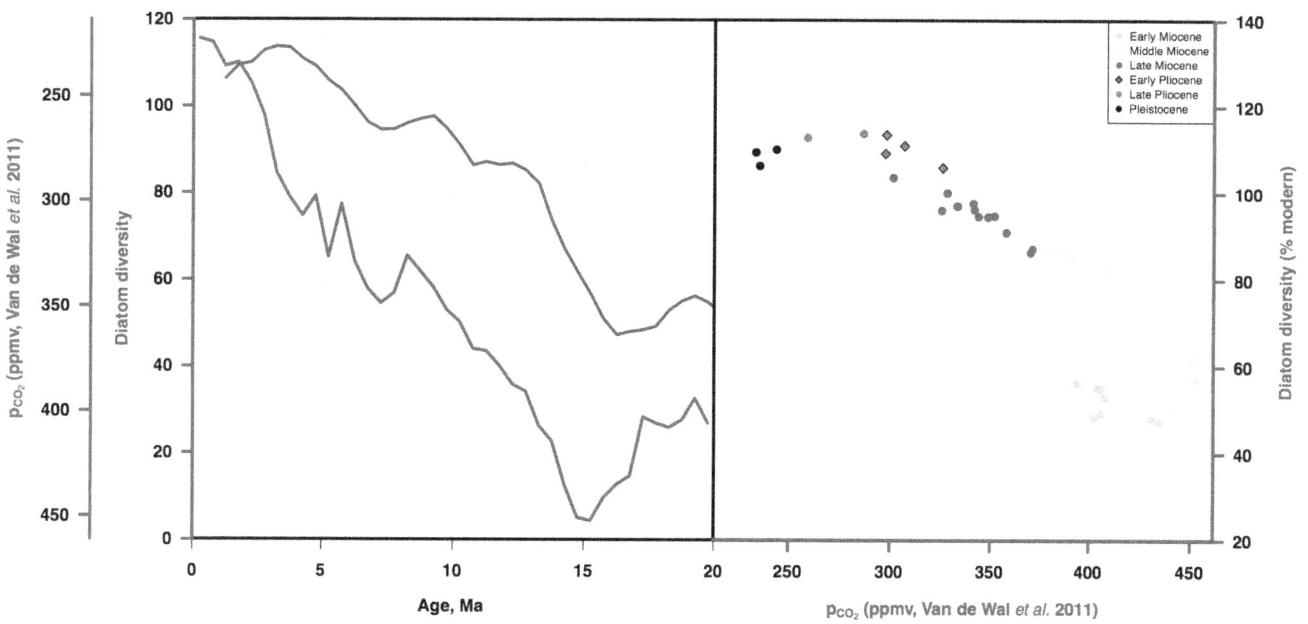

Figure 13. Diatom diversity vs estimated atmospheric pCO_2 for the last 20 Ma. a) Three point moving average of three diatom diversity time series estimates from figure 7 (blue line) vs time series of estimated pCO_2 [31] (red line). b) Diatom diversity vs pCO2. Symbols as in figure 11. Age scale: [43].

Table 1. Correlations between diatom diversity and selected parameters.

	Pearson r	p value	Significance
LBR Diversity:			
Diversity/∂^{18}O	0.818	<2.2E-16	***
Diversity/∂^{18}O detrended	0.198	2.65E-02	*
Diversity/∂^{18}O 5-pt average	0.823	<2.2E-16	***
Diversity/∂^{18}O 5-pt average detrended	0.199	2.52E-02	*
Diversity/∂^{18}O 0-40 Ma detrended	0.423	1.14E-04	***
Diversity/∂^{18}O 5-15 Ma	0.975	2.83E-13	***
Diversity/∂^{18}O 5-15 Ma detrended	0.938	1.02E-09	***
Diversity/∂^{13}C	−0.401	3.30E-06	***
Diversity/∂^{13}C detrended	−0.293	8.52E-04	***
Diversity/∂^{13}C 0-40 Ma detrended	−0.331	3.06E-03	**
Diversity/∂^{13}C (15Ma onwards)	−0.922	3.26E-12	***
Diversity/∂^{13}C (15Ma onwards) detrended	−0.649	6.41E-10	***
Diversity/∂^{13}C 5-pt average	−0.440	2.53E-07	***
Diversity/∂^{13}C 5-pt average detrended	−0.318	2.90E-04	***
Diversity/pCO_2 (15Ma onwards)	−0.925	2.04E-12	***
Diversity/pCO_2 (15Ma onwards) detrended	−0.633	2.96E-04	***
Diversity/No of NSB samples	0.839	<2.2E-16	***
Diversity/No of NSB samples detrended	0.687	<2.2E-16	***
BDR Diversity:			
Diversity/∂^{18}O	0.848	<2.2E-16	***
Diversity/∂^{18}O detrended	0.242	5.21E-03	**
Diversity/∂^{13}C	−0.477	7.35E-09	***
Diversity/∂^{13}C detrended	−0.332	9.94E-05	***
Diversity/∂^{13}C (15Ma onwards)	−0.928	1.68E-13	***
Diversity/∂^{13}C (15Ma onwards) detrended	−0.888	6.15E-11	***
Diversity/∂^{13}C 5-pt average	−0.457	5.88E-08	***
Diversity/∂^{13}C 5-pt average detrended	−0.326	1.76E-04	***
Diversity/pCO_2 (15Ma onwards)	−0.868	5.41E-10	***
Diversity/pCO_2 (15Ma onwards) detrended	−0.519	3.30E-03	**

***p<0.001.
**0.001<p<0.01.
*0.01<p<0.05.

rapid rise beginning at the base of the mid Miocene, reaching diversity levels, at their peak in the early Pliocene, ca 50% higher than the E/O peak, while the RS curve shows only a very gradual increase in diversity throughout the mid Miocene-Recent, reaching values only ca 75% of the E/O peak diversity in the Recent. As we have shown, the differences are due to the failure of the assumptions underlying the RS analysis in the mid-Miocene to Recent, e.g. no major changes in biogeographic endemism, or changes in relative abundance structure in the sampled populations.

The CSC estimate is a range through estimate of raw NSB data. It is similar in overall shape to the LBR curve, though this varies with time interval. The Eocene and older parts of the curves differ substantially: compared to the LBR curve the CSC curve shows no major peak in the late Eocene vs diversity fall in the late Oligocene, and a much steeper rise in the mid Eocene from low diversity values of the Paleocene-early Eocene. However over the last ca 30 Ma both curves show a plateau in diversity between ca 30 and 20 Ma, a rapid rise beginning ca 18-15 Ma reaching a brief peak in diversity in the early Pliocene slightly more than twice the prior plateau values, and a slight decline into the later Pliocene-Recent. The two curves differ substantially however in estimates of absolute diversity. For much of the Cenozoic, and in particular the last 30 Ma, the CSC diversity values are as much as 50% greater than those in the LBR curve.

Higher absolute diversity values in the CSC vs the LBR curves are difficult to explain as a simple calculation artifact: the LBR curve derives its absolute values via calibration to the BDR catalog curve, which is in principle also a range through estimate of diversity, derived even in part from the same deep-sea occurrence data, even if using different literature sources and additional onshore section data. The absolute values for the LBR curve should be thus similar, or, given the use of additional sources, somewhat higher than the CSC curve; the results however are the opposite. We speculate that the CSC curve contains substantial amounts of erroneous diversity due to data outliers and incomplete identification of synonyms in the taxonomy. The diatom taxonomy used in the Neptune database was only provisional when first compiled, and recent revision (Iwai et al, in prep; [55], not yet incorporated into NSB) has identified many additional synonyms. These may have inflated diversity in the CSC calculation in comparison to the absolute estimates in the more completely revised BDR catalog. Data outliers, due to age model errors, reworking and mis-identifications, may individually contribute only a small amount to inflated diversity estimates but cumulatively, particularly in range through calculations, may have also increased diversity by significant amounts. The use of range through methods at least tends to distribute local data errors of the sort described above over several time bins, so the net result of such error might well be to inflate local absolute values but leave the overall relative shape of the curve largely unchanged.

Why should however the relative pattern of diversity change be so similar between the LBR and CSC curves? One possibility is that the CSC curve is derived from a completely sampled record, so that changes in sampling had no effect on the diversity estimate. This is not in accord with the strong overall and detrended correlation between raw NSB diversity and number of samples (Fig. 5). Nor is this in accordance with our simulations of diversity vs sample size (Fig. 3) - 90% sampling completeness of diversity for a simulated diversity of 100 is only reached in the Neogene diatom data with sample sizes of ca 10,000 - values of data density not even approximated in the NSB data except for a few bins >5,000 but <7,000 in the Plio-Pleistocene. At more typical Neogene sample densities/bin of ca 1,000 (Fig. 5a), only ca 2/3 of the true diversity is recovered (Fig. 3a), and there is a significant slope showing that diversity does vary with sample size.

Given the very strong correlation of CSC diversity and sample size at all intervals over the Cenozoic, the correlation seen between our best estimate of diversity (LBR) and CSC curves over the last

30 Ma can only be due to a strong correlation between LBR diversity and sample availability over the last 30 Ma. This is shown in Fig. 14, where a very strong correlation can be seen for this time interval ($r = .84$, $p < .001$), in contrast to the Paleocene-Eocene, where no correlation is apparent. Why should this be, since the LBR diversity has been computed to be independent of sample size? Although the first order trend may well be coincidence (due to both series increasing rapidly towards the Recent), the correlation of detrended series (Fig. S2) is significant ($r = .69$, $p < .001$) and requires explanation. We suggest that diatom diversity may have influenced sample availability, thereby to some extent inverting the assumed relationship between variables. This is because diatom diversity has likely been causally correlated to the relative abundance of siliceous sediments in each sampled interval (Renaudie and Lazarus, in prep; see also 'Diversity and Climate' below). Unlike most fossils, where abundances in fossil-bearing sediments depend primarily on external geologic factors such as sea-level that regulate creation of sediment and its preservation after deposition, deep-sea sediments that have reasonably well-preserved fossil diatoms (diatom oozes and admixtures of siliceous ooze with carbonate and clay) are to a large extent created by the abundances of diatoms themselves in the overlying water column ([56]). The relationship to diversity can either be a direct influence of diversity on the relative abundance of diatoms in the plankton (the effect of more diverse evolutionary adaptations on relative nutrient capture), or indirect via the response of both diatom abundance and diatom diversity to changing amounts of nutrient input into the oceans due to changing Cenozoic climate. These hypotheses are not mutually exclusive and a mixture of direct and indirect response is feasible. Testing these ideas are however beyond the scope of our current study.

Diatom diversity and evolution of other groups of organisms

We group our discussion of implications into those related to the diversity history and its correlation to the evolution of other groups of organisms; and those related to the correlation between diversity and climate. In the first category we cite three examples.

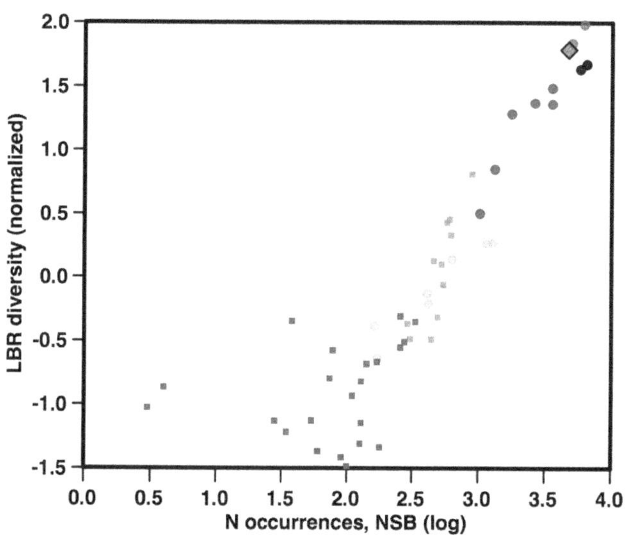

Figure 14. Diatom occurrences per time interval in the NSB database (log scale) vs LBR consensus diatom diversity estimate from figure 7. Symbols as in figure 11. Age scale: [43].

Grasslands. Rabosky and Sorhannus [18] suggested that the relatively low post-Eocene diversity of diatoms in their reconstruction argued against both a grassland expansion-driven increase of silica to the oceans, with consequent diversification of marine diatoms; and against co-evolutionary, inverse relationships between Cenozoic diatom and coccolithophore diversification. Our results do not support their arguments, although there are other reasons to question a grassland-marine diatom hypothesis, as proposed by Falkowski et al. [7]. Because grasses can alter the extent to which weathering products are stored in soils prior to their being dissolved and transported by water, it is reasonable to propose relatively short-term, local effects of different vegetation on local freshwater bodies [57]. It is not clear however if this scales to the global oceanic silica cycle over millions of years. Regardless of the silica content of standing grassland biomass, significant effects on the marine silica budget can only come from long-term ($>10^3$ year) differences in net rates of silicate weathering, yet grasslands may not actually increase long-term weathering rates vs. other vegetation types [32]. Grasses also were evolutionarily well developed before their late Neogene ecologic expansion [58]. Grassland evolution and ecologic expansion may thus not have driven diversification of marine diatoms. It is more likely both were influenced by tectonically driven Neogene increases in silicate weathering surfaces with consequent increase in marine nutrients, CO2 drawdown, cooling of continental climate and cooling climate forcing of enhanced ocean circulation [32,59,60].

Cetaceans. Our results weaken support for the hypothesis that Cenozoic cooling and the consequent concentration of high export productivity in upwelling regions, dominated by efficient, short food-chain diatoms, provided high levels of large zooplankton/nekton food supply that supported the radiation of cetaceans [60,61]. Although the Cenozoic-scale pattern seems plausible [60], and multiple factor models may still be valid [61], we did not, as in [61], find a significant correlation between latest Oligocene-Recent Cetacean and diatom diversity (maximum r values of ca. 0.2, $p > 0.1$: Fig. S3 and Table S2). We note however that the number of data points is small, and our method of averaging different curves means that, although the overall data series is robust, the precise values of a small number of points is not likely to be constant in alternate computations. Neither our negative, or the earlier [61] study's positive results should thus be considered a strong test of the hypothesis.

Radiolarians. Lazarus et al. [48] proposed that expanded Cenozoic diatom export productivity, together with increasing water column stratification, led to reduced silica availability in low latitude waters and an evolutionary reduction in radiolarian silica use, but did not statistically test diatom-radiolarian correlations. Our results support their inferred influence on radiolarians of an increase in diatom productivity over the Cenozoic, with a strong negative correlation of Cenozoic radiolarian silicification to diatom diversity ($r = .86$, $p < .001$; Fig. S4), although, as silicification data are few and noisy, detrended tests are not significant ($p > .1$).

Diversity and Climate

Environmental context of Cenozoic diatom evolution. As we wish to consider how the diversity history of diatoms is correlated to changing environments it is useful to review the most relevant aspects of Cenozoic ocean change. The main features are well known [62] and need be only briefly mentioned here. Early Cenozoic oceans were relatively warm, with little biogeographic differentiation. After reaching maximum warmth in the early/mid Eocene, ocean surface temperatures began to cool, and polar regions and tropical regions began to be more strongly differentiated from each other in the late Eocene. Cold deep-water circulation between the poles, underlying a more strongly isolated warm low to mid latitude surface water region developed along with Antarctic glaciation at the Eocene-Oligocene boundary. The late Oligocene to mid-Miocene were marked by a general warming trend, although punctuated by brief cooling episodes such as the M1 glacial event at the Oligocene-Miocene boundary. Renewed cooling near the end of the mid Miocene and associated expansion of the Antarctic ice-sheets was associated with the development of mid-latitude coastal upwelling systems. Cooling was briefly interrupted by a relatively modest warming interval in the early Pliocene, before being resumed in the late Pliocene-Recent, in association with widespread northern hemisphere glaciation. In parallel to this ocean climate history, both atmospheric pCO_2 and rates of weathering (which provide nutrients for new diatom productivity) were changing. Although still very poorly constrained, with different proxies often yielding very different estimates, Cenozoic pCO_2 showed a largely parallel trend, from low initial values in the Paleocene-early Eocene, maximum values of ca 1000 ppm in the mid-late Eocene, and declining values in the Oligocene-Recent (with modern pre-industrial pCO_2 values of ca 280 ppm) [63]. Weathering rates are even more poorly constrained, and, when not computed directly from pCO_2, are argued, based on various geochemical proxies, to either have increased substantially in the Oligocene-Recent [64] or have remained, at least for the Late Miocene-Recent, relatively constant [65].

Cenozoic climatic-ocean history was long believed to have been primarily driven by the tectonic opening of gateways that altered ocean circulation and polar heat transport, such as the development of a circumpolar Antarctic current near the Eocene-Oligocene boundary [56], but more recently changing concentrations of atmospheric pCO_2 have been implicated as the primary driver of Cenozoic climate change [66]. The role of ocean gateways however was not negligible, and this affects in particular our ability to interpret the causes of late Miocene to Recent changes in diatom diversity. Our understanding of Late Miocene and early Pliocene oceans is still very incomplete and rapidly evolving, and current understanding is thus worth summarizing.

Late Miocene climate was probably only moderately warmer than the Present (2–4°C globally: [67], however with several degrees additional warming in high latitudes: [67,68]), and thus not much warmer than the early Pliocene [69,70]. The early Pliocene is specifically noted as it is often used as a past proxy for projections of future climate warming [69], although the pCO_2 concentrations of the early Pliocene may not be representative of more extensive future warming, particularly beyond 2100 [66,71]. Late Miocene-early Pliocene climate was driven by an as yet not fully resolved combination of pCO_2 (via atmospheric temperature and wind fields) and unique, non-recurring past events such as the tectonically driven partial or complete closure of several oceanic gateways. The Panama gateway in particular closed gradually in the late Miocene-early Pliocene, which redirected tropical Atlantic waters northwards, warming the North Atlantic, and increasing pole-to-pole (Norwegian-Greenland Sea to Southern Ocean) surface-to-deep circulation and Southern Ocean diatom productivity [60,72,73]. Paleoceanographic studies suggest that the initial partial closure of the Panama gateway had already established the modern pattern of circulation by the early late Miocene [72–74], although final closure did not occur until the earliest Pliocene. Biogeographic, abundance frequency and evolutionary rate data for siliceous plankton [36,75,76] also suggest that an essentially modern global biogeographic pattern was already established by the early late Miocene. Moderately warmer than present late Miocene ocean climates, despite very low late Miocene pCO_2 estimates from epsilon-p plankton alkenones (ca 250+/− 50 ppm - equal to or even below modern pre-industrial values) [68], also have suggested that these past tectonic controlled circulation patterns rather than pCO_2 were primarily responsible for late Miocene ocean conditions.

More recent work however calls this tectonic 'gateway' model of late Miocene oceanography into question. Several ocean-atmosphere models have found Cenozoic gateways to have had (compared to pCO_2 or non-tectonic paleoceanographic mechanisms) only a secondary impact on global climate [68,77–80], while new studies suggest higher, if still moderate (ca 250–450 ppm) late Miocene pCO_2 [31,78,81]. It is important also to note that, while modeling studies provide valuable insight into the relative importance of causal factors in climate change, their ability to reconstruct absolute values of past climate is still limited. In particular, climate models tend to systematically and substantially underestimate the extent to which polar conditions warm, either as the result of changing pCO_2 or due to changes in gateways or other boundary conditions [77,82].

The late Miocene surface ocean conditions that these different controlling factors created are known to have differed from the modern, including for diatoms relevant environmental controls such as circulation, fronts, and distribution of nutrients. As documented by numerous deep-sea drilling sediment sections, biosiliceous sediments were more widespread in polar regions, indicating a reduced extent of permanent sea ice, even in regions near the Antarctic coast [83,84], while in lower latitudes the surface-deep ocean density contrast was lower, but with a deeper thermocline [68,85]. Surface water temperatures were also higher, although only by at most a few degrees [82]. Surface water temperature estimates from polar regions are however sparse, due to the general scarcity of carbonate-shelled plankton in high-latitude late Neogene pelagic sediments. How these conditions might relate to diatom evolution are discussed below.

Abiotic controls on large scale patterns of evolution. The relative importance of abiotic vis biologic interactions as shapers of large-scale patterns of evolution is a central question in evolutionary biology, but available data are few and difficult to interpret [86]. Most studies of plankton have so far shown a rather complex, episodic or threshold correlation of diversity to environment. This includes correlations to extreme events such as the abrupt mass extinction at the K/T boundary due to meteor impact, as well as many more minor climatic events (cooling, anoxia and others) in the Cretaceous and Cenozoic [38,87,88]. Only a few studies using general statistical comparisons of entire time series of diversity and environmental data have been done (e.g. Ezard [27] on calcareous zooplankton), and these have found significant if also complex and intermittent control by environmental factors, with single factor correlation coefficients of ca 0.4 in detrended data [27], similar to coefficients seen for Phanerozoic fossil invertebrates [89]. Our results for Cenozoic marine plankton diatom floras confirm an abiotic control of diversity, but differ from prior results in that: the correlation between diversity and climate is remarkably strong (r of ca 0.6 in detrended data; Spearman's rho >0.9 in the raw data), and unlike prior results e.g. [27,89], in diatoms colder climate, not warmer, is correlated to higher diversity.

We speculate that the higher degree of correlation between diversity and climate in diatoms, vs calcareous zooplankton or marine invertebrate benthos, may reflect the more direct impact that the physical environment has on phytoplankton growth, via regulation of temperature, light and nutrients. Zooplankton or invertebrate benthos by contrast are higher in the food chain and nutrient-climate correlations are reduced. This idea however ignores other aspects of phytoplankton growth such as grazing by zooplankton, and other pathways by which climate can influence invertebrate and zooplankton diversity, such as temperature effects on the energetics of carbonate shell formation. Comparative studies (e.g. using similar methods) of other microfossil records with complementary ecologies and shell mineralogy (calcareous phytoplankton: coccolithophores; siliceous zooplankton: radiolaria) might provide more insight. Currently, the best available estimate for Cenozoic calcareous phytoplankton diversity [38] shows a largely dissimilar diversity history to Cenozoic diatoms. Despite sharing an interval of low diversity in the late Oligocene-early Miocene, the overall trend in calcareous phytoplankton diversity has been downwards over the Cenozoic, from a diversity maximum in the earliest Eocene, which is regarded as the warmest interval of the Cenozoic. The different trajectories of diatoms vs calcareous phytoplankton has generally been interpreted to reflect the groups' contrasting ecologic adaptations to mesotrophic-oligotrophic (calcareous plankton) vs eutrophic (siliceous plankton) water conditions. These shifted from broadly mesotrophic conditions in the earlier Cenozoic towards regionally partitioned oligotrophy and eutrophy as cooling climate increased localized upwelling [7]. Regional upwelling specifically increased in the Southern Ocean after the Eocene-Oligocene boundary, in equatorial open ocean systems, and in mid-latitude coastal upwelling systems, such as the Benguela and California systems, in the late Miocene. The oceans thus came to be more strongly divided into a broad, geographically interconnected, low to mid latitude warm surface water, more strongly oligotrophic ocean dominated by calcareous phytoplankton, and a series of distinct, separated regional areas in both the tropics and high latitudes of more eutrophic waters dominated by differentiated endemic floras of siliceous plankton. This pattern has long been known [13,14] but our results (e.g. Fig. 8) provide a quantitative evaluation of this phenomenaon, and show that diversification in Cenozoic diatoms was due primarily to the increased diversity resulting from geographic endemism between these distinct regions. Since the mid Eocene tropical diversity has increased from ca 35 species to a maximum in the Pliocene of 60 species (ca 25 species, $<2\times$ relative regional increase), while overall diversity, e.g. including endemic polar floras, increased from ca 55 to 145 species over the same time interval (ca 90 species, $3.6\times$ the tropical species increase, $>2.6\times$ relative regional increase). There is thus a strong geographic cause in addition to the physiologic/ecologic cause (oligo- vs eutrophy) in Cenozoic diatom diversification. It can be argued that the primary reason for Cenozoic diatom diversification is not that diatoms are better adapted to increasing Cenozoic eutrophic environments, but that they are better adapted to Cenozoic environments that are geographically distinct.

Diatoms and Cenozoic geochemical cycles. Much of the earth-sciences interest in global Cenozoic diatom diversity lies in the presumed correlation of increasing Cenozoic diatom diversity to increasing Cenozoic export productivity by diatoms in the oceans, which in turn may have substantially affected the Cenozoic evolution of the global carbon cycle, including atmospheric pCO_2. Until quite recently [26] no direct sedimentary proxy for global diatom export productivity has been available, but it is generally thought that biogenic opal in marine sediments (largely consisting of diatoms) is a good, if rough, approximation of high diatom export productivity in the overlying water column [90]. The relative abundance of siliceous deep-sea sediments (vs carbonate or terrigenous) over the Cenozoic should thus indicate, at least approximately, the relative importance of diatoms to global oceanic export production. Our first order trend in diatom diversity is of major increase over the Cenozoic, suggesting a similar large increase in diatom export productivity and the relative abundance of biogenic opal in sediments. Prior compilations of biogenic opal abundance in Cenozoic deep-sea sediments [23–25] suggest that siliceous sediments have indeed become more common over the Cenozoic, although as these compilations were of a qualitative nature no numeric comparison is possible. Studies over shorter time intervals exist but are often difficult to evaluate. Cortese et al. [91] quantitatively studied details of the shifting geographic patterns of opal deposition in the late Neogene, but did not estimate opal relative abundance on a global basis, making comparison to our results difficult. Although no relative abundance estimates of siliceous sediment were given, recent studies [47,92] calculated absolute Cenozoic diatom sediment abundance histories from, respectively, the MRC and Neptune databases. These estimates are likely to be affected by biases in the relative amounts of siliceous vs non-siliceous data entered per time interval by the database compiler (in both instances overseen by the senior author of this paper). Lastly, although only a regional estimate, dissolved ocean silica usage, estimated from silicon isotope gradients in the water column, increased in the Southern Ocean in the late Eocene between ca 37–35 Ma, which is interpreted to reflect increasing diatom export productivity [26]. This matches well our calculated increase in both global and Southern Ocean diatom diversity at this time. Thus, available data on changing Cenozoic opal export is very limited, but tends to parallel the diversity increase in diatom diversity seen in our study, supporting, albeit only weakly, the use of diversity as a proxy for opal export productivity.

While export productivity is important to understanding many aspects of the marine ecosystem, of particular interest is the effect that marine export may have had on the global carbon cycle, and thus on atmospheric pCO_2 and climate. As noted earlier, $\partial^{13}C$ is often used as a proxy for global carbon cycle behavior. Late Neogene $\partial^{13}C$ shows a substantial shift (ca 1 per mil), which has been interpreted as reflecting an increase in the fraction of organic carbon sequestered by less ^{13}C depleted groups of organisms, specifically C4 terrestrial plants and marine diatoms [30]. Our diatom diversity curve shows a close correlation to this carbon isotopic shift beginning in the mid Miocene. More generally, diatom diversity changes (in detrended time series) are positively correlated with coeval changes in detrended $\partial^{13}C$ since the early Oligocene. These comparisons thus provide qualitative support for the idea that increasing Cenozoic diatom diversity is correlated to increasing fractions of diatom export productivity, and increasing relative abundance of diatom opal in sediment, since the mid Miocene. Together with evidence from living ecosystems and the limited independent evidence for Cenozoic opal productivity (reviewed above) our results suggest that diatom diversity can be used to some degree as a proxy for the influence of diatoms on marine export productivity and the carbon cycle. Diatom diversity is further strongly correlated to pCO_2 over the last 15 my. Since the pCO_2 record is derived from the $\partial^{18}O$ record [31], this is not too surprising, but it is interesting that the correlation between diversity and pCO_2 is stronger than between diversity and $\partial^{18}O$, implying that the observed correlation to pCO_2 is not simply one inherited from the observed correlation of diversity and $\partial^{18}O$. Our analysis thus also suggests that increasing diversity and abundance of diatoms are closely linked to late Neogene changes in the global carbon cycle. Whether this latter link is due to both diatom diversity and pCO_2 being driven by a third factor, such as increasing late Neogene silicate weathering, is however beyond the scope of our current study (see also below).

Implications of correlations for response to future anthropogenic global warming

Our analysis shows that warmer oceans are associated with lower diatom diversity, suggesting the possibility that future warmer oceans due to anthropogenic warming may result in lower diatom diversity, i.e. extinctions, with possibly substantial consequences for the functioning of the ocean carbon pump. This is suggested both by the correlation of increasing diversity with increasingly cold climate in the Neogene, and the reduction of diversity that occurred with warming ocean conditions in the late Oligocene. The broad uniformity of response over such a long range of time, and under such a wide variety of ocean conditions (as reviewed above) suggest that response will be similar also in future climate change. There are however significant problems in converting these general correlations into concrete predictions of diatom response to future warming. Oligocene floras were, as we document, almost entirely composed of species not present today, and even many genera were different. It is thus difficult to be sure that the response behavior of the modern, taxonomically different flora would be the same. The diatom floras of the Late Miocene ocean were much more similar taxonomically to those of the present, and the diatom diversity-climate relationship is particularly sensitive in this time interval. Diatom diversity in the late Miocene was, in our estimate, up to ca. 20% lower than modern values (and up to 50% lower in the middle Miocene, ca 15 Ma), and important cold-water polar species, such as F. kerguelensis and Neodenticula seminae, which play a major role in export productivity in the modern ocean [5] were absent. Given the potential importance of this issue, and the lack of any prior estimate of marine diatom extinction risk, it is important to consider this question. Evaluating the probability of future extinction however requires evaluating a chain of cause-effect processes that links past pCO_2->global climate->marine ocean conditions->diatom diversity->the carbon pump.

Although still uncertain, current paleoceanographic studies suggest that late Miocene ocean conditions might be a relevant analog for future marine plankton diatoms due to global climate warming, particularly on longer (post 2100) time scales. Reconstructions indicate higher temperatures and reduced sea-ice extent in polar regions, factors that are known to play a major role in the distribution of living diatom species and other phytoplankton, and diatom export productivity [50,93,94]. It thus seems reasonable to presume these factors, which, from our results are temporally correlated with lower diatom diversity, might also affect future diatom diversity, even if the proximal mechanisms are largely still unknown. Unlike the late Miocene however, where diversity-productivity relationships were in at least approximate evolutionary equilibrium, future climate change is expected to be orders of magnitude more rapid, so that rapid species loss would not be compensated for by significant evolutionary response, e.g. the result would be a perturbed system.

The effect of loss of diatom diversity on future ocean productivity is unknown but potentially significant. Bopp et al. [2] model a substantial drop in diatom abundance, particularly in high latitudes, in a 4X pCO_2 climate change scenario, although only a moderate impact on the ocean biologic pump. In Bopp et al. [95], this moderate degree of global change in pump efficiency is in part due to increased pump activity in polar regions, which partially compensates for an a broader drop in pump functioning in lower latitudes. These models however assume future diatom plankton response, particularly in polar regions, can be extrapolated from the living plankton flora. Would this still be true if important diatom species, particularly in polar regions, were to become extinct? Would extinct species of diatoms be replaced by fully functionally equivalent (e.g. for export productivity) other diatom species, by functionally less efficient taxa, or functionally very different coccolithophores or non-skeletal plankton? Our results, together with evidence that rapid diversity loss is linked to reduced productivity [53,54], suggest that loss of diatom diversity in future oceans, if it occurred, would indeed affect diatom abundance behavior, diatom-carbon export and thus further modify the global ocean pump, with consequences for future ocean regulation of pCO_2.

While the above presents arguments why the correlation of diatom diversity to climate state should be considered in thinking about future climate change, there are limits as well. We argue that while our results provide a useful perspective on eventual system response, a variety of issues, including temporal scaling and other potential casual factors, limit our ability to use our results to make any direct estimate of extinction risk due to anthropogenic change. As reviewed above, Late Miocene oceans, although largely very similar to the modern ocean, were still dissimilar in details of circulation and climatic context. The proximate conditions that affect diatom distributions - nutrient concentrations, water column stability, frontal structures; and the variability of these on seasonal to multi-decadal scales, could have been significantly different: not only from those of the modern ocean, but also to the oceans that anthropogenic global warming will produce. Our understanding of these proximate controls in Late Miocene oceans is very sketchy, particularly in high latitude regions which are most important to understanding diversity and productivity response. Also important is the temporal scaling between our data and future global change in how global nutrient are input to the system, which at the time scales of our study may exert a rate limiting effect on climate change, since the residence times of key elements in the oceans are much less than the scale of our analyses (Silica ca 20 kyr; Carbon ca 200 kyr: [30,90]). Although the proximate controls on past diatom diversity might have been temperature or other rapidly acting factor, if nutrients were the limiting factor, changes in weathering may well only affect oceans over time scales of many hundreds or thousands of years. It is thus difficult to use only a generic proxy such as $\partial^{18}O$ to identify truly analogous conditions in the past that can be used to predict future diatom behavior in a warmer world. In order to do this, we need much more information on the proximate controls on individual species distributions, both for now extinct species, and for modern species, whose ecology also is often only poorly known.

Conclusions

Cenozoic diatom diversity and its correlation to climate change is important to understanding how evolution in pelagic systems functions, and how biotic diversity may regulate climate change itself over various time scales - from millions of years, to those expected for future anthropogenic global warming. Existing estimates of diatom diversification have been contradictory and have methodologic problems. We derive a new, internally coherent estimate using multiple methodologies and datasets, using a newly created, first-ever comprehensive catalog of Cenozoic marine diatom species ranges as well as a new version of the Neptune marine microfossil database, while explicitly controlling for prior problems in diversity reconstruction such as changing sample sizes, evenness of population frequencies - by use of a novel modeling-based correction function, and other factors. We show that diatoms diversified strongly over the Cenozoic, with major increases near the Eocene/Oligocene boundary, and in the mid Miocene. Diversification occurs primarily by increasing diversity of endemic high latitude floras, and the disjunct development of high productivity regions in Cenozoic oceans is suggested to be a primary reason for diversification of the group. Diatom diversity is strongly correlated to evolutionary characteristics of other groups of organisms such as radiolaria; and to the oxygen isotope proxy record of global climate change: diversity not only increases with intervals of cooling, but also significantly decreases during intervals of past warming such as the late Oligocene-early Miocene. Over the last 15 my diatom diversity is strongly correlated to both the global carbon isotope record and to estimated past atmospheric pCO_2. These correlations suggest that diatoms have played an important role in the evolution of mid-Miocene to Recent climate, via their prominent role in the oceanic carbon pump. The correlation of warmer climate to lower diversity also suggests that global warming could potentially place a significant fraction of diatom diversity at risk of extinction, particularly as we show that important export productivity species originated only in the last few million years in association with the development of cold polar oceans. Both the time resolution of our study (0.5–1.0 my resolution) and the complexity of cause-effect relationships however mean that we cannot evaluate from our data alone the likelihood of future extinction over the next decades or centuries as a possible consequence of anthropogenic global warming.

Supporting Information

Figure S1 Number of diatom species vs biomass in water column samples from Atlantic ocean transect, from data presented in [**50**]. Age scale: [43]**.**

Figure S2 Detrended numbers of occurrences in time intervals in the NSB database vs detrended LBR diatom diversity estimate. Data point symbols age coded according to main paper figure 11. Age scale: [43].

Figure S3 Comparison of species diversity in Neogene cetaceans (red: sampled in bin and range-through) to diatom diversity of this study (blue: 3-point moving average of raw data and Neogene portion of Cenozoic detrended values as reported in table ST2). Cetacean data from [61]. Age scale: [42].

Figure S4 Diatom diversity (average of the three estimates for a bin, not smoothed) vs radiolarian shell silicification, from [**48**]**.** Silicification data binned to 1 my intervals, intervals with no silicification data excluded. Age scale: [43].

Table S1 The BDC (Barron Diatom Catalog). a) Data. b) List of publications/sources used.

Table S2 Pearson correlation coefficient r and color coded p value intervals for diatom diversity, Neoceti diversity (from [**61**]**) and geologic age.**

Table S3 Main data and results of this study in numeric form.

Acknowledgments

We thank M. Aberhan, L. Armand, W. Kiessling, B. Kotrc, H. Oberhänsli, M. Thomas, M. A. Barcena and two anonymous reviewers for constructive comments on earlier drafts of this paper. The BDC list is included in the Supporting Information. The NSB database can be accessed online: please contact the senior author for details.

Author Contributions

Conceived and designed the experiments: DL JR. Analyzed the data: DL JR AT PD. Wrote the paper: DL. Compiled data used in study: DL JB JR PD.

References

1. Cervato C, Burckle L (2003) Pattern of first and last appearance in diatoms: Oceanic circulation and the position of polar fronts during the Cenozoic. Paleoceanography 18: web.
2. Bopp L, Aumont O, Cadule P, Alvain S, Gehlen M (2005) Response of diatoms distribution to global warming and potential implications: A global model study. Geophysical Research Letters 32: 1–4 (web).
3. Armbrust EV (2009) The life of diatoms in the world's oceans. Nature 459: 185–192.
4. Nelson DM, Tréguer P, Brzezinski M, Leynaert A, Quéguiner B (1995) Production and dissolution of biogenic silica in the ocean: revised global estimates, comparison with regional data and relationship to biogenic sedimentation. Global Biogeochemical Cycles 9: 359–372.
5. Smetacek V (1999) Diatoms and the global carbon cycle. Protist 150: 25–32.
6. Falkowski P, Schofield O, Katz ME, Van de Schootbrugge B, Knoll A (2004) Why is the land green and the ocean red? In: Thierstein H, Young JR, editors. Coccolithophores: From Molecular Processes to Global Impact. Berlin: Springer. pp. 429–453.
7. Falkowski PG, Katz ME, Knoll AH, Quigg A, Raven JA, et al. (2004) The evolution of modern eukaryotic phytoplankton. Science 305: 354–360.
8. IPCC Core Writing Team (2007) Climate Change 2007: Synthesis Report. 104.
9. Richardson AJ, Schoeman DS (2004) Climate impact on plankton ecosystems in the northeast Atlantic. Science 305: 1609–1612.
10. Hays GC, Richardson AJ, Robinson CR (2005) Climate change and marine plankton. Trends Ecology Evolution 20: 337–344.
11. Thomas MK, Kremer CT, Klausmeier CA, Litchman E (2012) A global pattern of thermal adaptation in marine phytoplankton. Science 338: 1085–1088.
12. Halpern BS, Kappel CV (2012) Extinction risk in a changing ocean. In: Hannah L, editor. Saving a Million Species: Extinction Risk from Climate Change. Washington: Island Press. pp. 285–307.
13. Barron JA, Baldauf JG (1995) Cenozoic marine diatom biostratigraphy and applications to paleoclimatology and paleoceanography. In: Blome CD, Whalen PM, Reed KM, editors. Siliceous Microfossils. The Paleontological Society. pp. 107–118.
14. Scherer RP, Gladenkov AY, Barron J (2007) Methods and applications of Cenozoic marine diatom biostratigraphy. In: Starratt S, editor. Pond Scum to Carbon Sink: Geological and Environmental Applications of the Diatoms. The Paleontological Society. pp. 61–83.
15. Lazarus D (2011) The deep-sea microfossil record of macroevolutionary change in plankton and its study. In: Smith A, McGowan A, editors. Comparing the Geological and Fossil Records: Implications for Biodiversity Studies. London: The Geological Society. pp. 141–166.
16. Spencer-Cervato C (1999) The Cenozoic deep sea microfossil record: explorations of the DSDP/ODP sample set using the Neptune database. Palaeontologica Electronica 2: web.
17. Lazarus DB (1994) The Neptune Project - a marine micropaleontology database. Mathematical Geology 26: 817–832.
18. Rabosky DL, Sorhannus U (2009) Diversity dynamics of marine planktonic diatoms across the Cenozoic. Nature 247: 183–187.
19. Dornelas M, Magurran AE, Buckland ST, Chao A, Chazdon RL, et al. (2012) Quantifying temporal change in biodiversity: challenges and opportunities. Proceedings of the Royal Society London, B 280: 1–11 (web).
20. Lloyd GT, Pearson PN, Young JR, Smith AG (2012) Sampling bias and the fossil record of planktonic foraminifera on land and in the deep sea. Paleobiology 38: 569–584.
21. Marshall CR (2010) Marine biodiversity dynamics over deep time. Science 329: 1156–1158.
22. Zachos J, Dickens GR, Zeebe RE (2008) An early Cenozoic perspective on greenhouse warming and carbon-cycle dynamics. Nature 451: 279–283.
23. Leinen M (1979) Biogenic silica accumulation in the central Equatorial Pacific and its implications for Cenozoic paleoceanography. Geological Society of America Bulletin 90: 1310–1376.
24. Barron JA, Baldauf JG (1989) Tertiary cooling steps and paleoproductivity as reflected by diatoms and biosiliceous sediments. In: Berger WH, Smetacek VS,

Wefer G, editors. Productivity of the Ocean: Present and past. Wiley, John and Sons Limited. pp. 341–354.
25. Baldauf JG, Barron JA (1990) Evolution of biosiliceous sedimentation patterns - Eocene through Quaternary: paleoceanographic response to polar cooling. In: Bleil U, Thiede J, editors. Geological History of the Polar Oceans: Arctic vs Antarctic. Dordrecht: Kluwer. pp. 575–607.
26. Egan KE, Rickaby REM, Hendry KR, Halliday AN (2013) Opening the gateways for diatoms primes Earth for Antarctic glaciation. Earth and Planetary Science Letters 375: 34–43.
27. Ezard THG, Aze T, Pearson PN, Purvis A (2011) Interplay between changing climate and species' ecology drives macroevolutionary dynamics. Science 332: 349–351.
28. Elderfield H, Ferretti P, Greaves M, Crowhurst S, McCave N, et al. (2012) Evolution of ocean temperature and ice volume through the mid-Pleistocene climate transition. Science 337: 704–709.
29. Kump LR, Arthur MA (1999) Interpreting carbon-isotope excursions: carbonates and organic matter. Chemical Geology 161: 181–198.
30. Katz ME, Wright JD, Miller KG, Cramer BS, Fennel K, et al. (2005) Biological overprint of the geological carbon cycle. Marine Geology 217: 323–338.
31. van de Wal RSW, de Boer B, Lourens L, Köhler P, Bintanja R (2011) Reconstruction of a continuous high-resolution CO_2 record over the past 20 million years. Climate of the Past 7: 1459–1469.
32. Beerling DJ, Taylor LL, Bradshaw CDC, Lunt DJ, Valdes PJ, et al. (2012) Ecosystem CO_2 starvation and terrestrial silicate weathering: mechanisms and global-scale quantification during the late Miocene. Journal of Ecology 100: 31–41.
33. Fils D, Cervato C, Reed J, Diver P, Tang X, et al. (2009) CHRONOS architecture: Experiences with an open-source services-oriented architecture for geoinformatics. Computers and Geosciences 35: 774–782.
34. Lloyd GT, Smith AG, Young JR (2011) Quantifying the deep-sea rock and fossil record bias using coccolithophores. In: McGowan AJ, Smith AG, editors. Comparing the Geological and Fossil Records: implications for biodiversity studies. London: Geological Society. pp. 167–178.
35. Lazarus D, Weinkauf M, Diver P (2012) Pacman profiling: a simple procedure to identify stratigraphic outliers in high density deep-sea microfossil data. Paleobiology 38: 144–161.
36. Renaudie J, Lazarus D (2013) On the accuracy of paleodiversity reconstructions: a case study in Antarctic Neogene radiolarians. Paleobiology 39: 491–509.
37. Sepkoski JJ, Jr. (1978) A kinetic model of Phanerozoic taxonomic diversity. I. Analysis of marine orders. Paleobiology 4: 223–251.
38. Bown PR, Lees JA, Young JR (2004) Calcareous nannoplankton evolution and diversity through time. In: Thierstein HR, Young JR, editors. Coccolithophores: From Molecular Processes to Global Impact. Berlin: Springer. pp. 481–508.
39. Aze T, Ezard THG, Purvis A, Coxall H, Stewart DRM, et al. (2011) A phylogeny of Cenozoic macroperforate planktonic foraminifera from fossil data. Biological Reviews DOI: 10.1111/j.1469-185X.2011.00178.x:
40. Lazarus DB, Spencer-Cervato C, Pianka-Biolzi M, Beckmann JP, von Salis K, et al. (1995) Revised chronology of Neogene DSDP holes from the world ocean. College Station: Ocean Drilling Program. ca. 250 p.
41. Spencer-Cervato C, Thierstein HR, Lazarus DB, Beckmann JP (1994) How synchronous are Neogene marine plankton events? Paleoceanography 9: 739–763.
42. Berggren WA, Kent DV, Swisher CC, Aubry M-P (1995) A revised Cenozoic geochronology and chronostratigraphy. In: Berggren WA, Kent DV, Aubry MP, Hardenbol J, editors. Geochronology, Time Scales and Stratigraphic Correlation: Framework for an Historical Geology. Tulsa: Society of Economic Paleontologists and Mineralogists. pp. 129–212.
43. Gradstein FM, Ogg JG, Smith AG (2004) A Geologic Time Scale 2004. 589.
44. Alroy J (2010) Fair sampling of taxonomic richness and unbiased estimation of origination and extinction rates. In: Alroy J, Hunt G, editors. Quantitative Methods in Paleobiology. The Paleontological Society. pp. 55–80.
45. Hannisdal B (2011) Detecting common-cause relationships with directional information transfer. In: McGowan A, Smith AG, editors. Comparing the Geological and Fossil Records: Implications for Biodiversity Studies. pp. 19–30.
46. Hannisdal B, Peters SE (2011) Phanerozoic earth system evolution and marine biodiversity. Science 334: 1121–1124.
47. Muttoni G, Kent DV (2007) Widespread formation of cherts during the early Eocene climatic optimum. Palaeogeography Palaeoclimatology Palaeoecology 253: 348–362.
48. Lazarus D, Kotrc B, Wulf G, Schmidt DN (2009) Radiolarians decreased silicification as an evolutionary response to reduced Cenozoic ocean silica availability. Proceedings National Academy of Sciences USA 106: 9333–9338.
49. Chase JM, Leibold MA (2002) Spatial scale dictates the productivity-biodiversity relationship. Nature 461: 427–430.
50. Cermeño P, Dutkiewicz S, Harris RP, Follows M, Schofield O, et al. (2008) The role of nutricline depth in regulating the ocean carbon cycle. Proceedings National Academy of Sciences, USA 105: 20344–20349.
51. Ptacnik R, Solimini AG, Andersen T, Tamminen T, Brettum P, et al. (2008) Diversity predicts stability and resource use efficiency in natural phytoplankton communities. Proceedings National Academy of Sciences 105: 5134–5138.
52. Barton AD, Dutkiewicz S, Flierl G, Bragg J, Follows M (2010) Patterns of diversity in marine phytoplankton. Science 327: 1509–1511.
53. Cardinale BJ, Matulich KL, Hooper DU, Byrnes JE, Duffy E, et al. (2011) The functional role of producer diversity in ecosystems. American Journal of Botany 98: 572–592.
54. Hooper DU, Adair EC, Cardinale BJ, Byrnes JEK, Matulich KL, et al. (2012) A global synthesis reveals biodiversity loss as a major driver of ecosystem change. Nature 486: 105–108.
55. Lazarus D, Diver P, Suzuki N, IODP Paleontology Coordination Group (2011) Tools for synthesis of the deep-sea microfossil record - recent advances in developing the Neptune database, taxonomic dictionaries, and age-model library. Geobiology and Environments of Siliceous Biomineralizers (abs). p. 20.
56. Kennett JP (1982) Marine Geology. Englewood Cliffs, N.J.: Prentice-Hall. 813 pp.
57. Kidder DL, Gierlowski-Kordesch EH (2005) Impact of grassland radiation on the nonmarine silica cycle and Miocene diatomite. Palaios 20: 198–206.
58. Strömberg CAE (2005) Decoupled taxonomic radiation and ecological expansion of open-habitat grasses in the Cenozoic of North America. Proceedings National Academy of Sciences, USA 102: 11980–11984.
59. Raymo M (1994) The Himalayas, organic carbon burial, and climate in the Miocene. Paleoceanography 9: 399–404.
60. Berger WH (2007) Cenozoic cooling, Antarctic nutrient pump, and the evolution of whales. Deep-Sea Research II 54: 2399–2421.
61. Marx FG, Uhen MD (2010) Climate, critters, and cetaceans: Cenozoic drivers of the evolution of modern whales. Science 327: 993–996.
62. Zachos J, Pagani M, Solan L, Thomas E, Billups K (2001) Trends, rhythms and aberrations in global climate 65 Ma to present. Science 292: 686–693.
63. Beerling DJ, Royer DL (2011) Convergent Cenozoic CO_2 history. Nature Geoscience 4: 418–420.
64. Lear CH, Elderfield H, Wilson PA (2003) A Cenozoic seawater Sr/Ca record from benthic foraminiferal calcite and its application in determining global weathering fluxes. Earth and Planetary Science Letters 208: 69–84.
65. Willenbring JK, Blanckenberg Fv (2010) Long-term stability of global erosion rates and weathering during late-Cenozoic cooling. Nature 465: 211–214.
66. Hansen J, Sato M, Kharecha P, Beerling DJ, Berner RA, et al. (2008) Target atmospheric CO_2: Where should humanity aim? Open Atmospheric Science Journal 2: 217–231.
67. Micheels A, Bruch AA, Eronen J, Fortelius M, Harzhauser M, et al. (2011) Analysis of heat transport mechanisms from a Late Miocene model experiment with a fully-coupled atmosphere-ocean general circulation model. Palaeogeography Palaeoclimatology Palaeoecology 304: 337–350.
68. LaRiviere JP, Ravello AC, Crimmins A, Dekens PS, Ford HL, et al. (2012) Late Miocene decoupling of oceanic warmth and atmospheric carbon dioxide forcing. Nature 486: 97–100.
69. Dowsett HJ, Robinson MM, Haywood AM, Hill DJ, Dolan AM, et al. (2012) Assessing confidence in Pliocene sea surface temperatures to evaluate predictive models. Nature Climate Change 2: 365–371.
70. Fedorov AV, Brierley CM, Lawrence KT, Liu Z, Dekens PS, et al. (2013) Patterns and mechanisms of early Pliocene warmth. Nature 496: 43–49.
71. Sanderson BM, O'Neill BC, Kiehl JT, Meehl GA, Knutti R, et al. (2011) The response of the climate system to very high greenhouse gas emission scenarios. Environmental Research Letters 6: 1–11 (web).
72. Nisancioglu KH, Raymo ME, Stone PH (2003) Reorganization of Miocene deep water circulation in response to the shoaling of the Central American Seaway. Paleoceanography 18: web.
73. Butzin M, Lohmann G, Bickert T (2011) Miocene ocean circulation inferred from marine carbon cycle modeling combined with benthic isotope records. Paleoceanography 26: 1–19 (web).
74. Lear CH, Rosenthal Y, Wright JD (2003) The closing of a seaway: ocean water masses and global climate change. Earth and Planetary Science letters 210: 425–436.
75. Barron J (2003) Planktonic marine diatom record of the past 18 m.y.: appearances and extinctions in the Pacific and Southern Ocean. Diatom Research 18: 203–224.
76. Kamikuri S, Motoyama I, Nishi H, Iwai M (2009) Evolution of Eastern Pacific Warm Pool and upwelling processes since the middle Miocene based on analysis of radiolarian assemblages: Response to Indonesian and Central American Seaways. Palaeogeography Palaeoclimatology Palaeoecology 280: 469–479.
77. Huber M, Caballero R (2011) The early Eocene equable climate problem revisited. Climate of the Past Discussions 7: 241–304.
78. Bradshaw CD, Lunt DJ, Flecker R, Salzmann U, Pound MJ, et al. (2012) The relative roles of CO_2 and palaeogeography in determining Late Miocene climate: results from a terrestrial model-data comparison. Climate of the Past Discussions 8: 715–786.
79. Lefebvre V, Donnadieu Y, Sepulchre P, Swingedouw D, Zhang Z-S (2012) Deciphering the role of southern gateways and carbon dioxide on the onset of the Antarctic Circumpolar Current. Paleoceanography 27: 1–9 (web).
80. Lunt DJ, Haywood AM, Schmidt GA, Salzmann U, Valdes PJ, et al. (2012) On the causes of mid-Pliocene warmth and polar amplification. Earth and Planetary Science Letters 321–2: 128–138.
81. Micheels A, Bruch AA, Uhl D, Utescher T, Mosbrugger V (2007) A Late Miocene climate model simulation with ECHAM4/ML and its quantitative validation with terrestrial proxy data. Palaeogeography Palaeoclimatology Palaeoecology 253: 251–270.

82. Goldner A, Herold N, Huber M (2013) The challenge of simulating warmth of the mid-Miocene climate optimum in CESM1. Climate of the Past Discussions 9: 3489–3518.
83. Burckle LH, Robinson D, Cooke D (1982) Reappraisal of Sea-Ice distribution in Atlantic and Pacific sectors of the southern Ocean at 18,000 yr BP. Nature 299: 435–437.
84. Levy R, Cody RD, Crampton JS, Fielding C, Gollege N, et al. (2012) Late Neogene climate and glacial history of the Southern Victoria Land coast from integrated drill core, seismic and outcrop data. Global and Planetary Change 80–81: 61–84.
85. Finkel ZV, Sebbo J, Feist-Burkhardt S, Irwin AJ, Katz ME, et al. (2007) A universal driver of macroevolutionary change in the size of marine phytoplankton over the Cenozoic. Proceedings National Academy Sciences USA 104: 20416–20420.
86. Benton MJ (2009) The Red Queen and the Court Jester: species diversity and the role of biotic and abiotic factors through time. Science 323: 728–732.
87. Wei KY, Kennett JP (1986) Taxonomic evolution of Neogene planktonic foraminifera and paleoceanographic relations. Paleoceanography 1: 67–84.
88. Stanley SM, Whetmore KL, Kennett JP (1988) Macroevolutionary differences between the two major clades of Neogene planktonic foraminifera. Paleobiology 14: 235–249.
89. Mayhew PJ, Bell MA, Benton TG, McGowan AJ (2012) Biodiversity tracks temperature over time. Proceedings National Academy Sciences USA 109: 15141–15145.
90. Ragueneau O, Tréguer P, Leynaert A, Anderson RF, Brzezinski MA, et al. (2000) A review of the Si cycle in the modern ocean: recent progress and missing gaps in the application of biogenic opal as a paleoproductivity proxy. Global and Planetary Change 26: 317–365.
91. Cortese G, Gersonde R, Hillenbrand C-D, Kuhn G (2004) Opal sedimentation shifts in the World Ocean over the last 15 Myr. Earth and Planetary Science Letters 224: 509–527.
92. Lazarus D (2006) The Micropaleontological Reference Centers network. Scientific Drilling 3: 46–49.
93. Longhurst AR (1998) Ecological Geography of the Sea. San Diego: Academic Press. 398 p.
94. Boyd PW, Doney SC (2002) Modelling regional responses by marine pelagic ecosystems to global climate change. Geophysical Research Letters 29: 1–4 (web).
95. Bopp L, Monfray P, Aumont O, Dufresne J-L, Le Treut H, et al. (2001) Potential impact of climate change on marine export productivity. Global Biochemical Cycles 15: 81–99.

18

An Ensemble Weighting Approach for Dendroclimatology: Drought Reconstructions for the Northeastern Tibetan Plateau

Keyan Fang[1,2,3]*, Martin Wilmking[4], Nicole Davi[5,6], Feifei Zhou[1], Changzhi Liu[2]

1 Key Laboratory of Humid Subtropical Eco-Geographical Process (Ministry of Education), College of Geographical Sciences, Fujian Normal University, Fuzhou, Fujian Province, China, **2** Key Laboratory of Western China's Environmental Systems (Ministry of Education), Research School of Arid Environment and Climate Change, Lanzhou University, Lanzhou, Gansu Province, China, **3** Department of Geosciences and Geography, University of Helsinki, Helsinki City, Helsinki, Finland, **4** Institute of Botany and Landscape Ecology, Greifswald University, Greifswald, Mecklenburg-Vorpommern, Germany, **5** Tree-Ring Lab, Lamont-Doherty Earth Observatory, Columbia University, Palisades, New York, United States of America, **6** Department of Environmental Science, William Paterson University, Wayne, New Jersey, United States of America

Abstract

Traditional detrending methods assign equal mean value to all tree-ring series for chronology developments, despite that the mean annual growth changes in different time periods. We find that the strength of a tree-ring model can be improved by giving more weights to tree-ring series that have a stronger climate signal and less weight to series that have a weaker signal. We thus present an ensemble weighting method to mitigate these potential biases and to more accurately extract the climate signals in dendroclimatology studies. This new method has been used to develop the first annual precipitation reconstruction (previous August to current July) at the Songmingyan Mountain and to recalculate the tree-ring chronology from Shenge site in Dulan area in northeastern Tibetan Plateau (TP), a marginal area of Asian summer monsoon. The ensemble weighting method explains 31.7% of instrumental variance for the reconstructions at Songmingyan Mountain and 57.3% of the instrumental variance in the Dulan area, which are higher than those developed using traditional methods. We focus on the newly introduced reconstruction at Songmingyan Mountain, which showsextremely dry (wet) epochs from 1862–1874, 1914–1933 and 1991–1999 (1882–1905). These dry/wet epochs were also found in the marginal areas of summer monsoon and the Indian subcontinent, indicating the linkages between regional hydroclimate changes and the Indian summer monsoon.

Editor: Eryuan Liang, Chinese Academy of Sciences, China

Funding: This research is financed by National Basic Research Program of China (2012CB955301), the National Science Foundation of China (41001115 and 41171039), and the Minjiang Special-term Professor fellowship. Support for Davi comes from NSF AGS # 1137729. The funders had no role in study design, data collection and analysis, decision to publish, or preparation of the manuscript.

Competing Interests: The authors have declared that no competing interests exist.

* E-mail: kujanfang@gmail.com

Introduction

Global warming has brought long-term climate data inferred from proxies into focus for both the scientific and public communities. These proxies enable us to assess recent climate change in the context of hundreds to thousands of years and to evaluate changes prior to any anthropogenic influences [1,2]. In addition, the availability of large-scale paleoclimate data increases the robustness of the analyses of regional climate regimes in relation to external forcings and internal feedback loops [3]. Because tree-ring records are climate sensitive and can be exactly dated, they have been widely used to extend short-term instrumental climate by centuries to millennia from regional to global scales [4]. A variety of detrending methods have been developed to isolate and extract the climate signal from tree-ring series, such as negative exponential or straight line splines [5]. However, the too flexible forms result in the inevitable loss of longer-timescale climate signal. In addition, the medium-frequency (e.g. decadal/multi-decadal scales) variations can bias the final chronology [6]. This is referred to as the "trend distortion" problem, which can be mitigated by the "signal-free" method [6]. In addition, potential bias in the traditional methods can arise when setting the mean value of the tree-ring indices to 1 for different tree-ring indices covering different time intervals (Figure 1). The mean values of individual tree-ring series can be different in different temporal intervals, therefore low (high) tree-ring indices can be increased (decreased) when assigning the same value to these indices (Figure 1). Third, the tree-ring chronology can be biased when including some less climate-sensitive tree-ring series or those showing a different climate-growth relationship [7,8]. In order to mitigate the three potential biases, we propose a method, termed the "ensemble weighting method", to iteratively weight individual tree-ring series according to their mean climate values and by the sensitivity of each series to climate.

We use this new ensemble weighting method to develop the first tree-ring chronology from the Songmingyan Mountains in the Linxia district of the northeastern Tibetan Plateau (TP) and to a shoulder region of Asian summer monsoon [9]. Understanding hydroclimate dynamics in regions that are only marginally affected by Asian summer monsoon is highly needed due to the sensitivity of these regions to large-scale atmospheric

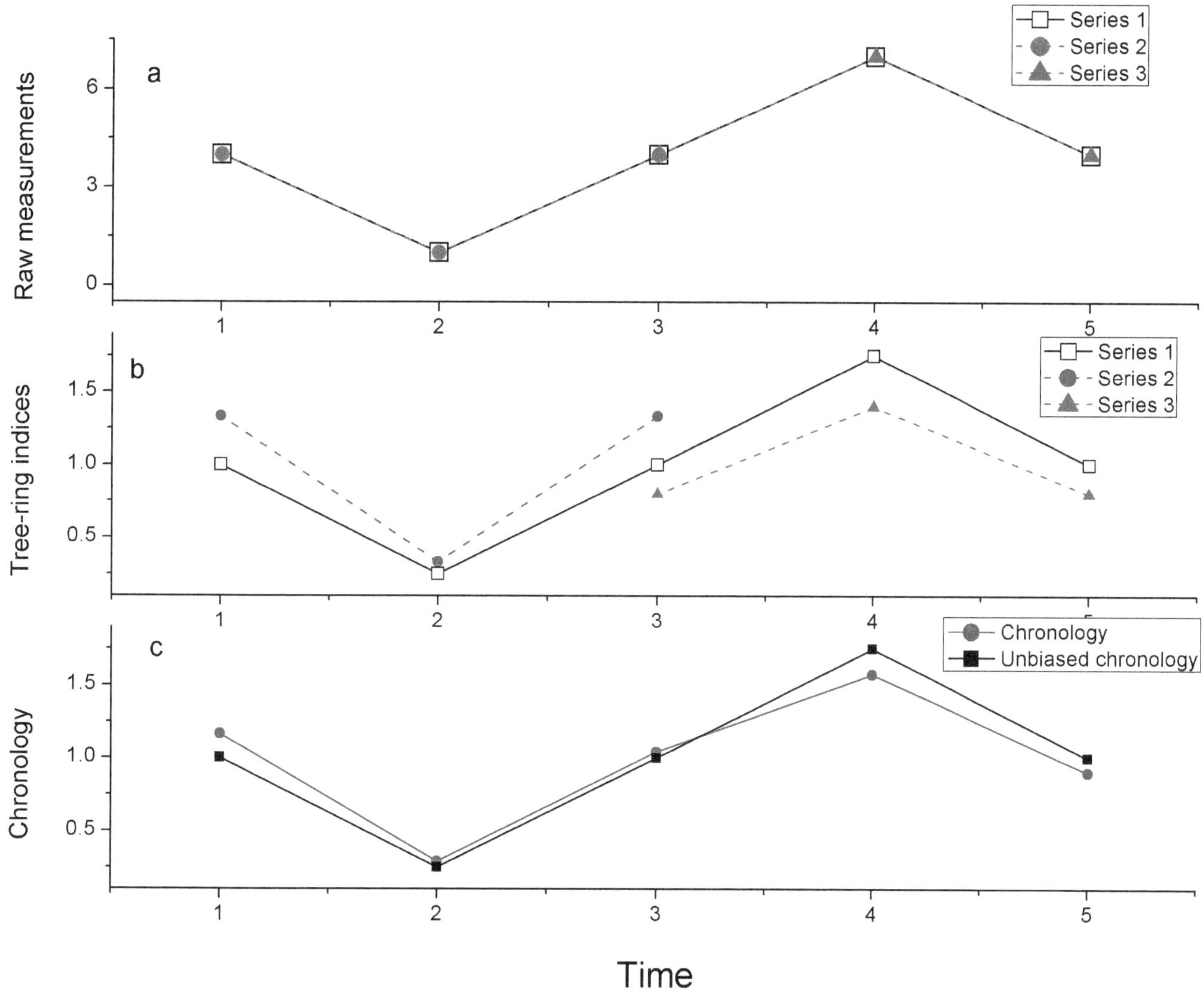

Figure 1. Schematic illustration of the distortion in chronology developments. (a) raw tree-ring measurements, (b) the standardized tree-ring indices calculated between the raw measurements and their mean values, and (c) the comparisons between chronology calculated as the mean among the tree-ring indices and the unbiased chronology.

changes and the potential impacts of water variability. Most instrumental records from these regions begin after the 1950s and tree-ring based reconstructions are sparse, limiting our ability to understand large-scale monsoon dynamics. Although there are many climate-sensitive tree-ring chronologies available for neighboring regions in northeastern TP [10,11,12,13,14,15], there are still no chronologies published from the Linxia district largely due to the limited availability of old-growth forests. In addition, we also test the use of this method for the development of a 1835-year tree-ring chronology in a nearby region in Dulan, which has been used for a hydroclimate reconstruction in the previous study [12,13,14,15]. Therefore, the goals of this study are to (1) introduce the "ensemble weighting method", (2) test its applicability by developing the first tree-ring chronology in the Linxia area and the tree-ring chronology in Dulan area, and (3) provide climate reconstructions for the northeastern TP and to detect its linkages with monsoon dynamics.

Data and Methods

The Instrumental Data

Monthly mean temperature peaks in July and monthly total precipitation peaks in August when the monsoon front reaches its northern shoulder region in the northeastern TP [16]. Old-growth forests in this region can only be found on the TP and on some of the mountain ranges on the arid or semi-arid Chinese Loess Plateau [17] (Figure 2). The newly introduced sampling site is located in the Songmingyan Mountain area (35.23°N, 103.39°E, 2589 m a.s.l.) in Linxia (Figure 2), in the northeastern TP. Total annual precipitation is 531 mm and annual mean temperature is 7.2°C according to the nearest meteorological station at Lintao (35.33°N, 103.82°E, 1891 m a.s.l., WMO NO. 52986). Pines and/or spruces growing near these temples are often well protected and can grow very old and can be used to develop long-term tree-ring chronologies [10,18,19]. The tree-ring samples used here are from Wilson's spruce (*Picea wilsonii*) growing in the sacred green island region on Songmingyan Mountain, the biggest national forest park in the Linxia district (Figure 2). The second tree-ring site at Dulan area is located near the Shenge town, northern to the Songmingyan Mountain area with a drier and colder climate than Songmingyan Mountain [15]. Total annual precipitation is 188 mm and the mean annual temperature is 3°C in Dulan meteorological station (36.00°N, 98.00°E, 3800 m a.s.l., WMO NO. 52986).

Tree-ring Data and Traditional Methods

We collected 30 cores from 20 *Picea wilsonii* trees at the Songmingyan Mountain site with permission from the Songmingyan Forest Park. 132 cores of 60 *Sabina przewalskii* trees were from Shenge site were downloaded from the International Tree-ring Data Bank (http://www.ncdc.noaa.gov/paleo/treering.html) [15]. These samples were mounted, air dried, polished and crossdated according to a skeleton plotting scheme by visually comparing the extremely narrow and wide rings [20]. The crossdated tree rings were measured to 0.001 mm accuracy and quality checked with moving correlations using the program COFECHA [21]. The growth trend in the raw measurements is removed by fitting an age-related growth curve (herein a cubic smoothing spline with a 50% cutoff at around 67% of the mean segment length) [22]. The dimensionless tree-ring indices are calculated as ratios between raw measurements and the fitted growth values, which are then averaged to produce a chronology based on a robust mean methodology [23]. In the signal-free method, the signal-free measurements are indexed as ratios between raw measurements and the initial robust mean chronology indices, which are again fitted with a growth curve (herein a spline) to create the "signal-free curve" representing the age-related growth trend. Then the tree-ring indices are calculated as ratios between raw measurements and the signal-free curve, which were used to create a new chronology. The final chronology is produced by iterating the aforementioned steps until the two latest versions of the chronologies showed only limited differences [6].

Ensemble Weighting Method

As shown in Figure 3, the ensemble weighting method for chronology development contains 3 stages and 6 steps, including:

Stage 1. Developing an initial chronology using traditional methods.

Step 1–2. Fitting growth curves and developing the chronologies using traditional methods detailed in the section above.

Stage 2. Developing the first version of the ensemble weighting chronology.

Step 3. The mean values of the initial chronology for the time period (segment length) the same as the duration of a given tree-ring index are calculated. These mean values are then assigned to the associated tree-ring index by multiplying the ratio between the mean chronology value corresponding to the length of the tree-ring index and the mean chronology value during the entire period.

Step 4. The target climate variable used for reconstruction is selected based on climate-growth relationships of the initial chronology. The Pearson correlation between resulting tree-ring indices and the target climate variable is calculated for individual tree-ring series. Each tree-ring index is weighted by its correlation

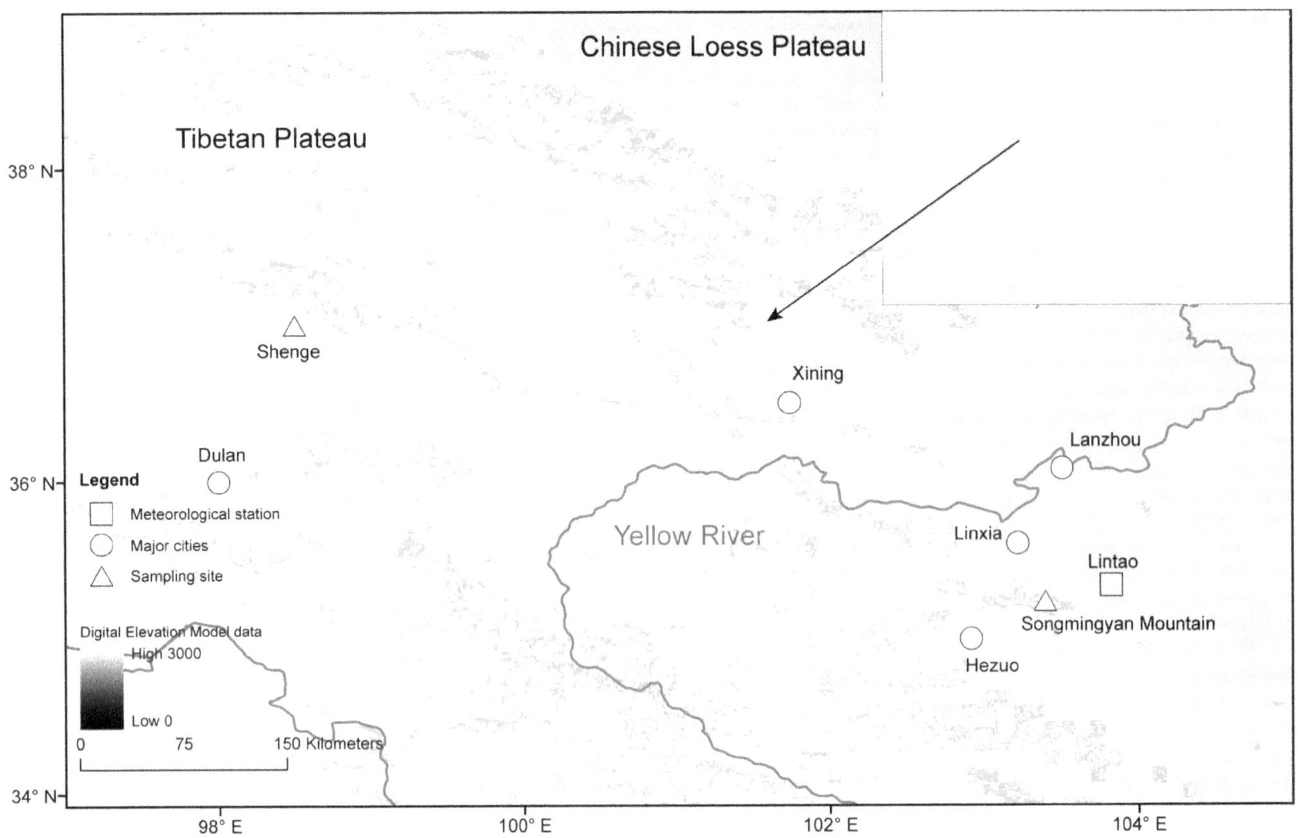

Figure 2. Map of the sampling site at the Songmingyan Mountain at Linxia area, the Shenge site at Dulan area, the meteorological station and major cities in the study region. The topographic features are indicated by digital elevation model data in grey colors and the boreal forests are shown in green color as well as the position of the study region in East Asia.

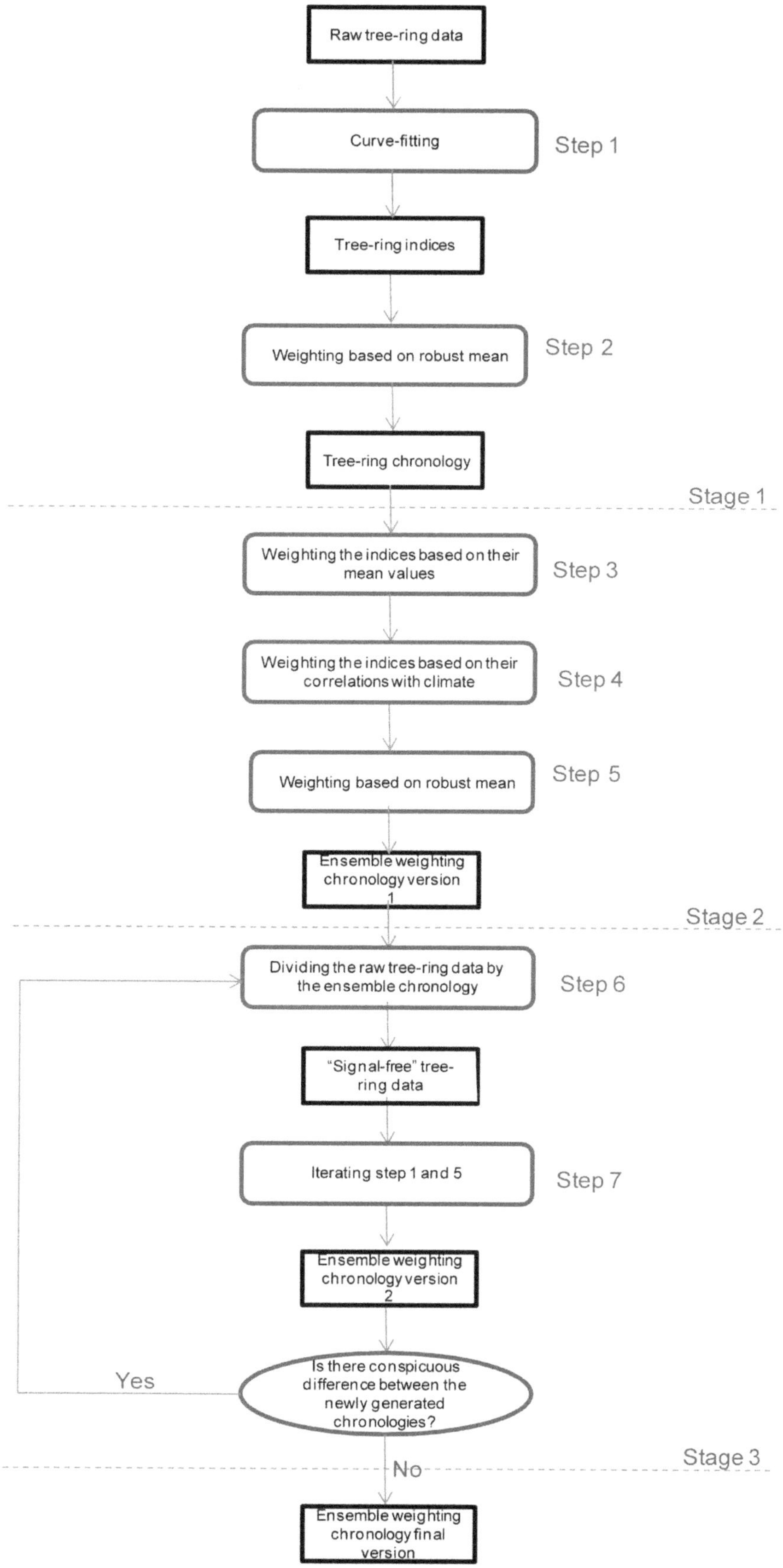

Figure 3. Flow diagram illustrating the development of an ensemble weighting chronology.

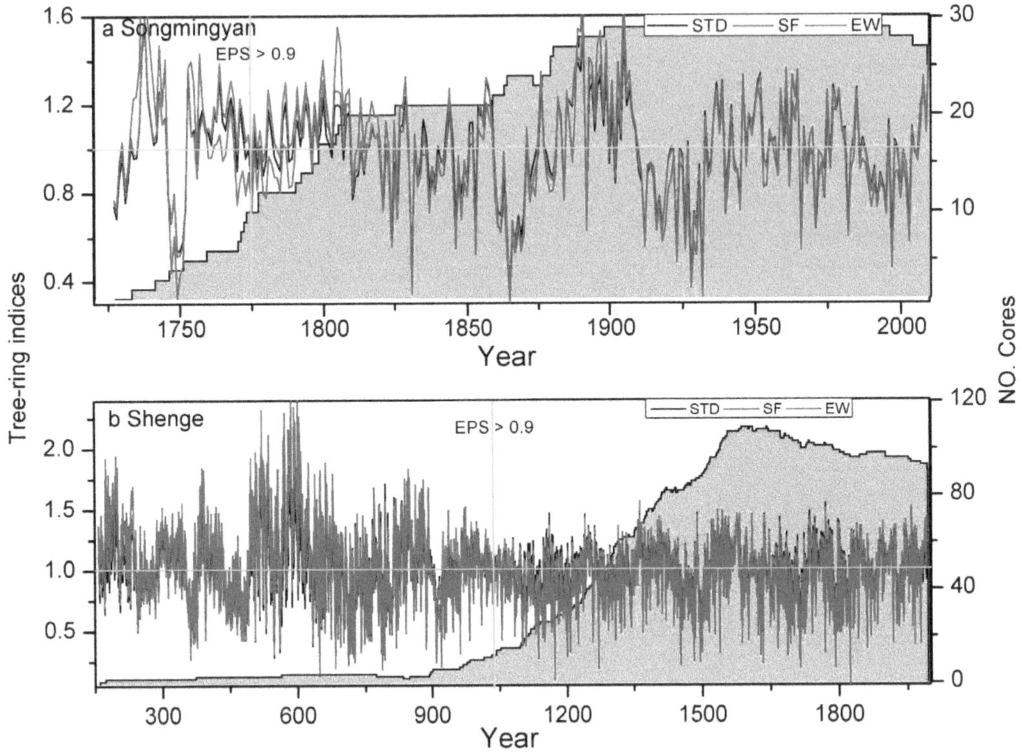

Figure 4. Comparisons of tree-ring chronologies based on the traditional method (STD), the signal-free (SF) method and the ensemble weighting (EW) method for (a) the Songmingyan Mountain site and (b) the Shenge site.

with climate at different powers (herein 0, 0.5, 1, 1.5 and 2). A power of 0 equals to an unweighted series. That is,

$$wTR = m \times r^p \times TR$$

where wTR is the weighted tree-ring index, TR is the unweighted tree-ring index, m is the mean chronology index in the sub-period of a given tree-ring series, r^p is the climate-growth correlations in different powers of p.

Step 5. The robust mean method is applied to the weighted tree-ring index to produce the chronologies (herein 5 chronologies) weighted by climate-growth correlations [8] at given powers (herein power of 5). The ensemble weighting chronology is the arithmetic mean of these chronologies, because an ensemble approach can dampen the influences of spurious climate-growth correlations due to the noise in both climate and tree ring data.

Stage 3. Developing the final chronology using signal-free iterations.

Step 6. The signal-free tree-ring measurements are produced as ratios between raw measurements and the original ensemble weighting chronology.

Step 7. The iteration procedures are the same as in the traditional signal-free method developed by Melvin and Briffa (2008) introduced above. The major difference from traditional applications of signal-free method is that the tree-ring indices during the iterations process are weighted. The iterations stop when only a minor difference between chronologies is found. The first and third stages, discussed above are similar to traditional methods. Major improvements occur in the second stage when the two weighting procedures are used to produce an ensemble of chronologies.

One limitation to this method, however, is the difficulty in determining the climate-growth correlations for sub-fossil samples that do not have any overlap with instrumental data. For such sub-fossil samples, we herein weight them based on their correlations with the master chronology and the correlations between the master chronology and climate, i.e. using a weighted multiplier between the series-chronology correlations and the chronology-climate correlations. The strength of the reconstructions was tested by linearly regressing the chronologies with the instrumental climate variable and evaluating the variance explained by each. The robustness of the reconstruction was further examined by split calibration-verification procedure [24], which calibrates the instrumental data from one sub-period and verifies the reconstruction using the remaining instrumental data. The verification sub-periods are the 1980–2008 and 1952–1979 for the Songmingyan Mountain reconstruction. Keeping in mind with the relatively short common period (1954–1993) between instrumental and tree-ring data at the Shenge site, we used a slightly longer sub-period for calibration (1954–1974 and 1973–1993) to maintain the robustness the split calibration-verification. Attention is also required for tree-ring series showing unstable climate-growth associations through the instrumental period [25], which could either be excluded or receive less weights. The samples used here generally show stable responses to climate through time as indicated by acceptable split calibration-verification statistics (detailed below).

Results

In this study, we applied 3 signal-free iterations for the Songmingyan Mountain site and 5 iterations for the Shenge site as suggested in previous studies [6]. We truncated the chronology

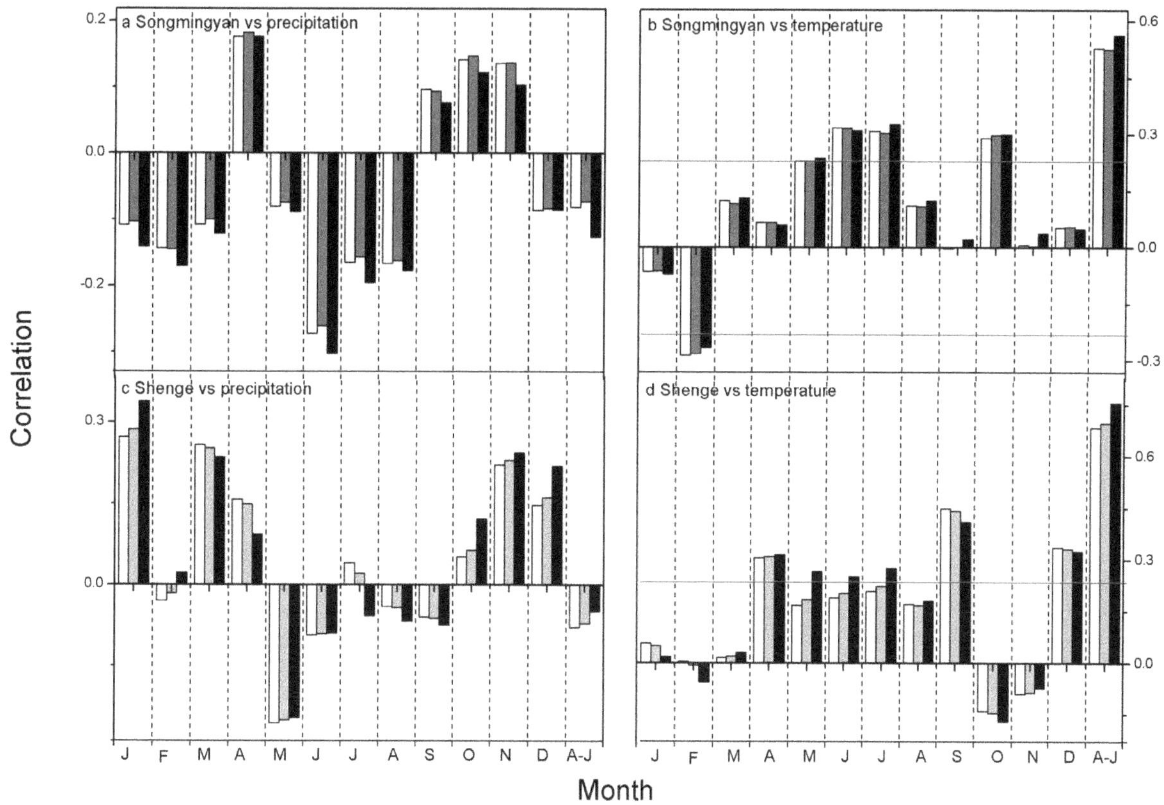

Figure 5. Correlations between tree-ring chronologies derived from traditional method (white bar), signal-free method (grey bar) and the ensemble weighting method (black bar) and the (a) monthly precipitation and annual precipitation from previous August to current July (A–J) for the Songmingyan Mountain, the (b) monthly temperature and annual temperature for the Songmingyan Mountain, the (c) monthly precipitation and annual precipitation from previous August to current July (A–J) for the Shenge site and the (d) monthly temperature and annual temperature for the Shenge site. The significance level of 0.1 is indicated by horizontal line.

at Songmingyan Mountain from 1773–2010 and Shenge chronology from 1041 to 1993 as there are sufficient replications indicated by an expressed population signal (EPS) over 0.90 [26]. Only the robust portions of the chronologies were employed in the following reconstructions. At the Songmingyan site, the signal-free method successfully mitigates the trend distortion problem by increasing (decreasing) the high (low) chronology indices in the latter half of the 18th century (during the ~1830s–1870s) (Figure 4a). At the Shenge site, the signal-free method adds back the climate signals removed in the traditional method and thus generally increase (decrease) the chronology indices when they are high (low) (Figure 4b). We additionally plotted the chronology indices derived from the three methods in the appendix (Figure S1) in a roughly 300-year interval to better illustrate the changes between them. Based on the ensemble weighting method, the chronology indices at Songmingyan Mountain increases (decreases) in the ~1790s–1820s and near 1900s (1870s–1880s and 1900s–1950s) (Figure 4a). At the Shenge site, 4 tree-ring samples (DU07A, DU12A1, DU70A and DU70B) spanning before ~800 show above average chronologies indices, which can be artificially lowered down in the traditional chronology. This problem has been mitigated in the ensemble weighting method that generally increases the chronology indices before ~800 and decreases the indices afterwards (Figure 4b). Both the signal-free and ensemble weight methods are more efficient and result in larger differences during the early periods with larger variance due to the availability of only a few tree-ring samples.

All the chronologies show positive, significant correlations with precipitation and negative correlations with temperature in the growing season (Figure 5). The highest climate-growth correlations are found with annual precipitation from the previous August to the current July for the Songmingyan Site and from previous July to currently June for the Shenge site (Figure 5). The ensemble weighting chronology shows higher correlations with precipitation than the chronologies developed using traditional and the signal-free methods (Figure 5). Therefore our later analyses are based on the reconstruction using ensemble weighting chronology since it contains a more "pure" precipitation signal and explains a higher percent of the variance. The precipitation reconstructions derived from the ensemble weighting chronologies explain 31.7% and 57.3% of the instrumental variance for Songmingyan Mountain and the Shenge site, respectively (Figure 6). The explained variance of the Shenge reconstruction is higher than that in previous reconstruction (47.8%) using traditional methods [15].

In the split calibration-verification method, the above-zero values of reduction of error (RE) and coefficient of efficiency (CE) statistics indicate that the reconstruction model has acceptable reliability in reproducing the climate signal in both sub-periods [27]. The RE and CE are acceptable for the verification sub-periods 1980–2008 (RE = 0.38, CE = 0.19) and 1952–1979 (RE = 0.31, CE = 0.06) for the Songmingyan Mountain reconstruction. For the Shenge site, RE and CE for the verifications are acceptable for verification sub-periods of 1975–1993 (RE = 0.45, CE = 0.25) and 1954–1972 (RE = 0.73, CE = 0.68). We additionally tested the robustness of the newly introduced reconstruction at

Songmingyan Mountain by comparing its extreme values with nearby reconstructions. Extremely dry years (< mean −2SD) in the reconstruction at Songmingyan Mountain are 1831, 1865, 1923, 1928, 1932 and 1997, and extremely wet (> mean +2SD) years are 1891 and 1905 (Figure 6). Some of the extreme dry years (1865, 1928 and 1997) reconstructed were also found in yearly dryness/wetness chats derived from historical documents from neighboring Lanzhou [28]. Although the extreme droughts in 1923 and 1932 were not found in the documentary records, these two extremely dry years were found in hydroclimate reconstructions in other nearby regions [10,19,29]. The extreme drought of 1997 was the most extreme in the Guiqing Mountain area [18], and is east of our study region.

Discussion

Performances of the Ensemble Weighting Method

The ensemble weighting chronologies contains two weighting procedures apart from the robust mean weighting in traditional methods [23]. This weighting procedure may have larger efficiency in modulating the chronologies for samples with both living and sub-fossil cores. This is because the living and fossil cores generally have less overlapping periods and can thus show larger differences in mean chronology values. Weighting procedure can better retain the low-frequency variations and tends to increase (decrease) the chronology indices in periods with high (low) climate signals.

Our weighting procedure is designed to put more emphasis on the most climate-sensitive samples, similar to fieldwork where scientists select sites and collect samples at sites with a high degree of climatic signal (determined by ecological conditions), at, for example, the treeline locations (a *priori* knowledge) [20]. The weighting of individual tree-ring series based on their correlations with climate provides a "quality control" to test whether individual tree-ring series contains a pure climate signal (a *posteriori* knowledge) [8]. This weighting procedure may be necessary in regions that have both arid and cold climate, such as northeastern TP. In these regions, tree growth may be sensitive to both precipitation and temperature [20] and thus it is possible to include tree-ring series with a different climate-growth relationships. In addition, this method is a more justified approach than completely excluding tree-ring samples from a chronology that have a limited climate signal. For example, the weighting of a few temperature-sensitive tree-ring series in the ensemble weighting chronology has increased (decreased) the chronology values in the warm (cold) period of the latter half of the 18th century and 19th century (early 19th century).

This ensemble weight procedure incorporates the principles of the signal-free method that has a number of iterations to mitigate the trend distortion problem. The key difference the original signal-free method gives equal weights to the tree-ring indices. From this point of view, we can consider this ensemble weighting method as a updated version of the signal-free method. Similar to the signal-free method, our method can be improve the regional curve standardization (RCS) method, a specific technique to overcome the segment length curse problem [30]. Apart from the trend distortion problem, our method can aid in producing a climate sensitive RCS chronology. However, there is no need to adjust the mean values of individual tree-ring indices in RCS, because their mean values are the ratio between mean growth measurements and the regional growth curve and thus can be different [31,32]. Therefore in the application of the ensemble weighting method in RCS, we only need to weight the tree-ring indices according to their associations with climate.

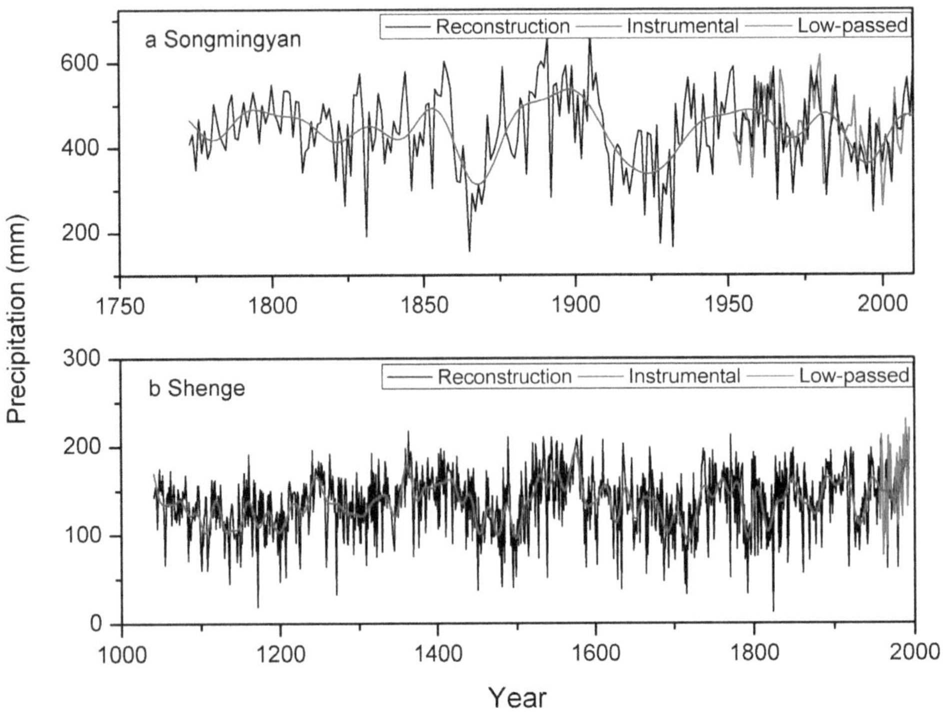

Figure 6. The reconstructed precipitations using the ensemble weighting chronology for the (a) Songmingyan and the (b) Shenge sites, and their low-passed values and the instrumental data.

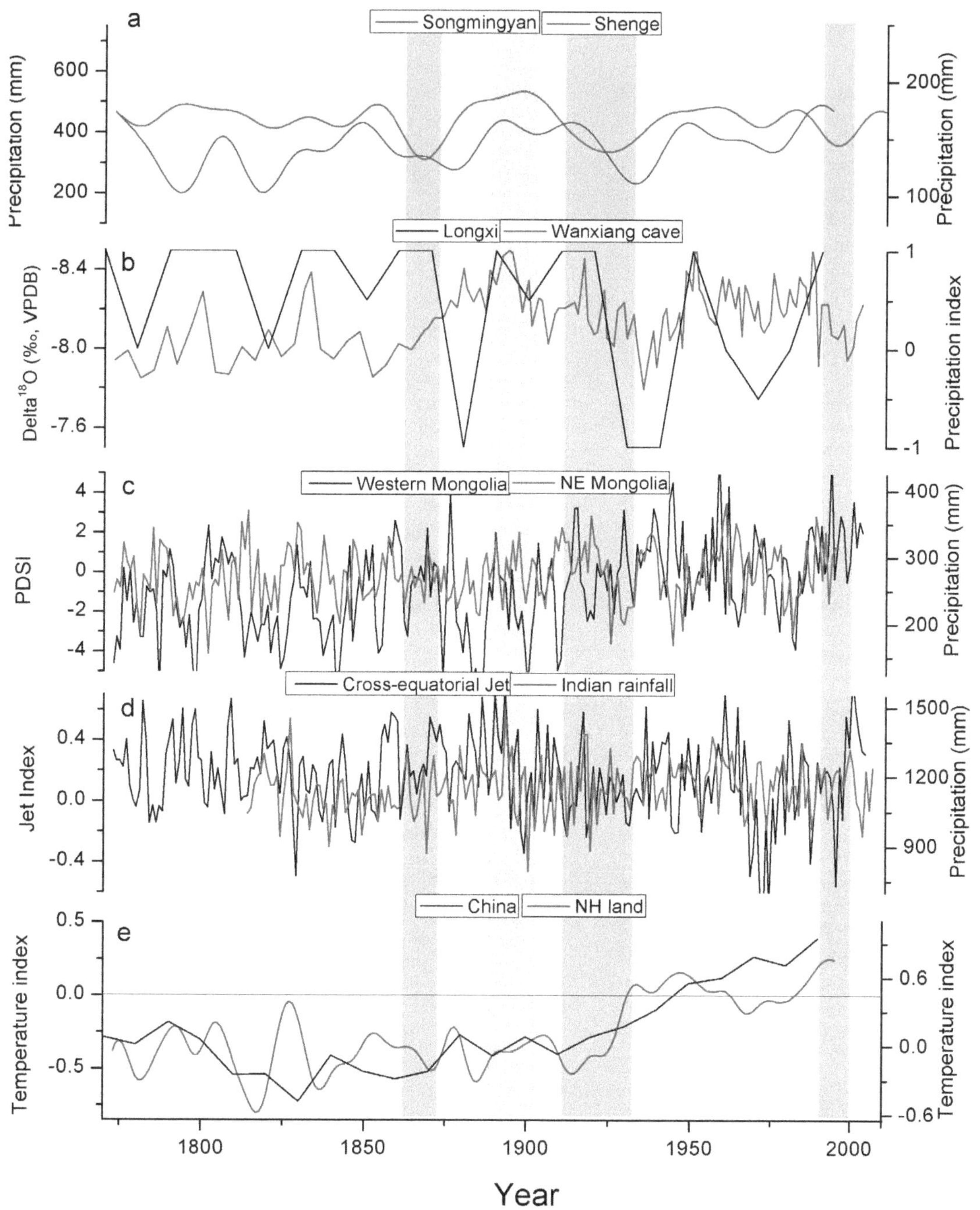

Figure 7. Comparisons among (a) precipitation reconstructions in the Songmingyan Mountain and the Shenge site, and (b) a speleothem-based monsoon record from the Wanxiang cave [38] southeastern to our study region and a precipitation reconstruction using historical documents from Longxi [39] eastern to our study region, (c) the hydroclimate reconstruction in far-western Mongolia [41] and the northeastern Mongolia [37], (d) coral-inferred variations of the low-level cross-equatorial jet in the western Indian Ocean [43] and the longest instrumental precipitation for all the India [42], and (e) the reconstructed temperature index in China derived from proxies of ice cores, tree rings, peat and historical documents [45] and the land temperature of northern hemisphere reconstructed from multiple proxies using composite-plus-scale method [1].

Precipitation Reconstruction and Monsoon Dynamics

A drought-sensitive growth pattern, negative correlations with temperature and positive correlations with precipitation, has been widely seen in arid regions [10,11,12,13,15,19], and is also found in our study region. Drought stress and less annual growth can occur when there are increases in evapotranspiration in warmer temperatures. We have found that tree growth at our study site correlates highest with annual averaged hydroclimate conditions rather than monthly or seasonal climate data like in our study region and some sites in southeastern TP [33,34] and in the northeastern TP [12,15,35], north central China [36] and eastern Mongolia [37]. This is because tree rings can integrate monthly precipitation of consecutive months in the growing season and the precipitation from the dormant season can "compensate" for monthly water shortages

[10]. This phenomenon is more conspicuous in regions with deep soil that can retain the water in winter (non-growing season) and facilitate tree growth in the following year. Tree growth at sites with very shallow soil tends to be more sensitive only to growing season hydroclimate conditions, such as the Xiaolong Mountain [10] and the Guiqing Mountain areas [18].

Dry epochs with more than 5 continuously dry years (< mean −SD) in the 20-year low-pass Songmingyan reconstruction are from 1862–1874, 1914–1933 and 1991–1999, and the wet epoch with over 5 continuously wet years (> mean −SD) is from 1882–1905 (Figure 7a). For the reconstruction at Shenge site, the dry epochs are found in periods of 1100–1111, 1127–1152, 1170–1178, 1192–1207, 1449–1459, 1470–1481, 1493–1506, 1686–1698, 1709–1725, 1788–1797, 1813–1822 and 1929–1935, and the wet epochs are found in the 1241–1251, 1355–1367, 1400–1421, 1525–1538, 1543–1557, 1566–1582, 1844–1851, 1888–1899, 1906–1918, 1945–1953 and 1982–1993. Our discussions only focus on the dry and wet epochs at Songmingyan Mountain, since most of these epochs have been mentioned in previous tree-ring reconstructions at Shenge site in the Dulan area [12,13,14,15]. For the reconstruction at Shenge site, we herein paid special attention to the most extreme drought over the entire reconstruction period in 1824 (Figure 6b). This extreme drought was also recorded by locally historical documents the major city of Xining near the study site [28]. This drought was centered in the Dulan area in northeastern TP, which was the driest region in 1824 in the Monsoonal Asia Drought Atlas (Figure S2). The drought reconstruction for Songmingyan Mountain does not record this drought, because it is located outside of this drought center. The reconstructed precipitation dropped sharply from 100.5 mm in 1823 to 12.8 mm in 1824, and increased to 144.7 mm in 1825. Similarly, this extreme drought was not observed in 1823 and 1825 (Figure S2). This drought might indicate that an abnormally high pressure controlled the Dulan area in 1824, which requires future modeling studies to examine the occurrence of this anomalous high and its associations with large-scale circulation anomalies. Along the waveguide of westerlies, this anomalous high might be related to the abrupt shift from positive phase of Pacific/North America teleconnection to its negative phase in 1824 ().

The dry and wet epochs in the Songmingyan reconstruction are seen in hydroclimate reconstructions from neighboring regions (Figure 7a and 7b), e.g. the reconstruction from Shenge site in Dulan region and the speleothem records from Wanxiang cave [38]. However, some dry epochs in Songmingyan reconstruction, for example, in 1862–1874 and 1914–1933 are not found in a document-based precipitation reconstruction from Longxi [39], east of our study region. The Longxi reconstruction shows similar variations with the Dulan reconstruction. These differences may be related to varying hydroclimate regimes in different regions and the different seasons for reconstruction (e.g. summer or annual precipitation).

These dry/wet epochs were not only found in marginal areas of the Asian summer monsoon in the northeastern TP region, but also in northeastern China and northeastern Mongolia, such as the dry epoch from 1914–1933 [10,18,29,37,38,40] (Figure 7c). Our study adds additional proxy evidence that this dry epoch reached the southwestern boundary of the Songmingyan Mountain. In the northeastern TP region this drought began in the 1910s, nearly a decade before other regions, including north central China and eastern Mongolia, suggesting that this persistent drought initiated from the west. However, the hydroclimate variations have limited resemblance with the hydroclimate changes from the westerlies-dominated regions, such as in western Mongolia [41] (Figure 7c). This indicates that hydroclimate changes in the past two centuries are more likely dominated by the monsoon and are different from the hydroclimate changes in westerlies-dominated regions, which is in agreement with previous studies [9]. Additional evidence of the influences of the monsoon in this region comes from the similar variations between hydroclimate changes and a long instrumental precipitation record from India [42] and the coral-inferred low-level cross-equatorial jet in the western Indian Ocean [43] (Figure 7d). Previous studies documented that the Asian summer monsoon, particularly the Indian summer monsoon, can reach northeastern TP along the eastern boundary of TP [44]. The dryness (wetness) in the northeastern TP and associated weakened (strengthened) Indian summer monsoon often corresponds to cold (warm) periods of the reconstructed land temperature of the northern hemisphere [10] and China [45], except for the recent monsoon failure since around the 1980s (Figure 7e). The positive relationship between temperature and the monsoon is likely a result of increased land-ocean temperature gradients [9,38] and the northward shifts of the intertropical convergence zone (ITCZ) [46]. The monsoon failure in recent decades may be caused by intensified human activities such as increases in aerosols emissions, which can weaken the land-ocean temperature gradients and the gradients between the northern and southern hemispheres [47] and the cooling of the upper troposphere [48].

Conclusions

We introduced an ensemble weighting method to alleviate two potential biases in traditional methods of chronology development. This method allows the mean value of tree-ring series to vary at different time intervals, instead of assigning a value of 1. In addition, this ensemble weighting method assigns weights to individual series depending on the strength of the climate-growth relationship. The resulting chronology is then averaged from an ensemble of chronologies with weighted individual tree-ring indices. The chronology development is iterated to adjust the mean values of individual tree-ring indices and to alleviate the trend distortion problem, similar to signal-free methods. We tested the efficiency of this method by developing a new tree-ring chronology at the Songmingyan Mountain and by recalculating a tree-ring chronology at Shenge site from a marginal area of the Asian summer monsoon in the northeastern TP. These reconstructions explain higher instrumental variance, 31.7% for the Songmingyan reconstruction and 57.3% for the Shenge reconstruction, than the reconstructions based on traditional methods. The reconstructed dry epochs range from the marginal area of the Asian summer monsoon from the northeastern TP to eastern Mongolia, as well as the monsoon dominated Indian subcontinent, indicating the linkages between regional hydroclimate changes and the Asian summer monsoon.

Supporting Information

Figure S1 Indices of the tree-ring chronologies developed from traditional method (black), the signal-free method (red) and the ensemble weighting method (blue) for the Shenge site at a roughly 300-year interval.

Figure S2 The reconstructed summer (June-August) Palmer Drought Severity Indices in years of 1823, 1824 and 1825 from the Monsoon Asia Drought Atlas (Cook et al. 2010).

Acknowledgments

We kindly thank Mingshi Zhao for his assistance in the field and laboratory. We acknowledge the Environmental and Ecological Science Data Center, West China for sharing the plant function data from boreal forests.

Author Contributions

Conceived and designed the experiments: KF. Performed the experiments: KF CL FZ. Analyzed the data: KF MW ND. Contributed reagents/materials/analysis tools: KF MW ND. Wrote the paper: KF.

References

1. Mann ME, Zhang Z, Hughes MK, Bradley RS, Miller SK, et al. (2008) Proxy-based reconstructions of hemispheric and global surface temperature variations over the past two millennia. Proceedings of the National Academy of Sciences 105: 13252–13257.
2. Cook E, Anchukaitis KJ, Buckley BM, D'Arrigo RD, Jacoby GC, et al. (2010) Asian Monsoon Failure and Megadrought During the Last Millennium. Science 328: 486–489.
3. Jones PD, Briffa KR, Osborn TJ, Lough JM, Van Ommen TD, et al. (2009) High-resolution palaeoclimatology of the last millennium: a review of current status and future prospects. Holocene 19: 3–49.
4. IPCC (2007) Climate Change 2007: The Physical Science Basis: IPCC.
5. Cook E, Briffa K (1990) Data analysis. In: Cook E, Kairiukstis L, editors. Methods of Dendrochology: Application in the Environmental Sciences. Dordrecht: Kluwer Academic Publisher.
6. Melvin TM, Briffa KR (2008) A signal-free approach to dendroclimatic standardisation. Dendrochronologia 26: 71–86.
7. Wilmking M, Juday GP, Barber VA, Zald HSJ (2004) Recent climate warming forces contrasting growth responses of white spruce at treeline in Alaska through temperature thresholds. Global Change Biology 10: 1724–1736.
8. Bunn AG, Hughes MK, Salzer MW (2011) Topographically modified tree-ring chronologies as a potential means to improve paleoclimate inference. Climatic Change 105: 627–634.
9. Chen FH, Chen JH, Holmes J, Boomer I, Austin P, et al. (2010) Moisture changes over the last millennium in arid central Asia: a review, synthesis and comparison with monsoon region. Quat Sci Rev 29: 1055–1068.
10. Fang K, Gou X, Chen F, Frank D, Liu C, et al. (2012) Precipitation variability of the past 400 years in the Xiaolong Mountain (central China) inferred from tree rings. Clim Dyn 39: 1697–1707.
11. Gou XH, Deng Y, Chen FH, Yang MX, Fang KY, et al. (2010) Tree ring based streamflow reconstruction for the Upper Yellow River over the past 1234 years. Chinese Science Bulletin 55: 4179–4186.
12. Liu Y, An Z, Ma H, Cai Q, Liu Z, et al. (2006) Precipitation variation in the northeastern Tibetan Plateau recorded by the tree rings since 850 AD and its relevance to the Northern Hemisphere temperature. Science in China Series D: Earth Sciences 49: 408–420.
13. Shao X, Huang L, Liu H, Liang E, Fang X, et al. (2005) Reconstruction of precipitation variation from tree rings in recent 1000 years in Delingha, Qinghai. Sci China 48: 939–949.
14. Zhang QB, Cheng GD, Yao TD, Kang XC, Huang JG (2003) A 2,326-year tree-ring record of climate variability on the northeastern Qinghai-Tibetan Plateau. Geophys Res Lett 30: 1739–1742.
15. Sheppard PR, Tarasov PE, Graumlich LJ, Heussner KU, Wagner M, et al. (2004) Annual precipitation since 515 BC reconstructed from living and fossil juniper growth of northeastern Qinghai Province, China. Clim Dyn 23: 869–881.
16. Ramage CS (1971) Monsoon Meteorology. New York: Academic Press.
17. Ran YH, Li X, Lu L, Li ZY (2012) Large-scale land cover mapping with the integration of multi-source information based on the Dempster–Shafer theory. International Journal of Geographical Information Science 26: 169–191.
18. Fang K, Gou X, Chen F, D'Arrigo R, Li J (2010) Tree-ring based drought reconstruction for the Guiqing Mountain (China): linkages to the Indian and Pacific Oceans. Int J Climatol 30: 1137–1145.
19. Li J, Chen F, Cook ER, Gou X, Zhang Y (2007) Drought reconstruction for north central China from tree rings: the value of the Palmer drought severity index. Int J Climatol 27: 903–909.
20. Fritts HC (1976) Tree rings and climate. New York: Academic Press.
21. Holmes RL (1983) Computer-assisted quality control in tree-ring dating and measurement. Tree-ring bull 43: 69–78.
22. Cook E, Kairiukstis L (1990) Methods of Dendrochronology Netherlands: Kluwer Academic Press.
23. Cook ER (1985) A time series analysis approach to tree ring standardization. Tucson: The University of Arizona.
24. Meko D, Graybill DA (1995) Tree-ring reconstruction of upper Gila river discharge. J Am Water Resour As 31: 605–616.
25. D'Arrigo RD, Kaufmann RK, Davi N, Jacoby GC, Laskowski C, et al. (2004) Thresholds for warming-induced growth decline at elevational tree line in the Yukon Territory, Canada. Global Biogeochem Cy 18: GB3021 doi: 3010.1029/2004GB002249.
26. Wigley TML, Briffa KR, Jones PD (1984) Average value of correlated time series, with applications in dendroclimatology and hydrometeorology. J Appl Meteorol Clim 23: 201–234.
27. Cook E, Meko DM, Stahle DW, Cleaveland MK (1999) Drought reconstructions for the continental United States. Journal of Climate 12: 1145–1162.
28. Zhang D, Li X, Liang Y (2003) Complementary Data of the Yearly Charts of Dryness/Wetness in China for the Last 500 years Period. Journal of Applied Meteorological Science 14: 379–384.
29. Liang E, Liu X, Yuan Y, Qin N, Fang X, et al. (2006) The 1920s drought recorded by tree rings and historical documents in the semi-arid and arid areas of northern China. Climatic Change 79: 403–432.
30. Cook E, Briffa KR, Meko DM, Graybill DA, Funkhouser G (1995) The'segment length curse'in long tree-ring chronology development for palaeoclimatic studies. Holocene 5: 229–237.
31. Esper J, Cook ER, Schweingruber FH (2002) Low-frequency signals in long tree-ring chronologies for reconstructing past temperature variability. Science 295: 2250–2253.
32. Briffa KR, Melvin TM (2009) A Closer Look at Regional Curve Standardization of Tree-Ring Records: Justification of the Need, a Warning of Some Pitfalls, and Suggested Improvements in Its Application. In: Hughes IMK, Diaz HF, Swetnam TW, Progress D, editors. Dendroclimatology: progress and prospects. New York City: Springer Verlag.
33. Fang K, Gou X, Chen F, Li J, D'Arrigo R, et al. (2010) Reconstructed droughts for the southeastern Tibetan Plateau over the past 568 years and its linkages to the Pacific and Atlantic Ocean climate variability. Clim Dyn 35: 577–585.
34. Fan ZX, Bräuning A, Cao KF (2008) Tree-ring based drought reconstruction in the central Hengduan Mountains region (China) since AD 1655. International Journal of Climatology 28: 1879–1887.
35. Gou X, Chen F, Cook E, Jacoby G, Yang M, et al. (2007) Streamflow variations of the Yellow River over the past 593 years in western China reconstructed from tree rings. Water Resour Res 43: W06434.
36. Fang K, Gou X, Chen F, Yang M, Li J, et al. (2009) Drought variations in the eastern part of Northwest China over the past two centuries: evidence from tree rings. Clim Res 38: 129–135.
37. Pederson N, Jacoby GC, D'Arrigo RD, Cook ER, Buckley BM, et al. (2001) Hydrometeorological Reconstructions for Northeastern Mongolia Derived from Tree Rings: 1651–1995. J Climate 14: 872–881.
38. Zhang P, Cheng H, Edwards RL, Chen F, Wang Y, et al. (2008) A test of climate, sun, and culture relationships from an 1810-year Chinese cave record. Science 322: 940–942.
39. Tan L, Cai Y, Yi L, An Z, Ai L (2008) Precipitation variations of Longxi, northeast margin of Tibetan Plateau since AD 960 and their relationship with solar activity. Climate of the Past 4: 19–28.
40. Davi N, Pederson N, Leland C, Nachin B, Suran B, et al. (2013) Is eastern Mongolia drying? A long-term perspective of a multi-decadal trend. Water Resources Research doi:10.1029/2012WR011834.
41. Davi N, Jacoby G, D'Arrigo R, Baatarbileg N, Li J, et al. (2009) A tree-ring-based drought index reconstruction for far-western Mongolia: 1565–2004. Int J Climatol 29 1508–1514.
42. Sontakke NA, Singh N, Singh HN (2008) Instrumental period rainfall series of the Indian region (AD 1813–2005): revised reconstruction, update and analysis. Holocene 18: 1055–1066.
43. Gong DY, Luterbacher J (2008) Variability of the low-level cross-equatorial jet of the western Indian Ocean since 1660 as derived from coral proxies. geophysical research letters 35: L01705, doi:01710.01029/02007GL032409.
44. An Z, Colman SM, Zhou W, Li X, Brown ET, et al. (2012) Interplay between the Westerlies and Asian monsoon recorded in Lake Qinghai sediments since 32 ka. Scientific reports 2: doi:10.1038/srep00619.
45. Yang B, Braeuning A, Johnson KR, Yafeng S (2002) General characteristics of temperature variation in China during the last two millennia. Geophys Res Lett 29: 1324. 1310.1029/2001GL014485.
46. Sachs J, Sachse D, Smittenberg R, Zhang Z, Battisti D, et al. (2009) Southward movement of the Pacific intertropical convergence zone AD 1400–1850. Nature-Geoscience: doi: 10.1038/NGEO1554.
47. Bollasina MA, Ming Y, Ramaswamy V (2011) Anthropogenic aerosols and the weakening of the South Asian summer monsoon. Science 334: 502–505.
48. Ding Y, Wang Z, Sun Y (2008) Inter-decadal variation of the summer precipitation in East China and its association with decreasing Asian summer monsoon. Part I: Observed evidences. International Journal of Climatology 28: 1139–1161.

Leafcutter Bee Nests and Pupae from the Rancho La Brea Tar Pits of Southern California: Implications for Understanding the Paleoenvironment of the Late Pleistocene

Anna R. Holden[1]*, Jonathan B. Koch[2], Terry Griswold[3], Diane M. Erwin[4], Justin Hall[5]

1 Entomology Section, Natural History Museum of Los Angeles County, Los Angeles, California, United States of America, **2** Department of Biology and Ecology Center, Utah State University, Logan, Utah, United States of America, **3** USDA-ARS Pollinating Insect Research Unit, Utah State University, Logan, Utah, United States of America, **4** Museum of Paleontology, University of California, Berkeley, Berkeley, California, United States of America, **5** Dinosaur Institute, Natural History Museum of Los Angeles County, Los Angeles, California, United States of America

Abstract

The Rancho La Brea Tar Pits is the world's richest and most important Late Pleistocene fossil locality and best renowned for numerous fossil mammals and birds excavated over the past century. Less researched are insects, even though these specimens frequently serve as the most valuable paleoenvironemental indicators due to their narrow climate restrictions and life cycles. Our goal was to examine fossil material that included insect-plant associations, and thus an even higher potential for significant paleoenviromental data. Micro-CT scans of two exceptionally preserved leafcutter bee nest cells from the Rancho La Brea Tar Pits in Los Angeles, California reveal intact pupae dated between ~23,000–40,000 radiocarbon years BP. Here identified as best matched to *Megachile* (*Litomegachile*) *gentilis* Cresson (Hymenoptera: Megachilidae) based on environmental niche models as well as morphometrics, the nest cells (LACMRLP 388E) document rare preservation and life-stage. The result of complex plant-insect interactions, they offer new insights into the environment of the Late Pleistocene in southern California. The remarkable preservation of the nest cells suggests they were assembled and nested in the ground where they were excavated. The four different types of dicotyledonous leaves used to construct the cells were likely collected in close proximity to the nest and infer a wooded or riparian habitat with sufficient pollen sources for larval provisions. LACMRLP 388E is the first record of fossil *Megachile* Latreille cells with pupae. Consequently, it provides a pre-modern age location for a Nearctic group, whose phylogenetic relationships and biogeographic history remain poorly understood. *Megachile gentilis* appears to respond to climate change as it has expanded its distribution across elevation gradients over time as estimated by habitat suitability comparisons between low and high elevations; it currently inhabits mesic habitats which occurred at a lower elevation during the Last Glacial Maximum ~21,000 years ago. Nevertheless, the broad ecological niche of *M. gentilis* appears to have remained stable.

Editor: David Frayer, University of Kansas, United States of America

Funding: The Annie M. Alexander Endowment provided funding for Diane M. Erwin. The funders had no role in study design, data collection and analysis, decision to publish, or preparation of the manuscript.

Competing Interests: The authors have declared that no competing interests exist.

* E-mail: aholden@nhm.org

Introduction

Megachile Latreille [1] is a large, worldwide genus of approximately 1,500 species of largely leafcutting, solitary bees. In the Western Hemisphere they inhabit temperate, arid, and tropical regions extending from Alaska to Tierra del Fuego [2]. There are 118 species native to North America [2]. The abundance of megachilids in California is not surprising given the wide diversity of habitats and microclimates [3,4].

Leafcutter bees are named for their use of leaf pieces in nest building. They constitute *Megachile* belonging to Michener's Group 1 [5] in which bees frequently construct two or more cells in a linear series. Their nesting sites are found under the bark of dead trees, in stems, in the burrows of wood-boring insects or in burrows self-dug in loose soil or those made by other animals [2,5,6]. The females use their sharp, serrated, scissor-like mandibles to cut oblong and circular leaf pieces, most likely from plant sources near the nest [2,5]. They line the nest cavities with overlapping layers of the oblong-shaped leaf disks. The leaf edges are compressed to extrude sap that, in combination with saliva, creates a glue-like substance that keeps the cells sealed and intact [2]. Each cell is provisioned with pollen and nectar by the female before she deposits a single egg on the food mass. After depositing the egg, she seals the cell with one to several circular leaf "caps" [2,5,6]. After a few weeks, depending upon species, the eggs hatch, and the larvae develop through multiple instars and feed on the provisions. Mature larvae spin cocoons of two or more layers of silk and diapause as prepupae. Cocoons are sturdy structures [7] made increasingly airtight by the larva's secretion of a brown liquid that fills and hardens the interstices of the silk layers [6]. This application binds the silk mesh and makes the cocoon extremely durable. Simultaneously, this fastens the cocoon's outer surface to the surrounding leaf disks that firmly hold the structure together. The larvae subsequently pupate and emerge as adults by chewing their way out through the cap.

That females may spend the majority of their time collecting pollen and nectar to provision their young [2] and construct intricate nests with specific materials indicates a very complex and highly evolved plant-insect interaction, and strongly suggests a long evolutionary history [8]. The use of leaf disks of various sizes, shapes, and textures also reflects highly complex and evolved behavior [2,8,9,10].

As currently known, the megachilid fossil record is restricted to the Cenozoic based on body fossils preserved as compressions and three-dimensionally preserved in amber, as well as trace evidence from fossil angiosperm leaves whose margins show smooth-edged oblong and circular cutouts [8,11,12]. Engel [13–15] and Engel and Perkovsky [11] have compiled the evolutionary history and an overview of the body fossil record, respectively. Morphological data (body fossils and leaf cutouts attributed to *Megachile*) and molecular data do not always agree on the time divergence of the genus [8,11–14,16–22]. Although the phylogenetic relationships and evolutionary history of the genus have become clarified as more studies incorporate molecular data [23], the fossil record remains incomplete and some specimens assigned to Megachilidae may need revision. For example, molecular data suggest that Megachilidae arose in the Cretaceous about 140–100 mya, but the genus *Megachile* is estimated to have originated only 22 mya [23]. However, leafcutters are derived species of *Megachile* and therefore, the fossil record based on leaf cutouts from the Early to middle Eocene in North American and Europe [16–20] suggests that basal divergences in the Megachilini occurred earlier in the Paleocene or Latest Cretaceous [11].

Here we report on fossil *Megachile* nest cells with pupae (LACMRLP 388E) recovered from the Late Pleistocene Rancho La Brea Tar Pits in southern California. Though geologically young, this is the first report of three-dimensionally preserved *Megachile* nest cells that shows rare preservation and life-stages. The pupal morphology, nesting behavior, and cell construction of LACMRLP 388E best match *M. gentilis,* a member of the native Nearctic *Litomegachile* Mitchell [24], a subgenus that today ranges from southern Canada to Cuba and southern Mexico [5,25,26].

The discovery of LACMRLP 388E provides valuable information for better understanding the environmental conditions of southern California in the Late Pleistocene. By setting specimens within a geological as well as an ecological context, Quaternary fossils are shown to be valuable precursors to modern biota [27]. Although the asphaltic deposits at Rancho La Brea are most often associated with vertebrate remains from saber-toothed cats and mammoths, the insects and plants found there are also significant fossils because they are original material, and thus, intact, three-dimensional, and structurally complex. As such they can provide the most valuable paleoenvironmental information for the richest Ice Age fossil locality. Our goal was to synchronize data by identifying both nest cell insect and plant material in order to make the significant paleoenvironmental inferences possible. These efforts resulted in environmental data, that a mesic environment occurred at a lower elevation than today at Rancho La Brea, the well-established provenance for LACMRLP 388E. This research also resulted in new information on *M. gentilis*, including diagnostic features of its nest cell architecture and insight regarding its relatively conserved ecological niche since the Late Pleistocene.

In additional to its role as a sensitive paleoenvironemental indicator and providing new information on *M. gentilis*, LACMRLP 388E is of rare value because its remarkable preservation, especially of the intact pupae, is of a standard unusual even for fossil material from Rancho La Brea.

Results

LACMRLP 388E

Provenance and age. LACMRLP 388E was excavated in 1970 from the northeast corner of Pit 91, one of approximately 100 asphaltic deposits at Rancho La Brea in Los Angeles. It was found as one intact specimen (LACMRLP 388E) in an area with asphalt-impregnated sediment and bone at a depth of 205 cm. However, on subsequent handling the two nest cells separated (LACMRLP 388Ea, Fig. 1*A–C;* LACMRLP 388Eb, Fig. 1*D–G, J*) along with a portion of the outer leaf layers (LACMRLP 388Ec) (Fig. 1*H, I*).

Fifty fossil bones from Pit 91 have been dated between 23,000–40,000 years BP based on radiocarbon dated bone collagen [28]. This is with the exception of one, anomalous bone on the opposite side of the pit from LACMRLP 388E, which was dated to 14,000 years BP [28]. Furthermore, LACMRLP 388E was recovered from a depth of less than 243 centimeters that included a cluster of bones dated to ages older than 35,000 years BP, increasing the likelihood that LACMRLP 388E is between 35,000 and 40,000 radiocarbon years BP [28]. That the nest cells were constructed underground but close to the surface before becoming embalmed in asphalt-rich matrix indicates that they may have been near an asphalt pipe from which oil soaked into the surrounding sediment [28].

Many aspects of LACMRLP 388E nest cell construction increased the probability of fossilization in asphalt. An underground burrow provided a protected environment from the elements, as well as many predators. Glue-like saliva containing anti-bacterial and anti-fungal properties likely prevented decay before fossilization [5,29,30]. Water-resistance was created from sap extruded from the cell leaf pieces, tight assembly of multiple leaf layers, the hydroscopic properties of the leaf epidermis, and the internal silk cocoon which was hardened and airtight from larval secretions plus lipids from the pollen provisions. Finally, the life stage–a bee in the pupal stage–would be less vulnerable to desiccation than a larva whose exoskeleton is less strongly sclerotized.

Pupae. Originally identified on-site as a bud (Obermayr, Pit 91 field notes p. 1770, 1970), LACMRLP 388E was later tentatively identified to the bee family Apidae. Micro-CT scans of LACMRLP 388E reveal a pupa within each cell, indicating that the specimen was preserved sometime during flowering season (Fig. 2*A, C* and Fig. 3*A, C*, Videos S1-6) since *Megachile* do not overwinter as pupae. A comparison of LACMRLP 388E to a modern leafcutter bee (*Megachile rotundata*) show obvious similarity in general morphology at pupal stage (Fig. 4). While nest cells are designed to house *Megachile* offspring, the megachilid cleptoparasite *Coelioxys* may usurp nest cells. Images of the pupae in dorsal and ventral views (Fig. 2*A, B, F* and Fig. 3*A, B, F,* Videos S3, 4) have the same general body morphology, presence of setal pits on scutum, the long labrum, and oval shape of the metasoma of *Megachile*, confirming the identity of the nests based on cell construction and structure. The shape of the metasoma and posteriorly rounded mesosoma eliminates the possibility of *Coelioxys*, which would have a more elongate, tapered metasoma in dorsal and ventral views and pointed axillae. One pupa is present in each cell and both sexes are represented (Fig. 2 and Fig. 3, Videos S1-6), the male being distinguished by a dorsal horizontal ridge on metasomal segment six (Fig. 2. *A, B*). Because

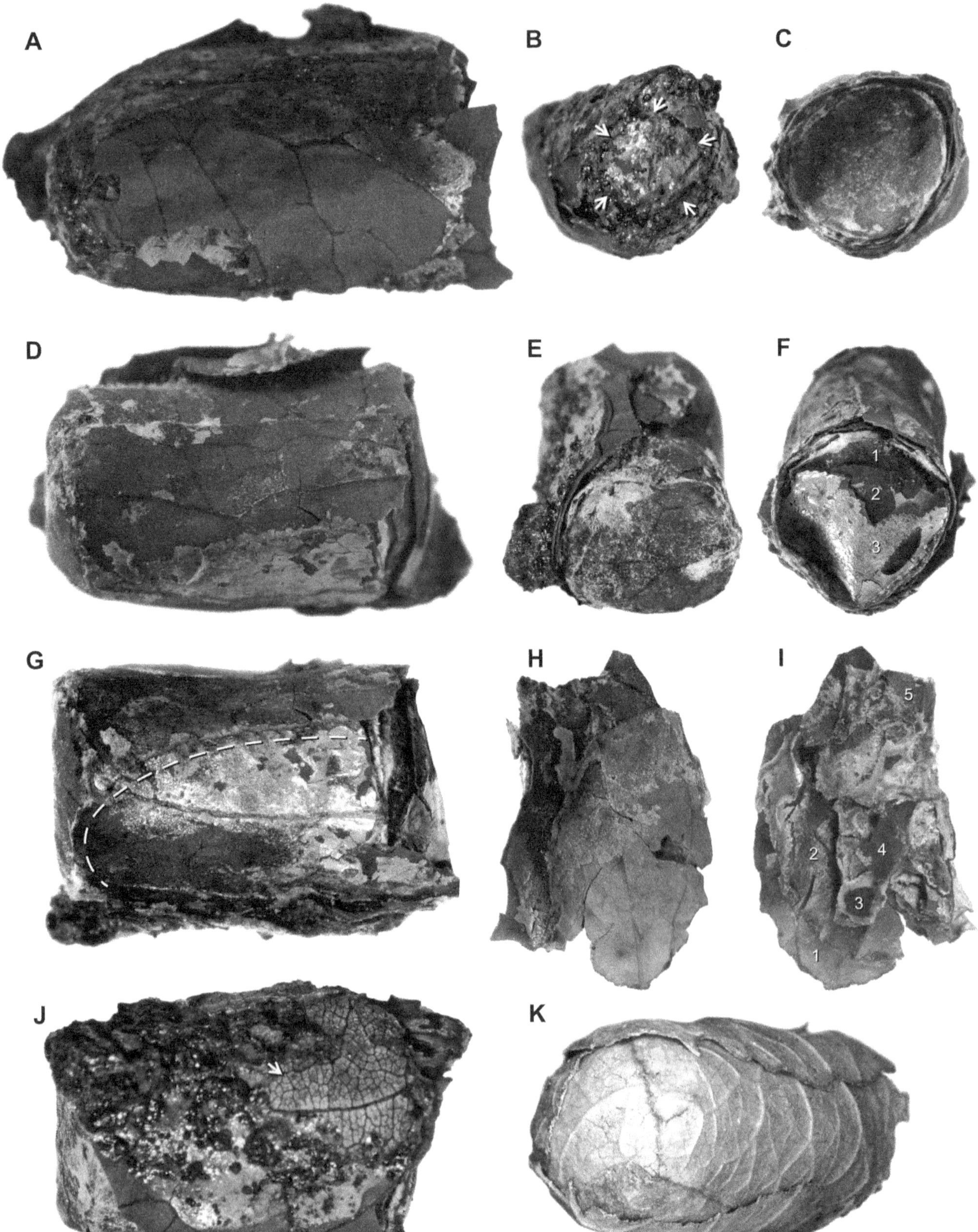

Figure 1. Images showing nest cell construction and modern nest cell of *Megachile gentilis* for comparison. A–C = LACMRLP 388Ea; D–G, J = LACMRLP 388Eb; H, I = LACMRLP 388Ec. (**A**) Nest cell containing male pupa showing cylindrical shape, tapered, rounded bottom at left typical of the first constructed cell, and remains of oblong leaf disc with Type 1 venation. (**B**) Bottom of first constructed cell (containing male pupa) with

possible portion of bottom circular leaf disk visible and outlined with arrows. (**C**) Cap of nest cell containing male pupa. (**D**) Nest cell containing female pupa. Arrow shows margin of oblong leaf disc with Type 2 venation. (**E**) Circular, bottom disc of nest cell (containing female pupa). In life, this end abutted the anterior end of LACMRLP 388Ea. (**F**) Cap of nest cell containing female pupa. (**G**) Nest cell of female pupa showing oblong side wall leaf cutout which does not reach bottom of cell and is instead supported by circular bottom disc. (**H**) Remains of oblong leaf disc with relatively smooth-cut margins and Type 3 venation. (**I**) View showing five overlapping oblong disks (1–5) comprising the sidewalls and circular bottom disk. discs. (**J**) Nest cell containing female pupa showing Type 4 venation on upper, right corner. (**K**) Nest cell of modern *M. gentilis*, showing circular disc bottom and oblong, sidewall leaf.

the cells were connected *in situ*, and are not cleptoparasitized, they can only be the same species.

Nest Cell Construction

Nest cell morphology. The two nest cells that constitute LACMRLP 388E are approximately 10.5 mm in length and 4.9 mm in diameter, constructed of oblong and circular leaf cutouts (Fig. 1*A*–*J*). The nest cells when found were joined end to end, the male cell in Figure 1*A* being the cell first assembled in the ground, while the female cell (Fig. 1*D*, *G*, *J*) would have been closer to the nest entrance. They were held together by additional cutouts that lined the nest cells but some dislodged after excavation and were lost. With subsequent handling another small portion sloughed off but was conserved. It consists of an additional five oblong-shaped disks (Fig. 1*H*, *I*). Outward morphology (Fig. 1*A*–*G*, *J*) and cross-sectional view of the cells (Fig. 2*E* and Fig. 3*E*, Video S7) show the sides and inner lining consist of 5–7 overlapping smooth-edged, oblong-shaped disks, the ends capped by small

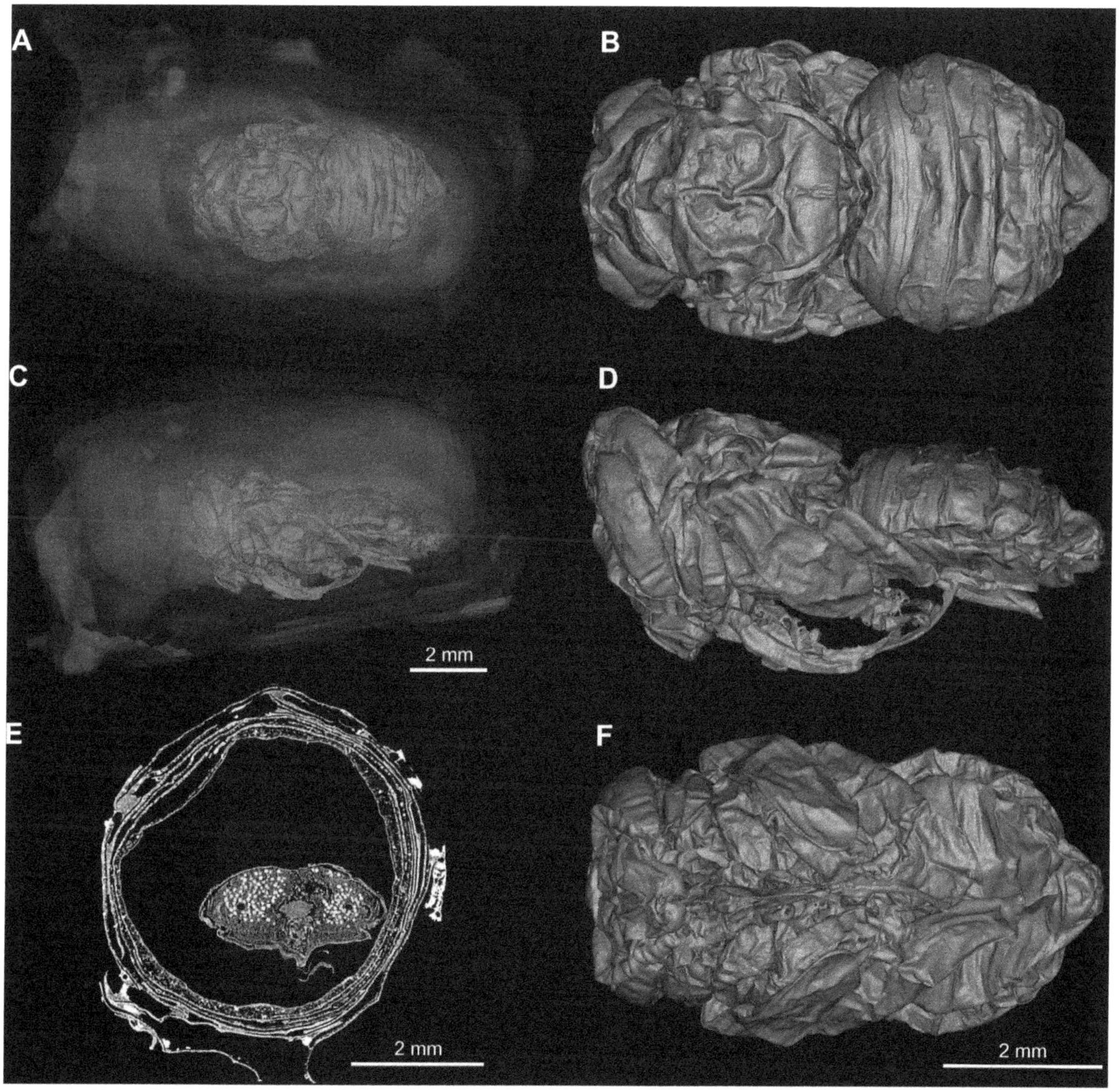

Figure 2. Micro-CT scans of LACMRLP 388Ea showing male pupa and its position within the nest cell. (A) Dorsal view of pupa within nest. (B) Dorsal view of pupa. (C) Lateral view of pupa within nest. (D) Lateral view of pupa. (E) Cross-section of nest and pupae. (F) Ventral view of pupa.

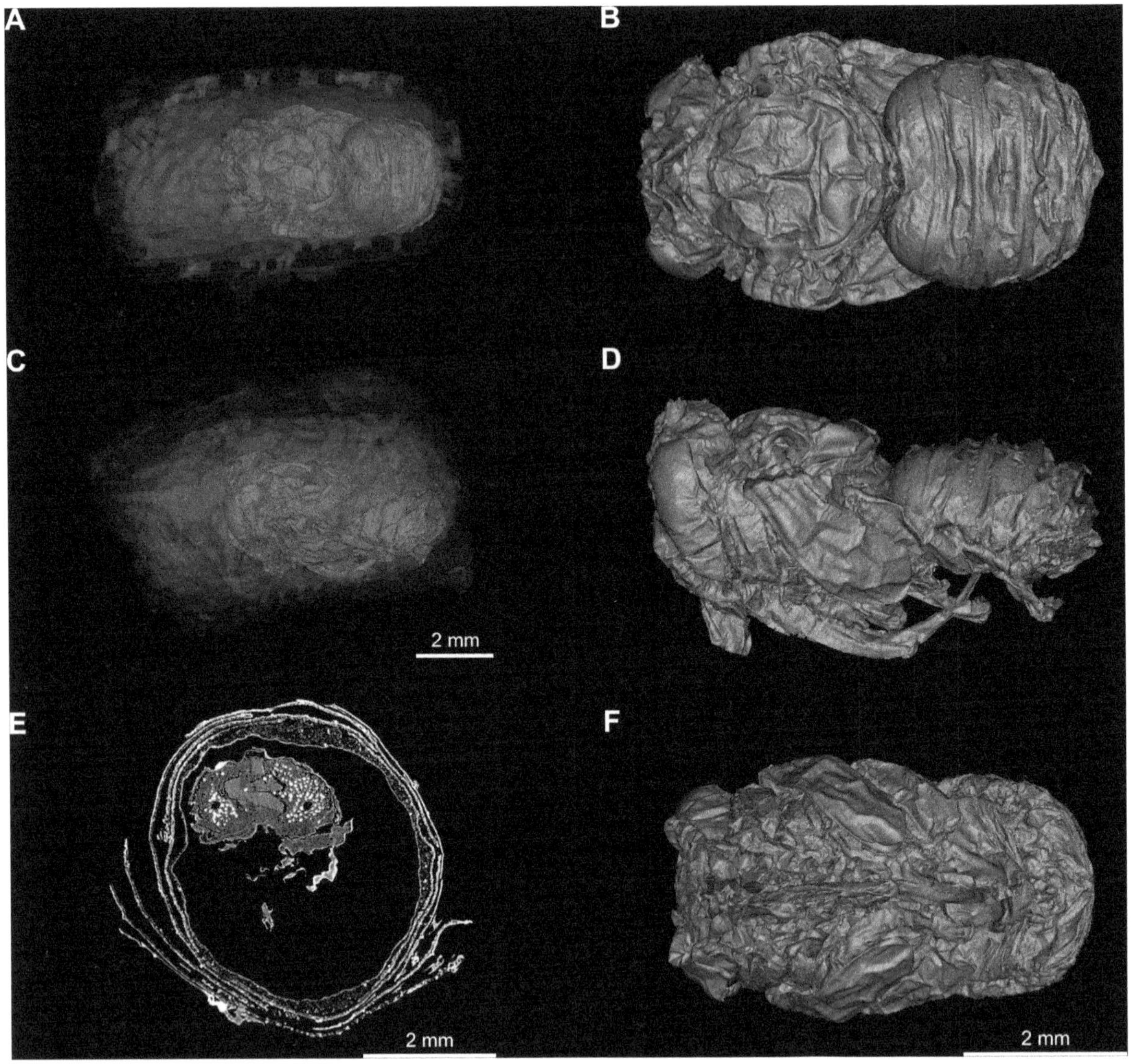

Figure 3. Micro-CT scans of LACMRLP 388Eb showing female pupa and its position within the nest cell. (A) Dorsal view of pupa within nest. (B) Dorsal view of pupa. (C) Lateral view of pupa within nest. (D) Lateral view of pupa. (E) Cross-section of nest and pupae. (F) Ventral view of pupa.

circular disks (Fig. 1*C, F*, Video S7). At least three discs cap the end of the male cell (Fig. 1*F*).

Most often, oblong leaves form the side walls and are bent into a cup at the proximal end (furthest from the nest entrance) that is glued with saliva and extruded leaf sap. A cap consisting of a series of layered circular discs is placed on the distal end (closest to the nest entrance). However, the separation of the two cells reveals the nest cell containing the female pupa includes a circular cap (Fig. 1*E*), as well as a circular bottom (Fig. 1*F*). Certain *Megachile* species insert a circular disc to provide an internal brace for the nest cell if the cup pieces are looser and less glued together than as constructed in other species [9,10,31]. Therefore, we presume that the nest cell containing the male pupa was the first cell in the cavity because it is hard to distinguish if the circular base is formed from the folded edges of the larger leaf fragments or the insertion of a smaller internal disc (Fig. 1*A, B* arrows). Literature on *Megachile* nest cell construction informs that the inclusion of a circular base is an uncommon construction for the genus [2,5].

Among extant Nearctic *Megachile* the small size of the cells and the use of leaves in nest construction limit candidate species to small members of the subgenus *Litomegachile*, specifically *M. brevis* Say, *M. gentilis* (Fig. 1*K*), *M. onobrychidis* Cockerell and *M. pseudobrevis* Mitchell. *Megachile brevis* and *M. pseudobrevis* are excluded from consideration because the former uses petals in cell construction [32] and the latter's current distribution is restricted to a limited range in the southeastern United States [25]. Exclusion of *M. pseudobrevis* is further supported by the incorporation of petals in two nests reported from Florida [33]. Further examination of the distinct nest cell architecture and exclusive use of leaves as cell material for LACMRLP 388E (Video S7), in combination with pupal morphology (Videos S1-6), indicate that these cells belong to *M. gentilis*, a species whose current distribution is concentrated in the southern portion of the Pacific coast, the Southwest, northern Mexico, and Texas [25, Discover Life *Megachile gentilis*, http://www.discoverlife.org/mp/

20m?kind = Megachile+gentilis, 2013] rather than *M. onobrychidis* (see below).

Plant material used. Since *Megachile* typically cut segments from the leaf margin, we can identify at least four, possibly five different taxa used in construction of the two cells based on the different patterns of venation (Figs. 1*A, D, G, J*). Dicotyledonous leaves typically have one or more large primary veins, with a reticulate network of successively narrower second, third, fourth, and fifth order veins in a pattern consistent within a species [34]. Although leaf venation is especially useful in fossil plant identification, venation characters are most reliable in combination with other morphological details such as leaf shape and size, shape of the leaf apex and base, and margin and petiole features [34], characters unavailable in the small areas of the leaf cutouts. Regrettably, this makes it difficult to identify with confidence the fragments to a specific species or even family. However, leaf texture, thickness, and the venation that is present are consistent with the cutouts being from woody trees, shrubs, or vines rather than herbaceous taxa, the leaves of which do not preserve well as fossils. Some *Megachile* include petals in their cells but venation of the cutouts of LACMRLP 388E is not consistent with petal venation [*35–37*], nor is their thickness (Video S7).

Historic and Contemporary Environmental Niche Space

We further researched the best species match for the La Brea bees by examining the historic environmental niche space of the La Brea *Litomegachile* relative to the contemporary environmental niche of *M. gentilis* and *M. onobrychidis,* the species whose nest cell morphology, distribution, size, and nesting behavior is closest to *M. gentilis*. To estimate the contemporary environmental niche space of *M. gentilis* and *M. onobrychidis* we associated a suite of informative bioclimatic variables with georeferenced natural history records (see Materials and Methods). We then constructed the historic environmental niche space of the La Brea bees by associating bioclimatic variables values of the Last Glacial Maximum (LGM, ~21,000 years ago) with georeferenced coordinates of the Rancho La Brea Tar Pits (latitude = 34.063068, longitude = −118.355412). While both species are broadly sympatric throughout California at present (Fig. 5), we found that the historic environmental niche of the La Brea bees in the Los Angeles Basin during the LGM fell within the 95% data concentration ellipse of the contemporary environmental niche space for *M. gentilis* in a principal component (PC) analysis (PC 1 = 32.7%, PC 2 = 26.5%) (Fig. 6). Furthermore, LGM habitat suitability estimates based on contemporary habitat suitability models (HSM) suggest that *M. gentilis* had a higher probability of being distributed at low elevations in the Los Angeles Basin than *M. onobrychidis* (Fig. 5, Fig. S2) (see Material and Methods).

Due to broad overlap in geography and environmental niche space between *M. gentilis* and *M. onobrychidis*, we further tested the assignment of the historic environmental niche of the La Brea bees using cluster analysis. Ninety percent of the contemporary *M. gentilis* distribution records, along with the La Brea fossil distribution record and associated historic environmental space values, were assigned to the contemporary environmental niche space of *M. gentilis* (Fig. 6). Sixty-five percent of the contemporary *M. onobrychidis* distribution records were assigned to the contemporary environmental niche space of *M. onobrychidis* (Fig. 6). The high assignment of the contemporary *M. gentilis* records to the *M. gentilis* environmental niche space is likely due to the narrow environmental distribution of the species. In comparison to *M. onobrychidis*, *M. gentilis* is distributed across mesic environments, where temperatures rarely fall below freezing during the coldest months (Fig. S1), with ample precipitation during the wettest months (Fig. S1). The ability for *M. onobrychidis* to exist in highly variable environments may have allowed that species to endure prolonged, freezing temperatures during overwintering periods, and disperse across a broad range of environments, whereas the environmental specificity of *M. gentilis* supports its narrow distribution in the Pacific US (Fig. 5). Cluster and principal component analysis, as well as estimates of habitat suitability during the LGM are consistent with our species identification of the La Brea bees unearthed in the Rancho La Brea Tar Pits in southern California as *M. gentilis*.

Discussion

The provenance of specimens recovered from Rancho La Brea cannot always be discerned because alluvial wash may have deposited some of the recovered material at a distance from their original location [38]. Nevertheless, the remarkable preservation and lack of water damage to LACMRLP 388E indicates with near certainty that it was assembled in the ground on the site of Pit 91 where it was found. The provenance of the specimen, in combination with environmental niche models and morphometric investigations, constitutes strong and comprehensive evidence to support confident fossil identifications and subsequent paleoenvironmental inferences for the Late Pleistocene, an epoch of particular climatic variation.

Because the leaves used in nest construction were collected in close proximity to each other and at the site of deposition, we can infer from their identification that a woody and riparian habitat existed at Rancho La Brea ~23,000–40,000 radiocarbon years BP. This inference is supported by Maxent habitat suitability models which suggest that *M. gentilis* was distributed in a mesic environment that likely occurred at a lower elevation during the Last Glacial Maximum (LGM) ~21,000 radiocarbon years BP (Fig. S1, Fig. S2). Presently, *M. gentilis* occurs in higher elevation habitats that constitute the perimeter of its projected habitat suitability during the LGM (Fig. 4). These areas surrounds the Los Angeles Basin in which Rancho La Brea is situated (Fig. 4).

Fossil plants from Pit 91 excavations have been grouped into four categories including a riparian association which includes plants from flood plain and canyon habitats [39]. Aquatic and moisture-loving plants indicate a permanent water source near Rancho La Brea [39,40]. But additional plants from Pit 91 indicate other associations such as coastal sage scrub, chaparral, and deep canyon [39–41]. Most of the excavated plants probably inhabited Rancho La Brea, but others may have been transported by streams or floods [39,41]. LACMRLP 388E suggests for the immediate Rancho La Brea area the presence of a gallery forest within a riparian zone. These forests exist along watercourses often in arid regions and standout in contrast to e.g. an adjoining grassland or open woodland habitat. The indication of a mesic environment at Rancho La Brea is relatively consistent with coastal run-off records [42] which indicate increased precipitation, possibly associated with increased glaciation [43], at ~25,000–20,000 years ago. Relative comparison of the geographic distribution of habitat suitability between the contemporary and LGM models reveals that *M. gentilis* has remained virtually unique to California and southwestern Arizona (Fig. 5) where the average temperature of the coldest quarter rarely falls below freezing (0°C). The contemporary distribution of *M. gentilis* is limited to mesic environments where the difference between the maximum of the warmest month and the minimum of the coldest month is centered around 33°C (Temperature Annual Range, Fig. S1). That *M. gentilis* has retained its abiotic niche since the LGM provides a

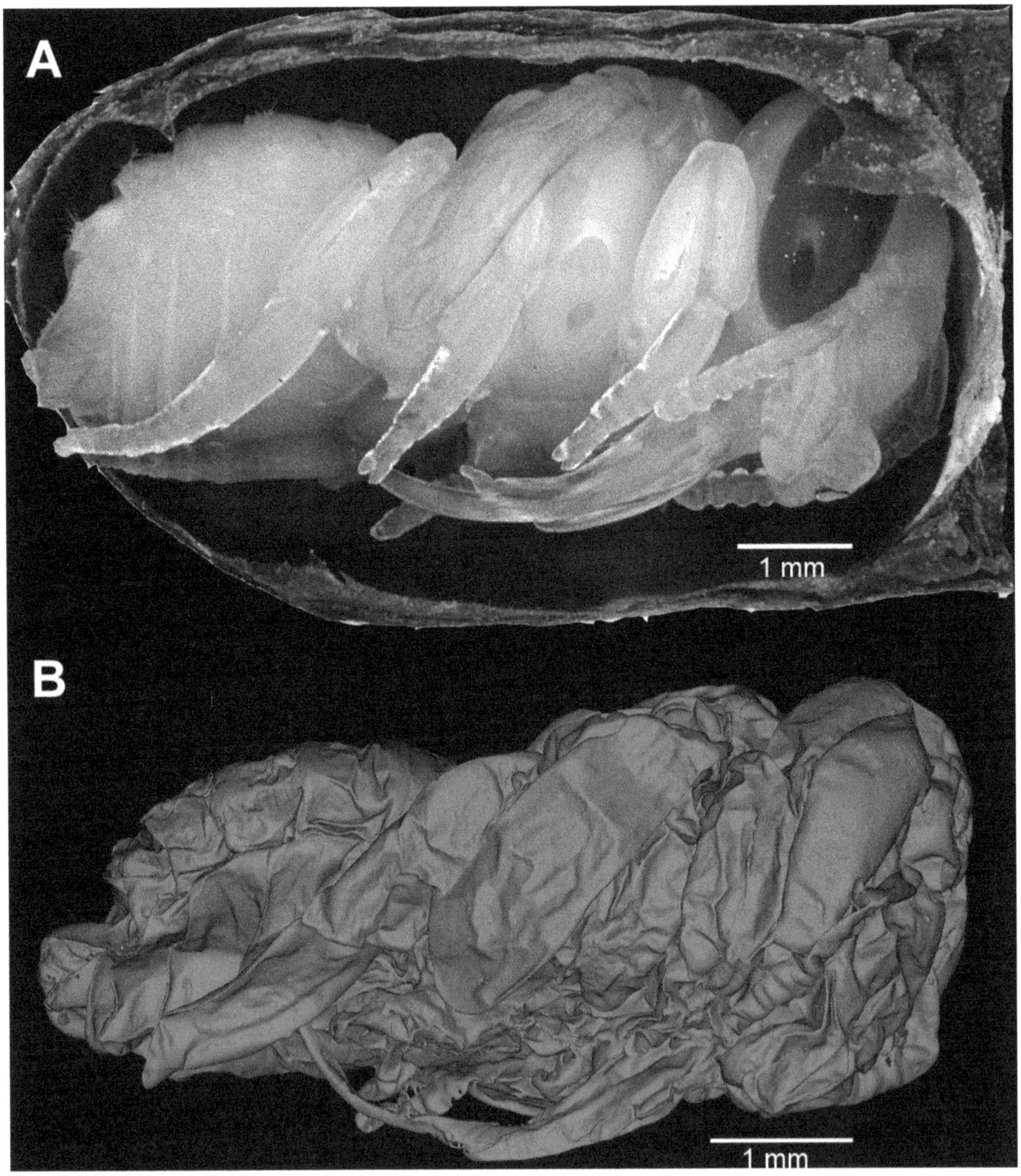

Figure 4. Comparison between (A) a female modern leafcutter bee (*Megachile rotundata*) and (B) LACMRLP 388E male and in order to display obvious similarity in general morphology at pupal stage.

strong benchmark with which to compare the climate restrictions of other fossils from Rancho La Brea.

Most animals and plants excavated from Rancho La Brea are extant, and those that have become extinct are mostly mammals such as common, larger carnivores (e.g. saber-toothed cats and dire wolves), common, medium-to-large herbivores (e.g. the western horse, mastodonts and mammoths), and some birds [39]. Only two species of insects recovered from Rancho La Brea, scarabs that relied on dung from mammals that became extinct, may have died out as well [39,44]. The current geographic distribution of most of the insect species from Rancho La Brea occurs in warmer parts of the United States and Mexico. Insect damage to the fossil bones attributed to tenebrionid and dermestid beetles is consistent with a warm period of at least 4–5 months at the time that the fossils were trapped [45]. This differs from interpretations of Late Pleistocene climatic conditions in southern California based on pollen from deep sea cores which indicate a shift to a cooler environment, as indicated by a transition from predominantly hardwood scrub-oak vegetation to coniferous pines and juniper [46–48] during the Last Glacial Interval from 24,000–14,000 years BP [46]. However, offshore palynological records may not reveal short-term increases in temperature that would result in both the colonization of insects with warm climate restrictions for brief periods due to their higher mobility, and in increased asphalt entrapment from more active seeps [45]. The current geographic distribution and climate restrictions of insect species preserved at Rancho La Brea, along with the physical properties of natural asphalt, indicate that the fossil insects represent warmer intervals of the Late Pleistocene [45]. But this does not resolve the discrepancy that some fossils recovered from Rancho La Brea such as *M. gentilis* inhabit mesic environments, while most of the tenebrionid beetles preserved at Rancho La Brea

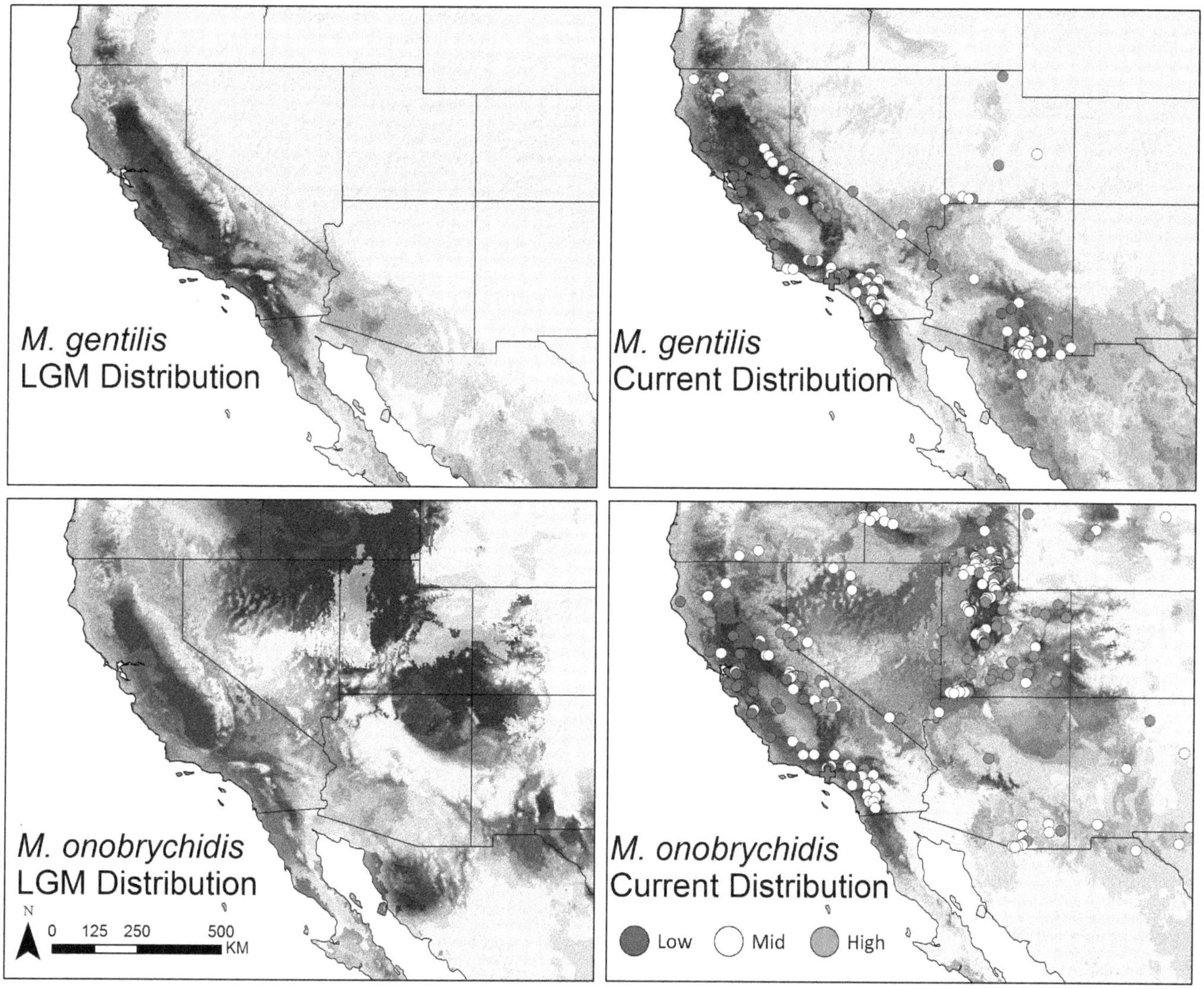

Figure 5. Estimated geographic distribution of *M. gentilis* and *M. onobrychidis* at present and during the last glacial maximum (LGM; ~21 000 years before present) in the western United States. Darker colors represent high MaxEnt habitat suitability (values closer to H = 1), whereas lighter colors represent low HS (values closer to HS = 0). Black squares represent georeferenced point localities of the respective species (Table S1). Red crosshairs represent the location of the La Brea Tar Pits in Los Angeles, California. Habitat suitability distributions of both species are based on eight bioclimatic variables.

inhabit dry scrub and woodland habitats [49]. Further research on insects that can be dated to a specific time period, and that also have a clear provenance for Rancho La Brea, as well as narrow climate restrictions, will greatly enrich our understanding of the paleoenvironment, as well as climate variation during the Late Pleistocene in southern California.

Aside from providing paleoenvironmental data, LACMRLP 388E enhances the fossil record of megachilids and the genus *Megachile* by providing the first reported specimen from the Pleistocene. This contributes to a better understanding of a genus whose phylogenetic relationships are still being researched.

The examination of LACMRLP 388E also presents neontologists with important information. O'Toole and Raw [2] state that, despite the diversity of nest sites and habitats exploited by leafcutter bees, their nest cells "differ little in structure and nesting behavior," a trait that makes the cells often difficult to identify. Yet the nest cell morphology of LACMRLP 388E provides one of the most important diagnostic tools to species-level. Nest cell construction of leafcutter bees is rarely closely examined or recorded. Where this has occurred [9,10,31], species-specific aspects of nest cell architecture have been documented. We surmise that there is indeed great morphological variation among *Megachile* species as has been demonstrated for another megachilid genus, *Osmia* [50]. A record of species-specific materials and nest structure may provide diagnostic tools useful in the field and for an organism that is often difficult to identify at an immature stage.

Materials and Methods

Excavation and Preparation

The cells, which were nested in chunky asphalt-impregnated matrix, were easier to excavate than if they had been preserved in more asphalt-concentrated matrix and therefore, were not exposed to solvent used in other preparation techniques. In addition, the immediate coating of the specimens with glyptol, an alkyd resin, upon discovery may have prevented further breakage during

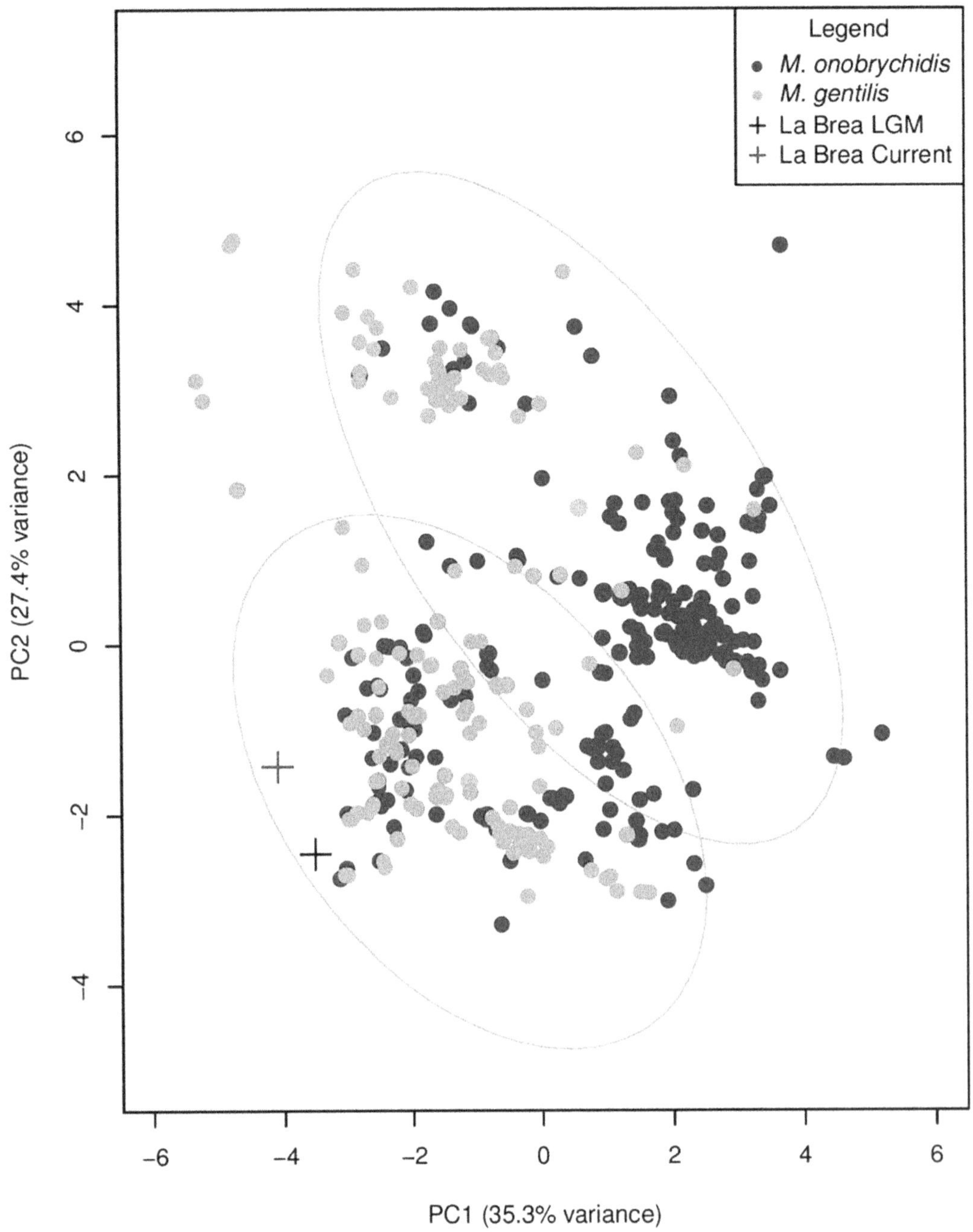

Figure 6. Principal components analysis of environmental space use by *M. gentilis* and *M. onobrychidis*. Estimates of environmental space use include eight bioclimatic variables, as well as, Cartesian (*i.e.*, latitude and longitude) and elevation variables. Clusters reflect 99% data concentration ellipses determined with the *scatterplot* (*car* library) functions in R v3.3.0 (R Core Team 2012).

handling and preparation (Obermayr, Pit 91 field notes p. 1770, 1970).

Micro-CT Scanning and Reconstruction

A 3D model was built for each pupae using Micro-CT data (Videos S1-6). Specimens were scanned on a SCANCO Micro-CT Scanner Model V1.2a. Linear Attenuation was 1/cm, 7.000000e+01 kVp with a slice thickness of 5 microns. Micro-CT slices were analyzed and 3D reconstructions of each bee built using Amira 5.4. LACMRLP 388Ea (male) was output as 2172 slices and reconstructed using 3,021,012 triangles (Videos S3, 5). LACMRLP 388Eb (female) was output as 2849 slices and reconstructed using 4,380,432 triangles (Videos S4, 6). To build 3D models of the nests, we scanned each nest on a MicroCAT II small animal CT system (Siemens Preclinical Solutions, Knoxville, TN, USA). Exposure settings were 70 kVp, 500 mAs, with a slice thickness of 2 mm and output as 436 slices. The female nest was built and output at 18,540 triangles. The male nest was built and output at 11,012 triangles. All models were cleaned and smoothed in Geomagic 10.

Use of Comparative Material

The fossil specimens were also compared to material from the comprehensive collection at the USDA-ARS Pollinating Insect Research Lab. The plant materials used and specific architecture was examined from the nest cells of other *Megachile* species of similar, diminutive size that include western United States distribution.

Estimation of Habitat Suitability at Present and During the LGM

Environmental space use. To investigate differences in environmental space use between *M. gentilis* and *M. onobrychidis*, we performed principal components analysis of bioclimatic variables, altitude, and Cartesian coordinates (*i.e.*, decimal degrees latitude and longitude) associated with distribution records. Contemporary environmental space use was estimated from recent specimen locality records (Dataset S1; National Pollinating Insect Database, Logan, UT) for eight informative bioclimatic variables: mean temperature of warmest quarter, mean temperature of coldest quarter, mean temperature of wettest quarter, mean temperature of driest quarter, precipitation of wettest quarter, precipitation of driest quarter, precipitation of warmest quarter, temperature annual range (Fig. S1; [51], http://worldclim.org). The contemporary (~1,950–2,000 AD) bioclimatic variables were selected based on their ability to capture seasonal trends and species-specific differences between *M. gentilis* and *M. onobrychidis*. Differences in the distribution of the study species for each bioclimatic variable were tested with the Wilcoxon rank sum test with significance (*P*) based on an alpha of 0.95 and visualized with violin plots (Fig. S1). Data concentration ellipses were estimated using the k-means clustering method ($K = 2$) to visualize contemporary record assignment, as well as assign the environmental space of the La Brea Tar Pits during the LGM to either *M. gentilis* or *M. onobrychidis*.

Geographic distribution of habitat suitability. To estimate the geographic distribution of habitat suitability in the western U.S. at present and during the LGM (~21,000 y BP), habitat suitability models (HSM) were constructed under the principle of maximum entropy with MaxEnt v3.3.1 [52]. Each species' HSM was constructed from the specimen locality records and bioclimatic variables selected for the principal component and cluster analysis (see Environmental Space Use). Under the principle of niche conservatism [53], we estimated the geographic distribution of habitat suitability of both species during the LGM with analog bioclimatic variables of their contemporary bioclimatic niche. To provide a robust estimate of LGM habitat suitability, we allowed for the HSM to exceed contemporary bioclimatic extremes in the event that maximum entropy was achieved during model construction. Constructing LGM HSMs that are younger than the fossil records of the study species guarantees that both *M. gentilis* and *M. onobrychidis*, or a common *Litomegachile* ancestor were already in existence. The spatial resolution of all bioclimatic variables used in the models is 2.5 arc-minutes. Data processing and geographic visualization of the HSM were implemented in ArcGIS 10.1 (ESRI, Redlands, CA). Model performance was evaluated in MaxEnt using the area under curve (AUC) statistic and a jackknife of variable importance (Table S1). The AUC is constrained between zero and one where values closer to one suggest better model performance. Models were averaged over 50 replicates, with 10-fold cross validation for each replicate for model validation.

Statistical analysis. Outside of MaxEnt, all analyses were conducted in R v3.3.0 [54]. Principal components were visualized with the *scatterplot* function in the *car* library. The k-means clustering method was performed with the *kmeans* function in the *cluster* library. Violin plots were visualized with the *vioplot* library.

Supporting Information

Figure S1 Violin plots of eight bioclimatic variables associated with the distributions of *M. gentilis* and *M. onobrychidis*. The width along each violin plot represents the frequency of specimen records associated with a measurement of the bioclimatic variable under observation. Wider widths reflect a higher frequency of specimen records, whereas thinner widths reflect a lower frequency of specimen records. Two-sample Wilcoxon tests were performed to test for precipitation/temperature differences between the species for each bioclimatic variable. Alpha levels for each test were set at alpha = 0.95. $P < 0.05$ suggests significant differences in bioclimatic space use between the two species. M. gen = M. gentilis and M. ono = M. onobrychidis.

Figure S2 Contemporary habitat suitability distribution of *M. gentilis* and *M. onobrychidis* across an elevation gradient in southern California. Kendall τ rank correlation coefficient estimates for *M. gentilis* revealed a positive and significant correlation between HS and Elevation ($\tau = 0.31$, $P < 0.05$), and a negative and non-significant correlation for *M. onobrychidis* ($\tau = 0.31$, $P = 0.52$).

Table S1 Analysis of bioclimatic variable importance for the MaxEnt habitat suitability models (HSM) of *M. gentilis* and *M. onobrychidis*. Values represent averages over 50 replicate runs of each species' habitat suitability model (HSM).

Dataset S1 Natural history records of *M. gentilis* and *M. onobrychidis* utilized to construct environmental space use and habitat suitability models (HSM). All specimens of each species present in the U.S. National Pollinating Insect Collection Database are presented in the table. *Megachile brevis onobrychidis* has been synonymized as *M. onobrychidis*. The former name is retained in the data table. Duplicate specimen records for each species per unique locality were removed for the final analyses.

Video S1 LACMRLP 388Ea. Rotating video nest of opaque to transparent nest cell containing male pupa.

Video S2 LACMRLP 388Eb. Rotating video nest of opaque to transparent nest cell containing female pupa.

Video S3 LACMRLP 388Ea. Rotating dorsal to ventral video of male pupa.

Video S4 LACMRLP 388Ea. Rotating dorsal to ventral video of female pupa.

Video S5 LACMRLP 388Ea. Rotating lateral video of male pupa.

Video S6 LACMRLP 388Eb Rotating lateral video of female pupa.

Video S7 LACMRLP 388Ea. Rotating quarter edge of nest cell. This video shows cross-sections of leaves from different angles.

Acknowledgments

We wish to thank the following individuals and institutions: from the Natural History Museum of Los Angeles County–L. Chiappe provided support for this project and considerable discussion of the manuscript. G. Kung assisted in micro CT scans, B. Brown provided review and guidance, R. Hulser assisted with references, and S. Abramowicz assisted with figures; from the George C. Page Museum–J. Harris for great support of paleoentomological research, C. Shaw for review and with S. Cox, for information on Pit 91 excavation, A. Farrell, G. Takeuchi expedited the preparation and documentation of specimens for the study, and C. Howard assisted with images; from the Berkeley Natural History Museums–J. Gross provided additional images; from the American Museum of Natural History–J. Rozen assisted in bee pupae identifications; from the Molecular Imaging Center at the University of Southern California–G. Dagilyan, A. Krishnan, and M. Aladadyan provided micro-CT scans and advice; from the Department of Cellular and Neurobiology at the University of Southern California–B. Patel offered guidance on the usage of large datasets and provided access to computer and software for micro-CT scan reconstructions; from the University of California, Los Angeles–A. Friscia offered information on the dating of specimens from Pit 91, M. Mc. Nitt-Gray and D. Stout provided nest cell micro-CT scans; and from the USDA-ARS Pollinating Insect Research Unit at Utah State University–H. Ikerd reared and imaged comparative bee material. We also thank D. W. Frayer and two anonymous reviewers for their constructive advice.

Author Contributions

Conceived and designed the experiments: ARH JBK TG DME. Performed the experiments: ARH JBK TG DME. Analyzed the data: ARH JBK TG DME. Contributed reagents/materials/analysis tools: ARH JBK TG DME JH. Wrote the paper: ARH JBK TG DME JH.

References

1. Latreille PA (1802–1803) Histoire naturelle, générale et particulière, des Crustacés et des Insectes, (Third edition). Paris: Dufart. 468 p.
2. O'Toole C, Raw A (1991) Bees of the world. London: Blandford Publishing. 192 p.
3. Munz PA, Keck DD (1959) A California flora. Berkeley and Los Angeles: University of California Press. 681 p.
4. Baldwin BG, Goldman DH, Keil DJ, Patterson R, Thomas JR (2012) The Jepson manual. Vascular plants of California (Second edition). Berkeley, Los Angeles, and London: University of California Press. 1600 p.
5. Michener CD (2007) The bees of the world (Second edition). Baltimore: The Johns Hopkins University Press, Baltimore. 953 p.
6. Hurd PD, Michener CD (1955) The megachiline bees of California (Hymenoptera: Megachilidae). In: Linsley EG, Essig EO, Freeborn SB, Usinger RL editors. The Bulletin of the California insect survey (Volume 3). Berkeley and Los Angeles: University of California Press. 1–247 p.
7. Krombein KV (1967) Trap-nesting wasps and bees: Life histories, nests, and associates (Volume 6). Washington, DC: Smithsonian Institution Press. 570 p.
8. Michez D, Vanderplanck M, Engel M (2012) Fossil bees and their plant associates, In: Patiny S editor. Evolution of plant-pollinator relationships. Cambridge: Cambridge University Press. pp. 103–164.
9. Kim JY (1992) Nest dimensions of two leaf-cutter bees (Hymenoptera: Megachilidae). Ann Entomol Soc Am 85: 85–90.
10. Kim JY (2007) Disc size regulation in the brood cell building behavior of leaf-cutter bee, *Megachile tsurugensis*. Naturwissenschaften 94: 981–990.
11. Engel MS, Perkovsky EE (2006) An Eocene bee in Rovno Amber (Hymenoptera: Megachilidae). Am Mus Novit 3506: 1–12.
12. Sarzetti L, Labandeira CC, Genise JF (2008) A leafcutter bee tracefossil from the Middle Eocene of Patagonia Argentina and a review of megachilid (Hymenoptera) ichnology. Palaeontol 51: 933–941.
13. Engel MS (1999) *Megachile glaesaria*, the first megachilid bee fossil from amber (Hymenoptera: Megachilidae). Am Mus Novit 3276: 1–13.
14. Engel MS (2001) A monograph of the Baltic amber bees and evolution of the Apoidea (Hymenoptera). Bull Am Mus Nat Hist 259: 1–192.
15. Engel MS, Grimaldi DA (2004) New light shed on the oldest insect. Nature 427: 627–630.
16. Berry EW (1916) The Lower Eocene floras of southeastern North America. U.S.G.S. Prof Pap 91: 1–469.
17. Berry EW (1931) An insect-cut leaf from the Lower Eocene. Am J Sci 21: 301–304.
18. Brooks HK (1955) Healed wounds and galls on fossil leaves from the Wilcox deposit (Eocene) of western Tennessee. Psyche 62: 1–9.
19. Lewis SE (1994) Evidence of leaf-cutting bee damage from the Republic sites (Middle Eocene) of Washington. J Paleontol 68: 172–173.
20. Wappler T, Engel MS (2003) The middle Eocene bee faunas of Eckfeld and Messel, Germany (Hymenoptera: Apoidea). J Paleontol 77: 908–921.
21. Cockerell TDA (1908) Fossil insect from Florissant, Colorado. Am Mus Nat Hist 24: 59–69.
22. Wedmann S, Wappler T, Engel MS (2009) Direct and indirect fossil records of megachilid bees from the Paleogene of Central Europe (Hymenoptera: Megachilidae). Naturwissenschaften 96: 703–712.
23. Litman JR, Danforth BN, Eardley CD, Praz CJ (2011) Why do leafcutter bees cut leaves? New insights into the early evolution of bees. Proc R Soc B 278: 3593–3600.
24. Mitchell TB (1935) A revision of the genus *Megachile* in the Nearctic Region. Part II. Morphology of the male sternites and genital armature and the taxonomy of the subgenera *Litomegachile*, *Neomegachile* and *Cressoniella* (Hymenoptera: Megachilidae). Trans Am Entomol Soc (1890-) 61: 1–44.
25. Bzdyk EL (2012) A revision of the *Megachile* subgenus *Litomegachile* Mitchell with an illustrated key and description of a new species (Hymenoptera, Meagchilidae, Megachilini). Zookeys 221: 31–61.
26. Sheffield CS, Genaro JA (2013) A new species of *Megachile* (*Litomegachile*) from Cuba, the Antilles (Hymenoptera: Megachilidae). J Melittology 19: 1–17.
27. Coope GR (2004) Several million years of stability among insect species because of, or in spite of, Ice Age climate instability? Philos Trans R Soc Lond B Biol Sci 359: 2009–214.
28. Friscia AR, Van Valkenburgh B, Spencer L, Harris JM (2008) Chronology and spatial distribution of large mammal bones in Pit 91, Rancho La Brea. Palaios 23: 35–42.
29. Cane JH, Gerdin SG, Wife G (1983) Mandibular gland secretions of solitary bees (Hymenoptera: Apoidea): Potential for nest cell disinfection. J Kan Entomol Soc 56: 199–204.
30. Cane JH, Carlson RG (1984) Dufour's gland triglycerides from *Anthophora*, *Emphoropsis* (Anthophoridae) and *Megachile* (Megachildae) bees (Hymentoptera: Apoidea). Comp Biochem Physiol 78B: 769–772.
31. Strickler K, Scott VL, Fisher RL (1991) Comparative nesting ecology of two sympatric leafcutting bees that differ in body size (Hymenoptera: Megachilidae). J Kans Entomol Soc 69: 26–44.
32. Michener CD (1953) The biology of a leafcutter bee (*Megachile brevis*) and its associates. Univ Kansas Sci Bull 35: 1659–1748.
33. Packer L (1987) The triungulin larva of *Nemognatha (Pauronemognatha) punctulata* LeConte (Coleoptera: Meloidae) with a description of the nest of its host-*Megachile brevis pseudobrevis* Say (Hymenoptera: Megachilidae). J Kans Entomol Soc 60: 280–287.
34. Ellis B, Daly DC, Hickey LJ, Johnson KR, Mitchell JD, et al. (2009) Manual of leaf architecture. Ithaca: Comstock Publishing Associates. 216 p.
35. Arnott HJ, Tucker SC (1963) Analysis of petal venation in Ranunculus. I. Anastomoses in *R. repens*, *V. pleniflorus*. Am J Bot 50: 821–830.
36. Arnott HJ, Tucker SC (1964) Analysis of petal venation in Ranunculus. II. Number and position of dichotomies in *R. repens var. pleniflorus*. Bot Gaz 125: 13–26.
37. Gustafsson MH (1955) Petal venation in the Asterales and related orders. Bot J Linn Soc 118: 1–18.
38. Shaw CA (1982) Techniques used in excavation, preparation, and curation of fossils from Rancho La Brea. Curator 25: 63–77.
39. Stock CS, Harris JM (1992) Rancho La Brea: A record of Pleistocene life in California. Los Angeles: Natural History Museum of Los Angeles County. 113 p.
40. Warter JK (1979) The environment of Pit 91, Rancho La Brea, as interpreted by plant remains. Abs, Ann Mtg S Ca Acad Sci 87: 44.
41. Shaw CS, Quinn JP (1986) Rancho La Brea: A look at coastal southern California's past. Ca Geo 39: 123–133.
42. Robert C (2004) Later Quaternary variability of precipitation in southern California and climatic implications: clay mineral evidence from the Santa Barbara Basin, ODP Site 893. Quaternary Sci Rev 23: 1029–1040.
43. Bischoff JL, Cummins K (2001) Wisconsin glaciation of the Sierra Nevada (79,000–15,000 yr BP) as recorded by rock flour in sediments of Owens Lake, California. Quaternary Res 55: 14–24.
44. Miller SE, Gordon RD, Howden HF (1981) Reevaluation of Pleistocene scarab beetles from Rancho La Brea, California (Coleoptera: Scarabaeidae). Proc Entomol Soc Wash 83: 625–630.
45. Holden AR, Harris JM, Timm RM (2013) Paleoecological and taphonomic implications of insect-damaged Pleistocene vertebrate remains from Rancho La Brea, southern California. PloS one, 8(7): e67119.
46. Heusser L (1998) Direct correlation of millennial-scale changes in western North American vegetation and climate with changes in the California current system over the past ~60 kyr. Paleoceanogr 13: 252–262.

47. Coltrain JB, Harris JM, Cerling TE, Ehleringer JR, Dearing MD, et al. (2004) Rancho La Brea stable isotope biogeochemistry and its implications for the palaeoecology of Late Pleistocene, coastal southern California. Palaeogeogr, Palaeoclimatol, Palaeoecol 205: 199–219.
48. Jacobs DK, Haney TA, Louie KD (2004) Genes, diversity, and geologic process on the Pacific Coast. Annu Rev Earth Planet Sci 32: 602–652.
49. Doyen JT, Miller SE (1980) Review of Pleistocene darkling ground beetles of the California asphalt deposits (Coleoptera: Tenebrionidae, Zopheridae). Pan-Pac Entomol 56: 1–10.
50. Cane JH, Griswold T, Parker FD (2007) Substrates and materials used for nesting by North American *Osmia* bees (Hymenoptera: Apiformes: Megachilidae). Ann Entomol Soc Am 100: 350–358.
51. Hijmans RJ, Cameron SE, Parra JL, Jones PG, Jarvis A (2005) Very high resolution interpolated climate surfaces for global land areas. Int J Climatol 25: 1965–1978.
52. Phillips SJ, Anderson RP, Schapire RE (2006) Maximum entropy modeling of species geographic distributions. Ecol Model 190: 231–259.
53. Peterson AT, Soberón J, Sánchez-Cordero V (1999) Conservatism of ecological niches in evolutionary time. Science 285: 1265–1267.
54. R Core Team (2012) R A language and environment for statistical computing. R Foundation for Statistical Computing, Vienna, Austria. ISBN 3-900051-07-0. Available: http://www.R-project.org.

PERMISSIONS

All chapters in this book were first published in PLOS ONE, by The Public Library of Science; hereby published with permission under the Creative Commons Attribution License or equivalent. Every chapter published in this book has been scrutinized by our experts. Their significance has been extensively debated. The topics covered herein carry significant findings which will fuel the growth of the discipline. They may even be implemented as practical applications or may be referred to as a beginning point for another development.

The contributors of this book come from diverse backgrounds, making this book a truly international effort. This book will bring forth new frontiers with its revolutionizing research information and detailed analysis of the nascent developments around the world.

We would like to thank all the contributing authors for lending their expertise to make the book truly unique. They have played a crucial role in the development of this book. Without their invaluable contributions this book wouldn't have been possible. They have made vital efforts to compile up to date information on the varied aspects of this subject to make this book a valuable addition to the collection of many professionals and students.

This book was conceptualized with the vision of imparting up-to-date information and advanced data in this field. To ensure the same, a matchless editorial board was set up. Every individual on the board went through rigorous rounds of assessment to prove their worth. After which they invested a large part of their time researching and compiling the most relevant data for our readers.

The editorial board has been involved in producing this book since its inception. They have spent rigorous hours researching and exploring the diverse topics which have resulted in the successful publishing of this book. They have passed on their knowledge of decades through this book. To expedite this challenging task, the publisher supported the team at every step. A small team of assistant editors was also appointed to further simplify the editing procedure and attain best results for the readers.

Apart from the editorial board, the designing team has also invested a significant amount of their time in understanding the subject and creating the most relevant covers. They scrutinized every image to scout for the most suitable representation of the subject and create an appropriate cover for the book.

The publishing team has been an ardent support to the editorial, designing and production team. Their endless efforts to recruit the best for this project, has resulted in the accomplishment of this book. They are a veteran in the field of academics and their pool of knowledge is as vast as their experience in printing. Their expertise and guidance has proved useful at every step. Their uncompromising quality standards have made this book an exceptional effort. Their encouragement from time to time has been an inspiration for everyone.

The publisher and the editorial board hope that this book will prove to be a valuable piece of knowledge for researchers, students, practitioners and scholars across the globe.

LIST OF CONTRIBUTORS

Daniel DeMiguel
Institut Català de Paleontologia Miquel Crusafont, Universitat Autònoma de Barcelona, Barcelona, Spain

David M. Alba
Institut Català de Paleontologia Miquel Crusafont, Universitat Autònoma de Barcelona, Barcelona, Spain

Dipartimento di Scienze della Terra, Università di Torino, Torino, Italy

Salvador Moyà -Solà
ICREA at Institut Català de Paleontologia Miquel Crusafont and Unitat d'Antropologia Biològica (Dept. BABVE), Universitat Autònoma de Barcelona, Barcelona, Spain

Neil H. Landman
Division of Paleontology American Museum of Natural History, New York, New York, United States of America

Isabelle Kruta
Division of Paleontology American Museum of Natural History, New York, New York, United States of America
Department of Geology and Geophysics, Yale University, New Haven, Connecticut, United States of America

J. Kirk Cochran
School of Marine and Atmospheric Sciences, Stony Brook University, Stony Brook, New York, United States of America

Hao Wang, Fan Lu, Yang Fang, Wenju Zhang, Qianhong Wu, Mei Yang and Jiakuan Chen
Institute of Biodiversity Science, School of Life Sciences, Fudan University, Shanghai, China

Lu Xu
Institute of Biodiversity Science, School of Life Sciences, Fudan University, Shanghai, China
College of Life Sciences, Northwest Normal University, Lanzhou, China

Kun Sun
College of Life Sciences, Northwest Normal University, Lanzhou, China

Qiong La and Yang Zhong
Institute of Biodiversity Science, School of Life Sciences, Fudan University, Shanghai, China
Department of Biology, Tibet University, Lhasa, China

H. John B. Birks
Department of Biology, University of Bergen, Bergen, Norway
Environmental Change Research Centre, University College London, London, United Kingdom
School of Geography and the Environment, University of Oxford, Oxford, United Kingdom

Jan Douda
Department of Ecology, Faculty of Environmental Sciences, Czech University of Life Sciences Prague, Prague, Czech Republic

Jana Doudová, Věroslava Hadincová, Karol Krak and Petr Zákravský
Institute of Botany, Academy of Sciences of the Czech Republic, Průhonice, Czech Republic

Alena Drašnarová and Bohumil Mandák
Department of Ecology, Faculty of Environmental Sciences, Czech University of Life Sciences Prague, Prague, Czech Republic
Institute of Botany, Academy of Sciences of the Czech Republic, Průhonice, Czech Republic

Petr Kuneš
Department of Botany, Faculty of Science, Charles University in Prague, Prague, Czech Republic

Carlos E. González-Orozco, Nunzio J. Knerr, Alexander N. Schmidt-Lebuhn, Christine C. Cargill, Mark Clements and Joseph T. Miller
Centre for Australian National Biodiversity Research, Commonwealth Scientific and Industrial Research Organisation, Plant Industry, Canberra, Australian Capital Territory, Australia

Malte C. Ebach and Shawn Laffan
School of Biological, Earth and Environmental Sciences, University of New South Wales, Sydney, New South Wales, Australia

Andrew H. Thornhill
Centre for Australian National Biodiversity Research, Commonwealth Scientific and Industrial Research Organisation, Plant Industry, Canberra, Australian Capital Territory, Australia
Australian Tropical Herbarium, James Cook University, Cairns, Queensland, Australia

Nathalie S. Nagalingum
National Herbarium of New South Wales, Botanic Gardens Trust, Sydney, New South Wales, Australia

Brent D. Mishler
University and Jepson Herbaria, Department of Integrative Biology, University of California, Berkeley, Berkeley, California, United States of America

Wenxiang Zhang, Qingzhong Ming, Zhengtao Shi, Guangjie Chen, Jie Niu, Fengqin Chang and Hucai Zhang
Key Laboratory of the Plateau Surface Process and Environment Changes of Yunnan Province, Key Laboratory of Plateau Lake Ecology and Global Change, Yunnan Normal University, Kunming, China

Guoliang Lei
Key Laboratory of Humid Subtropical Eco-geographical Process, Ministry of education, Fuzhou, China

Yu Li, Nai'ang Wang and Chengqi Zhang
College of Earth and Environmental Sciences, Center for Hydrologic Cycle and Water Resources in Arid Region, Lanzhou University, Lanzhou, China

Yuyini Licona-Vera and Juan Francisco Ornelas
Departamento de Biología Evolutiva, Instituto de Ecología, AC, Xalapa, Veracruz, Mexico

Olev Vinn
Department of Geology, University of Tartu, Tartu, Estonia

Mark A. Wilson
Department of Geology, The College of Wooster, Wooster, Ohio, United States of America

Mari-Ann Mõtus
Institute of Geology, Tallinn University of Technology, Tallinn, Estonia

Nicholas A. Famoso and Edward Byrd Davis
Department of Geological Sciences and Museum of Natural and Cultural History, University of Oregon, Eugene, Oregon, United States of America

Allyson L. Carroll, Stephen C. Sillett and Russell D. Kramer
Department of Forestry and Wildland Resources, Humboldt State University, Arcata, California, United States of America

Qiang Li
The State Key Laboratory of Loess and Quaternary Geology, Institute of Earth Environment, Chinese Academy of Sciences, Xi'an, China

Yu Liu
The State Key Laboratory of Loess and Quaternary Geology, Institute of Earth Environment, Chinese Academy of Sciences, Xi'an, China
Department of Environmental Science and Technology, School of Human Settlements and Civil Engineering, Xi'an Jiaotong University, Xi'an, China

Huiming Song, Yongyong Ma and Yanhua Zhang
The State Key Laboratory of Loess and Quaternary Geology, Institute of Earth Environment, Chinese Academy of Sciences, Xi'an, China
University of Chinese Academy of Sciences, Beijing, China

Na Gao
Institute of Geology, China Earthquake Administration, Beijing, China

Ye-Na Tang, Yi-Feng Yao and Cheng-Sen Li
State Key Laboratory of Systematic and Evolutionary Botany, Institute of Botany, Chinese Academy of Sciences, Beijing, China

Xiao Li
School of Chinese Classics, Renmin University of China, Beijing, China

David Kay Ferguson
Department of Paleontology, University of Vienna, Vienna, Austria
State Key Laboratory of Systematic and Evolutionary Botany, Institute of Botany, Chinese Academy of Sciences, Beijing, China

Věra Pavelková Řičánková, Jan Robovský and Jan Riegert
Department of Zoology, Faculty of Science, University of South Bohemia, České Budě jovice, Czech Republic

Sandra Ravnsbæk Holm and Jens-Christian Svenning
Ecoinformatics and Biodiversity, Department of Bioscience, Aarhus University, Aarhus C, Denmark

Alberto Rodriguez-Ramirez and Jian-xin Zhao
Radiogenic Isotope Facility, School of Earth Sciences, The University of Queensland, Brisbane, Queensland, Australia

Craig A. Grove
NIOZ Royal Netherlands Institute for Sea Research, Department of Marine Geology, Den Burg, Texel, The Netherlands

Jens Zinke
School of Earth and Environment, The University of Western Australia and the UWA Oceans Institute, Australia and the Australian Institute of Marine Science, Perth, Western Australia, Australia

John M. Pandolfi
Australian Research Council Centre of Excellence for Coral Reef Studies, Centre for Marine Science, School of Biological Sciences, The University of Queensland, Brisbane, Queensland, Australia

David Lazarus and Johan Renaudie
Museum für Naturkunde, Berlin, Germany

John Barron
United States Geological Survey, Menlo Park, California, United States of America

Patrick Diver
Divdat Consulting, Wesley, Arkansas, United States of America

Andreas Türke
Museum für Naturkunde, Berlin, Germany
Department of Geosciences, University of Bremen, Bremen, Germany

Feifei Zhou
Key Laboratory of Humid Subtropical Eco-Geographical Process (Ministry of Education), College of Geographical Sciences, Fujian Normal University, Fuzhou, Fujian Province, China

Keyan Fang
Key Laboratory of Humid Subtropical Eco-Geographical Process (Ministry of Education), College of Geographical Sciences, Fujian Normal University, Fuzhou, Fujian Province, China
Key Laboratory of Western China's Environmental Systems (Ministry of Education), Research School of Arid Environment and Climate Change, Lanzhou University, Lanzhou, Gansu Province, China
Department of Geosciences and Geography, University of Helsinki, Helsinki City, Helsinki, Finland

Changzhi Liu
Key Laboratory of Western China's Environmental Systems (Ministry of Education), Research School of Arid Environment and Climate Change, Lanzhou University, Lanzhou, Gansu Province, China

Martin Wilmking
Institute of Botany and Landscape Ecology, Greifswald University, Greifswald, Mecklenburg-Vorpommern, Germany

Nicole Davi
Tree-Ring Lab, Lamont-Doherty Earth Observatory, Columbia University, Palisades, New York, United States of America
Department of Environmental Science, William Paterson University, Wayne, New Jersey, United States of America

Anna R. Holden
Entomology Section, Natural History Museum of Los Angeles County, Los Angeles, California, United States of America

Jonathan B. Koch
Department of Biology and Ecology Center, Utah State University, Logan, Utah, United States of America

Terry Griswold
USDA-ARS Pollinating Insect Research Unit, Utah State University, Logan, Utah, United States of America

Diane M. Erwin
Museum of Paleontology, University of California, Berkeley, Berkeley, California, United States of America

Justin Hall
Dinosaur Institute, Natural History Museum of Los Angeles County, Los Angeles, California, United States of America

Index

Lightning Source UK Ltd.
Milton Keynes UK
UKHW050825261119
354107UK00010B/33/P

9 781682 864487